Lecture Notes in Computer Science 6986

Commenced Publication in 1973
Founding and Former Series Editors:
Gerhard Goos, Juris Hartmanis, and Jan van Leeuwen

Advanced Research in Computing and Software Science

Subline of Lectures Notes in Computer Science

Petr Kolman   Jan Kratochvíl (Eds.)

Graph-Theoretic Concepts in Computer Science

37th International Workshop, WG 2011
Teplá Monastery, Czech Republic, June 21-24, 2011
Revised Papers

 Springer

Volume Editors

Petr Kolman
Jan Kratochvíl
KAM MFF UK
Charles University
Malostranské nám 25, 11800 Praha 1, Czech Republic
E-mail: {kolman, honza}@kam.mff.cuni.cz

ISSN 0302-9743   e-ISSN 1611-3349
ISBN 978-3-642-25869-5   ISBN 978-3-642-25870-1 (eBook)
DOI 10.1007/978-3-642-25870-1
Springer Heidelberg Dordrecht London New York

Library of Congress Control Number: Applied for

CR Subject Classification (1998): G.2.2, I.2.8, E.1, F.2, I.3.5, C.2

LNCS Sublibrary: SL 1 – Theoretical Computer Science and General Issues

Typesetting: Camera-ready by author, data conversion by Scientific Publishing Services, Chennai, India

Printed on acid-free paper

Springer is part of Springer Science+Business Media (www.springer.com)

Preface

The 37th International Workshop on Graph-Theoretic Concepts in Computer Science (WG 2011) took place in Teplá Monastery, Czech Republic, during June 21–24, 2011. It was attended by 80 participants who came from all over the world not only to deliver or listen to interesting talks, but also to celebrate the 65th birthday of Luděk Kučera, a long-time member of the WG Steering Committee.

The conference series has a long tradition. Since 1975, WG has been organized 21 times in Germany, four times in The Netherlands, twice in Austria, France and the Czech Republic, and once in Greece, Italy, Norway, Slovakia, Switzerland, and the UK. WG aims at merging theory and practice by demonstrating how concepts from graph theory can be applied to various areas in computer science, and by extracting new graph theoretic problems from applications. Its goal is to present emerging research results and to identify and explore directions of future research. As always, this year's conference was well-balanced with respect to established researchers and young scientists.

This year's conference received 54 submissions, two of which were withdrawn before entering the review process. Each submission was carefully reviewed by at least three, and on average four, members of the Program Committee. The Committee decided to accept 28 papers for presentation at the conference and publication in the proceedings. The conference program was further enriched by two invited talks presented by Dániel Marx and Alberto Marchetti-Spaccamela and whose extended abstracts are also included in the proceedings.

The site of the conference was the Premonstratensian abbey at Teplá, near Mariánské lázně in western Bohemia. As mentioned above, the annual WG workshops have a long tradition going back to 1975 when the first WG took place in Berlin. The Premonstratensian Order has an incomparably longer tradition of, among other virtues, promoting science, technology and knowledge in general. The abbey at Teplá used to be a center of art and science for centuries, its library was one of the largest in the country, the abbey used to have its own observatory. Among the Premonstratensians from Teplá one can find rectors of the Charles University in Prague (Chrysostomus Pfrogner, Alois David), the vice-rector of the Naval Academy at Rijeka (Vojtěch Knight Kuneš) or the director of the Polytechnical Institute in Budapest (Lambert Mayer). Thus, though the vocation of a friar and a scientist may seem very different at the first sight, there are some similarities, too, the bottom line of which is a quest for truth. The abbey turned out to be a great venue for the conference.

The scientific program of the conference was complemented by a social program that contributed to a friendly and relaxing atmosphere. This included an organ concert given by a world-class organist, Aleš Bárta, in the baroque church of the abbey, a wine tasting of the finest selection of wines from Znovín Znojmo, and an excursion to Bečov Castle with a display of the unique reliquary of St. Maurus.

We wish to thank all who contributed to the success of WG 2011: the authors for submitting high-quality papers, the external reviewers for their timely reports, the Program Committee members for their excellent and responsible work, P. Augustin Ján Kováčik, O. Praem., Administrator, and Ms. Novotná from the abbey for their help and hospitality, Aleš Bárta for an excellent concert, Znovín Znojmo for sponsoring WG with top-quality wines from South Moravian vineyards, Luděk Kučera for suggesting the venue, our colleagues and students Jiří Fiala, Tomáš Vyskočil, and Martin Koutecký for their help with the organization, and last but not least Anna Kotěšovcová and her team from Conforg for a very smooth organization of WG 2011.

August 2011   Petr Kolman
                                                                  Jan Kratochvíl

Organization

Program Committee

Tetsuo Asano	Japan Advanced Institute of Science and Technology, Nomi, Japan
Andreas Brandstädt	University of Rostock, Germany
Leizhen Cai	Chinese University of Hong Kong, SAR China
Sunil Chandran	Indian Institute of Science, Bangalore, India
Jianer Chen	Texas A&M University, College Station, USA
Derek Corneil	University of Toronto, Canada
Michel Habib	LIAFA-Université Paris 7, France
Pinar Heggernes	University of Bergen, Norway
Juraj Hromkovič	Federal Institute of Technology Zurich, Switzerland
Petr Kolman	Charles University in Prague, Czech Republic
Jan Kratochvíl - Chair	Charles University in Prague, Czech Republic
Ludek Kučera	Charles University in Prague, Czech Republic
Ernst Mayr	Technical University Munich, Germany
Haiko Müller	University of Leeds, UK
Sang-Il Oum	Korea Advanced Institute of Science and Technology, Daejeon, Korea
Christophe Paul	National Center for Scientific Research, Montpellier, France
Andrzej Proskurowski	University of Oregon, Eugene, USA
Michal Stern	Caesarea Rothschild Institute and Academic College of Tel Aviv-Jaffa, Israel
Dimitrios Thilikos	National and Kapodistrian University of Athens, Greece
Dorothea Wagner	University of Karlsruhe, Germany

Additional Reviewers

Adler, Isolde	Bousquet, Nicolas
Aravind, N.R.	Chaplick, Steven
Asinowski, Andrei	Cibulka, Josef
Bachmaier, Christian	Daligault, Jean
Barat, Janos	Fernau, Henning
Basavaraju, Manu	Fiala, Jiří
Beck, Nili	Fleiner, Tamas
Bodlaender, Hans	Fomin, Fedor V.
Böckenhauer, Hans-Joachim	Fotakis, Dimitris

Francis, Mathew
Gemsa, Andreas
Giannopoulou, Archontia
Gioan, Emeric
Golovach, Petr
Gregor, Petr
Gurski, Frank
Görke, Robert
Hartmann, Tanja
Havet, Frederic
Hicks, Illya
Hunter, Paul
Jampani, Krishnam Raju
Kaminski, Marcin
Kanj, Iyad
Keller, Lucia
Klavík, Pavel
Knop, Dušan
Kolliopoulos, Stavros
Komm, Dennis
Koutsonas, Athanassios
Kratsch, Dieter
Krause, Philipp Klaus
Krug, Marcus
Krugel, Johannes
Král, Daniel
Královič, Richard
Kurur, Piyush
Le, Van Bang
Leveque, Benjamin
Lieber, Tobias
Limouzy, Vincent
Lin, Min Chih
Mamcarz, Antoine
McConnell, Ross
Meister, Daniel
Mertzios, George B.
Misra, Neeldhara
Mnich, Matthias
Molitor, Paul

Monaco, Gianpiero
Müller, Tobias
Nanongkai, Danupon
Nasre, Meghana
Nevries, Ragnar
Norine, Serguei
Obdržálek, Jan
Pajor, Thomas
Paulusma, Daniel
Perez, Anthony
Pergel, Martin
Raffinot, Mathieu
Rao, Michael
Rautenbach, Dieter
Rossmanith, Peter
Rotics, Udi
Rutter, Ignaz
Sau, Ignasi
Sawada, Joe
Schiermeyer, Ingo
Sereni, Jean-Sebastian
Sikdar, Somnath
Sivadasan, Naveen
Sprock, Andreas
Sritharan, R.
Stanton, Brendon
Steffen, Björn
Steinová, Monika
Tancer, Martin
Telle, Jan Arne
Tůma, Vojtěch
Täubig, Hanjo
Villanger, Yngve
Volec, Jan
Völker, Markus
Wahlström, Magnus
Weihmann, Jeremias
Wolff, Alexander
Zwols, Yori

Table of Contents

Structures and Hyperstructures in Metabolic Networks 1
  Alberto Marchetti-Spaccamela

Important Separators and Parameterized Algorithms 5
  Dániel Marx

Split Clique Graph Complexity 11
  *Liliana Alcón, Luerbio Faria, Celina M.H. de Figueiredo, and
  Marisa Gutierrez*

On Searching for Small Kochen-Specker Vector Systems.............. 23
  Felix Arends, Joël Ouaknine, and Charles W. Wampler

Characterizations of Deque and Queue Graphs...................... 35
  Christopher Auer and Andreas Gleißner

Graph Classes with Structured Neighborhoods and Algorithmic
Applications... 47
  Rémy Belmonte and Martin Vatshelle

Exact Algorithms for Kayles 59
  Hans L. Bodlaender and Dieter Kratsch

The Cinderella Game on Holes and Anti-holes 71
  *Marijke H.L. Bodlaender, Cor A.J. Hurkens, and
  Gerhard J. Woeginger*

On the Complexity of Planar Covering of Small Graphs.............. 83
  *Ondřej Bílka, Jozef Jirásek, Pavel Klavík, Martin Tancer, and
  Jan Volec*

Approximability of Economic Equilibrium for Housing Markets with
Duplicate Houses 95
  Katarína Cechlárová and Eva Jelínková

Planarization and Acyclic Colorings of Subcubic Claw-Free Graphs..... 107
  Christine Cheng, Eric McDermid, and Ichiro Suzuki

List Coloring in the Absence of a Linear Forest 119
  *Jean-François Couturier, Petr A. Golovach, Dieter Kratsch, and
  Daniël Paulusma*

Parameterized Complexity of Eulerian Deletion Problems   131
   Marek Cygan, Dániel Marx, Marcin Pilipczuk,
   Michał Pilipczuk, and Ildikó Schlotter

Restricted Cuts for Bisections in Solid Grids: A Proof via Polygons.   143
   Andreas Emil Feldmann, Shantanu Das, and Peter Widmayer

Maximum Independent Set in 2-Direction Outersegment Graphs   155
   Holger Flier, Matúš Mihalák, Peter Widmayer, and Anna Zych

Complexity of Splits Reconstruction for Low-Degree Trees   167
   Serge Gaspers, Mathieu Liedloff, Maya Stein, and Karol Suchan

Empires Make Cartography Hard: The Complexity of the Empire
Colouring Problem   179
   Andrew R.A. McGrae and Michele Zito

Alternation Graphs   191
   Magnús M. Halldórsson, Sergey Kitaev, and Artem Pyatkin

Improved Bounds for Minimum Fault-Tolerant Gossip Graphs   203
   Toru Hasunuma and Hiroshi Nagamochi

Parameterized Two-Player Nash Equilibrium   215
   Danny Hermelin, Chien-Chung Huang, Stefan Kratsch, and
   Magnus Wahlström

Counting Independent Sets in Claw-Free Graphs   227
   Konstanty Junosza-Szaniawski, Zbigniew Lonc, and
   Michał Tuczyński

On the Independence Number of Graphs with Maximum Degree 3   238
   Iyad A. Kanj and Fenghui Zhang

On Computing an Optimal Semi-matching   250
   František Galčík, Ján Katrenič, and Gabriel Semanišin

Planar k-Path in Subexponential Time and Polynomial Space   262
   Daniel Lokshtanov, Matthias Mnich, and Saket Saurabh

Approximability of the Path-Distance-Width for AT-free Graphs   271
   Yota Otachi, Toshiki Saitoh, Katsuhisa Yamanaka, Shuji Kijima,
   Yoshio Okamoto, Hirotaka Ono, Yushi Uno, and Koichi Yamazaki

Hanani-Tutte and Monotone Drawings.   283
   Radoslav Fulek, Michael J. Pelsmajer, Marcus Schaefer, and
   Daniel Štefankovič

On Collinear Sets in Straight-Line Drawings.   295
   Alexander Ravsky and Oleg Verbitsky

From Few Components to an Eulerian Graph by Adding Arcs   307
    Manuel Sorge, René van Bevern, Rolf Niedermeier, and
    Mathias Weller

Recognizing Some Subclasses of Vertex Intersection Graphs of 0-Bend
Paths in a Grid   319
    Steven Chaplick, Elad Cohen, and Juraj Stacho

A Polynomial Time Algorithm for Bounded Directed Pathwidth   331
    Hisao Tamaki

Author Index   343

Structures and Hyperstructures
in Metabolic Networks

Alberto Marchetti-Spaccamela*

Sapienza Università di Roma, Italy

1   Introduction

There has been an increasing interest by the computational biology community
in the study of chemical reactions within cells; indeed cells can be considered
as chemical factories that manufacture the various products of the cells and the
metabolic capacities of an organism are directly defined by the set of its possible
biochemical reactions. The links between reactions and compounds (or metabolites) that are used and produced by such reactions constitute the *metabolic
network* of an organism.

A metabolic network consists of a set of metabolites and a set of reactions.
Each reaction transforms a subset of metabolites, the *substrates*, into another
subset of metabolites, the *products* of the reaction. Such a network can be modelled as a directed hypergraph $G = (N, R)$ with N being the set of vertices
corresponding to *metabolites* (also called *compounds*) and R the set of hyper-
edges corresponding to *reactions*. A hyperedge $r \in R$ is directed away from a
compound $c \in N$ only if c is a substrate of r, and directed into c only if c is a
product of r. Clearly we can also represent a directed hypergraph using directed
bipartite graphs with node sets (A, B) and arcs in both direction. Namely, A denotes the set of compounds and B the set of reactions; if x is substrate (product)
of reaction R then there is a directed arc from x to R (from R to x).

Note that reactions can be reversible and that we need to take into account
stoichiometry. Namely, each reversible reaction can be modeled as two different
reactions of opposite direction and an integer weight associated to each sub-
strate/product of a reaction allows to represent the stoichiometry coefficient of
the compound in the reaction. Next to that we observe that a subset of the
compounds which are nutrients of the cell could be regarded as being available
in infinite supply (for instance, from the environment).

The above presentation does not exhaust all possible biological aspects that
could be taken into account (e.g. it does not model the role of proteins in cat-
alyzing the reactions). We refer to the survey [8] and to references therein for a
more thoroughly presentation.

2   Structural Characterization

The vast literature focusing on metabolic networks can be roughly classified in two
categories depending on whether networks are studied either from a structural

* Partially supported by project INRIA ARC project SIMBIOSI.

P. Kolman and J. Kratochvíl (Eds.): WG 2011, LNCS 6986, pp. 1–4, 2011.

perspective, or from a dynamic one. In this section we briefly refer to the first perspective.

In 1999 Barabasi and his group published two papers that had a strongly impact in the community. The papers' contribution can be summarized in two main claims; the first one was the apparently startling discovery that the distribution of connections of compounds (i.e. the number hyperlinks in which the compound is involved) follows a power law; namely, the frequency of nodes with connectivity k falls exponentially as $k^{-\alpha}$ where α denotes the power law coefficient [5].

The second main claim was that the finding of a power law distribution indicates that metabolic network are *scale-free* networks [11]. This second claim had a big impact in the biological community and in the media. In fact the claim was counter to traditional expectations of the community and for this reason its occurrence seems to imply some deep meaning as similar discoveries in physics (e.g. the occurrence of scale free networks had a deep impact in the study of phase transitions).

The above papers had a broad interest in the biological community for their impact: the papers proposed a model for metabolic networks that provides a proper probabilistic framework that could be used a reference model for biology.

More recently there has been a detailed analysis of Barabasi's claims (see for example [6,7,12]). Main criticisms are: there was a high rate errors in used data and available data can be also explained using other degree distributions (not scale-free); moreover the analysis was mainly done modeling the network using a graph that shows interaction of compounds but misses crucial aspects of metabolic reactions (e.g. direction of the reaction, conservation of mass). Finally it was also remarked that scale free networks are very general: if metabolic networks are scale free then this does not provide any clue on them.

Preliminary investigations using two well known structural characterization of graphs confirm that metabolic networks are not scale free. Namely, we have performed experiments in characterizing the structural properties of the directed bipartite graph representing the network using structural characterization as treewidth and Kelly width. Our results show the existence of a core subnetwork that involves only a fraction of the compounds of the metabolic network and that seems to characterize the structure of the network.

We finally mention that in [9] the authors show that, similarly to the Web, metabolic networks show a bow-tie structure. As a conclusion the structural characterization of metabolic networks that might allow to obtain a a suitable structural model that might provide a framework for the analysis of specific networks [4].

3   Dynamic Characterization

The dynamics of metabolic networks can be also based on graph-related formalisms on a constraint-based modelling in which the network may still be modelled as an edge-labelled hypergraph, but several types of constraints are added to restrict the possible fluxes through the network (to take into account other aspects such as stoichiometric and thermodynamic constraints).

The choice of a particular model heavily depends on the type of the specific question one wishes to address but also on the type of data that is available. An important aspect that must be taken into account is the computational cost of a given analysis, and therefore its scalability to large datasets (such as genome-scale metabolic networks). This might force the choice of tractable models.

Significant research has been carried on in studying suitably defined subnetworks of a given metaboic network. An important aspect is the study of admissible flux distributions that corresponds to a set of reactions, which, perform the transformation of available substrates into removable products with the special property that all intermediate compounds are balanced (steady-state assumption) and irreversible reactions are taken in the appropriate direction (thermodynamic constraint). Such an admissible flux distribution is called a mode; finding modes in a network is equivalent to finding suitably defined hyperpaths in the network.

A central concept in this methodology is the notion of an elementary mode which represents a minimal functional subsystem and that can be detected by finding minimal hyperpaths in the network. The computation of elementary modes still forms a limiting step in metabolic studies whose computational complexity has been characterized [1,2]. Moreover, several algorithms have been proposed to address this problem leading to increasingly faster methods. However, although a theoretical upper bound on the number of elementary modes that a network may possess has been established and enumeration algorithms have been proposed [13], the complexity of this enumeration problem is a main open problem in the area.

Since even relatively small networks have many elementary modes and given that their enumeration seems out of reach, it is important to suitably restrict the definition for which enumeration is possible either theoretically or in practice. This opens the possibility to define new problems that can be directly formulated as graph problems. As an example, in [3] the authors study the problem of finding minimal sets of metabolites (called precursors) that are sufficient to produce a set of target metabolites. The model takes into account self-regenerating metabolites involved in hypercycles, which may be used to generate target metabolites from potential precursors. Even if the problem of enumerating minimal precursors is intractable an algorithm to enumerate all minimal precursor sets for a set of target metabolites can be applied in real networks to identify a minimal medium necessary for a cell to ensure some metabolic functions.

The structural analysis of metabolic networks aims both at understanding the function and the evolution of metabolism. While it is commonly admitted that metabolism is modular, the identification of metabolic modules remains an open topic. Several definitions of what is a module have been proposed, and the research in the area is still very active. The notion of chemical organizations, aims to define sets of molecules which are closed and self-maintaining [10].

References

1. Acuña, V., Chierichetti, F., Lacroix, V., Marchetti-Spaccamela, A., Sagot, M.F., Stougie, L.: Modes and cuts in metabolic networks: Complexity and algorithms. Biosystems 95(1), 51–60 (2009)
2. Acuña, V., Marchetti-Spaccamela, A., Sagot, M.F., Stougie, L.: A note on the complexity of finding and enumerating elementary modes. Biosystems 99(3), 210–214 (2010)
3. Cottret, L., Milreu, P.V., Acuña, V., Marchetti-Spaccamela, A., Stougie, L., Charles, H., Sagot M.F.: Graph-based analysis of the metabolic exchanges between two co-resident intracellular symbionts, baumannia cicadellinicola and sulcia muelleri, with their insect host, homalodisca coagulata. PLOS Computational Biology 6 (2010)
4. Picard, F., Daudin, J.-J., Robin, S.: A mixture model for random graphs. Statistics and Computing 18(2), 173–183 (2008)
5. Jeong, H., Tombor, B., Albert, R., Oltvai, Z.N., Barabási, A.-L.: The large-scale organization of metabolic networks. Nature 407(5), 651–654 (2000)
6. Keller, E.F.: Revisiting scale-free networks. Bioessays 27(10), 1060–1068 (2005)
7. Khanin, R., Wit, E.: How scale-free are biological networks. Journal of Computational Biology 13(3), 810–818 (2006)
8. Lacroix, V., Cottret, L., Thébault, P., Sagot, M.F.: An introduction to metabolic networks and their structural analysis. IEEE/ACM Trans. Comput. Biology Bioinform. 5(4), 594–617 (2008)
9. Ma, H., Zeng, A.P.: Reconstruction of metabolic networks from genome data and analysis of their global structure for various organisms. Bioinformatics 19(2), 270–277 (2003)
10. Milreu, P.V., Acuña, V., Birmelé, E., Crescenzi, P., Marchetti-Spaccamela, A., Sagot, M.F., Stougie, L., Lacroix, V.: Enumerating Chemical Organisations in Consistent Metabolic Networks: Complexity and Algorithms. In: Moulton, V., Singh, M. (eds.) WABI 2010. LNCS, vol. 6293, pp. 226–237. Springer, Heidelberg (2010)
11. Ravasz, E., Somera, A.L., Mongru, D.A., Oltvai, Z.N., Barabási, A.L.: Hierarchical organization of modularity in metabolic networks. Science 297, 1551–1555 (2002)
12. Stumpf, M.P.H., Wiuf, C., May, R.M.: Subnets of scale-free networks are not scale-free: Sampling properties of networks. PNAS 102(12), 4221–4224 (2005)
13. Terzer, M., Stelling, J.: Large-scale computation of elementary flux modes with bit pattern trees. Bioinformatics 24(19), 2229–2235 (2008)

Important Separators and Parameterized Algorithms

Dániel Marx

Institut für Informatik, Humboldt-Universität zu Berlin, Germany
dmarx@cs.bme.hu

Abstract. The notion of "important separators" and bounding the number of such separators turned out to be a very useful technique in the design of fixed-parameter tractable algorithms for multi(way) cut problems. For example, the recent breakthrough result of Chen et al. [3] on the DIRECTED FEEDBACK VERTEX SET problem can be also explained using this notion. In my talk, I will overview combinatorial and algorithmic results that can be obtained by studying such separators.

1 Introduction

Problems related to cutting a graph into parts satisfying certain properties or separating different parts of the graph from each other form a classical area of graph theory and combinatorial optimization, with strong motivation coming from applications. The study of these problems revealed deep mathematical structures, such as connections to linear programming and semidefinite programming. In this talk, we explore an aspect of these problems that has been investigated and exploited only recently. It seems that understanding the extremal properties of small separators can be used to obtain combinatorial results and fixed-parameter tractability results. In particular, the notion of "important separators" has been used (implicitly or explicitly) in recent results on parameterized algorithms for separation and related problems [11,15,2,12,1,10,9].

An (X, Y)-separator is a set S of edges that separate X and Y for each other, that is, $G \setminus S$ has no component containing vertices from both X and Y (most of what we discuss here can be extended to vertex cutsets, but for simplicity we stick to edge cuts now). An (X, Y)-separator S is *inclusionwise minimal* if no subset $S' \subset S$ is an (X, Y)-separator. The main definition of the talk is the following:

Definition 1. Let $X, Y \subseteq V(G)$ be vertices, $S \subseteq E(G)$ be an (X, Y)-separator, and let R be the set of vertices reachable from X in $G \setminus S$. We say that S is an important (X, Y)-separator if it is inclusionwise minimal and there is no (X, Y)-separator S' with $|S'| \leq |S|$ such that $R \subset R'$, where R is the set of vertices reachable from X in $G \setminus S'$.

Note that an important (X, Y)-separator is not necessarily an important (Y, X)-separator. Intuitively, we want to minimize the size of the (X, Y)-separator and

P. Kolman and J. Kratochvíl (Eds.): WG 2011, LNCS 6986, pp. 5–10, 2011.

at the same time we want to maximize the set of vertices that remain reachable from X after removing the separator. The important separators are the separators that are Pareto-optimal with respect to these two objectives. Note that we do not want the number of vertices reachable from X to be maximal, we just want that this set of vertices is *inclusionwise maximal* (i.e., we have $R \subset R'$ and *not* $|R| < |R'|$ in the definition). The main observation of [11] is that the number of important (X,Y)-separators of size at most k can be bounded by a function of k; a better bound is implicit in [2].

Theorem 2. [11,2] *Let $X, Y \subseteq V(G)$ be two sets of vertices in graph G, let $k \geq 0$ be an integer, and let $\mathcal{S}_k$ be the set of all (X,Y)-important separators of size at most k. Then $|\mathcal{S}_k| \leq 4^k$ and $\mathcal{S}_k$ can be constructed in time $|\mathcal{S}_k| \cdot n^{O(1)}$.*

The following lemma clearly proves the bound in Theorem 2: if the sum is at most 1, then there cannot be more than 4^k important (X,Y)-separators of size at most k.

Lemma 3. [10] *Let $X, Y \subseteq V(G)$. If $\mathcal{S}$ is the set of all important (X,Y)-separators, then $\sum_{S \in \mathcal{S}} 4^{-|S|} \leq 1$.*

As an application, we can prove the following surprisingly simple, but still non-trivial combinatorial result:

Lemma 4. *Let $X, Y \subseteq V(G)$. The union of all inclusionwise minimal (X,Y)-separators of size at most k contains at most $k \cdot 4^k$ edges incident to Y.*

2   Multiway Cut

Let G be a graph and T be a set of terminals. A *multiway cut* is a set S of edges such that every component of $G \setminus S$ contains at most one vertex of T.

MULTIWAY CUT
Input: Graph G, set T of vertices, integer k
Find: A multiway cut S of size at most k

The MULTIWAY CUT problem is known to be NP-hard already for $|T| = 3$ terminals [6] (for two terminals, it is the classical minimum $s - t$ cut problem, hence it is polynomial-time solvable). For every fixed k, the problem is polynomial-time solvable: using brute force, we can try all possible subsets of k edges in time $n^{O(k)}$. Of course, for moderately large values of n, such a solution seems to be practically useless already for very small values of k, say for $k = 10$. Can we do anything significantly smarter than complete enumeration of these subsets? The main goal of parameterized complexity is to design algorithms where the combinatorial explosion is restricted to a well-defined parameter (such as the size k of the solution we are looking for). Recall that a problem with a parameter k is *fixed-parameter tractable* if it can be solved in time $f(k) \cdot n^{O(1)}$ for some function f depending only on the parameter k [7,8,13]. If $f(k)$ is "nice," say, $f(k)$ is c^k for some small constant c, then such an algorithm can be useful for small values of k even for large n.

We show that MULTIWAY CUT is FPT parameterized by k. The following observation connects MULTIWAY CUT and the concept of important separators:

Lemma 5 (Pushing Lemma). *Let* $t \in T$ *be a terminal that is not separated from* $T \setminus t$ *in* G. *If* G *has a multiway cut* S, *then it also has a multiway cut* S' *with* $|S'| \leq |S|$ *that contains an important* $(t, T \setminus t)$-*separator.*

Using this observation, we can solve the problem by branching on the choice of an important separator and including it into the solution:

Theorem 6. [2] MULTIWAY CUT *can be solved in time* $4^k \cdot n^{O(1)}$.

Proof. We solve the problem by a recursive branching algorithm. If all the terminals are separated from each other, then we are done. Otherwise, let $t \in T$ be a terminal not separated from the rest of the terminals. Let us use the algorithm of Theorem 2 to construct the set $\mathcal{S}_k$ consisting of every important $(t, T \setminus t)$-separator of size at most k. By Lemma 5, there is a solution that contains one of these separators. Therefore, we branch on the choice of one of these separators, and for every important separator $S' \in \mathcal{S}_k$, we recursively solve the MULTIWAY CUT instance $(G \setminus S', k - |S'|)$. If one of these branches returns a solution S, then clearly $S \cup S'$ is a multiway cut of size at most k in G.

The correctness of the algorithm is clear from Lemma 5. We claim that the search tree explored by the algorithm has at most 4^k leaves. We prove this by induction on k, thus let us assume that the statement is true for every value less than k. This means that we know that the recursive call $(G \setminus S', k - |S'|)$ explores a search tree with at most $4^{k-|S'|}$ leaves. Using Lemma 3, we can bound the number of leaves of the search tree by

$$\sum_{S' \in \mathcal{S}_k} 4^{k-|S'|} \leq 4^k \cdot \sum_{S' \in \mathcal{S}_k} 4^{-|S'|} \leq 4^{-k}.$$

$\square$

The running time can be improved from $4^k \cdot n^{O(1)}$ to $2^k \cdot n^{O(1)}$ with somewhat different techniques [15,5].

A natural generalization of MULTIWAY CUT can be obtained if, instead of requiring that all the terminals are separated from each other, we require that a specified set of pairs are separated from each other:

MULTICUT
Input: Graph G, pairs (s_1, t_1), ..., (s_ℓ, t_ℓ), integer k
Find: A set S of at most k edges such that $G \setminus S$ has no $s_i - t_i$ path for any i

---

Theorem 6 implies that MULTICUT is FPT jointly parameterized by k and ℓ, that is, can be solved in time $f(k, \ell) \cdot n^{O(1)}$. We can guess how the solution S partitions the 2ℓ vertices s_i, t_i ($1 \leq i \leq \ell$), identify those vertices that are supposed to be in the same component of $G \setminus S$, and solve the resulting MULTIWAY CUT instance. It is a more challenging question whether the problem is FPT parameterized by k (the size of the solution) only. Very recently, a positive answer was given to this question:

Theorem 7. [1,12] MULTICUT *is FPT parameterized by k.*

The proof in [12] introduces a new way of using important separators: with the "random sampling of important separators" technique we can significantly simplify the problem instance. This technique has found applications for other problems [4,10] and it is very likely that it will be of use in the future.

3   Directed Graphs

Problems on directed graphs are notoriously more difficult than problems on undirected graphs. This is phenomenon has been observed equally often in the area of polynomial-time algorithms, approximability, and fixed-parameter tractability. Let us see if the techniques based on important separators survive the generalization to directed graphs. First, important separators can be defined analogously for directed graphs and the bound of 4^k of Lemma 2 still holds. This gives us some hope that we would be able to use the technique on directed graphs and in particular to show that DIRECTED MULTIWAY CUT (Delete a set of at most k edges such that there is no $t_1 \rightarrow t_2$ path in $G \setminus S$ for any two distinct $t_1, t_2 \in T$) is fixed-parameter tractable. However, the Pushing Lemma (Lemma 5) is not true on directed graphs. This means that a straightforward generalization of Theorem 6 to directed graphs is not possible. Nevertheless, Chitnis et al. [4] showed, using the random sampling technique of [12], that the problem is FPT:

Theorem 8. [4] DIRECTED MULTIWAY CUT *is FPT.*

What about the more general DIRECTED MULTICUT problem? In contrast to the undirected version, the directed problem is W[1]-hard parameterized by k [12]. But the problem can be interesting even for small values of ℓ. The case $\ell = 2$ can be reduced to DIRECTED MULTIWAY CUT in a simple way, thus Theorem 8 implies that DIRECTED MULTICUT for $\ell = 2$ is FPT parameterized by k. The case of a fixed $\ell \geq 3$ and the case of jointly parameterizing with ℓ and k are still open.

Chen et al. [3] considered the following (slightly unnatural) variant of DIRECTED MULTICUT:

SKEW MULTICUT
Input: Graph G, pairs (s_1, t_1), ..., (s_ℓ, t_ℓ), integer k
Find: A set S of at most k edges such that $G \setminus S$ has no $s_i \rightarrow t_j$ path for any $i \leq j$

For this problem, the Pushing Lemma can be made to work: there is a solution that contains an important $(s_1, \{t_1, \ldots, t_\ell\})$-separator. Therefore, arguments analogous to Theorem 6 give:

Theorem 9. [3] SKEW MULTICUT *is* FPT *parameterized by* k.

The reason why Chen et al. [3] considered this problem is that it formed an important ingredient in their proof showing that DIRECTED FEEDBACK VERTEX SET is FPT.

DIRECTED FEEDBACK VERTEX SET
Input: A directed graph G, integer k
Find: A set S of at most k vertices such that $G \setminus S$ has no directed cycle

---

Using the technique of iterative compression (introduced by Reed et al. [14]), Chen et al. [3] gave a nice reduction from DIRECTED FEEDBACK VERTEX SET to SKEW MULTICUT. Together with Theorem 9, this reduction established the fixed-parameter tractability of DIRECTED FEEDBACK VERTEX SET, resolving a longstanding open problem.

Theorem 10. [3] DIRECTED FEEDBACK VERTEX SET *is* FPT *parameterized by* k.

4 Conclusions

The notion of important separators seems to be useful for a wide range of combinatorial and algorithmic problems. In a particular application, first we need to observe that important separators are relevant (an example of this is the Pushing Lemma for MULTIWAY CUT) and then we can try to apply the upper bound of Theorem 2. The random sampling technique of [12] raises the applicability of important separators to a new level. After the initial application for MULTICUT, randomized sampling turned out to be useful for DIRECTED MULTIWAY CUT [4], and, in a very different context, for a clustering problem [10]. Based on the recent surge of results using important separators, one can safely expect that it will find further uses.

References

1. Bousquet, N., Daligault, J., Thomassé, S.: Multicut is FPT. In: Proceedings of the 43rd ACM Symposium on Theory of Computing, pp. 459–468 (2011)
2. Chen, J., Liu, Y., Lu, S.: An Improved Parameterized Algorithm for the Minimum Node Multiway Cut Problem. In: Dehne, F., Sack, J.-R., Zeh, N. (eds.) WADS 2007. LNCS, vol. 4619, pp. 495–506. Springer, Heidelberg (2007)
3. Chen, J., Liu, Y., Lu, S., O'Sullivan, B., Razgon, I.: A fixed-parameter algorithm for the directed feedback vertex set problem. J. ACM 55(5) (2008)

4. Chitnis, R., Hajiaghayi, M., Marx, D.: Fixed-parameter tractability of directed multiway cut parameterized by the size of the cutset. Accepted to SODA (2012)
5. Cygan, M., Pilipczuk, M., Pilipczuk, M., Wojtaszczyk, J.: On multiway cut parameterized above lower bounds. Accepted to IPEC (2011)
6. Dahlhaus, E., Johnson, D.S., Papadimitriou, C.H., Seymour, P.D., Yannakakis, M.: The complexity of multiterminal cuts. SIAM J. Comput. 23(4), 864–894 (1994)
7. Downey, R.G., Fellows, M.R.: Parameterized Complexity. Monographs in Computer Science. Springer, New York (1999)
8. Flum, J., Grohe, M.: Parameterized Complexity Theory. Springer, Berlin (2006)
9. Heggernes, P.: van 't Hof, P., Lokshtanov, D., Paul, C.: Obtaining a bipartite graph by contracting few edges. CoRR abs/1102.5441 (2011)
10. Lokshtanov, D., Marx, D.: Clustering with Local Restrictions. In: Aceto, L., Henzinger, M., Sgall, J. (eds.) ICALP 2011. LNCS, vol. 6755, pp. 785–797. Springer, Heidelberg (2011)
11. Marx, D.: Parameterized graph separation problems. Theoret. Comput. Sci. 351(3), 394–406 (2006)
12. Marx, D., Razgon, I.: Fixed-parameter tractability of multicut parameterized by the size of the cutset. In: Proceedings of the 43nd ACM Symposium on Theory of Computing, pp. 469–478 (2011)
13. Niedermeier, R.: Invitation to fixed-parameter algorithms. Oxford Lecture Series in Mathematics and its Applications, vol. 31. Oxford University Press, Oxford (2006)
14. Reed, B., Smith, K., Vetta, A.: Finding odd cycle transversals. Operations Research Letters 32(4), 299–301 (2004)
15. Xiao, M.: Algorithms for Multiterminal Cuts. In: Hirsch, E.A., Razborov, A.A., Semenov, A., Slissenko, A. (eds.) CSR 2008. LNCS, vol. 5010, pp. 314–325. Springer, Heidelberg (2008)

Split Clique Graph Complexity

Liliana Alcón[1], Luerbio Faria[2],
Celina M.H. de Figueiredo[3], and Marisa Gutierrez[1,4]

[1] Universidad Nacional de La Plata, La Plata, Argentina
[2] Universidade do Estado do Rio de Janeiro, Rio de Janeiro, Brazil
[3] Universidade Federal do Rio de Janeiro, Rio de Janeiro, Brazil
[4] CONICET, Argentina

Abstract. A *complete set* of a graph G is a subset of vertices inducing a complete subgraph. A *clique* is a maximal complete set. Denote by $\mathcal{C}(G)$ the *clique family* of G. The *clique graph* of G, denoted by $K(G)$, is the intersection graph of $\mathcal{C}(G)$. Say that G is *a clique graph* if there exists a graph H such that $G = K(H)$. The clique graph recognition problem, a long-standing open question posed in 1971, asks whether a given graph is a clique graph and it was recently proved to be NP-complete even for a graph G with maximum degree 14 and maximum clique size 12. Hence, if P$\neq$NP, the study of graph classes where the problem can be proved to be polynomial, or of more restricted graph classes where the problem remains NP-complete is justified. We present a proof that given a split graph $G = (V, E)$ with partition (K, S) for V, where K is a complete set and S is a stable set, deciding whether there is a graph H such that G is the clique graph of H is NP-complete. As a byproduct, we prove that a problem about the Helly property on a family of sets is NP-complete. Our result is optimum in the sense that each vertex of the independent set of our split instance has degree at most 3, whereas when each vertex of the independent set has degree at most 2 the problem is polynomial, since it is reduced to check whether the clique family of the graph satisfies the Helly property. Additionally, we show three split graph subclasses for which the problem is polynomially solvable: the subclass where each vertex of S has a private neighbor, the subclass where $|S| \leq 3$, and the subclass where $|K| \leq 4$.

Keywords: clique graph, Helly property, NP-complete, split graphs.

1 Introduction

Consider finite, simple and undirected graphs. V and E denote the vertex set and the edge set of the graph G, respectively. A *complete set* of G is a subset of V inducing a complete subgraph. A *clique* is a maximal complete set. The *clique family* of G is denoted by $\mathcal{C}(G)$. The *clique graph* of G is the intersection graph of $\mathcal{C}(G)$.

The *clique operator*, K, assigns to each graph G its clique graph which is denoted by $K(G)$. On the other hand, say that G is *a clique graph* if G belongs to the image of the clique operator, i.e. if there exists a graph H such that $G = K(H)$.

P. Kolman and J. Kratochvíl (Eds.): WG 2011, LNCS 6986, pp. 11–22, 2011.
© Springer-Verlag Berlin Heidelberg 2011

Clique operator and its image were widely studied. First articles focused on recognizing clique graphs [7,11]. Graphs fixed under the operator K or fixed under the iterated clique operator, K^n, for some positive integer n; and the behavior under these operators of parameters such as number of vertices or diameter were studied. For several classes of graphs, the image of the class under the clique operator was characterized; and, in some cases, also the inverse image of the class. Results of the previous bibliography can be found in the survey [14]. Clique graphs have been much studied as intersection graphs and are included in several books [5,8,10].

The characterization of clique graphs given in [11] proposed the computational complexity of the recognition of clique graphs, a long-standing open question [5,10,11,14] just recently settled as NP-complete [1,2].

A graph is *split* if its vertex set can be partitioned into a complete set and a stable set. In this paper, we are concerned with the time complexity of the problem of recognizing split clique graphs, for which we establish NP-complete and polynomial results.

SPLIT CLIQUE GRAPH
INSTANCE: A split graph $G = (V, E)$.
QUESTION: Is there a graph H such that $G = K(H)$?

We prove that SPLIT CLIQUE GRAPH is NP-complete. As a byproduct, we prove that a problem about the Helly property is NP-complete. Given a *set family* $\mathcal{F} = (F_i)_{i \in I}$, the sets F_i are called *members* of the family. $F \in \mathcal{F}$ means that F is a member of $\mathcal{F}$. The family is *pairwise intersecting* if the intersection of any two members is not the empty set. The *intersection* or *total intersection* of $\mathcal{F}$ is the set $\bigcap \mathcal{F} = \bigcap_{i \in I} F_i$. The family $\mathcal{F}$ has the *Helly property*, if any pairwise intersecting subfamily has nonempty total intersection. Besides the theoretical interest, the Helly property has applications in many different areas such as optimization and location problems, semantics, coding, computational biology, data bases, image processing and, in special, graph theory where it has been a useful and a natural tool. We refer to [6] for a survey on the Helly property and its complexity aspects.

Given a family of sets $\mathcal{F}$, say that a family $\mathcal{F}'$ is a *spanning* family for $\mathcal{F}$ if: $\bigcup_{F' \in \mathcal{F}'} F' = \bigcup_{F \in \mathcal{F}} F$; for each $F' \in \mathcal{F}'$, $|F'| > 1$; for each $F' \in \mathcal{F}'$, there exists $F \in \mathcal{F}$ such that $F' \subseteq F$; and for each $F \in \mathcal{F}$, $\bigcup_{F' \subseteq F, F' \in \mathcal{F}'} F' = F$.

SPANNING HELLY FAMILY
INSTANCE: A family of sets $\mathcal{F}$.
QUESTION: Does $\mathcal{F}$ admit a spanning family $\mathcal{F}'$ that satisfies the Helly property?

Our NP-completeness result yields that SPANNING HELLY FAMILY is NP-complete even when restricted to the members of the input family $\mathcal{F}$ having cardinality 2 or 3. Note that the problem is polynomial when all members of $\mathcal{F}$ have cardinality 2, and we leave as open the problem when all members of $\mathcal{F}$ have cardinality exactly 3. Note that the problem 3SAT$_{\overline{3}}$, defined in Section 2, when restricted to having exactly three literals per clause is polynomial [9].

2   NP-Complete Split Clique Graph Classes

Theorem 1 is a well known characterization of Clique Graphs. The edge with end vertices u and v is represented by uv. We say that the complete set C *covers* the edge uv when u and v belong to C. A *complete set edge cover* of a graph G is a family of complete sets of G covering all edges of G.

Theorem 1 (Roberts and Spencer [11]). *G is a clique graph if and only if there exists a complete set edge cover of G satisfying the Helly property.*

Notice that for any graph G the clique family $\mathcal{C}(G)$ is a complete set edge cover of G, but, in general, this family does not satisfy the Helly property. Graphs such that $\mathcal{C}(G)$ satisfies the Helly property are called *clique-Helly* graphs. It follows from Theorem 1 that every clique-Helly graph is a clique graph. In [13], clique-Helly graphs are characterized and a polynomial-time algorithm for their recognition is presented. Lemma 2 extends that result and leads to a polynomial-time algorithm to check if a given complete set edge cover of a graph satisfies the Helly property which in turn yields that CLIQUE GRAPH is in NP [1,2].

A *triangle* is a complete set with exactly 3 vertices. The set of triangles of G is denoted $T(G)$. Let $\mathcal{F}$ be a complete set edge cover of G and T a triangle, and denote by $\mathcal{F}_T$ the subfamily of $\mathcal{F}$ formed by all the members containing at least two vertices of T.

Lemma 2 (Alcón and Gutierrez [3]). *Let $\mathcal{F}$ be a complete set edge cover of G. The following conditions are equivalent:*
i) $\mathcal{F}$ has the Helly property.
ii) For every $T \in T(G)$, the subfamily $\mathcal{F}_T$ has the Helly property.
iii) For every $T \in T(G)$, the subfamily $\mathcal{F}_T$ has nonempty intersection, this means $\bigcap \mathcal{F}_T \neq \emptyset$.

A graph admits a complete set edge cover with the Helly property if and only if the graph admits a complete set edge cover with the Helly property such that no member is contained in another; such cover is called an *RS-family* of the graph. Thus Theorem 1 is equivalent to the following simpler statement: G is a clique graph if and only if G admits an RS-family. The following properties are stated and proved by Roberts and Spencer [11].

Lemma 3 (Lemma 1 and Theorem 3 of [11]). *Let $\mathcal{F}$ be an RS-family of a graph G. Then $\mathcal{F}$ contains a complete set of size 2 if and only if this complete set is a clique of G. If a triangle T is a clique of G, then T is a member of $\mathcal{F}$.*

We show that SPLIT CLIQUE GRAPH is NP-complete by a reduction from the following version of the 3–satisfiability problem with at most 3 occurrences per variable [9]. Let $U = \{u_i, 1 \leq i \leq n\}$ be a set of boolean variables. A literal is either a variable u_i or its complement $\overline{u_i}$. A clause over U is a set of literals. Let $C = \{c_j, 1 \leq j \leq m\}$ be a collection of clauses over U. We say that variable u_i occurs in clause c_j (and then in C) if u_i or $\overline{u_i} \in c_j$. We say that variable u_i

occurs in clause c_j as literal u_i (or that literal u_i occurs in c_j) if $u_i \in c_j$, and as literal $\overline{u_i}$ (or that literal $\overline{u_i}$ occurs in c_j) if $\overline{u_i} \in c_j$.

$3\text{SAT}_{\overline{3}}$

INSTANCE: $I = (U, C)$, where $U = \{u_i, 1 \le i \le n\}$ is a set of boolean variables, and $C = \{c_j, 1 \le j \le m\}$ a set of clauses over U such that each clause has two or three variables, each variable occurs at most three times in C.

QUESTION: Is there a truth assignment for U such that each clause in C has at least one true literal?

In order to reduce $3\text{SAT}_{\overline{3}}$ to SPLIT CLIQUE GRAPH, we need to construct in polynomial time a particular instance G_I of SPLIT CLIQUE GRAPH from a generic instance $I = (U, C)$ of $3\text{SAT}_{\overline{3}}$, in such a way that the constructed graph G_I is a clique graph if and only if C is satisfiable. The particular instance G_I is a 3-split graph and we first characterize 3-split clique graphs.

3-Split Graphs

A split graph admits a *split partition* of its vertex set into a complete set K and a stable set S. The family of cliques of a split graph with split partition (K, S) is composed by the closed neighbourhood $N[s]$, for each $s \in S$, and the complete set K if it is not contained in $N[s]$, for $s \in S$. An *ℓ-cone* is an $(\ell + 1)$-clique containing a vertex of S that is called its *extreme* vertex and the remaining ℓ vertices are in K composing the *basis* of the cone. The *triangles of an ℓ-cone* are its ℓ triangles that contain the extreme vertex of the cone. The set of the remaining vertices of a triangle of an ℓ-cone are the *basis* of the triangle. Note that a 2-cone is a triangle that is a clique and so by Lemma 3 forced to belong to any RS-family of a split clique graph.

A *3-split* graph admits a split partition where each vertex of the stable set S has degree 2 or 3, in this case (K, S) is called a *3-split partition*.

Theorem 4. *Let G be a 3-split graph with 3-split partition (K, S). The following are equivalent:*

1. *G is a clique graph;*
2. *There exists an RS-family $\mathcal{F}$ of G composed by K, each 2-cone and exactly two triangles of each 3-cone;*
3. *There exists a family of complete sets of G containing each basis of a 2-cone and the bases of exactly two triangles of each 3-cone that satisfies the Helly property;*
4. *There exists a family of edges containing all the edges corresponding to the bases of the 2-cones and the edges of the bases of exactly two triangles of each 3-cone that induces a triangle-free subgraph of $G[K]$.*

Proof. 1. implies 2.: Let G be a 3-split graph with 3-split partition (K, S) and let $\mathcal{F}$ be an RS-family of G. Assume K is not a member of $\mathcal{F}$. Consider $\mathcal{F}'$ the family obtained from $\mathcal{F}$ by the addition of member K and by the removal of complete sets K' that satisfy K' is a member of $\mathcal{F}$ and $K' \subset K$. Suppose there exists a pairwise intersecting subfamily of $\mathcal{F}'$ without a common vertex. It is clear this subfamily must contain K, since the original RS-family $\mathcal{F}$ has the Helly property. Let $F_1, F_2, ..., F_\ell, K$ be the pairwise intersecting subfamily without a common vertex. Observe that $\ell \geq 2$. Since $F_1, F_2, ..., F_\ell$ are members of $\mathcal{F}$, they have a common vertex s. It is clear s is not in K, and so $s \in S$. In case $N(s) = \{x, y\}$, then $F_1 = \{s, x\}$ and $F_2 = \{s, y\}$ but this contradicts Lemma 3 since F_1 and F_2 are not cliques of G. Hence, $N(s) = \{x, y, z\}$ and the assumption that $F_1, F_2, ..., F_\ell$ have no common vertex in K forces $\ell = 3$, $F_1 = \{s, y, z\}$, $F_2 = \{x, s, z\}$ and $F_3 = \{x, y, s\}$, Note that F_1, F_2 and F_3 are the three triangles containing vertex s. Now we can eliminate one of these three triangles from $\mathcal{F}'$, the remaining two triangles have a common vertex in K and cover the same set of edges as $\mathcal{F}'$. Observe that in case we have another intersecting subfamily in $\mathcal{F}'$ without a common vertex, it must be the three triangles of another 3-cone. We repeat the same reasoning for each such pairwise intersecting subfamily to obtain an RS-family containing K.

So we may assume that K is a member of the RS-family $\mathcal{F}$. Observe that each 2-cone is a clique and must be a member of $\mathcal{F}$. Let $C_s = \{s, x, y, z\}$ be a 3-cone with extreme s and basis $T = \{x, y, z\}$. In order to cover the edges incident to s, note that $\mathcal{F}$ must contain exactly two triangles of C_s or must contain the 3-cone C_s itself. Suppose $C_s \in \mathcal{F}$. Note that no other member of $\mathcal{F}$ contains s. By Lemma 2, let $u_T \in \bigcap \mathcal{F}_T$. Since $u_T \in K \cap C_s = T$, we may assume $u_T = y$. Consider $\mathcal{F}'$ obtained from $\mathcal{F}$ by the removal of cone C_s and the addition of triangles $\{y, x, s\}$ and $\{y, z, s\}$. Now suppose $F_1, F_2, ..., F_\ell, \{y, x, s\}$ is a pairwise intersecting subfamily of $\mathcal{F}'$ without a common vertex. Since $F_i \cap \{y, x, s\} \neq \emptyset$ and $F_i \cap \{y, x, s\} \neq s$, we may assume $x \in F_1$ and $y \notin F_1$, $x \notin F_2$ and $y \in F_2$. Since F_1, F_2, C_s are pairwise intersecting members of $\mathcal{F}$, we must have $z = F_1 \cap F_2 \cap C_s$. Now $z, x \subset F_1$ implies $F_1 \in \mathcal{F}_T$, so $y \in F_1$, a contradiction. Suppose $F_1, F_2, ..., F_\ell, \{y, x, s\}, \{y, z, s\}$ is a pairwise intersecting subfamily of $\mathcal{F}'$ without a common vertex. We have $y \notin F_1$ but $F_1 \cap \{y, x, s\} \neq \emptyset$ and $F_1 \cap \{y, z, s\} \neq \emptyset$, which implies $F_1 \in \mathcal{F}_T$, again leading to a contradiction.

4. implies 1.: Let $\mathcal{E}$ be a family of edges containing all the edges corresponding to the bases of the 2-cones and the edges of the bases of exactly two triangles of each 3-cone that induces a triangle-free subgraph of $G[K]$. Let $e = xy$ be an edge of the family $\mathcal{E}$. Call $S_e = \{s \in S | \{x, y\} \subseteq N(s)\}$. Observe that: (1) if s is the extreme vertex of a 2-cone then s belongs to exactly one set S_e; (2) if s is the extreme vertex of a 3-cone then s belongs to exactly two sets S_e. Consider the complete set family $\mathcal{F}$ whose members are K and the triangles $T_{e,s}$, where $e \in \mathcal{E}$ and $s \in S_e$. By (1) and (2) if a subfamily of triangles $T_{e,s}$ is pairwise intersecting then the corresponding family of edges e is pairwise intersecting. Since by hypothesis the family of edges do not contain a triangle, then they have a common vertex in K, which implies $\mathcal{F}$ is an RS-family.

The remaining implications are simpler to establish and omitted in the extended abstract.   □

The family of edges defined in statement 4 of Theorem 4 is called an *RS-basis* of a 3-split clique graph.

Construction of G_I from $I = (U, C)$

Let $I = (U, C)$ be any instance of $3SAT_{\overline{3}}$. We assume with no loss of generality that each variable occurs two or three times in C, and no variable occurs twice in the same clause. In addition, if variable u_i occurs twice in C, then we assume it is once as literal u_i and once as literal $\overline{u}_i$; and if variable u_i occurs three times in C, then we assume it is once as literal u_i and twice as literal $\overline{u}_i$.

For each variable u_i, let j_i be the subindex of the unique clause where variable u_i occurs as literal u_i; and $\overline{J}_i = \{j \mid \text{literal } \overline{u}_i \text{ occurs in } c_j\}$.

For each clause c_j with $|c_j| = 3$, let $I_j = \{i \mid \text{variable } u_i \text{ occurs in } c_j\}$; and for each clause c_j with $|c_j| = 2$, let $I_j = \{i \mid \text{variable } u_i \text{ occurs in } c_j\} \cup \{n+1\}$. Notice that in any case $|I_j| = 3$. Given $I_j = \{i_1, i_2, i_3\}$, with $i_1 < i_2 < i_3$, let $i_1^* = i_2$, $i_2^* = i_3$ and $i_3^* = i_1$.

From instance $I = (U, C)$, we construct a graph $G_I = (V, E)$ as follows.

The vertex set V is the union:

$$V = \bigcup_{1 \leq i \leq n} \{a^i_{j_i}, b^i_{j_i}, c^i_{j_i}, d^i_{j_i}, e^i_{j_i}, f^i_{j_i}, g^i_{j_i}, h^i_{j_i}\} \cup$$

$$\bigcup_{1 \leq i \leq n} \bigcup_{j \in \overline{J}_i} \{a^i_j, b^i_j, c^i_j, d^i_j, e^i_j, f^i_j, p^i_j, q^i_j\} \cup$$

$$\bigcup_{1 \leq j \leq m, |c_j| = 2} \{a^{n+1}_j, c^{n+1}_j, d^{n+1}_j\}.$$

In order to have the property that $G_I = (V, E)$ is a split graph, the edge set E is composed so that:

$$K = \bigcup_{1 \leq i \leq n} \{a^i_{j_i}, d^i_{j_i}, g^i_{j_i}, h^i_{j_i}\} \cup \bigcup_{1 \leq i \leq n} \bigcup_{j \in \overline{J}_i} \{a^i_j, d^i_j\} \cup \bigcup_{1 \leq j \leq m, |c_j| = 2} \{a^{n+1}_j, d^{n+1}_j\}.$$

is a complete set and the remaining vertices $S = V \setminus K$ compose the set:

$$S = \bigcup_{1 \leq i \leq n} \{b^i_{j_i}, c^i_{j_i}, e^i_{j_i}, f^i_{j_i}\} \cup \bigcup_{1 \leq i \leq n} \bigcup_{j \in \overline{J}_i} \{b^i_j, c^i_j, e^i_j, f^i_j, p^i_j, q^i_j\} \cup \bigcup_{1 \leq j \leq m, |c_j| = 2} \{c^{n+1}_j\}.$$

and is a stable set.

We finish the definition of the edge set by defining the edges incident to the vertices of the stable set S: For $1 \leq i \leq n$, $N(b^i_{j_i}) = \{a^{i^*}_{j_i}, d^i_{j_i}\}$, $N(c^i_{j_i}) = \{a^{i^*}_{j_i}, a^i_{j_i}, d^i_{j_i}\}$, $N(e^i_{j_i}) = \{d^i_{j_i}, h^i_{j_i}\}$, $N(f^i_{j_i}) = \{a^{i^*}_{j_i}, g^i_{j_i}\}$. For $1 \leq i \leq n, j \in \overline{J}_i$, $N(b^i_j) = \{a^{i^*}_j, d^i_j\}$, $N(c^i_j) = \{a^{i^*}_j, a^i_j, d^i_j\}$, $N(e^i_j) = \{d^i_j, h^i_{j_i}\}$, $N(f^i_j) = \{a^{i^*}_j, g^i_{j_i}\}$, $N(p^i_j) = \{a^i_{j_i}, g^i_{j_i}, a^i_j\}$, $N(q^i_j) = \{a^i_{j_i}, h^i_{j_i}, a^i_j\}$. For $1 \leq j \leq m, |c_j| = 2$, $N(c^{n+1}_j) = \{a^{n+1}_j, a^{n+1^*}_j\}$.

Note that the constructed instance G_I is a 3-split graph. Notice that for each variable u_i, graph G_I contains as induced subgraph, Truth Setting component T_i, the graph depicted in Figure 1 for the case variable u_i has 3 occurrences. Throughout the paper, we shall use the convention in the figures: vertices of K

are black, vertices of S are white; only edges between vertices of the same cone are drawn which means all other edges between black vertices are omitted. Note further that our NP-completeness result yields that SPANNING HELLY FAMILY, defined in the Introduction, is NP-complete even when restricted to the members of the input family $\mathcal{F}$ having cardinality 2 or 3.

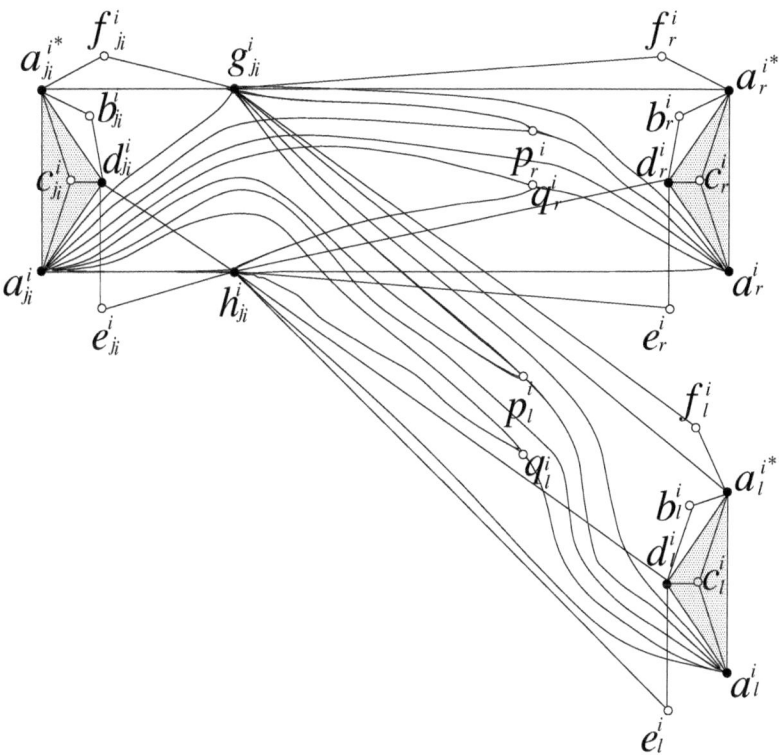

Fig. 1. Graph T_i corresponding to a variable u_i, with $\overline{J}_i = \{r, l\}$

We refer to Figure 2 for the proof of Lemma 5.

Lemma 5. (*True edge–False edge*) *Suppose $\mathcal{F}$ be an RS-basis of the constructed graph G_I. For each $j, 1 \leq j \leq m$, and for each $i \in I_j, i \neq n+1$, exactly one of the edges $a^i_j a^{i^*}_j$, $a^i_j d^i_j$ belongs to $\mathcal{F}$. For each $i, 1 \leq i \leq n$, and for each $j \in \overline{J}_i$, if $a^i_j d^i_j \in \mathcal{F}$ then $a^i_{ji} a^{i^*}_{ji} \in \mathcal{F}$, and if $a^i_j a^{i^*}_j \in \mathcal{F}$ then $a^i_{ji} d^i_{ji} \in \mathcal{F}$.*

Proof. Consider any $j, 1 \leq j \leq m$, and $i \in I_j, i \neq n+1$. Assume with no loss of generality, $j = j_i$. By considering the 2-cone $N[b^i_{j_i}]$, notice that edge $a^{i^*}_{j_i} d^i_{j_i}$ must belong to the RS-basis $\mathcal{F}$ which implies that both edges $a^i_{j_i} a^{i^*}_{j_i}$ and $a^i_{j_i} d^i_{j_i}$ cannot belong to $\mathcal{F}$, which implies that exactly one of the edges $a^i_{j_i} a^{i^*}_{j_i}$, $a^i_{j_i} d^i_{j_i}$ belongs to $\mathcal{F}$.

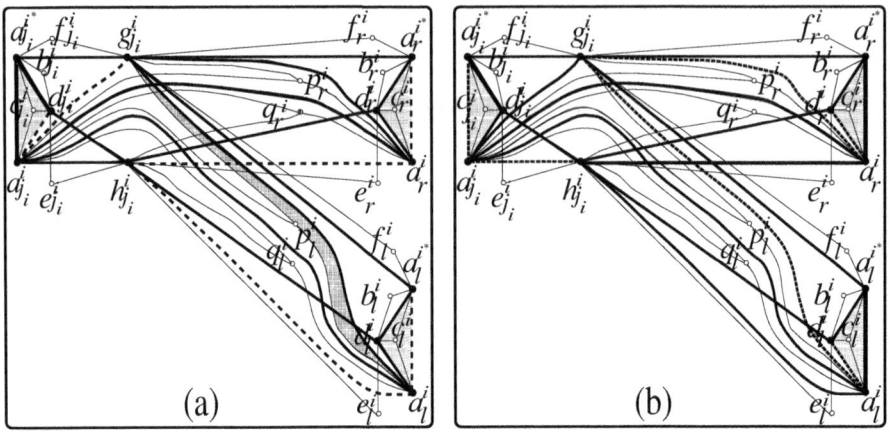

Fig. 2. (a) RS-basis for T_i containing edge $a_r^i d_r^i$ is depicted in bold edges. Dashed edges are the edges of the bases of the 3-cones that are not members of the RS-basis. (b) Respectively for edge $a_r^i a_r^{i^*}$.

Consider any $i, 1 \le i \le n$, and $j \in \overline{J}_i = \{r, l\}$. Say $j = r$ and refer to Figure 2(a). Notice that, edge $h_{j_i}^i d_r^i$ must belong to the RS-basis $\mathcal{F}$. Assume that $a_r^i d_r^i \in \mathcal{F}$. Then $a_r^i h_{j_i}^i \notin \mathcal{F}$, and so by considering the 3-cone $N[q_r^i]$, edges $a_r^i a_{j_i}^i, h_{j_i}^i a_{j_i}^i \in \mathcal{F}$. Notice that edge $h_{j_i}^i d_{j_i}^i$ must belong to the RS-basis $\mathcal{F}$. Hence $a_{j_i}^i d_{j_i}^i \notin \mathcal{F}$, and so by the first statement, $a_{j_i}^i a_{j_i}^{i^*} \in \mathcal{F}$. Assume that $a_r^i a_r^{i^*} \in \mathcal{F}$ and refer to Figure 2(b) to obtain an analogous reasoning. $\square$

Lemma 5 is the key for the NP-completeness result. Given any variable u_i and any clause c_j where u_i occurs, any RS-basis of G_I is forced to choose exactly one of the edges $a_{j_i}^i a_{j_i}^{i^*}$, $a_{j_i}^i d_{j_i}^i$. If $r \in \overline{J}_i$, then any RS-basis of G_I is forced to choose different types of edges incident to vertices a_r^i and $a_{j_i}^i$, respectively. If $r, \ell \in \overline{J}_i$, then any RS-basis of G_I is forced to choose the same type of edges incident to vertices a_r^i and a_ℓ^i, respectively. The correspondence between the two possible truth assignments of variable u_i and the two possible edges incident to vertex $a_{j_i}^i$ is clear.

Theorem 6. SPLIT CLIQUE GRAPH *is NP-complete.*

Proof. As mentioned in the Introduction, SPLIT CLIQUE GRAPH belongs to NP.

Let G be the constructed 3-split graph obtained from an instance $I = (U, C)$ of 3SAT$_{\overline{3}}$. Suppose G is a clique graph, and we exhibit a truth assignment for U such that C is satisfied. By Theorem 4, let $\mathcal{F}$ be an RS-basis for G. Let $u_i \in U$ be a variable. Set u_i equal to true if and only if edge $a_{j_i}^i d_{j_i}^i \in \mathcal{F}$. To see that this truth assignment for U satisfies C consider a clause c_j and its corresponding triangle $\{a_j^i, a_j^{i^*}, a_j^{i^{**}}\}$. Since $\mathcal{F}$ induces a triangle-free subgraph of $G[K]$, there exists $i \in I_j$ such that the edge $a_j^i a_j^{i^*}$ is not a member of $\mathcal{F}$. Notice that $i \ne n+1$.

By Lemma 5, edge $a_j^i a_j^{i^*} \notin \mathcal{F}$ implies that edge $a_j^i d_j^i \in \mathcal{F}$. If $j = j_i$ then variable u_i is true and clause c_j is satisfied. If $j \neq j_i$, then $j \in \bar{J}_i$, by Lemma 5 edge $a_j^i d_j^i \in \mathcal{F}$ implies edge $a_{j_i}^i a_{j_i}^{i^*} \in \mathcal{F}$, and edge $a_{j_i}^i d_{j_i}^i \notin \mathcal{F}$. It follows that u_i is false, and then c_j is satisfied.

Conversely, given a truth assignment of U that satisfies C, by Theorem 4, it suffices to exhibit an RS-basis $\mathcal{F}$ in order to prove that G is a clique graph.

For each $j, 1 \leq j \leq m$, for each $i \in I_j$, the edges $a_j^{i^*} g_j^i, d_j^i h_j^i, a_j^{i^*} d_j^i$.

For each $j, 1 \leq j \leq m$, for $i = n + 1$, the edges $a_j^{n+1^*} a_j^{n+1}$.

For each $i, 1 \leq i \leq n$, such that variable u_i is true, the edges $d_{j_i}^i a_{j_i}^i, a_{j_i}^i g_{j_i}^i$; and for each $j \in \bar{J}_i$, the edges $h_{j_i}^i a_j^i, a_j^i a_j^{i^*}$.

For each $i, 1 \leq i \leq n$, such that variable u_i is false, the edges $a_{j_i}^i a_{j_i}^{i^*}, a_{j_i}^i h_{j_i}^i$; and for each $j \in \bar{J}_i$, the edges $g_{j_i}^i a_j^i, a_j^i d_j^i$.

The proof is completed by showing that the chosen set of edges indeed induces a triangle-free subgraph of $G[K]$ containing all the basis of 2-cones and two edges of the basis of each 3-cone. Details are omitted in the extended abstract.   □

3 Polynomially Solvable Split Clique Graph Classes

In the following three theorems we present non trivial split graph classes for which clique graphs can be recognized in polynomial time. Let G be a split graph with split partition (K, S), without loss of generality assume $K = \bigcup_{s \in S} N(s)$, to obtain a unique possible split partition.

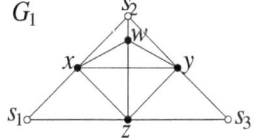 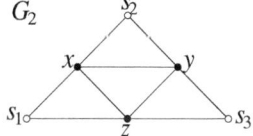

Fig. 3. (a) w is a private neighbour of s_2. (b) no vertex in S has a private neighbour.

We say that a vertex $x \in K$ is a *private neighbor* of $s \in S$, if s is the only vertex in S adjacent to x, i.e. $N(x) \cap S = \{s\}$. We refer to Figure 3.

Theorem 7. *If every vertex $s \in S$ has a private neighbor then G is a clique graph.*

Proof. Suppose every vertex $s \in S$ has a private neighbor h_s. Let x and y be vertices of K. We say that x is a twin of y when $N[x] = N[y]$. Observe this is an equivalence relation, and so the equivalence classes define a partition of K. Let R_s be the class of h_s for $s \in S$; and $R_1, R_2, \ldots, R_k$ the remaining classes, this means the classes that do not contain any vertex h_s for $s \in S$. We notice that

$((R_s)_{s \in S}, R_1, R_2, R_3, \ldots R_k)$ is a partition of K. Since h_s is a private neighbor of s, if $s' \in S$ and $s' \neq s$ then $R_s \neq R_{s'}$.

For every $s \in S$, we call I_s the set $\{i : 1 \leq i \leq k$ such that $R_i \subseteq N(s)\}$. Let $\mathcal{F}$ be the family of complete sets of G whose members are: K; $F_{s,i} = R_s \cup R_i \cup \{s\}$, for each $s \in S, I_s \neq \emptyset$ and $i \in I_s$; $F_s = R_s \cup \{s\}$, for each $s \in S, I_s = \emptyset$. We claim that $\mathcal{F}$ is an RS-family of G, and so G is a clique graph.

Details are omitted in the extended abstract.   □

Theorem 8. *Let G be a split graph with $|S| \leq 3$. Graph G is a clique graph if and only if G is not the Hajós graph depicted in Figure 3.(b).*

Proof. It is well known that if G is a clique graph then G is not the Hajós graph. Let us prove the reciprocal implication. Assume G is a graph with split partition (K, S), $| S | \leq 3$ and G is not the Hajós graph. By Theorem 1, if the clique family of G has the Helly property then G is a clique graph. If the clique family does not satisfy the Helly property, then there exists a subfamily of cliques pairwise intersecting without a common vertex.

It is clear that such subfamily must contain $N[s_1]$, $N[s_2]$ and $N[s_3]$ as members, where s_1, s_2 and s_3 are the vertices in S.

For $1 \leq i < j \leq 3$, let $x_{i,j}$ be three vertices of K such that $x_{i,j} \in N[s_i] \cap N[s_j]$. Since G is not the Hajós graph, then K must contain at least one more vertex.

Call it u and suppose u is a private neighbor, for instance of s_1, then $u \in N[s_1] \setminus (N[s_2] \cup N[s_3])$. In this case it is easy to check that the complete set family $\mathcal{F}$ $N[s_1] \setminus N[s_2]$, $N[s_1] \setminus N[s_3]$, $N[s_2]$, $N[s_3]$ and K satisfies the conditions given by Theorem 1, so G is a clique graph. We depict in Figure 4 such family. Details are omitted in the extended abstract.   □

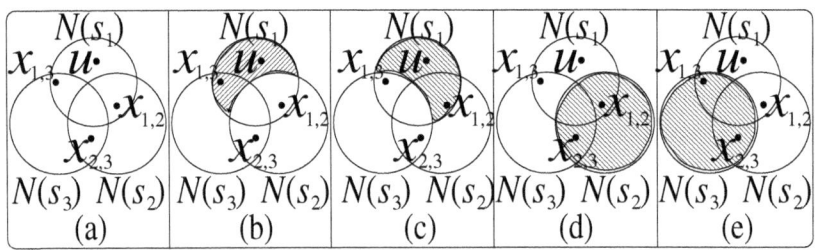

Fig. 4. Case in which u is a private neighbor, assumed of s_1

Theorem 9. *Let G be a split graph with $|K| \leq 4$. Graph G is a clique graph if and only if: (1) There are no three bases of 2-cones forming a triangle; and (2) There are no four bases of cones satisfying: one is the basis $B = \{a, b, d\}$ of a 3-cone, the other three bases $B_1 = \{a, c\}$, $B_2 = \{b, c\}$, and $B_3 = \{d, c\}$ are bases of 2-cones.*

4   Open Related Problems

We summarize in a table the results and open problems we have managed to state about the complexity of the problem of recognizing clique graphs when restricted to split graphs. Denote by $3split_2$ the class of 3-split graphs, where the vertices of the independent set have degree at least 2 and at most 3, and by $3split_3$ the subclass of 3-split graphs, where the vertices of the independent set have degree exactly 3.

	$3split_3$	$3split_2$	$\forall s \in S$, s has a private neighbor.	$|S|$ bounded		$|K|$ bounded	
				$|S| \leq 3$	general	$|K| \leq 4$	general
Split graph $G = (V,E)$ partition (K,S).	?	NPC	P	P	?	P	?

The present work presents three distinct sufficient conditions for a split graph to be a clique graph that lead to three non trivial polynomial split clique graph classes. The complexity of recognizing split clique graphs with $|K|$ or $|S|$ bounded remains open.

Several subclasses of clique graphs have been studied for which polynomial-time recognition is known. In particular, for several classes of graphs the corresponding class of clique graphs is known [14]. Note that it is well known that the clique graph of a chordal graph is a dually chordal graph [4,12] but the complexity of deciding whether a chordal graph is a clique graph was a challenging open problem. We have proved that deciding whether a given split graph is a clique graph is an NP-complete problem. Note that the class of split graphs is the intersection of chordal graphs and complements of chordal graphs.

The NP-completeness of CLIQUE GRAPH [1,2] suggested the study of the problem restricted to classes of graphs not properly contained in the class of clique graphs. One such class is the class of split graphs, the object of the present paper, and the recognition of split clique graphs is proved NP-complete. Another challenging still open problem is the recognition of planar clique graphs [3].

Let G be a split graph with split partition (K, S). In case G is a 3-split graph, Theorem 4 says G admits an RS-family containing K. We leave as open the complexity of deciding if a split clique graph with split partition (K, S) admits an RS-family containing K.

Our NP-completeness result for split clique graph recognition is optimum in the sense that each vertex of the independent set of our split instance has degree at most 3, whereas when each vertex of the independent set has degree at most 2 the problem is polynomial, since it is reduced to check whether the clique family of the graph satisfies the Helly property. Actually, by Theorem 4 the problem is polynomial when the input is a 3-split graph such that the number of 3-cones is bounded, which implies that 3-split clique graph recognition when $|K|$ is bounded or when $|S|$ is bounded is in P. We leave as open the complexity of recognizing split clique graphs such that every vertex of the independent set has degree exactly 3. Note that the problem $3SAT_{\overline{3}}$ when restricted to having exactly three literals per clause is polynomial [9].

Acknowledgment. Two of the polynomial cases were presented at the 4th Latin-American Workshop on Cliques in Graphs and the corresponding abstract appeared in the proceedings published by Matemática Contemporânea.

References

1. Alcón, L., Faria, L., de Figueiredo, C.M.H., Gutierrez, M.: Clique Graph Recognition is NP-Complete. In: Fomin, F.V. (ed.) WG 2006. LNCS, vol. 4271, pp. 269–277. Springer, Heidelberg (2006)
2. Alcón, L., Faria, L., de Figueiredo, C.M.H., Gutierrez, M.: The complexity of clique graph recognition. Theoretical Computer Science 410, 2072–2083 (2009)
3. Alcón, L., Gutierrez, M.: Cliques and extended triangles. A necessary condition for planar clique graphs. Discrete Appl. Math. 141, 3–17 (2004)
4. Brandstädt, A., Dragan, F.F., Chepoi, V.D., Voloshin, V.: Dually chordal graphs. SIAM J. Discrete Math. 11, 437–455 (1998)
5. Brandstädt, A., Le, V.B., Spinrad, J.P.: Graph Classes: A survey. SIAM Monographs on Discrete Mathematics and Applications (1999)
6. Dourado, M.C., Protti, F., Szwarcfiter, J.L.: Complexity Aspects of the Helly Property: Graphs and Hypergraphs. Electron. J. Combin. 17, 1–53 (2009)
7. Hamelink, R.C.: A partial characterization of clique graphs. J. Combin. Theory Ser. B 5, 192–197 (1968)
8. McKee, T.A., McMorris, F.R.: Topics in Intersection Graph Theory. SIAM Monographs on Discrete Mathematics and Applications (1999)
9. Papadimitriou, C.H.: Computational Complexity. Addison-Wesley (1994)
10. Prisner, E.: Graph Dynamics. Pitman Research Notes in Mathematics 338, Longman (1995)
11. Roberts, F.S., Spencer, J.H.: A characterization of clique graphs. J. Combin. Theory Ser. B 10, 102–108 (1971)
12. Szwarcfiter, J.L., Bornstein, C.F.: Clique graphs of chordal graphs and path graphs. SIAM J. Discrete Math. 7, 331–336 (1994)
13. Szwarcfiter, J.L.: Recognizing clique-Helly graphs. Ars Combin. 45, 29–32 (1997)
14. Szwarcfiter, J.L.: A survey on Clique Graphs. In: Linhares-Sales, C., Reed, B. (eds.) Recent Advances in Algorithms and Combinatorics. CMS Books Math./Ouvrages Math. SMC, vol. 11, pp. 109–136. Springer, New York (2003)

On Searching for Small
Kochen-Specker Vector Systems*

Felix Arends[1], Joël Ouaknine[2], and Charles W. Wampler[3]

[1] Google Germany GmbH
felix.arends@gmx.de
[2] Department of Computer Science, Oxford University, UK
joel@cs.ox.ac.uk
[3] Department of Mathematics, University of Notre Dame, USA
charles.w.wampler@gm.com

Abstract. Kochen-Specker (KS) vector systems are sets of vectors in $\mathbb{R}^3$ with the property that it is impossible to assign 0s and 1s to the vectors in such a way that no two orthogonal vectors are assigned 0 and no three mutually orthogonal vectors are assigned 1. The existence of such sets forms the basis of the Kochen-Specker and Free Will theorems. Currently, the smallest known KS vector system contains 31 vectors. In this paper, we establish a lower bound of 18 on the size of any KS vector system. This requires us to consider a mix of graph-theoretic and topological embedding problems, which we investigate both from theoretical and practical angles. We propose several algorithms to tackle these problems and report on extensive experiments. At the time of writing, a large gap remains between the best lower and upper bounds for the minimum size of KS vector systems.

Keywords: Kochen-Specker vector systems, topological graph embedding problems, constraint solving, graph enumeration algorithms.

1 Introduction

In a recent, thought-provoking paper, John H. Conway and Simon Kochen demonstrate that *"if [...] there exist any experimenters with a modicum of free will, then elementary particles must have their own share of this valuable commodity"* [10]. More precisely, Conway and Kochen consider so-called 'spin-1' particles (such as photons) whose 'spin' (a physical property) can be measured along any given direction. The squared outcome of such measurements is always either 0 or 1. Conway and Kochen's Free Will theorem asserts that, if an experimenter can choose the direction along which to perform a spin-1 experiment freely (i.e., in a way that is not determined by the past), then the response of the spin-1 particle to such an experiment is also not determined by the past.

---

* This work is based on the first author's Master's thesis [1]; an extended version of this paper is also available as [2].

P. Kolman and J. Kratochvíl (Eds.): WG 2011, LNCS 6986, pp. 23–34, 2011.
© Springer-Verlag Berlin Heidelberg 2011

This theorem rests on three basic axioms of quantum mechanics and relativity, the most crucial of which (for our purposes) is the following:

The SPIN Axiom [10]. *Measurements of the squared components of spin of a spin-1 particle in three orthogonal directions always yield the outcomes 1, 0, 1 in some order.*[1]

The SPIN axiom not only follows from the postulates of quantum mechanics, but has also been verified experimentally [12]. This axiom alone already gives rise to what is known as the 'Kochen-Specker paradox' [11]: if the response of a spin-1 particle to any conceivable spin measurement were predetermined prior to the actual measurement, then those responses would define a function from the unit sphere in three dimensions to the set $\{0, 1\}$, satisfying the so-called 101-property: any three points on the sphere with mutually orthogonal position vectors must be assigned the values 1, 0, 1 in some order. The Kochen-Specker paradox—which is in fact a mathematical theorem—is that no such function exists.

The impossibility of such '101-functions' can be proved by exhibiting a finite set of points on the sphere on which such functions cannot be defined. The first such set, discovered by Kochen and Specker more than forty years ago, contained 117 points [13]. Subsequent sets, usually referred to as 'records' [17], cut this number down to 33 and then 31 [20]. The latter is the size of the smallest known 'Kochen-Specker vector system', discovered approximately 20 years ago by Conway and Kochen.

As pointed out in [17], finding small Kochen-Specker vector systems has both theoretical and practical motivations. Conway himself has stressed the problem on several occasions whilst giving public lectures on the Free Will theorem. The work we describe here reports on some partial progress in this endeavour; our main result is that any Kochen-Specker vector systems must contain at least 18 vectors (Thm. 10). Achieving this bound required us to consider a mix of graph-theoretic and topological embedding problems, for which we devised and analysed a number of algorithms. In addition, we establish bounds on the theoretical complexity of some of the principal problems involved (Thms. 2 and 3), and also show that the key task of checking canonicity in the 30-year-old Colbourn-Read orderly graph enumeration algorithm [9] cannot belong to NP— and much less to P—unless NP = co-NP (Thm. 7).

Unfortunately, it would appear that narrowing the gap between the best lower and upper bounds for the minimum size of KS vector systems remains a formidable challenge, and significant progress in this area will likely require substantially new ideas.

The work most closely related to ours is that of Pavičić *et al.* [18, 17]. They consider higher-dimensional generalisations of the problem treated in this paper, but their formulation and results are incomparable to ours. We return to the differences between our approach and theirs in Sec. 2; we also refer the reader to [18, 1] for a more thorough discussion of the matter.

---

[1] As pointed out in [10], such measurements 'commute', so the order in which they are performed does not matter.

Different Standards of Proof. Due to the mixed discrete/continuous aspects of the problems considered in this paper, it is important to pay special attention to the nature of the proofs involved. The usual kind of proof is *mathematical*. Along with such proofs, we also present several results that have *computer-aided* proofs, in which extensive calculations were carried out by computer. A third category of results could be deemed to have *numerical* proofs, by which we mean that a computer program was used and floating-point arithmetic was involved in a way that cannot be guaranteed to be entirely accurate. While we have occasionally made use of the latter as heuristics, it should be stressed that all results presented in this paper are backed by fully rigorous mathematical or computer-aided proofs.

2   Kochen-Specker Vector Systems

Kochen-Specker vector systems can be represented in multiple ways. In the Introduction, we have implicitly described such systems as certain finite sets of points on the surface of the sphere $\mathbb{S}^2$. In fact, an immediate consequence of the SPIN axiom is that squared-spin measurements along opposite directions necessarily yield the same outcome, so that it is sensible to identify antipodal points. Accordingly, let us therefore define a **vector system** as a finite subset of the open northern hemisphere $\mathbb{H}^2 = \{(x, y, z) : x^2 + y^2 + z^2 = 1 \text{ and } z > 0\}$.[2] Alternatively, vector systems can be represented as finite subsets of the projective plane $\mathbb{P}^2$, and also as finite sets of points on the surface of a cube, where once again we identify antipodal points. We variously make use of all three of these representations in the rest of this paper.

A vector system $\mathcal{K} \subseteq \mathbb{H}^2$ is **101-colourable** if it is possible to assign either 0 or 1 to each vector in $\mathcal{K}$ such that (i) no two orthogonal vectors are both assigned 0, and (ii) no three mutually orthogonal vectors are all assigned 1.

Finally, a **Kochen-Specker (KS) vector system** is a vector system that is not 101-colourable. The *size* of such a system is simply the number of vectors it contains.

At the time of writing, the smallest known KS vector system is still the one discovered approximately twenty years ago by Conway and Kochen [20]. The 31 vectors of this system can be represented as lying on a cubic grid centered at the origin, as depicted in Fig. 1. Note that orthogonality relationships among the vectors are easily inferred through elementary geometry, thanks to the regularity of the grid. Non-101-colourability can be established by (somewhat tedious) case analysis.

Note that the colourability conditions (i) and (ii) as given above, although in appearance stronger than the SPIN axiom, are implicitly equivalent to it. For instance, while the SPIN axiom, strictly speaking, asserts nothing about a vector system consisting of exactly two orthogonal vectors, it implicitly requires one of

---

[2] Dispensing entirely with the equator simplifies somewhat our technical development later on; it is harmless since any finite set of points on the sphere can always be rigidly rotated so as to avoid the equator.

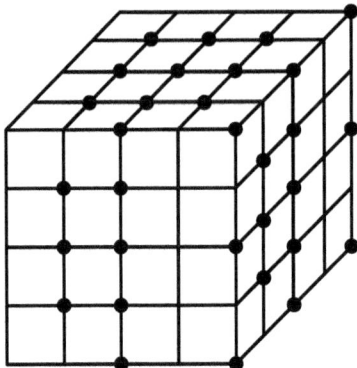

Fig. 1. A visual representation of Conway and Kochen's KS vector system of size 31

these vectors to be assigned 1, since if both were assigned 0 one could derive a contradiction by considering a third vector orthogonal to the other two.

This seemingly innocuous observation has consequences for the way in which KS vector systems are built and measured. Pavičić *et al.* [17] and Larsson [14], for example, require every pair of orthogonal vectors to belong to a triple of mutually orthogonal vectors, invoking a strict application of the SPIN axiom. Following this convention, they argue that the KS vector system depicted in Fig. 1 should be viewed as having size *51* rather than 31 (cf. [17]): indeed, this system as represented above contains 20 pairs of orthogonal vectors without a third orthogonal vector present.

In contrast, our own conventions—following, among others, [13, 19, 20, 7, 10, 11]—are predicated on colourability conditions (i) and (ii) as given earlier. For further discussion on the matter, we refer the reader to [17] and [1].

Vector Systems and Graphs. Any vector system $\mathcal{K}$ gives rise to an associated undirected graph $G_\mathcal{K}$, the vertices of which are the vectors of $\mathcal{K}$, with an edge between two vertices iff the corresponding vectors are orthogonal. In other words, $G_\mathcal{K} = (V, E)$, where $V = \mathcal{K}$ and $E = \{\{\boldsymbol{u}, \boldsymbol{v}\} : \boldsymbol{u}, \boldsymbol{v} \in \mathcal{K}$ and $\boldsymbol{u} \cdot \boldsymbol{v} = 0\}$.

We define 101-colourability for graphs in the obvious way: assignment of 0 or 1 to the vertices in such a way that (i) no two adjacent vertices are both assigned 0, and (ii) no triangle (3-clique) is assigned all 1s. Clearly, $\mathcal{K}$ is a KS vector system iff $G_\mathcal{K}$ is not 101-colourable.

Of course, an arbitrary graph H may not correspond to any realisable (3-dimensional) vector system: the orthogonality constraints corresponding to graph edges may fail to be simultaneously satisfiable. Let us define a graph H to be ***embeddable*** if there exists some vector system that it corresponds to. More precisely, we ask that there be a vector system $\mathcal{K}$ that can be put in one-to-one correspondence with the vertices of H in such a way that adjacent vertices are mapped to orthogonal vectors. Note that we do *not* require that

non-adjacent vertices should go to non-orthogonal vectors; this relaxation simplifies the embeddability-checking process, discussed in Sec. 3. However, it is
necessary for distinct vertices to go to distinct vectors. Formally, $H = (V, E)$
is embeddable if it has a supergraph $H' = (V, E')$ over the same set of vertices
such that H' is isomorphic to $G_\mathcal{K}$ for some vector system $\mathcal{K}$.

Finding a small KS vector system therefore corresponds to finding a small
graph that is both not 101-colourable and embeddable, and accordingly this is
the approach we have followed and report on in this paper.

Note that any graph containing a square (4-cycle) is unembeddable: indeed,
orthogonality constraints would force a pair of opposite vertices of the square
to be mapped to collinear (i.e., identical) vectors. Accordingly, we shall mainly
focus on *square-free* graphs in the remainder of this paper. This turns out to be a
fairly powerful restriction: all square-free graphs with 9 or fewer vertices are embeddable, while there are only two distinct (up to isomorphism) unembeddable
square-free graphs with 10 vertices [1].

A second interesting observation about embeddable graphs is the following:

Proposition 1. *Any embeddable graph is 4-colourable.*

To see this, consider an embeddable graph G and let $\mathcal{K}$ be an embedding of
it as a vector system in the hemisphere $\mathbb{H}^2$. Partition $\mathbb{H}^2$ into four quadrants
as delineated by the xz-plane and the yz-plane, and colour each vector of $\mathcal{K}$
according to the quadrant it lies in. Since vectors belonging to the same quadrant
cannot be mutually orthogonal, corresponding vertices of G cannot be adjacent.
Thus the quadrant colouring of $\mathcal{K}$ gives rise to a valid 4-colouring of G.   □

As the next result indicates, 101-colourability is in theory an expensive condition to check, even when restricting to graphs that are both square-free and
4-colourable. In practice, however, our SAT-based colourability checker (implemented using MiniSat 2.0 [16]) was systematically able to decide 101-colourability
of graphs having at most 30 vertices in microseconds.

Theorem 2. *Deciding whether a square-free 4-colourable graph is 101-colourable
is NP-complete.*

Membership in NP is obvious; for hardness, we refer the reader to [1].

Finally, note that every 3-colourable graph is automatically 101-colourable.

3   Embeddability

In this section, we examine the problem of determining whether a given (square-
free) graph is embeddable or not. See also [1, 2] in which we discuss some of the
practical approaches and algorithms we have used including homotopy continuation and interval arithmetic.

It is fairly straightforward to see that embeddability queries can be phrased
in the existential theory of the reals: given a graph G, postulate a triple of real

variables (x, y, z) for every vertex of G, and express the various constraints using polynomial equalities and inequalities. For example, $x^2 + y^2 + z^2 = 1$ and $z > 0$ together ensure that the corresponding vector should lie in the hemisphere $\mathbb{H}^2$. Orthogonality constraints are likewise expressed by setting the relevant dot products equal to zero, and so on. Embeddability of the graph G therefore corresponds to solvability of this constraint system over the reals.

It is plain that the constraints can be constructed in polynomial time. Since the existential theory of the reals has polynomial space complexity [8, 23], we have:

Theorem 3. *Graph embeddability can be decided in PSPACE.*

For a graph with 30 vertices, the corresponding constraint system requires 90 real variables (or rather, assuming the graph has at least one triangle, 81 variables since we can quotient out rotational symmetries by fixing the vectors associated with one of its triangles). Unfortunately, current real arithmetic solvers cannot in practice handle systems containing more than just a handful of real variables. Thm. 3 is therefore mainly of theoretical interest at the present time.

Our next observation is that, given a graph G, one can in fact construct a single multivariate polynomial P_G over the reals such that G is embeddable iff P_G has a root. We simply extend the above approach by transforming inequalities into equalities, through the use of auxiliary variables, and conjoining multiple equalities into a single one via a standard squaring trick. For example, the inequality $z > 0$ is equivalent to the conjunction of the equalities $uv = 1$ and $u^2 = z$, where u and v are implicitly existentially quantified. In turn, both equalities can be conjoined into a single one by writing $(uv - 1)^2 + (u^2 - z)^2 = 0$, etc. We therefore have:

Proposition 4. *A graph G is embeddable iff the polynomial P_G has a real root.*

It is easy to see that we can arrange for P_G to have degree four. Moreover, graphs with at most 30 vertices give rise to polynomials in fewer than 1000 variables (the bulk of which are required to ensure that all vectors are pairwise distinct). Unfortunately, deciding whether such polynomials have real roots is in general also well beyond the practical capabilities of today's algorithms and computers. For an in-depth account of relevant algorithms and results in this area, we refer the reader to [4].

Finally, let us remark that graph embeddability can alternatively be phrased in terms of *isometric* (i.e., *distance-preserving*) *embeddability*: a graph is embeddable (in the sense of this paper) iff its vertices can distinctly be placed in the upper half of $\mathbb{R}^3$ so as to lie at distance 1 from the origin, and such that adjacent vertices are precisely $\sqrt{2}$ units apart. More information on isometric embeddings and related topics in topological graph theory can be found in [5].

Cubic Grids. Conway and Kochen's KS vector system of size 31 lies on a regular cubic grid, as shown in Fig. 1. That grid can be viewed as consisting of all vectors with integer coordinates lying on the surface of the cube $[-2, 2]^3$, with antipodal points identified.

We can, of course, consider grids of different granularities by introducing a grid parameter N: the corresponding grid can be viewed as the set of vectors with integer coordinates lying on the surface of the cube $[-N, N]^3$, again with antipodal points identified.

One of the chief advantages of cubic grids is that all orthogonality relationships are inferable by straightforward inspection; thus grid embeddability provides a genuine mathematical proof of embeddability. Moreover, we have so far not encountered any graph which we believed to be embeddable (through the use of homotopy continuation or interval arithmetic) yet which was not found to be embeddable on some cubic grid. This leads us to formulate the following:

Conjecture 5. Every embeddable graph can be embedded on some cubic grid.

In our experiments, we found grid-solving to be consistently highly efficient. For example, embedding the 31-vertex Conway-Kochen graph took less than 10ms on the grid with parameter $N = 2$, approximately 250ms on the (N=8)-grid, and 26s on the (N=12)-grid. Naturally, all embeddability proofs carry mathematical certainty; however the absence of a grid embedding does not allow one to draw any conclusion regarding (proper) embeddability.

Let us conclude this section by noting that embeddability clearly remains a highly challenging problem at present. For example, a 12-vertex graph is given in [1] for which, despite our best efforts, we have not succeeded in proving or disproving embeddability (even numerically).

4   Lower Bounds

A natural strategy for finding small KS vector systems is first to search for small graphs that are not 101-colourable. Such graphs should moreover be square-free (otherwise they cannot be embeddable) and connected (otherwise a smaller instance would be available).

Our initial approach was to generate these graphs at random, subject to various parameters, and check whether they are embeddable. Several hundred millions of connected square-free graphs were generated, yielding thousands that were not 101-colourable. Unfortunately, for most graphs with 30 vertices or less, we were simply unable to determine embeddability; and the ones for which we did succeed were all found to be unembeddable. Interestingly, our random graph generator produced several isomorphic copies of the 31-vertex Conway-Kochen specimen.

We then turned to sub-systems of the various cubic grids, which are embeddable by construction. We were able to exhaustively search the grids with parameters $N = 2$ and $N = 4$; all vector systems of size 30 or less were found to be 101-colourable, whereas the only systems of size 31 that were not 101-colourable were all isomorphic to the 31-vector Conway-Kochen system. We also randomly sampled extensively from sub-systems of the grids with parameters

$N = 6$, $N = 8$, and $N = 12$.[3] Again, no smaller system was found, and all KS systems of size 31 were found to be isomorphic to the Conway-Kochen system.

Enumerating Connected Square-Free Graphs. At the time of writing, the On-Line Encyclopedia of Integer Sequences [24] lists the numbers of non-isomorphic connected square-free graphs with up to and including 17 vertices: there are 19,297,850,417 in total, and 17,992,683,043 on 17 vertices alone. We re-enumerated all these graphs and checked each one for 101-colourability, a task which required solving more than 19 billion instances of an NP-complete problem.

In [9], Colbourn and Read propose an 'orderly' procedure for graph enumeration. The key idea is to generate the adjacency matrices of graphs in unique canonical forms. More precisely, given an adjacency matrix, consider the bit-string obtained by concatenating the entries strictly above the diagonal, column by column (from top to bottom), left to right. The **canonical** representation of a given graph G is the unique adjacency matrix of G with the greatest bit-string value in lexicographic order.

As pointed out in [9], a crucial property of this particular notion of canonicity is the following: if $\mathbf{M}$ is the canonical adjacency matrix of a graph G on n vertices, then the $(n-1) \times (n-1)$ submatrix of $\mathbf{M}$ obtained by deleting the last column and the last row of $\mathbf{M}$ is also the canonical adjacency matrix of some subgraph of G on $n-1$ vertices. In the terminology of [21] (see also [15]), this enables the design of an *effective* graph enumeration algorithm: Generate the adjacency matrices of graphs on a fixed number of vertices by a depth-first search process which starts from the trivial 1×1 matrix and successively augments the matrix by adding a single column and row to it until the target number of vertices has been reached. In so doing, whenever a non-canonical matrix is encountered, immediately discard it and backtrack. This procedure guarantees that every canonical matrix will appear exactly once at some point in the search. Moreover, the number of non-canonical matrices that are produced (and immediately discarded) in the process is kept relatively low.

A second key advantage of the Colbourn-Read notion of canonicity is that one may use it to enumerate all non-isomorphic graphs with some hereditary property (i.e., any property of a graph which automatically holds for all its induced subgraphs). Note however that while square-freeness is clearly hereditary, connectedness is not. Nevertheless, the following result, whose proof can be found in [1, 2], shows that the Colbourn-Read algorithm is still suitable for our purposes:

Proposition 6. *Suppose that $\mathbf{M}$ is the canonical adjacency matrix of a connected graph G. Then the submatrix of $\mathbf{M}$ obtained by deleting the last column and the last row of $\mathbf{M}$ is also the (canonical) adjacency matrix of some connected subgraph of G.*

---

[3] For odd values of N, it turns out that the smallest grid which is not itself 101-colourable—and therefore a candidate for hosting KS vector systems—is the one with parameter $N = 15$.

A major attraction of orderly graph enumeration algorithms is that *"expensive isomorphism tests are replaced by relatively inexpensive verifications of canonicity"* [9]—see also [21, 22, 15]. Canonicity checking is indeed a pivotal component of the Colbourn-Read algorithm, yet somewhat surprisingly its precise complexity appears to have remained open since orderly algorithms were first introduced 30 years ago.[4] We provide some partial answers to this question below.

The first observation is that canonicity checking is clearly in co-NP: if a matrix is not canonical, then one needs only exhibit another one that is higher in lexicographic order together with an isomorphism between the two. We now establish a hardness result:

Theorem 7. *If the problem of canonicity checking were in NP, then NP = co-NP.*

We show that an NP algorithm for canonicity checking would entail the existence of an NP algorithm for proving that a graph has no clique of size a given integer k. Since the latter is well-known to be co-NP-complete, the conclusion that NP = co-NP would follow.

Note that the canonical adjacency matrix of a graph having a clique of size k will necessarily contain 1s in the upper-left triangle covering vertices up to k, for otherwise an adjacency matrix with a higher lexicographic order could immediately be obtained by re-labelling the vertices of the clique with the integers 1 to k. Conversely, any adjacency matrix with an upper-left triangle covering vertices up to k consisting entirely of 1s necessarily represents a graph having a clique of size k.

Let G be a graph that has no clique of size k. Guess the canonical adjacency matrix for G, guess and verify the adjacency mapping showing that the matrix does indeed represent G, and verify that the matrix is indeed canonical using the putative NP algorithm for canonicity checking. By the above observation, the upper-left triangle of this matrix covering vertices up to k cannot consist entirely of 1s, thereby proving that all cliques in G must have size strictly less than k. □

Thm. 7 strongly suggests that canonicity checking is unlikely to be in NP, and much less in P.

Note however that this hardness result is not obviously applicable to square-free graphs, since in particular the latter have no cliques of size greater than 3. We are nonetheless able to establish the following weaker statement, whose proof can be found in [1, 2]:

Theorem 8. *If the problem of canonicity checking for adjacency matrices of square-free graphs were in NP, then the graph isomorphism problem (for arbitrary graphs) would be in co-NP.*

---

[4] Note that the problem of *canonisation*—i.e., given a graph, construct its canonical adjacency matrix—is well-known to be both NP-hard and co-NP-hard [3]. Yet it is conceivable that merely *verifying* canonicity could be substantially easier.

In practice, notwithstanding Thms. 7 and 8, we have found that canonicity checking could be made extremely efficient. We implemented a fairly simple backtracking algorithm which enabled us to check canonicity of the vast majority of matrices with 17 vertices or fewer within microseconds. This led us to the following *computer-aided* result:

Theorem 9. *Every square-free graph with at most 16 vertices is 101-colourable. Moreover, there is a unique graph with 17 vertices that is not 101-colourable (shown on the left-hand side of Fig. 2).*

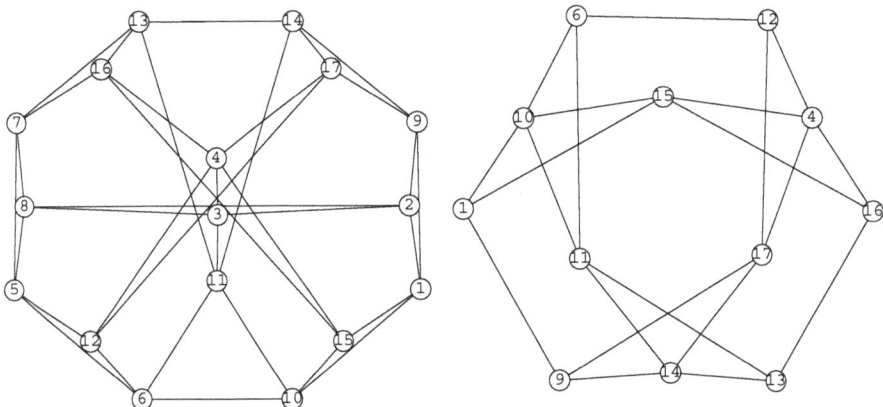

Fig. 2. The unique square-free graph on 17 vertices that is not 101-colourable (left), together with its smallest unembeddable subgraph

Unfortunately, this 17-vertex candidate turned out to be unembeddable. Using Bertini [6], a state-of-the-art software package for numerical algebraic geometry, we identified a 12-vertex subgraph (shown on the right-hand side of Fig. 2) whose associated embedding polynomial was shown to have precisely 12 distinct complex roots, none of which are purely real. This produces a *numerical* proof of the fact that our 17-vertex candidate cannot be embedded. Finally, we were able to upgrade this to a *computer-aided* proof using our interval arithmetic solver (see [1, 2]), yielding the following lower bound:

Theorem 10. *A Kochen-Specker vector system must contain at least 18 vectors.*

Unfortunately, Thm. 10 still leaves an astronomical gap to bridge before proving that Conway and Kochen's KS vector system of size 31 is the smallest possible (if indeed that is the case). Extrapolating from known data, the graph below suggests that there are some 10^{32} connected square-free graphs on 30 vertices or less, well out of the brute-force reach of current technology.

Number of Connected Square-Free Graphs (Log10)

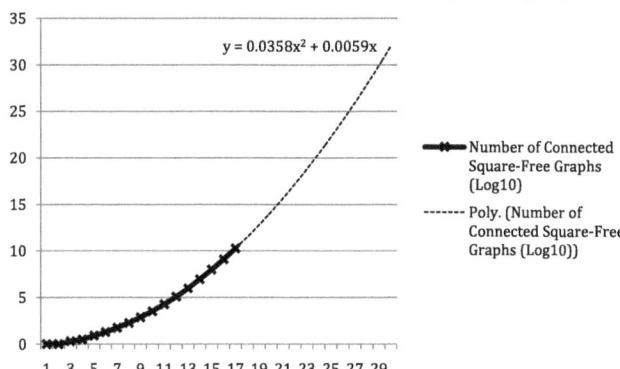

5   Conclusion

We have proposed the problem of finding small Kochen-Specker vector systems—or proving that none exist of size less than 31—as a difficult and worthwhile algorithmic challenge. Higher-dimensional generalisations of the problem have also been considered by others, notably Pavičić *et al.* [17].

The results we have obtained (greater details of which are available in [1]) can largely be summarised by listing a number of properties that any minimal KS vector system must enjoy. Such a system:

- has at most 31 vectors (Conway and Kochen's KS vector system);
- contains at least 18 vectors (Thm. 10);
- has associated graph that is square-free (Sec. 2);
- has associated graph that is not 101-colourable and not 3-colourable (Sec. 2);
- has associated graph that is 4-colourable (Prop. 1);
- has associated graph with minimum degree 3 [1];
- has associated graph in which each vertex belongs to a triangle [1];
- is not a subsystem of the cubic grid with grid parameter $N = 4$, unless it is the Conway-Kochen 31-vector system itself (Sec. 3).

In our view, two central challenges are to (i) devise more efficient algorithms for determining graph embeddability (in which respect Conjecture 5 could play a key role), and (ii) find efficient means to drastically cut down the number of candidate graphs that must be examined.

Acknowledgements. We thank Nick Trefethen for introducing the first two authors to the third, Jean-Pierre Merlet for sharing his experience on interval arithmetic, and Don Knuth for drawing our attention to [9]. The second author was supported by EPSRC and the third author was supported by NSF grant DMS-0712910.

References

[1] Arends, F.: A lower bound on the size of the smallest Kochen-Specker vector system. Master's thesis. Oxford University (2009),
www.cs.ox.ac.uk/people/joel.ouaknine/download/arends09.pdf

[2] Arends, F., Ouaknine, J., Wampler, C.W.: On searching for small Kochen-Specker vector systems (extended version). Technical report (2011),
www.cs.ox.ac.uk/people/joel.ouaknine/publications/ks11abs.html

[3] Babai, L., Luks, E.M.: Canonical labeling of graphs. In: Proc. STOC. ACM (1983)

[4] Basu, S., Pollack, R., Roy, M.-F.: Algorithms in Real Algebraic Geometry. Springer, Heidelberg (2006)

[5] Beineke, L.W., Wilson, R.J. (eds.): Topics in Topological Graph Theory. Encyclopedia of Mathematics and its Applications. Cambridge University Press (2009)

[6] http://www.nd.edu/~sommese/bertini/

[7] Bub, J.: Schütte's tautology and the Kochen-Specker theorem. Found. Phys. 26, 787–806 (1996)

[8] Canny, J.: Some algebraic and geometric computations in PSPACE. In: Proc. STOC. ACM (1988)

[9] Colbourn, C.J., Read, R.C.: Orderly algorithms for graph generation. Int. J. Comput. Math. 7, 167–172 (1979)

[10] Conway, J.H., Kochen, S.: The free will theorem. Found. Phys. 36(10), 1441–1473 (2006)

[11] Conway, J.H., Kochen, S.: The strong free will theorem. Notices of the AMS 56(2) (2009)

[12] Huang, Y.-F., Li, C.-F., Zhang, Y.-S., Pan, J.-W., Guo, G.-C.: Experimental test of the Kochen-Specker theorem with single photons. Phys. Rev. Lett. 90(250401) (2003)

[13] Kochen, S., Specker, E.P.: The problem of hidden variables in quantum mechanics. J. Math. Mech. 17, 235–263 (1967)

[14] Larsson, J.-Å.: A Kochen-Specker inequality. Europhys. Lett. 58, 799–805 (2002)

[15] McKay, B.D.: Isomorph-free exhaustive generation. J. Alg. 26, 306–324 (1998)

[16] http://www.minisat.se/

[17] Pavičić, M., Merlet, J.-P., McKay, B., Megill, N.D.: Kochen-Specker vectors. J. Phys. A: Math. Gen. 38(7), 1577–1592 (2005)

[18] Pavičić, M., Merlet, J.-P., Megill, N.D.: Exhaustive enumeration of Kochen-Specker vector systems. Research report RR-5388, INRIA (2004)

[19] Peres, A.: Two simple proofs of the Kochen-Specker theorem. J. Phys. A: Math. Gen. 24, L175–L178 (1991)

[20] Peres, A.: Quantum Theory: Concepts and Methods. Kluwer (1993)

[21] Read, R.C.: Every one a winner, or: How to avoid isomorphism search when cataloguing combinatorial configurations. Annals Discrete Math. 2, 107–120 (1978)

[22] Read, R.C.: A survey of graph generation techniques. Lecture Notes in Mathematics, vol. 884. Springer, Heidelberg (1981)

[23] Renegar, J.: On the computational complexity and geometry of the first-order theory of the reals. Parts I-III. J. Symb. Comput. 13(3), 255–352 (1992)

[24] Sloane, N.J.A.: The On-Line Encyclopedia of Integer Sequences (2010),
http://www.research.att.com/~njas/sequences/

Characterizations of Deque and Queue Graphs*

Christopher Auer and Andreas Gleißner

University of Passau, 94030 Passau, Germany
{auerc,gleissner}@fim.uni-passau.de

Abstract. In graph layouts the vertices of a graph are processed according to a linear order and the edges correspond to items in a data structure inserted and removed at their end vertices. Graph layouts characterize interesting classes of planar graphs: A graph G is a stack graph if and only if G is outerplanar, and a graph is a 2-stack graph if and only if it is a subgraph of a planar graph with a Hamiltonian cycle [2]. Heath and Rosenberg [12] characterized all queue graphs as the arched leveled-planar graphs. In [1], we have introduced linear cylindric drawings (LCDs) to study graph layouts in the double-ended queue (deque) and have shown that G is a deque graph if and only if it permits a plane LCD.

In this paper, we show that a graph is a deque graph if and only if it is the subgraph of a planar graph with a Hamiltonian path. In consequence, we obtain that the dual of an embedded queue graph contains a Eulerian path. We also turn to the respective decision problem of deque graphs and show that it is $\mathcal{NP}$-hard by proving that the Hamiltonian path problem in maximal planar graphs is $\mathcal{NP}$-hard. Heath and Rosenberg state [12] that queue graphs are "almost" proper leveled-planar. We show that bipartiteness captures this "almost": A graph is proper leveled-planar if and only if it is a bipartite queue graph.

1 Introduction

In a graph layout the vertices are processed according to a total order, which is called *linear layout*. The edges correspond to data items that are inserted to and removed from a data structure: Each edge is inserted at its end vertex that occurs first according to the linear layout and is removed at its other end vertex. These operations must obey the principles of the underlying data structure, such as "last-in, first-out" for a stack or "first-in, first-out" for a queue.

Queue and stack layouts, the latter also known as book embeddings, have been studied extensively in the past, e.g., in [2, 3, 6, 7, 9, 10, 11, 12, 17, 18, 20], and are used for 3D drawings of graphs [17, 18], in VLSI design [3] and in other application scenarios [12]. Graph layouts have also shed new light on Gauss codes and permutation networks [13]. A graph G is a stack graph, i.e., has a stack layout, if and only if it is outerplanar, and it is a 2-stack graph if and only if it is a subgraph of a planar graph with a Hamiltonian cycle (HC). Heath et al.

---

* Supported by the Deutsche Forschungsgemeinschaft (DFG), grant Br835/15-1.

P. Kolman and J. Kratochvíl (Eds.): WG 2011, LNCS 6986, pp. 35–46, 2011.
© Springer-Verlag Berlin Heidelberg 2011

[7, 12] characterize the class of queue graphs as the arched leveled-planar graphs. Such graphs have a planar drawing with vertices placed on levels and inter-level edges only between two adjacent levels or intra-level edges (the *arches*) from the left-most vertex to accessible vertices on the right side (see Fig. 3(a)).

In [1] we have studied graph layouts in the double-ended queue (*deque*): A deque has two ends, a head and a tail, to insert and remove items. A deque operates like a stack if an item is inserted and removed at the same side. It operates like a queue if an item is inserted at one and removed from the other side. In [1] we have introduced linear cylindric drawings (LCDs) of graphs. The property of planarity also plays an important role in LCDs: A graph is a deque graph if and only if it admits a plane LCD [1]. By further investigating linear cylindric drawings, we are able to prove a new characterization of deque graphs: A graph is a deque graph if and only if it is a subgraph of a planar graph with a Hamiltonian path (HP). Remember that a 2-stack graph is a subgraph of a planar graph with an HC. A single deque can emulate two stacks and additionally allows queue items. Intriguingly, the "surplus" of power a deque has in comparison to two stacks, i.e., the queue items, exactly captures the gap between HCs and HPs in planar graphs. Based on these observations, we will also give a new characterization of queue graphs. Specifically, we will see that the dual of an embedded queue graph contains a Eulerian path.

We will then turn to the question how hard the decision problem "Is a given graph a deque graph?" is. We will prove its $\mathcal{NP}$-completeness by showing that the HP problem in maximal planar graphs is $\mathcal{NP}$-complete.

Heath and Rosenberg called arched leveled-planar graphs, which characterize all queue graphs, "almost proper leveled-planar" [12]. Proper leveled-planar graphs, i.e., leveled planar graphs with edges only between adjacent levels, play an important role in the field of graph drawing, e.g., in [5, 14]. We will show that the property of bipartiteness exactly corresponds to the gap between arched and proper leveled-planar graphs: A graph is a bipartite queue graph if and only if it is proper leveled-planar.

The remainder of this paper is organized as follows: Sect. 2 introduces deque layouts and linear cylindric drawings. In Sect. 3.1, we prove that a graph is a deque graph if and only if it is a subgraph of a planar graph with an HP. We will also give an intuitive explanation for this theorem, which leads to interesting insights into deque and queue layouts. In Sect. 3.3 we show that the decision problem "Is a given graph a deque graph?" is $\mathcal{NP}$-complete. We then characterize all bipartite queue graphs as the proper leveled-planar graphs (Sect. 4). In Sect. 5 we give a conclusion and some pointers to future work.

2   Preliminaries

In this section we introduce deque layouts and LCDs. We consider simple undirected graphs $G = (V, E)$ with n vertices and m edges. Let $\prec$ be a total order on V, called *linear layout*. If $u \prec v$ ($u \succ v$), we say that u *precedes (succeeds)* v. In the following, we assume that the edges are directed according to the linear layout, i.e., each edge $\{u, v\}$ with $u \prec v$ is denoted by its directed version

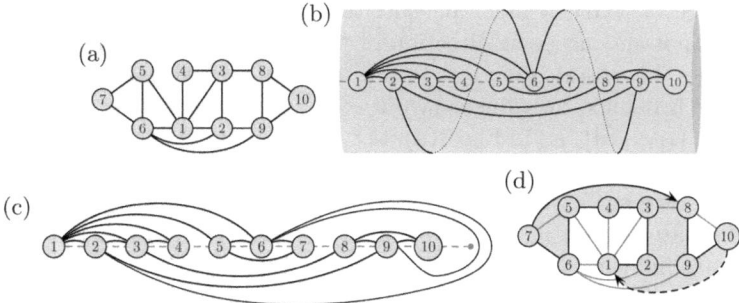

Fig. 1. A planar graph (see (a)) and its linear cylindric drawing on a 3D cylinder (see (b)), which can be obtained by "cutting" along the Hamiltonian path (see (d)). The linear cylindric drawing can be embedded in the plane (see (c)).

(u, v). A linear layout defines the order in which the vertices are processed in a graph layout, where at each vertex all incoming edges have to be removed from the data structure first and then all outgoing edges have to be inserted into the data structure. These operations must obey the principles of the underlying data structure, where we focus on the deque in the following.

A deque has two ends, a *head* $\mathtt{h}$ and a *tail* $\mathtt{t}$, to insert and remove edges. Let $\alpha/\omega : E \to \{\mathtt{h}, \mathtt{t}\}$ be two functions that assign to each edge the side of its insertion and removal, respectively. α/ω are called *input/output* assignments (*I/O assignments*). If $\alpha(e) = \omega(e)$, then e is called *stack edge*, otherwise *queue edge*, according to the manner e is processed in the deque. Note that a deque can emulate two stacks by restricting α/ω such that only stack edges are allowed. If all stack (queue) edges are inserted at the same side, a single stack (queue) is emulated. The content of a deque is denoted by $\mathcal{C} = (e_1, \dots, e_k)$, where e_1 is at the deque's head and e_k at the tail. At vertex v, at first all edges from preceding vertices have to be removed from the deque according to ω. Note that an edge can only be removed if it is situated at the deque's head or tail, i.e., if $\mathcal{C} = (e_1, \dots, e_k)$, then afterwards only a "kernel" $\mathcal{C}' = (e_r, \dots, e_s)$ with $1 \le r \le s \le k$ remains, where $\omega(e_t) = \mathtt{h}$ for $t = 1, \dots, r - 1$ and $\omega(e_t) = \mathtt{t}$ for $t = s + 1, \dots, k$. Afterwards, all edges to succeeding vertices of v are inserted according to α. That is, if $\mathcal{C}' = (e_r, \dots, e_s)$, then its content afterwards is $\mathcal{C}''_i = (e'_1, \dots, e'_{j'}, e_r, \dots, e_s, e_1, \dots, e''_{j''})$ where $\alpha(e'_i) = \mathtt{h}$ for all $i = 1, \dots, j'$ and $\alpha(e''_i) = \mathtt{t}$ for all $i = 1, \dots, j''$.

A linear layout with I/O assignments denoted by $\Delta(G) = (\prec, \alpha, \omega)$ is called *deque layout* if all edges can be processed in a deque according to $\prec$ and α/ω and a graph is a *deque graph* if and only if it has a deque layout. Note that the property of having a deque layout is hereditary, i.e., if a graph has a deque layout, then so does every subgraph. Also note that we can neglect the exact order in which the edges are inserted/removed at a vertex as there is always a canonical order (see also [1]).

In [1] we have introduced *linear cylindric drawings (LCDs)*: In an LCD of a graph G, the vertices are placed disjointly on a straight line, the *front line*, on the surface of a 3D cylinder parallel to the cylinder's axis. The edges are drawn as monotone curves in direction of the cylinder's axis and do not cross the front line. The total order according to which the vertices are placed on the front line is defined by a linear layout $\prec$. As an example consider the planar graph $G = (\{1, \ldots, 10\}, E)$ in Fig. 1(a). Let $\prec$ be the linear layout corresponding to the numbering of the vertices. The corresponding LCD of G with the linear layout $\prec$ is shown in Fig. 1(b): The vertices are placed on the front line (dashed) and all edges are either routed entirely above or below the front line or wrap at most once around the cylinder.

In [1] we have investigated the relationship between deque layouts and LCDs. We have found out that a graph is a deque graph if and only if it admits a plane linear cylindric drawing (cf. Theorem 1 in [1]). The key observation to obtain this result is as follows: The regions above and below the front line correspond to the head and tail of the deque, respectively. An edge (u, v), leaving u and entering v from above the front line, e.g., edge $(1, 5)$ in Fig. 1(b), is inserted to and removed from the head of the deque, i.e., it is a stack edge. Edges entering their end vertices from opposing sides of the front line are queue edges, e.g., edge $(6, 9)$ is inserted to the head and removed from the tail of the deque.

An LCD on the cylinder can be continuously transformed to a drawing in the plane by mapping the surface of the cylinder to a disk. Hence, all deque graphs are planar. Fig. 1(c) shows the result of the mapping when applied to the LCD in Fig. 1(b). Again, the front line is displayed by a dashed line and the region on the right side of the cylinder in Fig. 1(b) is symbolized by the gray dot at the right end of the front line in Fig. 1(c). All queue edges are now routed once around the disk.

3   Deque Graphs

3.1   Characterizing Deque Graphs

In this section we characterize deque graphs as the subgraphs of planar graphs with an HP.

Theorem 1. *A graph is a deque graph if and only if it is a subgraph of a planar graph with an HP.*

Proof. "$\Rightarrow$": Let G be a deque graph and, hence, G has a plane LCD according to Theorem 1 in [1]. Consider, for instance, the LCD in Fig. 1(c). Note that edges between immediate neighbors on the front line can always be inserted without destroying planarity. By inserting these edges, we obtain the super-graph G' of G which has an HP and is a deque graph since it allows a plane LCD. Since G' is also planar, G has a planar super-graph with an HP.

"$\Leftarrow$": Consider the planar super-graph $G' = (V', E')$ of G with an HP. Any plane drawing of G' can be continuously transformed to a plane LCD where the HP is a straight line. In the so obtained LCD, the vertices V' are placed on the front line according to the HP and the edges E' do not cross the front line. By Theorem 1 in [1], we can conclude that G' is a deque graph and so is G.   □

Since the deque can emulate two stacks, the surplus of the deque to additionally allow queue edges exactly bridges the gap between Hamiltonian *cycles* and Hamiltonian *paths* in planar graphs. With a slight abuse of notation, we get for planar graphs:

$$\frac{\text{Hamiltonian Path}}{\text{Hamiltonian Cycle}} = \frac{\text{Deque}}{2 \text{ Stacks}} .$$

3.2 Hamiltonian Paths in Deque and Queue Graphs

Remember that the stack graphs are exactly the outerplanar graphs. A graph is a 2-stack graph if and only if it has a planar super-graph with an HC, where the proof in [2] relies on the following idea: Consider a plane drawing of a planar graph with an HC and "cut" along this cycle. This divides the plane into two regions where the vertices lie along the cut. Hence, we have obtained two outerplanar graphs, each of which uses one stack in the 2-stack layout or, equivalently, both use the same deque with stack edges only.

We can apply a similar interpretation to Theorem 1: In Fig. 1(d) the graph from Fig. 1(a) is augmented by edges to form the HP $p = 1, \ldots, 10$. In the following, let V^* be the plane regions in G's embedding, then $G^* = (V^*, E)$ is the dual of G where the set of edges of G^* is equal to that of G, i.e., $e \in E$ is an edge between two regions $u, v \in V^*$ if e joins the regions u and v in G's embedding. Now, join start and end of HP p by a path from vertex 10 and 1 by a path in G^*. Hence, we obtain a closed curve that divides the plane into region A (shaded in Fig. 1(d)) and its complement $\complement A$. Like in the case of the 2-stack graphs, all edges that lie completely within A or $\complement A$ are stack edges, e.g., edge $(5, 7)$, situated completely within A, is inserted to and removed from the head of the deque (see Fig. 1(b)). Edges that go from one region to the other are exactly the queue edges in the deque layout, e.g., edges $(2, 6)$ and $(6, 9)$, where, for instance, $(2, 6)$ is inserted at the head and removed from the tail in the deque.

By applying the same interpretation, we obtain a new characterization of queue graphs. Note that a queue graph is essentially a deque graph, where all edges are queue edges. Hence, we obtain:

Corollary 1. $G = (V, E)$ *is a queue graph if and only if* $G = (V, E)$ *has a planar super-graph* $G' = (V', E')$ *with an HP* $p = v_1, \ldots, v_n$ *fulfilling the following properties: HP p does not use any edges of E and there is a path $p^* = v_1^*, \ldots, v_r^*$ in G''s dual graph G^* from a region at v_1 to a region at v_n that visits each edge $e \in E$ exactly once, i.e., p^* is a Eulerian path in G^*.*

Corollary 2. *Let $G = (V, E)$ be an embedded queue graph and G^* be its dual, then G^* contains a Eulerian path.*

3.3 Deciding If a Graph Is a Deque Graph Is $\mathcal{NP}$-Complete

The problem of deciding whether a maximal planar graph has an HC is $\mathcal{NP}$-complete [4, 16]. Consequently, the decision problem "Is a given graph a 2-stack

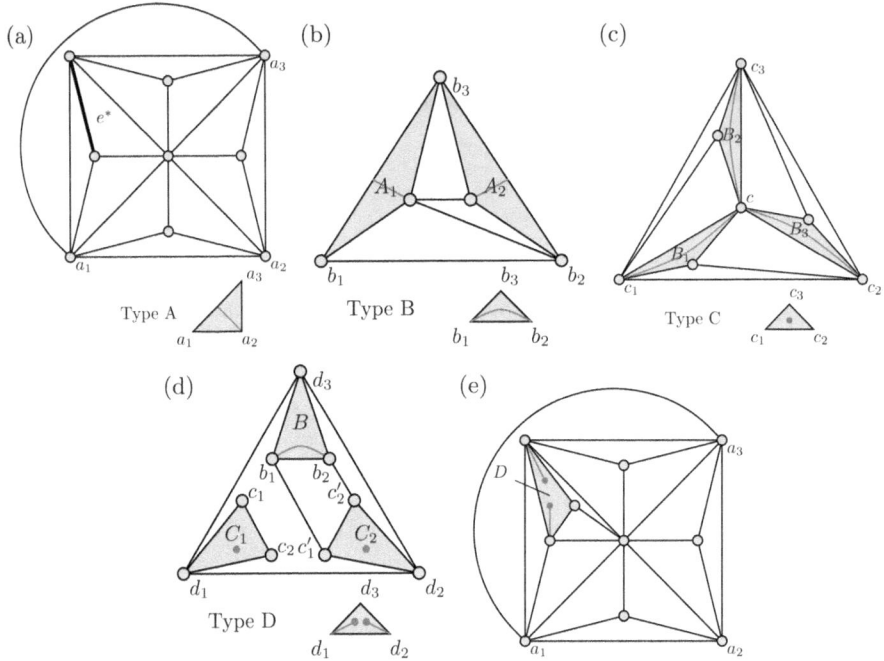

Fig. 2. Widgets needed for the $\mathcal{NP}$-completeness reduction in Theorem 2

graph?" is $\mathcal{NP}$-complete. In contrast, deciding if a graph permits a stack layout can be done in linear time [2, 15]. We show that deciding if a graph is deque graph is also $\mathcal{NP}$-complete. To do so we show the $\mathcal{NP}$-completeness of the following problem:

Theorem 2. *The decision problem "Given a maximal planar graph, does it contain a Hamiltonian path?" is $\mathcal{NP}$-complete.*

Proof. In [16], Wigderson uses an elegant construction to transform a 3-connected, cubic planar graph G to a maximal planar graph G' and shows that G contains an HC if and only if G' contains an HC. For 3-connected, cubic planar graphs the HC problem is known to be $\mathcal{NP}$-complete [8]. To the best of our knowledge (and surprise) the HP problem for maximal planar graphs has not been shown to be $\mathcal{NP}$-complete so far.

In the following, we will introduce only those widgets from [16] that we need for our purposes. For all widgets and proofs of their properties see [16]. The widget of *Type A* is the base building block of the construction and is depicted in Fig. 2(a): Wigderson has shown that if an HC enters and leaves, i.e., *traverses*, a Type-A widget A, then all vertices of A must be visited before the A can be left. He shows this by enumerating all possibilities of traversing a Type-A widget, where all possibilities use edge e^* (bold line) (see Fig. 2(a)); a fact which we will use later in our adaption. In the following, a Type-A widget is symbolized by a shaded triangle with a gray line (cf. Fig. 2(a); note that this widget is symmetric with respect to the gray line within the triangle).

The *Type-B* widget, shown in Fig. 2(b), is composed by two Type-A widgets A_1 and A_2. Wigderson has shown that an HC traversing a Type-B widget must enter and leave at vertices b_1 and b_2 and, again, all vertices in A_1 and A_2 have to be visited during this traversal. Hence, it is not possible to visit, for instance, b_3 and then visit the remaining vertices of the widget later (cf. [16]). A Type B widget is symbolized by a shaded triangle, where the corners symbolizing b_1 and b_2 are connected by an arc.

For our adaption we compose two additional widgets of Type-C and D: The *Type-C* widget is depicted in Fig. 2(c) and it consists of three Type-B widgets. A Type-C widget C has the property that any HP in a graph that contains C must have at least one of its end points in C, i.e., C cannot be traversed completely: Suppose that an HP enters at c_1, then it must visit all vertices of B_1 until it reaches *center vertex* c. Then either B_2 or B_3 have to be traversed entirely. W. l. o. g. it traverses B_2 and reaches c_3. Still, the vertices of B_3 have to be visited. The only possibility (by the properties of Type-A and B widgets) of doing this is to enter C at c_2 again and to visit all vertices of B_3 just before center vertex c is reached. There the HP ends. The same holds if the HP enters at c_2 or c_3. A Type-C widget is denoted by a triangle with a gray dot symbolizing the end point of an HP. Note that a Type-C widget is invariant with respect to rotation around vertex c.

Two Type-C widgets can be composed to obtain a *Type-D* widget, which assures that both ends of an HP end at a Type-D widget. The Type-D widget is displayed in Fig. 2(d). For the sake of clarity we did not triangulate the Type-D widget since it can be triangulated arbitrarily. Let D be a Type-D widget then an HP of a graph containing D must have its one end in C_1 and its other end in C_2. The Type-B widget B in D assures that any HP leaves D at d_1 and d_2: Since any HP has to end in C_1 and C_2, B has to traversed by an HP. By the property of Type-B widgets, the HP has to enter and leave B at b_1 and b_2 and during the traversal d_3 is traversed.

The construction in [16] replaces each vertex of the 3-connected, cubic graph G by a widget composed of three Type-B widgets (consisting of two Type-A widgets) and afterwards elegantly triangulates the whole graph to obtain maximal planar graph G'. In our adaption we apply the same construction and then pick an arbitrary Type-A widget A in G'. We then insert a Type-D widget D into A as displayed in Fig. 2(e): Edge e^* is replaced by the edge d_1 and d_2 of D and vertex d_3 is connected to the central vertex of A. We denote the so obtained graph by G''. Remember that e^* is the edge that always has to be used when traversing a Type-A widget.

Suppose that G contains an HC. Then, by the construction in [16], G' also contains an HC. This HC must use edge e^* in the Type-A widget A of G'. In G'' and the corresponding Type-D widget (cf. Fig. 2(d) and (c)), consider the path $d_2 \to c_2' \to b_2 \to b_1 \to c_1'$ and from c_1' to the vertex before the center vertex of C_2. This path visits all vertices of C_2 and B and ends in C_2. Similarly, the path

$d_1 \rightarrow c_1 \rightarrow c_2$ and, again, from c_2 to the vertex before the center vertex of C_1 visits all vertices of C_1 and ends in C_1. Hence, in G'' instead of using edge e^* we visit all vertices in D and let the HP end in C_1 and C_2. Consequently, G'' has an HP.

Conversely, let G'' contain an HP. Then, by construction, the HP must end in C_1 and C_2. Note that the HP must leave D at d_1 and d_2. In G', there is an HP starting and ending at d_1 and d_2. By adding edge e^* to the HP we get an HC and, hence, G also contains an HC.   □

Theorem 3. *The decision problem "Is a given graph a deque graph?" is $\mathcal{NP}$-complete.*

Proof. Let $G = (V, E)$ be a graph, then non-deterministically guess a deque layout $\Delta(G) = (\prec, \alpha, \omega)$ of G, which can be done in linear time. Insert all edges between immediate successors according to $\prec$ in order to construct a super-graph G' of G which contains an HP. G' is planar if and only if G is a deque graph (Theorem 1). Planarity of G' can be decided in polynomial time. Consequently, the decision problem is in $\mathcal{NP}$.

Now, let $G = (V, E)$ be maximal planar, i.e., if G' is a super-graph of G and G' is planar, then $G' = G$. Hence, $G = (V, E)$ is a deque graph if and only if it contains an HP. Thus, the decision problem "Does a maximal planar graph contain an HP?" reduces to "Is a given graph a deque graph?", which is then also $\mathcal{NP}$-complete by Theorem 2.   □

4   Queue Graphs

Now we turn to queue graphs. Heath and Rosenberg proved that a graph has a queue layout if and only if it has an arched leveled-planar embedding [12]. The leveling of a graph is an assignment of its vertices to positive integers. An arched leveled embedding is a leveling together with a total left-to-right order of the vertices on each level such that edges connect either vertices of adjacent levels (*inter-level edges*) or the left-most vertex on a level with an accessible vertex on the right side on the same level (*arches*).

We consider a special case by forbidding arches, i.e., ask for a *proper* leveled-planar embedding. We show that this restriction is exactly captured by bipartiteness. The vertices of different levels are usually considered unrelated. However, for sake of simplicity, we operate with a single total order on all vertices, which is a linear extension of the union of the levelwise orders. Additionally, we consider the arches as directed from the left-most vertex to the vertex on the right side.

Theorem 4. *A graph is proper leveled-planar if and only if it is bipartite and has a queue layout.*

Proof. "⇒": Let $G = (V, E)$ be a graph with a proper level-drawing, which is just a special case of an arched leveled-planar drawing, so that G, according to [12], has a queue layout. The vertices placed on even, and the vertices placed on odd levels of the drawing form a bipartition of V.

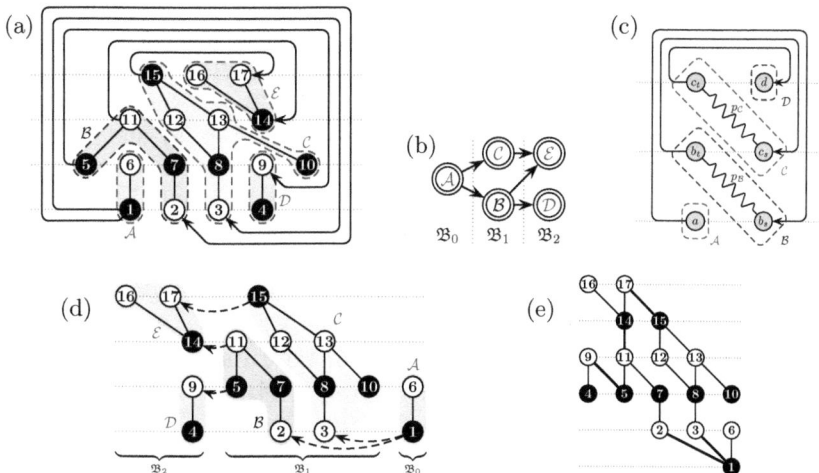

Fig. 3. Bipartite graph G with an arched leveled-planar embedding (a), which is transformed via (d) to a proper leveled-planar embedding (e). (b) is the component graph $\mathbb{G}$ of G. (c) illustrates the proof that any component graph has a proper leveling.

"$\Leftarrow$": Let $G = (V, E)$ be a bipartite graph with a queue layout and partition $V_w \dot\cup V_b = V$, i.e., $V_w \cap V_b = \emptyset$, see Fig. 3(a). We will refer to V_w (V_b) as the set of *white* (*black*) vertices. G has an arched leveled-planar embedding Σ consisting in a level assignment $\phi : V \to \mathbb{N}$ and a total order $\prec$ on V, which imposes the order of the vertices on each level. Let $A \subseteq E$ be the arches of G caused by Σ. We will transform Σ into a proper leveled-planar embedding Σ' without arches. W. l. o. g. assume that G is connected. Otherwise Σ can be transformed independently for each connected component of G.

We call a vertex *even* (*odd*) if it is placed on an even (odd) level by Σ and denote the set of all even (odd) vertices by V_{even} (V_{odd}). Obtain the graph $G' = (V, E \setminus A)$ by temporarily removing the arches from G. As Σ clearly is a proper leveled-planar embedding of G', $V_{\text{even}} \dot\cup V_{\text{odd}}$ form another bipartition of the vertices of G'. We call $V_+ = (V_{\text{even}} \cap V_w) \cup (V_{\text{odd}} \cap V_b)$ the set of *positive* vertices and $V_- = (V_{\text{even}} \cap V_b) \cup (V_{\text{odd}} \cap V_w)$ the set of *negative* vertices, respectively. Note that in G' none of the positive vertices is connected to any of the negative vertices as each edge in $E \setminus A$ connects vertices of simultaneously different parity and color.

Let $\mathbb{V}_+$ ($\mathbb{V}_-$) be the set of connected components of G' which consist of positive (negative) vertices and let $\mathbb{V} = \mathbb{V}_+ \cup \mathbb{V}_-$. Denote by $\mathfrak{c}(v)$ the connected component of G' containing vertex $v \in V$. We define the directed graph $\mathbb{G} = (\mathbb{V}, \mathbb{E})$ by asserting a so called *component connection* $(\mathcal{A}, \mathcal{B}) \in \mathbb{E}$ if and only if the components $\mathcal{A}, \mathcal{B} \in \mathbb{V}$ are connected by an arch in G, i.e., if there are vertices $u \in \mathcal{A}$ and $v \in \mathcal{B}$ with $(u, v) \in A$. Note that while each arch contributes to a single connection $(\mathcal{A}, \mathcal{B}) \in \mathbb{E}$, this connection may be caused by several arches.

As an intermediate result, we prove that there is a proper leveling of $\mathbb{G}$. In order to avoid confusion of said leveling with planar level embeddings of G, we speak of *buckets* instead of levels in the context of $\mathbb{G}$. In other words, we show that $\mathbb{V}$ can be partitioned into buckets $\mathfrak{B}_0, \mathfrak{B}_1, \mathfrak{B}_2, \ldots$ such that for each component connection $(\mathcal{A}, \mathcal{B}) \in \mathbb{E}$ there is an i with $\mathcal{A} \in \mathfrak{B}_i$ and $\mathcal{B} \in \mathfrak{B}_{i+1}$. Observe that each arch connects vertices of different color but the same parity, i.e., the connected vertices (and thereby, their components) are of different sign. Thus, $\mathbb{G}$ is bipartite with the partition $\mathbb{V}_+ \,\dot{\cup}\, \mathbb{V}_-$ and, in particular, contains no triangle. Assume there are components $\mathcal{A}, \mathcal{B}, \mathcal{C}, \mathcal{D} \in \mathbb{V}$ with $(\mathcal{A}, \mathcal{B}), (\mathcal{B}, \mathcal{C}), (\mathcal{C}, \mathcal{D}) \in \mathbb{E}$, see Fig. 3(c). It remains to show that there are no transitive edges $(\mathcal{A}, \mathcal{D}) \in \mathbb{E}$ and no directed cycles caused by $(\mathcal{D}, \mathcal{A})$ or $(\mathcal{B}, \mathcal{A}) \in \mathbb{E}$. The following argument applies analogously for the situation with more components involved, i.e., $(\mathcal{A}_1, \mathcal{A}_2), (\mathcal{A}_2, \mathcal{A}_3) \ldots, (\mathcal{A}_{k-1}, \mathcal{A}_k) \in \mathbb{E}$ $(k \in \mathbb{N})$, and for the case $|\mathbb{V}| \leq 3$. Let $a \in \mathcal{A}, b_s, \ldots, b_t \in \mathcal{B}, c_s, \ldots, c_t \in \mathcal{C}$, and $d \in \mathcal{D}$ and let $p = a, b_s, \ldots, b_t, c_s, \ldots, c_t, d$ be a simple path in G from $\mathcal{A}$ via $\mathcal{B}$ and $\mathcal{C}$ to $\mathcal{D}$ ignoring edge directions. $(a, b_s), (b_t, c_s), (c_t, d) \in A$ are arches of G with $\phi(a) = \phi(b_s) \neq \phi(b_t) = \phi(c_s) \neq \phi(c_t) = \phi(d)$ as arches on the same level would have to share the same source vertex. W. l. o. g. assume $\phi(b_t) > \phi(b_s)$. For any vertex $u \in V$ and a collection of vertices $S \subseteq V$, we say that u *is placed at the bottom left* of S if $\phi(u) < \min_{v \in S} \phi(v)$ or if $\phi(u) \leq \max_{v \in S} \phi(v)$ and for all $v \in S$ with $\phi(u) = \phi(v)$, $u \prec v$. Accordingly, we say that u *is placed at the top right* of S if $\phi(u) > \max_{v \in S} \phi(v)$ or if $\phi(u) \geq \min_{v \in S} \phi(v)$ and for all $v \in S$ with $\phi(u) = \phi(v)$, $v \prec u$. The subpath $p_\mathcal{B} = b_s, \ldots, b_t$ is a kind of "separator" of G' in the following sense. If a vertex u is placed at the bottom left and a vertex v is placed at the top right of $p_\mathcal{B}$, then, due to the planarity of Σ, any path q from u to v in G must either share a vertex with $p_\mathcal{B}$ or contain at least one arch. Thus, if q consists solely of inter-level edges, both u and v are connected to $p_\mathcal{B}$ and thereby are contained in $\mathcal{B}$. This implies that if $\mathfrak{c}(u) \neq \mathcal{B}$ or $\mathfrak{c}(v) \neq \mathcal{B}$, then $\mathfrak{c}(u) \neq \mathfrak{c}(v)$ since q contains an arch. Hence, $\phi(c_t) > \phi(c_s)$ as otherwise c_s and c_t would be placed at different sides of $p_\mathcal{B}$, but in fact $c_s, c_t \in \mathcal{C}$. Similarly, the subpath $p_\mathcal{C} = c_s, \ldots, c_t$ is a separator of G', too. Note that $p_\mathcal{B}$ and $p_\mathcal{C}$ cannot be bypassed simultaneously by a single arch because they share only the level $\phi(b_t) = \phi(c_s)$. Therefore, any path q that connects a and d and is vertex disjoint to p, must contain at least two arches so that neither $(\mathcal{A}, \mathcal{D}) \in \mathbb{E}$ nor $(\mathcal{D}, \mathcal{A}) \in \mathbb{E}$. Furthermore, assume by contradiction $(\mathcal{B}, \mathcal{A}) \in \mathbb{E}$ caused by an arch (b_*, a_*) with $b_* \in \mathcal{B}$ and $a_* \in \mathcal{A}$. Then there must be a path $p_\mathcal{A} = a_*, \ldots, a \in \mathcal{A}$ being a separator of G'. b_* must be a left-most vertex on the level $\phi(b_*) = \phi(a_*)$. But then $b_* \notin \mathcal{B}$ as b_s and b_* would be placed at different sides of $p_\mathcal{A}$, completing the proof for the proper leveling of $\mathbb{G}$.

Using the partition of $\mathbb{V}$ into buckets as described above, we can obtain a new total order $\prec'$ of the vertices without introducing an edge crossing as follows. Consider two vertices $u, v \in V$. If u and v belong to components of different buckets, then they are ordered by $\prec'$ in reverse order of their buckets. In other words, if there are $i > j$ with $\mathfrak{c}(u) \in \mathfrak{B}_i$ and $\mathfrak{c}(v) \in \mathfrak{B}_j$, then $u \prec' v$. If, however, $\mathfrak{c}(u)$ and $\mathfrak{c}(v)$ are in the same bucket, then the initial order $\prec$ is preserved, i.e.,

$u \prec' v \Leftrightarrow u \prec v$. The corresponding level embedding of G', see Fig. 3(d), is still planar as $\prec'$ does particularly not alter the order within a component and does also not introduce new interleavings of different components. Note that the vertices which where connected by an arch in G, now lie next to each other. Next we transform the leveling ϕ. For each vertex $v \in V$, let $\phi'(v) = \phi(v) + i$ if $\mathfrak{c}(v) \in \mathfrak{B}_i$. ϕ' and $\prec'$ constitute the new level embedding Σ' of G', see Fig. 3(e). Σ' is proper leveled-planar as the relative positions of the vertices belonging to the same bucket is the same as in Σ, while vertices of different buckets are horizontally separated by $\prec'$ and there is no edge between them.

Applying Σ' to G, each arch in $(u, v) \in A$ becomes an inter-level edge with $\phi'(v) - \phi'(u) = 1$. As the arches determined the order of the buckets and now the border vertices of neighboring buckets face each other in Σ', reinserting the arches does not cause any crossings in Σ'. Thus, Σ' is a proper leveled-planar embedding of G. □

Heath and Rosenberg have shown in [12] that the decision problem "Is a graph a queue graph?" is $\mathcal{NP}$-complete. They have also shown that deciding if a graph is proper leveled-planar is $\mathcal{NP}$-complete. Hence, we obtain the following:

Corollary 3. *The decision problem "Is a graph a queue graph?" is $\mathcal{NP}$-complete even for bipartite graphs.*

5 Conclusion

In this paper, we have proved that a graph is a deque graph if and only if it is a subgraph of a planar graph with a Hamiltonian path. Using this result, we have also characterized the gap between Hamiltonian paths and cycles in planar graphs with respect to graph layouts. We used our findings to obtain new insights into queue graphs, e.g., the dual of an embedded ququc graph contains a Eulerian path. We then showed that deciding if a graph can be laid out in a deque is $\mathcal{NP}$-complete. We have also found out that the property of bipartiteness of a queue graph exactly matches the difference between proper and arched leveled-planar graphs: A graph is a bipartite queue graph if and only if it is proper leveled-planar.

To layout an arbitrary planar graph four stacks are sufficient and necessary [19, 20]. However, there are non-planar graphs that permit a 4-stack layout, e.g., the complete graph K_5. We are currently in the process of proving that by a slight modification of the deque data structure we are able to exactly characterize all planar graphs. In [12], Heath and Rosenberg have conjectured that every planar graph is a stack-queue graph; a conjecture, which is still open. We have obtained indicators that this conjecture might not be true. We have the hope that our extended deque layouts will help to prove our conjecture.

References

1. Auer, C., Bachmaier, C., Brandenburg, F.J., Brunner, W., Gleißner, A.: Plane Drawings of Queue and Deque Graphs. In: Brandes, U., Cornelsen, S. (eds.) GD 2010. LNCS, vol. 6502, pp. 68–79. Springer, Heidelberg (2011)

2. Bernhart, F., Kainen, P.: The book thickness of a graph. J. Combin. Theory, Ser. B 27(3), 320–331 (1979)
3. Chung, F.R.K., Leighton, F.T., Rosenberg, A.L.: Embedding graphs in books: A layout problem with applications to VLSI design. SIAM J. Algebra. Discr. Meth. 8(1), 33–58 (1987)
4. Chvátal, V.: The Traveling Salesman Problem: A Guided Tour of Combinatorial Optimization. John Wiley and Sons, New York (1985)
5. Di Battista, G., Nardelli, E.: Hierarchies and planarity theory. IEEE Transactions on Systems, Man, and Cybernetics 18(6), 1035–1046 (1988)
6. Dujmović, V., Wood, D.R.: On linear layouts of graphs. Discrete Math. Theor. Comput. Sci. 6(2), 339–358 (2004)
7. Dujmović, V., Wood, D.R.: Stacks, queues and tracks: Layouts of graph subdivisions. Discrete Math. Theor. Comput. Sci. 7(1), 155–202 (2005)
8. Garey, M.R., Johnson, D.S.: Computers and Intractability; A Guide to the Theory of NP-Completeness. W. H. Freeman & Co., New York (1990)
9. Heath, L.S., Leighton, F.T., Rosenberg, A.L.: Comparing queues and stacks as mechanisms for laying out graphs. SIAM J. Discret. Math. 5(3), 398–412 (1992)
10. Heath, L.S., Pemmaraju, S.V.: Stack and queue layouts of directed acyclic graphs: Part II. SIAM J. Comput. 28(5), 1588–1626 (1999)
11. Heath, L.S., Pemmaraju, S.V., Trenk, A.N.: Stack and queue layouts of directed acyclic graphs: Part I. SIAM J. Comput. 28(4), 1510–1539 (1999)
12. Heath, L.S., Rosenberg, A.L.: Laying out graphs using queues. SIAM J. Comput. 21(5), 927–958 (1992)
13. Rosenstiehl, P., Tarjan, R.E.: Gauss codes, planar hamiltonian graphs, and stack-sortable permutations. J. of Algorithms 5, 375–390 (1984)
14. Sugiyama, K., Tagawa, S., Toda, M.: Methods for visual understanding of hierarchical system structures. IEEE Transactions on Systems, Man, and Cybernetics 11(2), 109–125 (1981)
15. Wiegers, M.: Recognizing Outerplanar Graphs in Linear Time. In: Tinhofer, G., Schmidt, G. (eds.) WG 1986. LNCS, vol. 246, pp. 165–176. Springer, Heidelberg (1987)
16. Wigderson, A.: The complexity of the Hamiltonian circuit problem for maximal planar graphs. Tech. rep., Department of EECS, Princeton University (1982)
17. Wood, D.R.: Bounded Degree Book Embeddings and Three-Dimensional Orthogonal Graph Drawing. In: Mutzel, P., Jünger, M., Leipert, S. (eds.) GD 2001. LNCS, vol. 2265, pp. 312–327. Springer, Heidelberg (2002)
18. Wood, D.R.: Queue Layouts, Tree-Width, and Three-Dimensional Graph Drawing. In: Agrawal, M., Seth, A.K. (eds.) FSTTCS 2002. LNCS, vol. 2556, pp. 348–359. Springer, Heidelberg (2002)
19. Yannakakis, M.: Four pages are necessary and sufficient for planar graphs. In: Proc. of the 18th Annual ACM Symposium on Theory of Computing, STOC 1986, pp. 104–108. ACM, New York (1986)
20. Yannakakis, M.: Embedding planar graphs in four pages. J. Comput. Syst. Sci. 38(1), 36–67 (1989)

Graph Classes with Structured Neighborhoods and Algorithmic Applications*

Rémy Belmonte and Martin Vatshelle

Department of Informatics, University of Bergen,
P.O. Box 7803, N-5020 Bergen, Norway
{remy.belmonte,martin.vatshelle}@ii.uib.no

Abstract. Boolean-width is a recently introduced graph width parameter. If a boolean decomposition of width w is given, several NP-complete problems, such as MAXIMUM WEIGHT INDEPENDENT SET, k-COLORING and MINIMUM WEIGHT DOMINATING SET are solvable in $O^*(2^{O(w)})$ time [6]. In this paper we study graph classes for which we can compute a decomposition of logarithmic boolean-width in polynomial time. Since $2^{O(\log n)} = n^{O(1)}$, this gives polynomial time algorithms for the above problems on these graph classes. For interval graphs we show how to construct decompositions where neighborhoods of vertex subsets are nested. We generalize this idea to neighborhoods that can be represented by a constant number of vertices. Moreover we show that these decompositions have boolean-width $O(\log n)$. Graph classes having such decompositions include circular arc graphs, circular k-trapezoid graphs, convex graphs, Dilworth k graphs, k-polygon graphs and complements of k-degenerate graphs. Combined with results in [1, 5], this implies that a large class of vertex subset and vertex partitioning problems can be solved in polynomial time on these graph classes.

1   Introduction

Two common ways of coping with NP-hard graph problems are to restrict instances to a certain graph class where the problem is polynomial, or to give FPT algorithms parameterized by a graph width parameter. In this paper we combine these two approaches by exploiting the fact that an FPT algorithm with running-time $2^{O(w)} \cdot poly(n)$ is polynomial whenever w is $O(\log n)$.

A theorem by Courcelle, Makovski and Rotics [10] states that every problem expressible in MSO_1 logic can be solved in linear time on graphs of bounded clique-width. Examples of graph classes with bounded clique-width can be found in Group I of Figure 1. However, many interesting classes of graphs have unbounded clique-width (see [4] and [16]). In order to obtain algorithms for larger classes of graphs, we have to compromise by considering a smaller range of problems or having less efficient running time. An example of such algorithms, related to the results in this paper, was shown by Kratochvíl, Manuel and Miller in [23],

---

* This project was partially supported by the Research Council of Norway.

P. Kolman and J. Kratochvíl (Eds.): WG 2011, LNCS 6986, pp. 47–58, 2011.
© Springer-Verlag Berlin Heidelberg 2011

where a large class of the (σ, ρ) vertex subset problems was shown to be solvable in polynomial time on interval graphs.

Boolean-width is a graph parameter recently introduced by Bui-Xuan, Telle and Vatshelle [6]. They present algorithms for solving MAXIMUM WEIGHT INDEPENDENT SET and MINIMUM WEIGHT DOMINATING SET in $2^{O(w)} \cdot poly(n)$ time, given a decomposition of boolean-width w. In this paper we study classes of graphs with boolean-width $O(\log n)$. We show that a large class of graphs including interval graphs, permutation graphs, convex graphs, circular k-trapezoid graphs, Dilworth k graphs and complements of planar graphs have boolean-width $O(\log n)$ (see Group II of Figure 1). Combining our results with the results in [6] leads to polynomial time algorithms for problems such as MINIMUM WEIGHT DOMINATING SET and MAXIMUM WEIGHT INDEPENDENT SET, for all the graph classes in Group I and II of Figure 1. To our knowledge, this is the first time an FPT algorithm parameterized by a graph parameter is used to give a polynomial time algorithm on a natural graph class where the parameter value is not bounded by a constant. Note that our result unifies and generalizes several previous polynomial time algorithms for MINIMUM WEIGHT DOMINATING SET. Interestingly, there is no graph class whose boolean-width is known *not* to be $O(\log n)$ for which MINIMUM WEIGHT DOMINATING SET can be solved in polynomial time. We are also able to prove that for most of the graph classes discussed in this paper the upper bounds we give on boolean-width are tight up to a constant factor, using the fact that they have clique-width $\Omega(\sqrt{n})$.

In the simple case of interval graphs we show how to construct decompositions such that every cut $(A, \overline{A})$ has nested neighborhoods, i.e. for every pair of vertices of A, the neighborhood of one is a subset of the neighborhood of the other when restricted to $\overline{A}$. We generalize the idea of a cut with nested neighborhoods to the notion of representative-size. We say a cut $(A, \overline{A})$ has representative-size r if every subset of A contains another subset of size at most r having the same neighborhood in $\overline{A}$. We also show that these decompositions have boolean-width $O(\log n)$, since there is only a polynomial number of subsets of constant size. Our proofs depend on having a certain representation of the input graph. For most of the graph classes discussed in this paper the required representation can be computed in polynomial time, meaning we can in polynomial time build a decomposition given a graph belonging to the graph class.

Telle and Proskurowski [30] introduced a framework covering a large class of vertex subset and vertex partitioning problems. This framework includes several well studied problems, among which are MAXIMUM INDEPENDENT SET and MINIMUM DOMINATING SET, but also PERFECT CODE, k-COLORING, H-COVER, H-HOMOMORPHISM and H-ROLE ASSIGNMENT. We use the algorithm Bui-Xuan et al. gave in [1, 5] to show that all the problems covered by this framework can be solved in polynomial time on all the graph classes in Group I and II of Figure 1.

In Section 2, we start by introducing standard graph theoretic notions and define boolean-width as well as some of the related terminology. We also define formally the notion of representing a neighborhood by a smaller set of vertices,

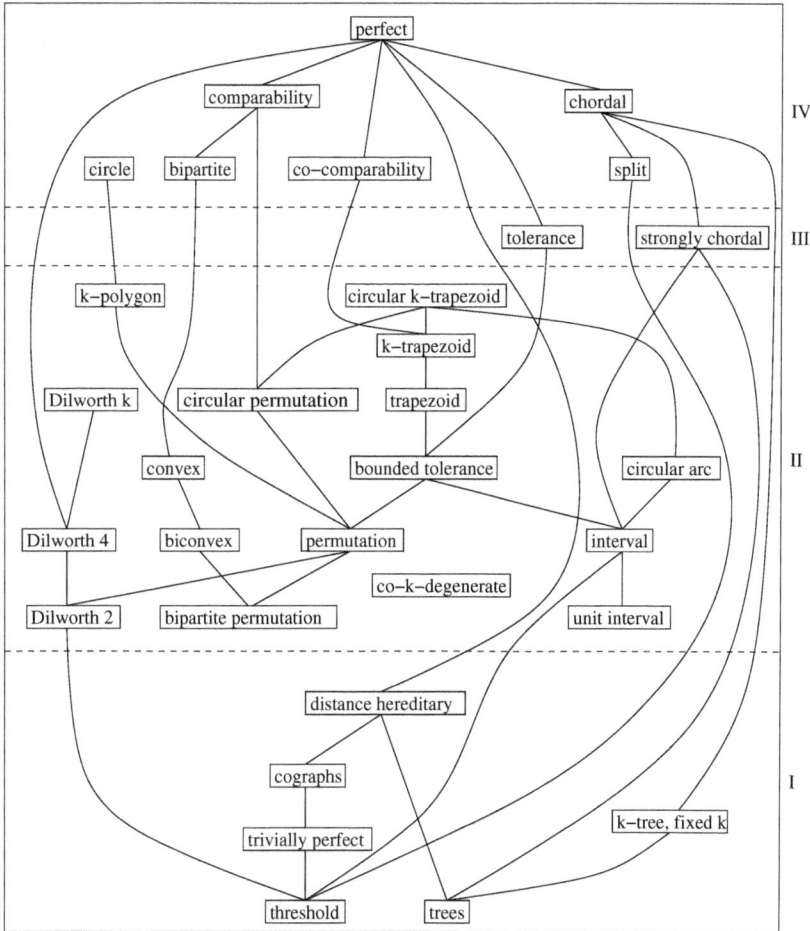

Fig. 1. Inclusion diagram of some well-known graph classes. A link between a higher class A and a lower class B means that B is a subclass of A. (I) Graph classes where boolean-width is bounded by a constant. (II) Graph classes having boolean-width $O(\log n)$. (III) It is unknown whether boolean-width is $O(\log n)$. (IV) There does not always exist a boolean-decomposition of value $O(\log n)$, or it is NP-complete to compute it. Many vertex subset and vertex partitioning problems can be solved in polynomial time on graph classes in Group I and II.

Main Result. We show that Dilworth k graphs, convex graphs, trapezoid graphs, circular permutation graphs, circular arc graphs and complements of k-degenerate graphs and circular k-trapezoid graphs have decompositions where neighborhoods can be represented by a constant number of vertices. This implies that a large class of vertex subset and vertex partitioning problems are solvable in polynomial time on these graph classes given their intersection model.

Many of these problems are well studied on many of these graph classes, see [2, 8, 11, 12, 13, 14, 17, 18, 24, 27, 31]. Our result implies many of these results.

whose size we call "representative-size" and relate this notion to boolean-width. In section 3, we show classes of graphs having representative-size bounded by a constant. In section 4, we show that constant representative-size allows to apply the results in [1, 5] and get polynomial time algorithms for the large class of vertex subset and vertex partitioning problems defined by Telle and Proskurowski [30]. Finally, in section 5 we show that our upper bounds are tight up to a constant factor and give evidence that a large class of graphs cannot have logarithmic boolean-width.

2   Framework

All graphs considered in this paper are undirected, finite and simple. A *graph* G is a pair (V, E) where V is the set of vertices of G and E is the set of edges. The *neighborhood* of a vertex u, denoted by $N(u)$, is the set of vertices u such that the edge $\{u, v\}$ is in E. The neighborhood of a set X is $N(X) = \bigcup_{x \in X} N(x)$. Given a set $A \subseteq V$, we denote by $\overline{A}$ the complement of A in V, i.e. $V \setminus A$. We call a bipartition $(A, \overline{A})$ of V a *cut* of G. Given a cut $(A, \overline{A})$ of G and $u \in A$, we call the set $N(u) \cap \overline{A}$ the *neighborhood of u across* $(A, \overline{A})$.

When applying divide-and-conquer to solve a graph problem, we first need to divide the input graph. A common way to store the information of how to divide a graph is to use a decomposition tree. The choice of a decomposition tree greatly influences the running time of any algorithm using the decomposition tree. In order to choose the best decomposition tree, we evaluate a decomposition tree by using a cut function. The following formalism is referred to as *branch decomposition* of a cut function and is standard in graph and matroid theory (see, e.g., [15, 26, 29]).

Definition 1. *A decomposition tree of a graph $G = (V, E)$ is a pair (T, δ) where T is a tree having internal nodes of degree three and $|V|$ leaves, and δ is a bijection between the vertices of G and the leaves of T. Every edge of T defines a cut $(A, \overline{A})$ of the graph via δ, by the leaves of the two subtrees of T we get by removing the edge. Let $f : 2^V \to \mathbb{R}$ be a symmetric function, i.e. $f(A) = f(\overline{A})$ for all $A \subseteq V$, also called a cut function. The f-width of (T, δ) is the maximum value of $f(A)$, taken over all cuts $(A, \overline{A})$ of G given by an edge of T. The f-width of G is the minimum f-width over all decomposition trees of G.*

The following equivalence relation on subsets of A was introduced in [6] and serves as a basis for the definition of boolean-width:

Definition 2. *Let $G = (V, E)$ be a graph and $A \subseteq V$. Two vertex subsets $X, X' \subseteq A$ are neighborhood equivalent with respect to A, denoted by $X \equiv_A X'$, if $N(X) \cap \overline{A} = N(X') \cap \overline{A}$. We denote by $nec(\equiv_A)$ the number of equivalence classes of $\equiv_A$.*

Definition 3. *[6] The cut-bool function of a graph G is defined as $cut\text{-}bool(A) = \log_2 nec(\equiv_A)$. Using Definition 1 with $f = cut\text{-}bool$ we define the boolean-width of a decomposition, denoted $boolw(T, \delta)$, and the boolean-width of a graph, denoted $boolw(G)$.*

It is known from boolean matrix theory that *cut-bool* is symmetric [21]. For more background on boolean-width, see the full version of [6].

Definition 4 (Representative-size). *Let $G = (V, E)$ be a graph and $(A, \overline{A})$ a cut of G. We say that the cut $(A, \overline{A})$ has representative-size r if r is the smallest integer such that for every subset S of A, there exists a set $S' \subseteq S$ with $|S'| \leq r$ and $S \equiv_A S'$. We denote by $rep\text{-}size(A)$ the representative-size of the cut $(A, \overline{A})$. Using Definition 1 with $f = rep\text{-}size$ we define the representative-size of a decomposition, denoted $rep\text{-}size(T, \delta)$, and the representative-size of a graph, denoted $rep\text{-}size(G)$.*

The next lemma relates representative-size and boolean-width:

Lemma 1. *Let $G = (V, E)$ be a graph, and (T, δ) a decomposition of G. If the representative-size of (T, δ) is r, then the boolean-width of (T, δ) is at most $r \log_2(|V|)$.*

Proof. For any cut $(A, \overline{A})$ of the decomposition (T, δ), we know that $rep\text{-}size(A)$ is at most r. This means that given any set $S \in A$, there exists a set S' such that $|S'| \leq r$ and $S \equiv_A S'$. Clearly, there are at most $\binom{|V|}{r} \leq |V|^r$ subsets of A of cardinality at most r. Hence we have that boolean-width is at most $\log_2 |V|^r = r \log_2 |V|$.   □

Caterpillar decompositions are decompositions where the underlying tree is a path with one leaf added as neighbor of each internal node of the path. Many of our proofs will construct caterpillar decompositions. To describe a caterpillar decomposition of a graph G, we only give an ordering $v_1, \ldots, v_n$ of the vertices of G. To construct the caterpillar decomposition (T, δ) from an ordering, first construct a caterpillar T from a path $u_1, \ldots, u_n$ of length $|V|$. Then let δ be a mapping of v_1 to u_1, v_n to u_n and for all $i \in \{2, \ldots, n-1\}$, of v_i to the leaf attached to u_i.

Many of the graph classes we study in this paper are special cases of intersection graphs. Let $\mathcal{F}$ be a family of nonempty sets. The *intersection graph* of $\mathcal{F}$ is obtained by representing each set in $\mathcal{F}$ by a vertex and connecting two vertices by an edge if and only if their corresponding sets intersect. The intersection model $\mathcal{F}$ usually consists of geometrical objects such as intervals of the real line.

3   Upper Bounds on Boolean-Width of Graph Classes

In this section we prove upper bounds on the boolean-width of several classes of graphs. Throughout the paper, when talking about a class of graphs, we denote by n the number of vertices $|V|$. We say that a class of graphs $\mathcal{C}$ has boolean-width $f(n)$ if every graph belonging to $\mathcal{C}$ has boolean-width at most $f(n)$. In particular, we focus on classes of graphs having boolean-width $O(\log n)$. We prove that the graph classes in Group II of Figure 1 have representative-size bounded by a constant. Combining this with Lemma 1 implies that they also have boolean-width $O(\log n)$.

First, we give a sketch of the proof for interval graphs showing that they have representative-size 1. We build the decomposition by ordering the vertices by the left endpoint of their intervals, then across each cut $(A, \overline{A})$ of the decomposition the neighborhood of the vertices are nested in order of right endpoint of their intervals. This means that, for every pair of vertices of A, the neighborhood of one is a subset of the neighborhood of the other when restricted to $\overline{A}$. Now we extend this idea to circular-arc graphs, which are the intersection graphs of arcs on a circle.

Lemma 2. *Given a circular-arc graph G we can, in polynomial time, compute a decomposition of G having representative-size at most 2 and boolean-width at most $2 \log n$.*

Proof. We compute the circular-arc intersection model of G in polynomial time using the algorithm of McConnell [25]. Let p be an arbitrary point on the circle. We define the distance of an arc from p as follows: if the arc contains p, then the distance is 0, otherwise it is the minimum distance between p and any point of the arc. For any vertex u, we denote by arc_u the arc corresponding to u. Note that since p is an arbitrary point then no pair of arcs have the same distance from p unless they intersect.

Build a caterpillar decomposition by adding the vertices in order of increasing distance of their associated arc from p, breaking ties arbitrarily. Note that this decomposition can be computed in polynomial time. We now consider any cut $(A, \overline{A})$ of this decomposition. By construction, for every $x \in A, y \in \overline{A}$, the distance of arc_x from p is less than or equal to the distance of arc_y from p.

Now, we prove that for any set $S \subseteq A$, there exists a subset $S' \subseteq S$ such that $|S'| \leq 2$ and $S \equiv_A S'$. Let d be the smallest distance from p to the arc of any vertex in $\overline{A}$. Let p^+ be the point on the circle which is at distance d going in clockwise direction from p. Likewise, p^- is the point at distance d going in counter-clockwise direction from p. We build S' starting from the empty set. If there exists a vertex in S whose arc contains p^+, then let u be one such vertex with arc_u extending furthest from p^+ in clockwise direction and add u to S'. Likewise, if there exists a vertex in S whose arc contains p^-, then let v be one such vertex with arc_v extending furthest from p^- in counter-clockwise direction and add v to S'. Now we prove that $N(S) \cap \overline{A} = N(S') \cap \overline{A}$.

Let z be some vertex of $N(S) \cap \overline{A}$, if no such z exists $S' = \emptyset$ satisfies the lemma. Assume for contradiction that $z \notin N(S')$. Let w be a vertex of $N(z) \cap S$ and p_i any point contained in both arc_w and arc_z. Since any arc of a vertex in A is at distance at most d from p and p_i is at distance at least d from p, then arc_w contains both p_i and a point of distance at most d from p. We can assume without loss of generality that arc_w contains all points from p^+ to p_i in clockwise direction. Since arc_u is the arc extending furthest in clockwise direction from p^+, arc_u will also contain p_i, contradicting the choice of p_i.

Therefore $S \equiv_A S'$, which implies that the decomposition we built has representative-size at most 2. By applying Lemma 1 it follows that circular-arc graphs have boolean-width at most $2 \log n$. □

We show a similar result for several other classes of graphs but their definitions and proofs are in the appendix due to space limitation. The proof for circular-arc graphs contains all the important ideas. The definitions of the graph classes can also be found in [3] or [28].

Theorem 1. *Convex graphs, circular-arc graphs, circular permutation graphs and trapezoid graphs have representative-size $O(1)$ and boolean-width $O(\log n)$.*

The graph classes in Group II of Figure 1 involving a parameter k are dealt with in Theorem 2. As an example, the proof showing that k-trapezoid graphs have representative-size at most k can be sketched as follows. A k-trapezoid is the polygon obtained by choosing an interval on each of k parallel lines in the plane and connecting the left and right endpoints of each neighboring interval. k-trapezoid graphs are intersection graphs of k-trapezoids. First, we build the caterpillar decomposition by ordering the k-trapezoids by their leftmost point. Then, for any cut $(A, \overline{A})$ of the decomposition and any set $S \subseteq A$, there is one k-trapezoid extending further to the right on each of the k lines. We call the set of vertices associated with these k-trapezoids S'. Moreover, for every vertex of S, any of its neighbors in $\overline{A}$ is also adjacent to at least one of the vertices in S'. Hence we have $S' \subseteq S, |S'| \leq k$ and $S' \equiv_A S$.

Theorem 2. *Complements of k-degenerate graphs, Dilworth k graphs, k-polygon graphs and circular k-trapezoid graphs have representative-size $O(k)$ and thus boolean-width $O(k \log n)$.*

Note that Theorem 1 and 2 encompass all graph classes in Group I and II of Figure 1. We find it interesting to note that some of these classes are seemingly unrelated to each other, but they all have decompositions sharing a common neighborhood structure, which allows for efficient dynamic programming approaches on a large class of problems. In particular, we combine these results with the following:

Theorem 3 (Bui-Xuan, Telle, Vatshelle [6]). *For any graph $G = (V, E)$, MINIMUM WEIGHT DOMINATING SET can be solved in $O(|V|^2 + |V| \cdot w \cdot 2^{3 \cdot w})$ time when given a decomposition of G having boolean-width w.*

Combining Theorem 3 with Theorem 1 and 2, we get:

Corollary 1. MINIMUM WEIGHT DOMINATING SET *can be solved in polynomial time on all the graph classes in Group I and II of Figure 1.*

The next section shows how to extend this result to a larger class of problems.

4   Vertex Partitioning Problems

In [30] Proskurowski and Telle introduced a generalized framework for handling many types of vertex subset and vertex partitioning problems in a unified manner. These problems can be described by a degree constraint matrix.

Definition 5. *A degree constraint matrix D_q is a q by q matrix with entries being finite or co-finite subsets of natural numbers. A D_q-partition in a graph G is a partition $\{V_1, V_2, ..., V_q\}$ of V such that for $1 \leq i, j \leq q$ we have $\forall v \in V_i :$ $|N(v) \cap V_j| \in D_q[i, j]$.*

A D_q vertex partitioning problem is the problem of finding a D_q partition satisfying a given D_q matrix and optionally maximizing or minimizing the weight of a given class of the D_q partition. This formalism was introduced by Telle and Proskurowski and encompass several well studied problems, such as MAXIMUM INDEPENDENT SET, MINIMUM DOMINATING SET, PERFECT CODE, k-COLORING, H-COVER, H-HOMOMORPHISM and H-ROLE ASSIGNMENT. The class of (σ, ρ) vertex subset problems is a subset of D_q vertex partitioning problems. For example, MAXIMUM INDEPENDENT DOMINATING SET is encoded by a 2 by 2 matrix with entries $[1, 1] = \{0\}, [1, 2] = \{0, 1, \dots\}, [2, 1] = \{1, 2, \dots\}$ and $[2, 2] = \{0, 1, \dots\}$, and maximizing the size of V_1. H-HOMOMORPHISM for a graph H on q vertices simply asks for the existence of a partition satisfying the q by q matrix constructed from the adjacency matrix of H by replacing entry 0 with $\{0\}$ and 1 with $\{0, 1, \dots\}$. Telle and Proskurowski showed that all D_q-problems are solvable in FPT time parameterized by tree-width [30]. Kobler and Rotics showed that D_q-problems are solvable on graphs of bounded clique-width [22], and with a little effort their algorithms can be made into FPT algorithms. Bui-Xuan et al. showed that D_q-problems are FPT when parameterized by boolean-width [1]. Kratochvíl et al. [23] showed that a subset of the D_q-problems are solvable in polynomial time on interval graphs. We generalize the results of [23] by showing that all D_q-problems are solvable in polynomial time on many well known graph classes, including interval graphs.

We will apply the algorithm of Bui-Xuan et al. [1], where the bottleneck for running time is the number of equivalence classes of d-neighborhoods. When solving a D_q-problem, the integer value $d(D_q)$ needed depends on the degree constraint matrix in the following way. Let $d(\{0, 1, \dots\}) = 0$. For every finite or co-finite non-empty set $\mu \subseteq \mathbb{N}$, let $d(\mu) = 1 + \min(\max x : x \in \mu, \max x : x \notin \mu)$. For a matrix D_q, the value $d(D_q)$ will be $\max_{i,j} d(D_q[i, j])$. When there is no ambiguity, we denote $d(D_q)$ by d. Note that d depends only on the problem and hence can be treated as a constant.

Definition 6 (d-neighbor equivalence). *Let $G = (V, E)$ be a graph, $A \subseteq V$ and d a positive integer. Two vertex subsets $X \subseteq A$ and $X' \subseteq A$ are d-neighbor equivalent with respect to A, denoted $X \equiv_A^d X'$ if:*
$$\forall v \in \overline{A}, (|N(v) \cap X| = |N(v) \cap X'|) \text{ or } (|N(v) \cap X| \geq d \text{ and } |N(v) \cap X'| \geq d)$$
We denote by $nec(\equiv_A^d)$ the number of equivalence classes of $\equiv_A^d$.

Note that X and X' are 1-neighborhood equivalent with regard to A if and only if $N(X) \cap \overline{A} = N(X') \cap \overline{A}$ and thus $nec(\equiv_A) = nec(\equiv_A^1)$. We show a connection between representative-size and d-neighbor equivalence.

Lemma 3. *Let $G = (V, E)$ be a graph and $(A, \overline{A})$ a cut of G. If $rep\text{-}size(A) = r$, then for every positive integer d and every set $X \subseteq A$, there exists $X_d \subseteq X$ such that $|X_d| \leq d \cdot r$ and $X_d \equiv_A^d X$.*

Proof. We prove the statement by induction on d. Let $R \subseteq X$ be an inclusion minimal set such that $N(R) \cap \overline{A} = N(X) \cap \overline{A}$. Since the representative-size of $(A, \overline{A})$ is r, we have that $|R| \leq r$. For $d \leq 1$ the lemma holds since $R \equiv_A^1 X$. Assume the induction hypothesis true up to $i - 1$, then we show it true for i. By induction hypothesis there exists $X_{i-1} \subseteq (X \setminus R)$ such that $X_{i-1} \equiv_A^{i-1} (X \setminus R)$ and $|X_{i-1}| \leq r \cdot (i-1)$. Thus it is enough to show $X_i \equiv_A^i X$, for $X_i = X_{i-1} \cup R$.

We partition the nodes of $\overline{A}$ into (P, Q) such that $\forall v \in P$, we have $|N(v) \cap (X \setminus R)| = |N(v) \cap X_{i-1}|$ and $\forall v \in Q$, we have $|N(v) \cap (X \setminus R)| \geq i - 1$ and $|N(v) \cap X_{i-1}| \geq i - 1$. Since $R \cap X_{i-1} = \emptyset$ and $R \subseteq X$, we know $|N(v) \cap (X \setminus R)| = |N(v) \cap X| - |N(v) \cap R|$ and $|N(v) \cap (X_{i-1} \cup R)| = |N(v) \cap X_{i-1}| + |N(v) \cap R|$. Hence for every vertex $v \in P$, we have $|N(v) \cap X| = |N(v) \cap X_{i-1}| + |N(v) \cap R| = |N(v) \cap (X_{i-1} \cup R)|$. Since $i > 1$, then for every vertex $v \in Q$ we have $N(v) \cap R \neq \emptyset$. Since $X \equiv_A R$, then for every vertex $v \in Q$ we have $|N(v) \cap X| \geq i$ and $|N(v) \cap X_i| \geq i$.

Since (P, Q) is a partition we get $X_i \equiv_A^i X$ and $|X_i| \leq r \cdot i$, thus by induction the lemma holds for all i.   □

For a decomposition (T, δ) of a graph G, let $nec_d(T, \delta)$ be the maximum $nec(\equiv_A^d)$ over all cuts $(A, \overline{A})$ of (T, δ).

Lemma 4. *Let $G = (V, E)$ be a graph, (T, δ) a decomposition of G and d a positive integer. If $rep\text{-}size(T, \delta) = r$, then $nec_d(T, \delta) \leq |V|^{d \cdot r}$.*

Proof. For any cut $(A, \overline{A})$ of the decomposition (T, δ), we know that $rep\text{-}size(A)$ is at most r. From Lemma 3 we know that for any $S \subseteq A$ there exists a set S' such that $|S'| \leq d \cdot r$ and $S \equiv_A^d S'$. Clearly, there are at most $\binom{|V|}{d \cdot r} \leq |V|^{d \cdot r}$ subsets of A of cardinality at most $d \cdot r$. Hence $nec_d(T, \delta) \leq |V|^{d \cdot r}$.   □

By combining Lemma 4 with Theorem 1 and Theorem 2 we get:

Theorem 4. *Let $G = (V, E)$ be a graph in Group I or II of Figure 1, then we can in polynomial time compute a decomposition (T, δ) such that $nec_d(T, \delta)$ is polynomial in $|V|$ assuming an intersection model of G is provided.*

Theorem 5 (Bui-Xuan, Telle, Vatshelle [5][1]). *For any graph $G = (V, E)$ and (T, δ) a decomposition of G, all D_q vertex partitioning problems can be solved in $O(nec_d(T, \delta)^{3 \cdot q} \cdot poly(|V|))$ time.*

Combining Theorem 4 with Theorem 5, we get:

Corollary 2. *All D_q vertex partitioning problems can be solved in polynomial time on all the graph classes in Group I and II of Figure 1 assuming an intersection model of the input graph is provided.*

5   Lower Bounds

We say that a class of graphs $\mathcal{C}$ has boolean-width $\Omega(f(n))$ if there exists an infinite family of graphs in $\mathcal{C}$ all having boolean-width $\Omega(f(n))$. In this section

---

[1] [5] is an arXiv version of [1] containing a more fitting version of this theorem.

we show that the upper bounds we gave on the boolean-width are tight in two senses. Firstly, for all graph classes (except Dilworth k graphs) in Group II of Figure 1, we are able to show that they have boolean-width $\Omega(\log n)$. Secondly, we show that for all graph classes in Group IV of Figure 1, it is highly unlikely that they have boolean-width $O(\log n)$. Note the following result on the relation between boolean-width and some other width parameters:

Theorem 6 (Bui-Xuan, Telle, Vatshelle [6]). *For any graph G it holds that* $\log rw(G) - 1 \leq \log cw(G) - 1 \leq boolw(G)$*, where $boolw(G), rw(G)$ and $cw(G)$ denote respectively the boolean-width, rank-width and clique-width of G.*

Hence if a graph class has rank-width or clique-width $\Omega(n^c)$ for some constant $c > 0$, then this graph class also has boolean-width $\Omega(\log n)$. We use this to show that the bounds we give in this paper are tight up to a constant factor.

Lemma 5. *All graph classes in Group II of Figure 1 (except Dilworth k graphs), have boolean-width $\Theta(\log n)$.*

Proof. Brandstädt and Lozin showed in [4] an infinite family of bipartite permutation graphs with clique-width $\Omega(\sqrt{n})$. Likewise, Golumbic and Rotics showed in [16] an infinite family of unit interval graphs with clique-width $\Omega(\sqrt{n})$. Moreover, Jelínek showed in [19] that $q \times q$ grids have rank-width exactly $q - 1$. Note that if a graph G has rank-width w, then its complement $\overline{G}$ has rank-width $w \pm 1$. Since all grids are 2-degenerate, then complements of 2-degenerate graphs have rank-width $\Omega(\sqrt{n})$. From Theorem 6, it follows that these three graph classes have boolean-width $\Theta(\log n)$. Hence the lemma follows since all graph classes in Group II of Figure 1 contain one these graph classes.   □

Another interesting question to ask is whether there exist more graph classes having logarithmic boolean-width. For some graph classes it is possible to provide examples of an infinite family of graphs having non-logarithmic boolean-width, for example the grid. However, for some classes of graphs, we do not know any example of infinite family of graphs having non-logarithmic boolean-width. We are nonetheless able to provide some lower bounds:

Lemma 6. *For all the classes in Group IV of Figure 1, either they do not have boolean-width $O(\log n)$, or a decomposition of boolean-width $O(\log n)$ cannot be computed in polynomial time, unless $P = NP$.*

Proof. Note first that for all the classes of graphs in Group IV of Figure 1, MINIMUM WEIGHT DOMINATING SET is NP-complete (see [9], [7] and [20]). Moreover, MINIMUM WEIGHT DOMINATING SET can be solved in time $O(2^{3 \cdot boolw} \cdot poly(n))$. Assume now that there exists a class $\mathcal{C}$ in Group IV of Figure 1 having boolean-width $O(\log n)$ and where such decompositions can be computed in polynomial time. Then MINIMUM WEIGHT DOMINATING SET can be computed in time $O(2^{O(\log n)} \cdot poly(n))$, which is a polynomial of n. Hence if a class of graphs on which MINIMUM WEIGHT DOMINATING SET is NP-complete has boolean-width $O(\log n)$, then decompositions of boolean-width $O(\log n)$ cannot be computed in polynomial time, unless $P = NP$.   ⊓

Note that this holds not only for MINIMUM WEIGHT DOMINATING SET, but as long as there exists a problem which can be solved in $O(2^{O(boolw)} \cdot poly(n))$ time. Finally, we get better lower bounds by working under a stronger hypothesis. The *E*xponential Time Hypothesis (ETH) states that there does not exists an algorithm for solving 3-SAT running in time $2^{o(n)}$. We can reformulate Lemma 6 as follows:

Lemma 7. *For all the classes in Group IV of Figure 1, either they do not have boolean-width $n^{o(1)}$, or a decomposition of boolean-width $n^{o(1)}$ cannot be computed in time $2^{o(n)}$, unless ETH fails.*

This means for instance that if split graphs have boolean-width $poly\text{-}log(n)$, then it is NP-hard to compute a decomposition of split graphs having boolean-width within a factor $log(n)$ of the optimum.

6 Conclusion

We have shown that all graph classes in Group II of Figure 1 have logarithmic boolean-width and we can compute such decompositions of logarithmic boolean-width, answering an open question from [6]. Applying the algorithm for vertex partitioning problems (as well as their weighted versions) in [1, 5], we show several graph classes for which a large class of vertex partitioning problems can be solved in polynomial time. What is the boolean-width of the graph classes in Group III of Figure 1? Is there any graph class not having boolean-width $O(\log n)$ where MINIMUM WEIGHT DOMINATING SET is polynomially solvable?

References

[1] Adler, I., Bui-Xuan, B.-M., Rabinovich, Y., Renault, G., Telle, J.A., Vatshelle, M.: On the Boolean-Width of a Graph: Structure and Applications. In: Thilikos, D.M. (ed.) WG 2010. LNCS, vol. 6410, pp. 159–170. Springer, Heidelberg (2010)
[2] Brandstädt, A., Kratsch, D.: On the Restriction of Some np-Complete Graph Problems to Permutation Graphs. In: Budach, L. (ed.) FCT 1985. LNCS, vol. 199, pp. 53–62. Springer, Heidelberg (1985)
[3] Brandstädt, A., Le, V.B., Spinrad, J.P.: Graph classes: a survey (1999)
[4] Brandstädt, A., Lozin, V.V.: On the linear structure and clique-width of bipartite permutation graphs. Ars Comb. 67 (2003)
[5] Bui-Xuan, B.M., Telle, J.A., Vatshelle, M.: Fast algorithms for vertex subset and vertex partitioning problems on graphs of low boolean-width. arXiv
[6] Bui-Xuan, B.-M., Telle, J.A., Vatshelle, M.: Boolean-Width of Graphs. In: Chen, J., Fomin, F.V. (eds.) IWPEC 2009. LNCS, vol. 5917, pp. 61–74. Springer, Heidelberg (2009)
[7] Chang, M.S.: Weighted Domination on Cocomparability Graphs. In: Staples, J., Katoh, N., Eades, P., Moffat, A. (eds.) ISAAC 1995. LNCS, vol. 1004, pp. 122–131. Springer, Heidelberg (1995)
[8] Chang, M.S.: Efficient algorithms for the domination problems on interval and circular-arc graphs. SIAM J. on Computing 27(6), 1671–1694 (1998)

[9] Corneil, D.G., Perl, Y.: Clustering and domination in perfect graphs. Discrete Applied Math. 9, 27–40 (1984)

[10] Courcelle, B., Makowsky, J.A., Rotics, U.: Linear time solvable optimization problems on graphs of bounded clique-width. Theory Comput. Syst. 33(2), 125–150 (2000)

[11] Damaschke, P., Müller, H., Kratsch, D.: Domination in convex and chordal bipartite graphs. Inf. Process. Lett. 36(5), 231–236 (1990)

[12] Díaz, J., Nešetřil, J., Serna, M., Thilikos, D.M.: H-Colorings of Large Degree Graphs. In: Shafazand, H., Tjoa, A.M. (eds.) EurAsia-ICT 2002. LNCS, vol. 2510, pp. 850–857. Springer, Heidelberg (2002)

[13] Elmallah, E.S., Stewart, L.K.: Independence and domination in polygon graphs. Discrete Appllied Math 44(1-3), 65–77 (1993)

[14] Farber, M., Keil, J.: Domination in permutation graphs. J. Algorithms 6, 309–321 (1985)

[15] Geelen, J.F., Gerards, B., Whittle, G.: Branch-width and well-quasi-ordering in matroids and graphs. J. Comb. Theory, Ser. B 84(2), 270–290 (2002)

[16] Golumbic, M.C., Rotics, U.: On the Clique-Width of Perfect Graph Classes Extended Abstract. In: Widmayer, P., Neyer, G., Eidenbenz, S. (eds.) WG 1999. LNCS, vol. 1665, pp. 135–147. Springer, Heidelberg (1999)

[17] van't Hof, P., Paulusma, D., van Rooij, J.M.M.: Computing role assignments of chordal graphs. Theor. Comput. Sci. 411(40-42), 3601–3613 (2010)

[18] Hsu, W.L., Tsai, K.H.: Linear time algorithms on circular-arc graphs. Inf. Process. Lett. 40(3), 123–129 (1991)

[19] Jelínek, V.: The rank-width of the square grid. Discrete Applied Math 158(7), 841–850 (2010)

[20] Keil, J.M.: The complexity of domination problems in circle graphs. Discrete Applied Math. 42(1), 51–63 (1993)

[21] Kim, K.H.: Boolean matrix theory and its applications. Marcel Dekker (1982)

[22] Kobler, D., Rotics, U.: Polynomial algorithms for partitioning problems on graphs with fixed clique-width (extended abstract). In: Proc. SODA, pp. 468–476 (2001)

[23] Kratochvíl, J., Manuel, P.D., Miller, M.: Generalized domination in chordal graphs. Nord. J. Comput. 2(1), 41–50 (1995)

[24] Liang, Y.: Dominations in trapezoid graphs. Inf. Process. Lett. 52(6), 309–315 (1994)

[25] McConnell, R.M.: Linear-time recognition of circular-arc graphs. Algorithmica 37, 93–147 (2003)

[26] Oum, S., Seymour, P.D.: Approximating clique-width and branch-width. J. Comb. Theory, Ser. B 96(4), 514–528 (2006)

[27] Rhee, C., Liang, Y., Dhall, S., Lakshmivarahan, S.: An $o(n + m)$-time algorithm for finding a minimum-weight dominating set in a permutation graph. SIAM J. on Computing 25(2), 404–419 (1996)

[28] de Ridder, H.N., et al.: Information System on Graph Classes and their Inclusions (ISGCI), http://wwwteo.informatik.uni-rostock.de/isgci

[29] Robertson, N., Seymour, P.D.: Graph minors. X. obstructions to tree-decomposition. J. Comb. Theory, Ser. B 52(2), 153–190 (1991)

[30] Telle, J.A., Proskurowski, A.: Algorithms for vertex partitioning problems on partial k-trees. SIAM J. Discrete Math. 10(4), 529–550 (1997)

[31] Tsai, K.H., Hsu, W.L.: Fast algorithms for the dominating set problem on permutation graphs. Algorithmica 9(6), 601–614 (1993)

Exact Algorithms for Kayles[*]

Hans L. Bodlaender[1] and Dieter Kratsch[2]

[1] Utrecht University, P.O. Box 80.089
3508 TB Utrecht, The Netherlands
hansb@cs.uu.nl
[2] Université Paul Verlaine – Metz
LITA, 57045 Metz Cedex 01, France
kratsch@univ-metz.fr

Abstract. In the game of Kayles, two players select alternatingly a vertex from a given graph G, but may never choose a vertex that is adjacent or equal to an already chosen vertex. The last player that can select a vertex wins the game. In this paper, we give an exact algorithm to determine which player has a winning strategy in this game. To analyse the running time of the algorithm, we introduce the notion of a K-set: a nonempty set of vertices $W \subseteq V$ is a K-set in a graph $G = (V, E)$, if $G[W]$ is connected and there exists an independent set X such that $W = V - N[X]$, where $N[X]$ is the union of X and the set of all vertices adjacent to at least one vertex of X. The running time of the algorithm is bounded by a polynomial factor times the number of K-sets in G. We show that the number of K-sets in a graph with n vertices is bounded by $O(1.6052^n)$, and thus we have an algorithm for KAYLES with running time $O(1.6052^n)$. We also show that the number of K-sets in a tree is bounded by $n \cdot 3^{n/3}$ and thus KAYLES can be solved on trees in $O(1.4423^n)$ time. We show that apart from a polynomial factor, the number of K-sets in a tree is sharp.

1 Introduction

When a problem is computationally hard, then there still are many situations in which the need can arise to solve it exactly. This motivates the field of exact algorithms, where exact, exponential-time algorithms whose running time is as small as possible are sought. Many such exact algorithms have been designed and analysed for problems that are NP-complete or #P-complete, see [6]. Of course, also problems that are complete for a 'harder' complexity class, e.g., PSPACE-complete often ask for exact solutions. Many PSPACE-complete problems arrive from the question which player has a winning strategy for a given position in a combinatorial game. Exact algorithms are of great relevance here, e.g., a program could use a heuristic to find a move, but once a position is simple enough, it switches to an exact algorithm to give optimal play in the endgame.

In this paper, we study exact algorithms for one such PSPACE-complete problem, namely the problem to determine which player has a winning strategy

---

[*] The work of the second author was supported by the ANR project AGAPE.

P. Kolman and J. Kratochvíl (Eds.): WG 2011, LNCS 6986, pp. 59–70, 2011.
© Springer-Verlag Berlin Heidelberg 2011

in an given instance of the game KAYLES. KAYLES is a two-player game that is played on a graph $G = (V, E)$. Alternatingly, the players choose a vertex from the graph, but players are not allowed to choose a vertex that already has been chosen or is adjacent to a vertex that already has been chosen. Thus, the player build together an independent set in G. The last player that chooses a vertex (i.e., turns the independent set into a maximal independent set) wins the game. Alternatively, one can describe the game as follows: the chosen vertex and its neighbors are removed and a player wins when his move empties the graph. The problem to determine the winning player for a given instance of the game is also called KAYLES. This problem was shown to be PSPACE-complete by Schaefer [9]. In an earlier paper [3], we showed that by exploiting Sprague-Grundy theory, KAYLES can be solved in polynomial time on several special graph classes, in particular graphs with a bounded asteroidal number (which includes well known classes of graphs like interval graphs, cocomparability graphs and cographs). Fleischer and Trippen [5] showed that KAYLES can be solved in polynomial time on stars of bounded degree, and also analysed this special case experimentally. For general trees, the complexity of KAYLES is a long standing open problem. Variants of the game on paths were shown to be linear time solvable by Guignard and Sopena [7]. For more background, the reader can consult [1,2,4].

It is not hard to find an algorithm that solves KAYLES in $O^*(2^n)$ time[1], by tabulating for each induced subgraph of G which player has a winning strategy from that position. In this paper, we improve upon this trivial algorithm, and give an algorithm that uses $O(1.6052^n)$ time. The algorithm uses ideas from [3], exploiting results from Sprague-Grundy theory. To analyze the running time of the algorithm, we introduce the notion of a K-set: a set of nonempty vertices $W \subseteq V$ is a K-set in a graph $G = (V, E)$, if $G[W]$ is connected and there exists an independent set X such that $W = V - N[X]$, where $N[X]$ is the set of vertices belonging to X or having a neighbour in X. With a nontrivial analysis we obtain that the number of K-sets of a graph with n vertices is bounded by $O(1.6052^n)$, which yields the bound on the running time of our algorithm. We also show that if G is a tree, then G has at most $n \cdot 3^{n/3}$ K-sets, and thus, KAYLES can be solved in $O^*(3^{n/3}) = O(1.4423^n)$ time on trees (and forests). We also give lower bounds for the number of K-sets. In particular, our bound of $3^{n/3}$ K-sets for trees is sharp except for polynomial terms.

2   Preliminaries

Graph terminology. Throughout this paper all graphs $G = (V, E)$ are undirected and simple. Let $S \subseteq V$. Then $N[S] = \cup_{s \in S} N[s]$ is the *closed neighborhood* of S, $N(S) = N[S] \setminus S$ is the *open neighborhood* of S, and $G[S]$ denotes the *subgraph of G induced by S*.

---

[1] We use the so called O^* notation: $f(n) = O^*(g(n))$ if $f(n) = O(g(n)p(n))$ for some polynomial $p(n)$. If a base α is obtained by rounding up a real then we may and shall write $O(\alpha^n)$ (instead of $O^*(\alpha^n)$). See also [6].

A nonempty set of vertices $W \subseteq V$ of a graph $G = (V, E)$ is called a *K-set* (*Kayles set*) of G, if it fulfills each of the following criteria:

- $G[W]$ is connected;
- there exists an independent set $X \subseteq V$ such that $W = V - N[X]$.

Sprague-Grundy theory. Next, we review some notions and results from Sprague-Grundy theory, and give some preliminary results on how this theory can be used for Kayles. For a good introduction to Sprague-Grundy theory, the reader is referred to [1,4].

A *nimber* is an integer belonging to $\mathbf{N} = \{0, 1, 2, \ldots\}$. For a finite set of nimbers $S \subseteq \mathbf{N}$, define the minimum excluded nimber of S as $mex(S) = \min\{i \in \mathbf{N} \mid i \notin S\}$. To each position in a two player game that is finite, deterministic, full-information, impartial, and with 'last player wins rule', one can associate a nimber in the following way. If no move is possible in the position (and hence the player that must move loses the game), the position gets nimber 0. Otherwise the nimber is the minimum excluded nimber of the set of nimbers of positions that can be reached in one move.

Theorem 1. *[1,4] There is a winning strategy for player 1 from a position, if and only if the nimber of that position is at least 1.*

Denote the nimber of a position p by $nb(p)$. Given two (finite, deterministic, impartial, ...) games $\mathcal{G}_1$, $\mathcal{G}_2$, the *sum* of $\mathcal{G}_1$ and $\mathcal{G}_2$, denoted $\mathcal{G}_1 + \mathcal{G}_2$ is the game where a move consists of choosing $\mathcal{G}_1$ or $\mathcal{G}_2$ and then making a move in that game. A player that cannot make a move in $\mathcal{G}_1$ nor in $\mathcal{G}_2$ loses the game $\mathcal{G}_1 + \mathcal{G}_2$. With (p_1, p_2) we denote the position in $\mathcal{G}_1 + \mathcal{G}_2$, where the position in $\mathcal{G}_i$ is p_i ($i - 1, 2$).

The binary XOR operation is denoted by $\oplus$, i.e., for nimbers i_1, i_2, $i_1 \oplus i_2 = \sum\{2^j \mid (\lfloor i_1/2^j \rfloor$ is odd$) \Leftrightarrow (\lfloor i_2/2^j \rfloor$ is even$)\}$.

Theorem 2. *[1,4] Let p_1 be a position in $\mathcal{G}_1$, p_2 a position in $\mathcal{G}_2$. The nimber of position (p_1, p_2) in $\mathcal{G}_1 + \mathcal{G}_2$ equals $nb((p_1, p_2)) = nb(p_1) \oplus nb(p_2)$.*

As Kayles is an impartial, deterministic, finite, full-information, two-player game with the rule that the last player that moves wins the game, we can apply Sprague-Grundy theory to Kayles, and we can associate with every graph G the nimber of the start position of the game Kayles, played on G. We denote this nimber $nb(G)$, and call it the *nimber* of G.

An important observation is the following: when $G = G_1 \cup G_2$ for disjoint graphs G_1 and G_2, then the game Kayles, played on G is the sum of the game Kayles, played on G_1, and the game Kayles, played on G_2. Hence, by Theorem 2, we have the following result.

Lemma 1. $nb(G_1 \cup G_2) = nb(G_1) \oplus nb(G_2)$.

Note that G_1 and G_2 might be disconnected graphs.

Our second observation shows how to express the nimber of a graph G in the nimbers of some subgraphs of G. Consider Kayles, played on $G = (V, E)$, and suppose that a vertex $v \in V$ is played. Then, the nimber of the resulting position is the same as the nimber of $G - N[v]$, as the effect of playing on v is the same as the effect of removing v and its neighbors from the graph. As the nimber of a position is the minimum nimber that is not in the set of nimbers of positions that can be reached in one move, we have:

Lemma 2. *(i) If $G = (V, E)$ is the empty graph, then $nb(G) = 0$.*
(ii) If $G = (V, E)$ is not the empty graph, then $nb(G) = mex(nb(\{G - N[v] \mid v \in V\})$.

3   An Upper Bound on the Number of K-sets

In this section, we will show an upper bound on the number of K-sets in a graph. This bound is needed for the analysis of our algorithm, see Section 5. Our main result is the following.

Theorem 3. *Let G be a graph with n vertices. Then G has $O(1.6052^n)$ K-sets.*

The proof of Theorem 3 is algorithmic: we give a branching procedure that generates all K-sets. By distinguishing different types of vertices, assigning these different weights, and considering the different branching vectors, we obtain a set of recurrences, whose solution gives us the desired bound. For information on branching algorithms and their analysis, in particular branching vectors and the corresponding recurrences we refer to [6].

We say that a K-set is *nontrivial*, if it has at least three vertices; otherwise we call it trivial. As each trivial set either consists of a single vertex or the two endpoints of an edge, the number of trivial K-sets is at most $n + m$, where m is the number of edges of the graph.

During our branching process, we decide at some points to put some vertices in an independent set X and forbid for some vertices to put them in the independent set. When placing a vertex in X, we say we *select* the vertex. The vertices in G are of four types:

- **White** or free vertices. Originally all vertices in G are white. We have not made any decision yet for a white vertex. All white vertices have weight one.
- **Red** vertices. Red vertices may not be placed in the independent set X: i.e., we already decided this during the branching. It still is possible that a red vertex becomes deleted later. Red vertices have a weight $\alpha = 0.5685$.
- **Green** vertices. A green vertex is 'safe': it never will be removed. I.e., we cannot place the green vertex nor any of its neighbors in the independent set X. Green vertices have weight zero.
- **Removed vertices**: these are either placed in the independent set or are a neighbor of a vertex in the independent set. All removed vertices have weight zero. Removed vertices are considered not existing, i.e., when discussing the neighbors of a vertex, these neighbors will be white, red, or green.

The *measure* of an instance G is the total weight of all vertices, and the difference in the measure from an instance to one of a subproblem often called *gain* is used to analyse the branching algorithm via branching vectors. Our branching process may be overcounting the number of K-sets (in particular, in some cases, we will not detect that a generated set is not connected), but the obtained bound nevertheless is valid as an upper bound.

The semantics of the colors imply that we can always perform the following actions:

- **Rule 1:** If a red vertex v has no white neighbors, we can color it green. This is valid, as we can no longer place a neighbor of v in X.
- **Rule 2:** If a green vertex v has a white neighbor w, we can color w red. This is valid, as placing w in X would remove v, which we are not allowed by the green color of v.

Rules 1 and 2 will always decrease the measure. They ensure that each red vertex will have a white neighbor, and that white vertices have no green neighbors.

The following action also can always be performed; the removed vertices can no longer be part of a nontrivial K-set.

- **Rule 3:** If $W \subseteq V$ is a set of white vertices that are not adjacent to non-deleted vertices not in W, and $|W| \leq 2$, then remove all vertices of W.

Before starting the main recursive branching, we first fix one vertex $v_0 \in V$, of which we will assume that it is an element of the K-set. In terms of colors, this means that we color v_0 green and all neighbors of v_0 red. Clearly, the total number of K-sets will be at most n times the bound on the number of K-sets that contain a specific vertex.

We obtain a fourth rule.

- **Rule 4:** If G has more than one connected component, then remove all vertices from components that do not contain v_0.

As a consequence, we have that while G has white vertices, a rule can be applied. Also, there may be a white vertex adjacent to a red vertex.

We consider two main types of branching. The first type of branching is a *vertex branch*. Let $v \in V$ be a white vertex. We consider two cases: v is placed in X, and v is not placed in X. In the former case, we remove and decrease the measure by the total weight of all white and red vertices in the closed neighborhood of v. In the latter case, we color v red and have a measure decrease of $1 - \alpha$. In some cases, we gain more by applications of Rules 1, 2, and 3.

In the second type of branching, we consider a number of cases, of which one must apply. Again, in some cases, we can gain more by applications of Rules 1, 2, and 3.

In the sequel we present all branching rules in a preference order. Hence when Case i branching is applied to an instance all earlier cases do not apply.

Case 1: There is a white vertex with at least three white neighbors. If v has three white neighbors, we can perform a vertex branch on v. The branching vector in this case will be $(4, 1 - \alpha)$, i.e., in one case, we decrease the measure by at least four, and in the other case, we decrease the measure by $1 - \alpha$.

Case 2: There is a white vertex with two white neighbors and at least one red neighbor. If v has two white neighbors and at least one red neighbor, then a vertex branch on v gives a branching vector of $(3 + \alpha, 1 - \alpha)$.

Suppose Cases 1 and 2 cannot be applied anymore. Then all white vertices have at most two white neighbors. Moreover, there cannot be a cycle of white vertices, as such a cycle would either be removed by Rule 4 or contains a vertex to which Case 2 applies. Similar for white vertices forming paths. Only the endpoints of such a path can be adjacent to a red vertex, and at least one endpoint is adjacent to a red vertex.

Case 3: The subgraph induced by white vertices contains a path of length at least two, with both endpoints incident to at least one red vertex. Suppose now we have a path of white vertices $v_1, \ldots, v_r$, $r \geq 2$, with v_1 and v_r incident to a red vertex. As Case 2 no longer applies, we can assume that $v_2, \ldots, v_{r-1}$ have no nondeleted neighbors outside the path.

Let R be the set of red vertices that are adjacent to v_1 and/or v_r.

Case 3.1: $r = 2$. We must either select v_1, or select v_2, or select neither v_1 nor v_2. In the latter case, both v_1 and v_2 can be colored green, so the measure is decreased by two in this case. Hence, we have a branching vector $(2 + \alpha, 2 + \alpha, 2)$.

Case 3.2: $r = 3$ and $|R| \geq 2$. We consider all cases of placing vertices from $\{v_1, v_2, v_3\}$ in X:

- Select v_1 and v_3: we decrease the measure by $3 + 2 \cdot \alpha$.
- Select v_1: we decrease the measure by $3 + \alpha$.
- Select v_2: we decrease the measure by 3.
- Select v_3: we decrease the measure by $3 + \alpha$.
- Choose none: we decrease the measure by 3. (All three vertices can be colored green.)

So, in this case, we obtain a branching vector $(3 + 2 \cdot \alpha, 3 + \alpha, 3, 3 + \alpha, 3)$.

Case 3.3: $r = 3$ and $|R| = 1$. The vertices v_1 and v_3 have a common red neighbor. Now, we can perform a vertex branch on v_1. If we select v_1, then v_3 becomes an isolated vertex, and thus we have a branching vector of $(3 + \alpha, 1 - \alpha)$.

Case 3.4: $r = 4$ and $|R| \geq 2$. Like in Case 3.2, we consider all cases of placing vertices from $\{v_1, v_2, v_3, v_4\}$ in X, and obtain a somewhat tedious case analysis. In each case, each vertex in $\{v_1, v_2, v_3, v_4\}$ either is removed or is green. If v_1 or v_4 is placed in X, we gain an additional α for the removal of the red neighbor of this vertex. In case we select both v_1 and v_4, we gain $2 \cdot \alpha$; here we use that

$|R| \geq 2$. This gives a branching vector of $(4+\alpha, 4+2\cdot\alpha, 4+\alpha, 4+\alpha, 4, 4, 4+\alpha, 4)$, corresponding to selecting $\{v_1, v_3\}$, $\{v_1, v_4\}$, $\{v_2, v_4\}$, $\{v_1\}$, $\{v_2\}$, $\{v_3\}$, $\{v_4\}$ or no vertex from this path for inclusion in X.

Case 3.5: $r = 4$ and $|R| = 1$. We do a vertex branch on v_1: if we select v_1, then Rule 3 will remove v_3 and v_4. So the branching vector is $(4 + \alpha, 1 - \alpha)$.

Case 3.6: $r \geq 5$. We branch as follows:

- v_1 is placed in X: we decrease the measure by $2 + \alpha$.
- v_2 is placed in X: we decrease the measure by 3.
- v_3 is placed in X and v_1 is not placed in X. v_1 can be colored green, and thus we decrease the measure by 4.
- v_4 is placed in X and v_1 and v_2 are not placed in X. v_1 and v_2 can be colored green, and thus we decrease the measure by 5.
- None of v_1, v_2, v_3, v_4 is placed in X. v_1, v_2, v_3 become green, and v_4 becomes red: a measure decrease of $4 - \alpha$.

Thus, the branching vector is $(2 + \alpha, 3, 4, 5, 4 - \alpha)$.

Case 4: The subgraph induced by white vertices contains a path of length at least two, with exactly one endpoint incident to a red vertex. Suppose $v_1, \ldots, v_r$ is a path of white vertices, and suppose $r \geq 2$ is maximal. Assume without loss of generality that v_1 has a red neighbor, say w.

Case 4.1: $r \geq 3$. We do a vertex branch on v_{r-2}. If we select v_{r-2} then we gain at least $3+\alpha$: if $r \geq 4$, then v_{r-2} has two white neighbors, and if $r = 3$, then v_{r-2} has a white neighbor (v_{r-1}) and a red neighbor (w). Moreover, v_r becomes an isolated vertex after v_{r-2} is placed in the independent set, and thus is removed by Rule 3. If we do not select v_{r-2}, we gain $1 - \alpha$, and thus we have a branching vector of $(3 + \alpha, 1 - \alpha)$.

Case 4.2: $r = 2$ and w has a white neighbor $x \neq v_1$. We can now perform a vertex branch on x. If we place x in the independent set, then w and x are removed, but also v_1, v_2 are removed as Rule 3 can be applied: they form a connected component of at most two white vertices. So, the measure is decreased by at least $3 + \alpha$. If we do not select x, we color x red so obtain a measure decrease of $1 - \alpha$. So, this case gives a $(3 + \alpha, 1 - \alpha)$ branching vector.

Case 4.3: $r = 2$ and w has no white neighbor. We either must select v_1, or we select v_2, or we select neither v_1 or v_2. If we select v_2, then w can be colored green, as its only white neighbor v_1 is removed. If we select neither v_1 nor v_2, then w, v_1 and v_2 can be colored green, so we decrease the measure $2 + \alpha$ in this case. So we obtain a branching vector of $(2 + \alpha, 2 + \alpha, 2 + \alpha)$.

If Cases $1 - 4$ cannot be applied, then there are no adjacent white vertices. The remaining cases thus deal with white vertices that have no white neighbors. If a white vertex has no red neighbors, then it is removed by Rule 3, so we assume that each white vertex has at least one red neighbor but no other neighbors.

Case 5: v_1 is a white vertex with no white but at least two red neighbors. We do a vertex branch on v_1. If we do not select v_1, it can be colored green, by Rule 1. So we obtain a branching vector $(1 + 2 \cdot \alpha, 1)$.

Case 6: v_1 is a white vertex with exactly one neighbor, which is red. Let w be the red neighbor of v_1.

Case 6.1: w has a white neighbor $x \neq v_1$. If a white neighbor of w has at least two red neighbors, then we can deal with it as in Case 5, and obtain a branching vector of $(1 + 2 \cdot \alpha, 1)$. So suppose all white neighbors of w have degree one, and thus w is their unique neighbor. We now have the following branch:

- w is a vertex in the K-set. In this case, w and all white neighbors of w are colored green. So, the measure decreases by at least $2 + \alpha$.
- w is not a vertex in the K-set. In this case, we must place *all* white neighbors of w in the independent set X. Again, the measure decreases by at least $2 + \alpha$.

So, we obtain a $(2 + \alpha, 2 + \alpha)$ branching vector.

Case 6.2: v_1 is the unique white neighbor of w. In this case, we do a vertex branch on v_1. If we do not place v_1 in the independent set, then both v_1 and w can be colored green. So, the branching vector is $(1 + \alpha, 1 + \alpha)$.

If no case applies, then there are no white, and hence also no red vertices left, so we found one (or zero, in case the green vertices are not connected) K-set. Our choice of $\alpha = 0.5685$ gives the best value for the base of the exponent for the given branching vectors, namely the claimed 1.6052. Thus, it follows that there are $O(1.6052^n)$ nontrivial K-sets that contain v_0. As the value 1.6052 is obtained by rounding, and there are at most $n + m$ trivial K-sets, the result follows.

4   A Bound on the Number of K-sets in Trees

In this section, we establish an upper bound on the number of K-sets in a tree. This bound is used in Section 5 to show a bound on the running time of our algorithm, when the input graph is a tree or a forest.

Theorem 4. *Let $T(n)$ be the maximum number of K-sets in a tree on n nodes. Then $T(n) \leq n \cdot 3^{n/3}$.*

Proof. We denote as a *rooted K-set* of a rooted tree T any K-set of T containing r, where r denotes the root of T. Let $R(n)$ be the maximum number of rooted K-sets in any rooted tree on n nodes. We claim that $R(n) \leq 3^{n/3} - 1$ for all $n \geq 2$.

We are going to prove this claim by induction. To see that the claim is true for the base case $n = 2$, note that the only K-set containing r is the one containing both nodes of the tree, and that $3^{2/3} - 1 > 1.08$.

As induction hypothesis let us assume that the claim is true for all $n' < n$ and consider any rooted tree T on $n > 2$ nodes. Let r be the root of the tree and $u_1, u_2, \ldots, u_p$ be the children of v. For every $i = 1, 2, \ldots, p$, let T_i be the subtree of T rooted at u_i. Furthermore for all $i = 1, 2, \ldots, p$, we denote by n_i the number of nodes of T_i.

Let W be any K-set of T containing its root r. Then for every i, the intersection of W with T_i is either empty or a K-set of T_i containing its root u_i. Note that $n_i = 1$ implies that W also contains u_i since $r \in W$ (and thus r cannot be taken into the independent set X generating W). Using the induction hypothesis and $\sum_{i=1}^{p} n_i = n - 1$, we establish the following upper bound for the number of rooted K-sets of a rooted tree on n nodes

$$R(n) \leq \prod_{i:n_i \geq 2} (R(n_i) + 1) \leq \prod_{i:n_i \geq 2} 3^{n_i/3}$$
$$\leq 3^{(n-1)/3} \leq 3^{n/3} - 1.$$

This completes the proof of our claim.

To complete the proof of the theorem simply note that any K-set is counted at least once as a rooted K-set for some vertex v chosen to be the root, and thus $T(n) \leq n \cdot R(n)$. □

The above proof can be used to obtain an algorithm to enumerate all K-sets of a tree in time $O^*(3^{n/3})$. This algorithm chooses any vertex r of maximum degree and branches into two subproblems: in one r is taken into W and in the other one r is discarded from W and thus all neighbors of r are discarded from S.

5 The Exact Algorithm

In this section we present our exact exponential-time algorithm solving KAYLES. The algorithm starts with a call to the procedure `compute_nimber` shown in Figure 1, with input $G = (V, E)$. If it returns a nimber that is at least one, then Player 1 has a winning strategy on G; if it otherwise returns nimber zero, then Player 2 has a winning strategy. Correctness of the procedure directly follows from the discussion in Section 2.

Note that the procedure `compute_nimber`$(G[W])$ is only called for K-sets, and thus $G[W]$ is always connected, with one possible exception: if G is not connected, then the first call to the procedure is for $G[V]$ with V not a K-set. As the overhead per recursive call is polynomial, the running time is a polynomial factor times the number of K-sets in G. The procedure computes the nimber $nb(W)$ of $G[W]$ for all K-sets W of G and stores the value in a table using Memorisation, i.e., computed values are stored in a table, and by look-up no value $nb(W)$ is computed more than once. It follows that the running time of the algorithm is $O^*(|\mathcal{K}(G)|)$ where $\mathcal{K}(G)$ is the set of K-sets of G.

Combining the bounds of the previous sections on the number of K-sets with the algorithm of this section, we establish the following result.

Theorem 5. KAYLES *can be solved in time* $O^*(1.6052^n)$ *for graphs on* n *vertices.* KAYLES *can be solved in time* $O^*(1.4423^n)$ *for trees on* n *nodes.*

Procedure compute_nimber(**G[W]**).

> **if** $nb(W)$ *already computed* **then**
>> | **return** $nb(W)$
>
> **else**
>> | $M := \emptyset$;
>> | **for** *all* $w \in W$ **do**
>>> | let $Z_1, Z_2, \ldots, Z_r$ $(r \geq 1)$ be the components of $G - N[w]$;
>>> | $nim := 0$;
>>> | **for** $i \leftarrow 1$ **to** r **do**
>>>> | $nim := nim \oplus$ compute_nimber$(G[Z_i])$;
>>>
>>> | $M := M \cup \{nim\}$;
>>
>> | $answer := \mathbf{mex}(M)$;
>> | $nb(W) := answer$;
>
> **return** $answer$

Fig. 1. Procedure compute_nimber

6   Lower Bounds

There are graphs on n vertices having $\Theta(3^{n/3})$ different K-sets. This implies a lower bound on the maximum number of K-sets of any graph on n vertices as well as a lower bound on the running time of any exact algorithm solving KAYLES by using all K-sets of the input graph. A similar lower bound can be achieved for trees.

Theorem 6. *For any $t \geq 1$, there is a graph on $n = 3t$ vertices having $3^{n/3} + 2n/3$ different K-sets.*

Proof. Consider the following family of (chordal) graphs G_t for all positive integers t on the vertex set $\{1, 2, \ldots, 3t\}$. The edge set of G_t is constructed as follows:

- $\{3i \; : \; i = 1, 2, \ldots, t\}$ is a clique of G_t, and
- for all $i = 1, 2, \ldots, t$, the vertex set $\{3i - 2, 3i - 1, 3i\}$ induces a path.

Let us count the K-sets W of G_n. For any K-set W let X be an independent set such that $W = V - N[X]$.

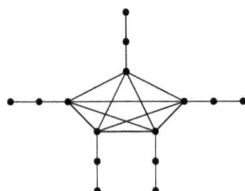

Fig. 2. Example of the construction of Theorem 6, with $t = 5$

Case 1: $W \cap \{3i : i = 1, 2, \ldots, t\} = \emptyset$, which implies $|X \cap \{3i : i = 1, 2, \ldots, t\}| = 1$. Say $X \cap \{3i : i = 1, 2, \ldots, t\} = \{3i_0\}$. Hence $W \subseteq \{3i - 1, 3i - 2\}$ for some i. Thus if $i \neq i_0$ then $W = \{3i - 1, 3i - 2\}$; and if $i = i_0$ then $W = \{3i - 2\}$. Hence there are $2t$ different K-sets in this case.

Case 2: $W \cap \{3i : i = 1, 2, \ldots, t\} \neq \emptyset$, which implies $X \cap \{3i : i = 1, 2, \ldots, t\} = \emptyset$. Then for any i, $X \cap \{3i - 2, 3i - 1, 3i\}$ may be one of the following sets: $\{3i - 2, 3i - 1, 3i\}$, $\{3i\}$, $\emptyset$. Thus there are 3^t different K-sets W in this case.

In total the graph G_t has at least $3^t + 2t$ K-sets.   □

Theorem 7. *For any $t \geq 1$, there is a tree on $n = 3t + 1$ nodes having $3^{(n-1)/3} + n - 1$ different K-sets.*

Proof. Consider the following family of trees T_t for all positive integers t. The node set of T_t is the set $\{0, 1, 2, \ldots, 3t + 1\}$. The edge set is constructed as follows:

- For all $i = 1, 2, \ldots, t$, the vertex set $\{3i - 2, 3i - 1, 3i\}$ induces a path, and
- the node 0 is adjacent to all nodes in the set $\{3i : i = 1, 2, \ldots, t\}$ and no others.

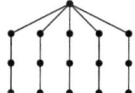

Fig. 3. Example of the construction of Theorem 7, with $t = 5$

For any K-set W let X be an independent set such that $W - V - N[X]$. To count the K-sets W of T_t we distinguish two cases.

Case 1: $0 \notin W$. Then $X \cap \{0, 3, 6, \ldots, 3t\} \neq \emptyset$. Hence $W \subseteq \{3i, 3i - 1, 3i - 2\}$ for some i. Thus $W = \{3i, 3i - 1, 3i - 2\}$, $W = \{3i\}$ or $W = \{3i - 2\}$. Hence there are $3t$ different K-sets.

Case 2: $0 \in W$. Then $X \cap \{0, 3, 6, \ldots, 3t\} = \emptyset$. For every i, consider $W \cap \{3i - 2, 3i - 1, 3i\}$. By connectedness of $G[W]$ and since $0 \in W$, we obtain that $W \cap \{3i - 2, 3i - 1, 3i\}$ is any of the following sets $\{3i - 2, 3i - 1, 3i\}$, $\{3i\}$, $\emptyset$. Thus there are 3^t different K-sets W in this case.

Summarizing, the tree T_t has at least $3^t + 3t$ K-sets.   □

7   Conclusions

In this paper, we gave an algorithm to determine which player has a winning strategy for the game KAYLES. To analyse the running time, we introduced the notion of K-sets, and obtained upper and lower bounds on the maximum number of K-sets that a graph can have. We also obtained such bounds for trees; up to a polynomial factor, the bounds are sharp for trees.

A number of interesting directions for further research remain. The complexity of KAYLES on trees remains a long standing open problem. But one can also ask if there exists a subexponential time algorithm for KAYLES on trees, e.g., with running time of the form $O(c^{\sqrt{n}})$.

Our algorithm uses exponential memory. It also is open if there exists a polynomial space algorithm with a running time of $O^*(2^n)$, and this may well be hard to obtain.

Our paper shows that a PSPACE-complete problem may be solvable by an exact algorithm of running time $O^*(c^n)$ with $c < 2$. It would be interesting to study exact algorithms for other PSPACE-complete problems, e.g., for other combinatorial games, or for a problem like QUANTIFIED 3-SATISFIABILITY [8]. An algorithm that solves QUANTIFIED (3-)SATISFIABILITY in $O^*(2^n)$ time is not hard to find, but it seems very hard (or impossible) to find an algorithm with a running time $O^*(c^n)$ with $c < 2$ for this problem.

References

1. Berlekamp, E.R., Conway, J.H., Guy, R.K.: Winning Ways for your mathematical plays. Games in General, vol. 1. Academic Press (1982)
2. Berlekamp, E.R., Conway, J.H., Guy, R.K.: Winning Ways for your mathematical plays. Games in Particular, vol. 2. Academic Press (1982)
3. Bodlaender, H.L., Kratsch, D.: Kayles and nimbers. Journal of Algorithms 43, 106–119 (2002)
4. Conway, J.H.: On Numbers and Games. Academic Press (1976)
5. Fleischer, R., Trippen, G.: Kayles on the Way to the Stars. In: van den Herik, H.J., Björnsson, Y., Netanyahu, N.S. (eds.) CG 2004. LNCS, vol. 3846, pp. 232–245. Springer, Heidelberg (2006)
6. Fomin, F.V., Kratsch, D.: Exact Exponential Algorithms. Springer, Heidelberg (2010)
7. Guignard, A., Sopena, É.: Compound Node-Kayles on paths. Theoretical Computer Science 410, 2033–2044 (2009)
8. Schaefer, T.J.: The complexity of satisfiability problems. In: Proceedings of STOC 1978, pp. 216–226 (1978)
9. Schaefer, T.J.: On the complexity of some two-person perfect-information games. Journal of Computer and System Sciences 16, 185–225 (1978)

The Cinderella Game on Holes and Anti-holes

Marijke H.L. Bodlaender[1], Cor A.J. Hurkens[2], and Gerhard J. Woeginger[2]

[1] Department of Information and Computing Sciences
Universiteit Utrecht, The Netherlands
[2] Department of Mathematics and Computer Science
TU Eindhoven, The Netherlands

Abstract. We investigate a two-player game on graphs, where one player (Cinderella) wants to keep the behavior of an underlying water-bucket system stable whereas the other player (the wicked Stepmother) wants to cause overflows. The bucket number of a graph G is the smallest possible bucket size with which Cinderella can win the game.

We determine the bucket numbers of all perfect graphs, and we also derive results on the bucket numbers of certain non-perfect graphs. In particular, we analyze the game on holes and (partially) on anti-holes for the cases where Cinderella sticks to a simple greedy strategy.

Keywords: Combinatorial game, on-line algorithms, perfect graphs.

1 Introduction

"Five empty buckets of capacity b stand in the corners of a regular pentagon. Cinderella and her wicked Stepmother play a game that goes through a sequence of rounds: at the beginning of every round, the Stepmother takes one liter of water from the nearby river, and distributes it arbitrarily over the five buckets. Then Cinderella chooses a pair of neighboring buckets, empties them into the river, and puts them back into the pentagon. Then the next round begins. The Stepmother's goal is to make one of these buckets overflow. Cinderella's goal is to prevent this. For which bucket sizes b can the Stepmother eventually enforce a bucket overflow? And for which bucket sizes can Cinderella keep the game running forever?"

This game has been proposed as a problem [5] for the 50th International Mathematical Olympiad for high-school students that took place in Germany in summer 2009. Bodlaender & al. [1] consider a generalization of this game where there are $n \geq 2$ buckets in a circle and where in every round Cinderella can empty an arbitrary group of c consecutive buckets; they construct optimal strategies and characterize optimal bucket sizes for many instances of this game. An earlier paper [2] by Chrobak & al. investigates another variant where the water does not arrive in rounds, but does arrive continuously over time; this earlier variant is built around independent sets in certain underlying graphs. In the current paper, we will study yet another variant where water arrives in rounds (as in [5] and [1]) and where the game board is an undirected graph (as in [2]).

P. Kolman and J. Kratochvíl (Eds.): WG 2011, LNCS 6986, pp. 71–82, 2011.

Bucket games and bucket numbers. The game is played on an undirected simple graph $G = (V, E)$. Every vertex in V contains a bucket which is empty at the beginning of the game. Every edge $[u, v] \in E$ indicates an incompatibility, and Cinderella must never touch two adjacent buckets u and v within the same round. In every round the Stepmother first distributes one liter of water over the buckets, and then Cinderella picks an independent set in the graph and empties the buckets in this independent set. The Stepmother wants to reach a bucket overflow, and Cinderella wants to avoid this. Note that the original puzzle [5] considers this game on the graph C_5, the cycle with five vertices a, b, c, d, e and the five edges $[a, c]$, $[c, e]$, $[e, b]$, $[b, d]$, and $[d, a]$.

We define $\mathtt{bucket}'(G)$ as the infimum of all bucket sizes for which Cinderella can keep the game running forever; furthermore we introduce the quantity $\mathtt{bucket}(G) = \mathtt{bucket}'(G) - 1$ which we call the *bucket number* of G. If Cinderella consistently avoids overflows for buckets of size $\mathtt{bucket}'(G)$, then at the end of every round she must leave buckets with contents $\mathtt{bucket}(G)$ or less. The Stepmother adds a liter and messes up the water levels, and then Cinderella restores order and brings the water levels back to at most $\mathtt{bucket}(G)$.

Perhaps the most natural strategy for Cinderella is the GREEDY strategy: *"Always remove as much water as possible from the system."* In other words, GREEDY empties the buckets in an independent set with currently largest total contents; ties are broken arbitrarily. In analogy to the value $\mathtt{bucket}(G)$, we define the so-called *greedy bucket number* $\mathtt{g\text{-}bucket}(G)$ of a graph G for games in which Cinderella works with the GREEDY strategy.

Summary of results. We relate the bucket number and the greedy bucket number of a graph to its clique number and chromatic number; roughly speaking, we show that both bucket numbers lie between the natural logarithm of the clique number and the natural logarithm of the chromatic number. These results (presented in Section 3) settle the game for all perfect graphs.

The main part of the paper then deals with the two simplest families of non-perfect graphs: with odd holes and with odd anti-holes. Odd holes are fully analyzed in Section 4. We prove that their bucket number alway equals 1, whereas their greedy bucket number always lies strictly above 1; see Theorem 10 for the exact statement of these greedy values. In Section 5 we investigate odd anti-holes. The bucket game on odd anti-holes is also discussed — in a slightly different guise — by [1] who conjecture that the bucket number of an odd anti-hole is essentially the natural logarithm of its clique number. We do not touch this (seemingly hard) conjecture, but manage to show that under this conjecture the greedy bucket number of every odd anti-hole is strictly larger than its bucket number.

Section 6 finally formulates some bold conjectures on bucket numbers that are in perfect agreement with our current knowledge. For instance, it could be true that the GREEDY strategy is sub-optimal if and only if the underlying game board graph is non-perfect.

2   Definitions and First Results

Consider a graph $G = (V, E)$. The set of all neighbors of a vertex $v \in V$ is denoted $N_G(v)$, or simply $N(v)$ if the graph is clear from the context. As usual, the chromatic number of G is denoted by $\chi(G)$ and its clique number by $\omega(G)$. For a subset $S \subseteq V$, we will often shortly write $\chi(S)$ to denote (in slight abuse of notation) the chromatic number of the sub-graph $G[S]$ induced by S.

The partial sums $\mathcal{H}\langle k \rangle$ of the harmonic series play a major role in our investigations.

$$\mathcal{H}\langle k \rangle \;=\; 1 + \frac{1}{2} + \frac{1}{3} + \frac{1}{4} + \;\ldots\ldots\; + \frac{1}{k} \tag{1}$$

We often summarize the contents of the buckets at a particular moment in time in a corresponding load vector $x = (x_v)_{v \in V}$ where x_v denotes the contents of bucket v. For a subset S of buckets, we denote $x(S) = \sum_{v \in S} x_v$. To keep the notation simple, we write $x(u, v)$ short for $x(\{u, v\})$ and $x(u, v, w)$ short for $x(\{u, v, w\})$.

The following three lemmas state simple but very useful observations.

Lemma 1. *Let H be a graph, and let G be a (not necessarily induced) sub-graph of H. Then* $\texttt{bucket}(G) \leq \texttt{bucket}(H)$ *and* $\texttt{g-bucket}(G) \leq \texttt{g-bucket}(H)$.

Lemma 2. *A disconnected graph G with connected components $G_1, \ldots, G_k$ satisfies* $\texttt{bucket}(G) = \max_{1 \leq i \leq k} \texttt{bucket}(G_i)$ *and furthermore* $\texttt{g-bucket}(G) = \max_{1 \leq i \leq k} \texttt{g-bucket}(G_i)$.

Lemma 3. *Let $G = (V, E)$ be a graph, and let u and v be two non-adjacent vertices with $N(u) \subseteq N(v)$. Then* $\texttt{bucket}(G) = \texttt{bucket}(G - u)$ *and* $\texttt{g-bucket}(G) = \texttt{g-bucket}(G - u)$.

3   The Game on General Graphs

In this section we relate the bucket number and the greedy bucket number of a graph to its clique number and chromatic number. The proofs of the following two Theorems 4 and 5 apply ideas that have been used before by Dietz & Sleator [3], Chrobak & al. [2], and Bodlaender & al. [1]. Theorem 4 substantially extends these ideas, whereas Theorem 5 follows quite easily by cosmetic modifications.

Theorem 4. *Every graph $G = (V, E)$ satisfies* $\texttt{g-bucket}(G) \leq \mathcal{H}\langle \chi(G) - 1 \rangle$.

Proof. Let x_v denote the contents of bucket $v \in V$ at the beginning of some round. We will show that the GREEDY strategy maintains the following system of invariants:

$$x(S) \;<\; \chi(S) \cdot \left(1 + \mathcal{H}\langle \chi(G) - 1 \rangle - \mathcal{H}\langle \chi(S) \rangle \right) \qquad \text{for all sets } S \subseteq V. \tag{2}$$

By applying the inequality in (2) to a single vertex set $S = \{v\}$, we get that all buckets satisfy the bound $x_v < \mathcal{H}\langle \chi(G) - 1 \rangle$; this implies the theorem.

Now let us show that the invariants (2) indeed are maintained. These invariants are trivially satisfied at the beginning of the game when all buckets are still empty. Let us inductively assume that GREEDY has maintained the invariants up to some fixed round. Next the Stepmother makes her move by raising the bucket contents to y_v for $v \in V$, and then GREEDY empties an independent set I with maximum y-weight. Consider an arbitrary set $S \subseteq V - I$. If $\chi(S) = \chi(G)$, then

$$y(S) \leq y(V - I) \leq \frac{\chi(G) - 1}{\chi(G)} y(V) \leq \frac{\chi(G) - 1}{\chi(G)} (x(V) + 1) < \chi(G) - 1. \quad (3)$$

To get the rightmost inequality in (3), we deduced $x(V) < \chi(G) - 1$ from the invariant. Note that (3) implies that S satisfies the invariant, exactly as desired. In the remaining cases we will assume $\chi(S) < \chi(G)$. Observe that the choice of I implies

$$y(S) \leq \chi(S) \cdot y(I). \quad (4)$$

Furthermore the chromatic number $\chi(S \cup I)$ is either $\chi(S)$ or $\chi(S) + 1$. Since $\chi(S) < \chi(G)$, in either case the inductive assumption yields

$$x(S \cup I) < (\chi(S) + 1) \cdot (1 + \mathcal{H} \langle \chi(G) - 1 \rangle - \mathcal{H} \langle \chi(S) + 1 \rangle). \quad (5)$$

By applying (4) and (5) we derive

$$
\begin{aligned}
y(S) &\leq \frac{\chi(S)}{\chi(S) + 1} (y(S) + y(I)) \leq \frac{\chi(S)}{\chi(S) + 1} (x(S \cup I) + 1) \\
&< \chi(S) \cdot \left(1 + \mathcal{H} \langle \chi(G) - 1 \rangle - \mathcal{H} \langle \chi(S) + 1 \rangle + \frac{1}{\chi(S) + 1} \right) \\
&= \chi(S) \cdot (1 + \mathcal{H} \langle \chi(G) - 1 \rangle - \mathcal{H} \langle \chi(S) \rangle).
\end{aligned}
$$

Therefore GREEDY maintains the invariant (2) for all sets $S \subseteq V - I$. Since all the buckets in I are emptied by GREEDY, this also yields the invariant for all the remaining sets $S \subseteq V$ that overlap with I. □

Theorem 5. *Every graph $G = (V, E)$ satisfies* $\mathtt{bucket}(G) \geq \mathcal{H} \langle \omega(G) - 1 \rangle$.

Proof. By Lemma 1, it is sufficient to prove the result for the complete graph K_n on n vertices. Note that on the game board K_n Cinderella can only empty a single bucket per round. In a first phase, the Stepmother always distributes her liter in such a way that all buckets in K_n are filled to the same level. This common filling level strictly increases over the rounds and converges to 1. The first phase ends when all buckets have contents $1 - \varepsilon$ (where ε is a small positive real number that can be chosen arbitrarily close to 0). Cinderella empties (at most) one of these buckets, whereas the remaining $n - 1$ buckets take their contents into the following round.

The second phase goes through $n - 2$ further rounds that we number from 1 to $n - 2$. At the beginning of the r-th such round, there are $n - r$ buckets that each contain at least $W := 1 - \varepsilon + \mathcal{H} \langle n - 1 \rangle - \mathcal{H} \langle n - r \rangle$ water. The Stepmother then fills these $n - r$ buckets to the same level

$$\frac{(n - r)\, W + 1}{n - r} = 1 - \varepsilon + \mathcal{H} \langle n - 1 \rangle - \mathcal{H} \langle n - (r + 1) \rangle.$$

Cinderella empties (at most) one of these buckets, whereas the remaining $n - (r + 1)$ buckets remain untouched and move on into the next round. At the end of round $n - 2$ there remains a bucket of contents $\mathcal{H} \langle n - 1 \rangle - \varepsilon$. □

By combining Theorems 4 and 5, we see that the bucket numbers of every graph G are sandwiched between the following bounds:

$$\mathcal{H} \langle \omega(G) - 1 \rangle \;\leq\; \mathtt{bucket}(G) \;\leq\; \mathtt{g\text{-}bucket}(G) \;\leq\; \mathcal{H} \langle \chi(G) - 1 \rangle. \qquad (6)$$

This fully settles the bucket game for all graphs G with $\omega(G) = \chi(G)$, and in particular yields the following corollary.

Corollary 6. *Every perfect graph G satisfies* $\mathtt{bucket}(G) = \mathtt{g\text{-}bucket}(G) = \mathcal{H} \langle \omega(G) - 1 \rangle.$

In the rest of this section, we settle the bucket game for all graphs with $n \leq 6$ vertices. Because of Corollary 6 we actually only need to check the non-perfect graphs in Figure 1.

Lemma 7. *The graph $C_5 + u$ satisfies* $\mathtt{bucket}(C_5 + u) = 3/2.$

Proof. We show that Cinderella can maintain the following system of invariants.

$$x_1, x_2, x_3, x_4, x_5, x_u \;<\; 3/2 \qquad\qquad\qquad\qquad\qquad (7a)$$

$$x_u + x_k + x_{k+1} \;<\; 2 \qquad \text{for } 1 \leq k \leq 5 \qquad\qquad (7b)$$

$$x(1, 2, 3, 4, 5, u) \;<\; 5/2 \qquad\qquad\qquad\qquad\qquad (7c)$$

The indices $k + 1$ in (7b) are taken modulo 5. The argument is an inductive proof with many case distinctions. We first show that the invariants can be maintained, if there is a bucket of contents at least $3/2$, and then settle the remaining cases. The (straightforward but tedious) details are omitted from this extended abstract. □

Theorem 8. *Every graph G on $n \leq 6$ vertices has* $\mathtt{bucket}(G) = \mathcal{H} \langle \omega(G) - 1 \rangle.$

Proof. Because of Corollary 6, it only remains to prove that every non-perfect graph G on $n \leq 6$ vertices satisfies $\mathtt{bucket}(G) \leq \mathcal{H} \langle \omega(G) - 1 \rangle$. Every non-perfect graph G on $n \leq 6$ vertices consists of an induced cycle C_5 on five vertices plus (perhaps) some sixth vertex u. The clique number of such a graph G is either 2 or 3.

First consider the case $\omega(G) = 2$. Then vertex u cannot be adjacent to two consecutive vertices on the cycle C_5. Every such graph G is a subgraph of the graph H that consists of a cycle C_5 on five vertices $1, 2, 3, 4, 5$ plus vertex u plus the two edges $[1, u]$ and $[3, u]$. By Lemma 1 we have $\texttt{bucket}(G) \leq \texttt{bucket}(H)$. By Lemma 3 we have $\texttt{bucket}(H) = \texttt{bucket}(C_5)$, as H contains the two non-adjacent vertices u and 1 with $N(u) \subseteq N(1)$. By Theorem 9 we have $\texttt{bucket}(C_5) = 1$. Putting things together yields the desired bound

$$\texttt{bucket}(G) \;\leq\; \texttt{bucket}(C_5) \;=\; 1 \;=\; \mathcal{H}\langle \omega(G) - 1 \rangle.$$

Next consider the case $\omega(G) = 3$. Every such graph G is a subgraph of the graph $C_5 + u$ that consists of a cycle C_5 on five vertices $1, 2, 3, 4, 5$ plus all five edges between vertex u and the vertices on the cycle. By Lemma 1 we have $\texttt{bucket}(G) \leq \texttt{bucket}(C_5 + u)$, and Lemma 7 shows that $\texttt{bucket}(C_5 + u) = 3/2 = \mathcal{H}\langle 2 \rangle$. This completes the proof of Theorem 8. $\qquad\square$

4   The Game on Holes

In this section the considered game board will be the cycle C_n on $n \geq 3$ vertices. The vertices/buckets correspond to the integers in $B = \{1, 2, \ldots, n\}$, and two buckets i and j are adjacent, if and only if $|i - j| \equiv 1 \pmod n$. Bucket numbers are always taken modulo n, so that k and $k + n$ refer to the same bucket.

The results in Section 3 yield $\texttt{bucket}(C_n) = \texttt{g-bucket}(C_n) = 1$ for all even n, as in these cases C_n is a perfect graph with clique number 2. Hence we will assume throughout this section that $n = 2m + 1$ is an odd integer; note that $\omega(C_{2m+1}) = 2$ and $\chi(C_{2m+1}) = 3$. We use $I_r = \{r+2, r+4, r+6, \ldots, r+2m\}$ to denote the (unique) independent set of m buckets that neither contains bucket r nor bucket $r + 1$.

Theorem 9. *The odd cycle C_{2m+1} satisfies* $\texttt{bucket}(C_{2m+1}) = 1$.

Proof. The lower bound $\texttt{bucket}(C_{2m+1}) \geq 1$ follows from Theorem 5. Hence we concentrate on the upper bound proof. Cinderella always completes a round by

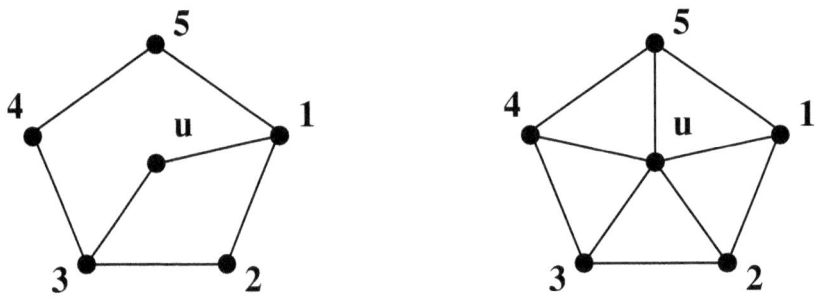

Fig. 1. The non-perfect graphs H (to the left) and graph $C_5 + u$ (to the right)

emptying an appropriate independent set I_r for some r, and she leaves buckets with contents $x_1, \ldots, x_{2m+1}$. As invariants she uses the inequalities

$$x_r + x_{r+1} < 1 \tag{8a}$$

$$x_k \leq 1 \qquad \text{for } 1 \leq k \leq 2m + 1. \tag{8b}$$

Then the Stepmother moves and increases the bucket contents to $y_1, \ldots, y_{2m+1}$. From $x_{r-1} = 0$, $x_{r+2} = 0$, and from (8a) we derive

$$y_{r-1} + y_r + y_{r+1} + y_{r+2} \leq x_{r-1} + x_r + x_{r+1} + x_{r+2} + 1 < 2,$$

which implies $y_{r-1} + y_r < 1$ or $y_{r+1} + y_{r+2} < 1$. First consider the case where $y_{r-1} + y_r < 1$. Then Cinderella empties I_{r-1} and maintains (8a). All buckets $k \in I_r$ were emptied in the preceding round and hence satisfy $y_k \leq 1$. All buckets $k \in I_{r-1}$ have been emptied in the current round and hence satisfy $y_k = 0$. The only bucket in $B - (I_r \cup I_{r-1})$ is bucket r, for which $y_{r-1} + y_r < 1$ implies $y_r < 1$. Hence Cinderella also maintains (8b). Next consider the case where $y_{r+1} + y_{r+2} < 1$. In this case Cinderella empties I_{r+1} and maintains the invariants by symmetric arguments. □

Theorem 10. *The odd cycle C_{2m+1} has* $\text{g-bucket}(C_{2m+1}) = 1 + \frac{1}{m} \cdot 2^{-m}$.

The proof of Theorem 10 will be presented in Section 4.1 (for the upper bound) and Section 4.2 (for the lower bound).

4.1   Proof of the Upper Bound for GREEDY

We show by induction that GREEDY always leaves configurations where the bucket contents $x_1, \ldots, x_{2m+1}$ satisfy the system (9a)–(9c) of invariants. In the next round, the Stepmother moves by raising the bucket contents to $y_1, \ldots, y_{2m+1}$. Then GREEDY reacts by emptying an independent set I with maximum y-weight, and thereby produces buckets with contents $z_1, \ldots, z_{2m+1}$. We assume without loss of generality that among any three consecutive buckets at least one is in I.

$$\sum_{i=1}^{2m+1} x_i < \frac{m+1}{m} \tag{9a}$$

$$\sum_{i=k}^{k+2t-1} x_i < 1 + \frac{1}{m} \cdot 2^{t-m} \qquad \text{for } 1 \leq k \leq 2m+1, \ 1 \leq t \leq m \tag{9b}$$

$$x_k < 1 + \frac{1}{m} \cdot 2^{-m} \qquad \text{for } 1 \leq k \leq 2m+1 \tag{9c}$$

The details of this proof can be found in the full version of the paper.

4.2   Proof of the Lower Bound for GREEDY

We present an adversarial strategy for the Stepmother that works against GREEDY in three phases.

In the first phase, the Stepmother always distributes her liter in such a way that all $2m+1$ buckets are filled to the same level. GREEDY always empties some independent set I_r with m buckets. This common filling level a_i after i moves of the Stepmother satisfies the recurrence $a_{i+1} = ((m+1)a_i + 1)/(2m+1)$. Hence these levels a_i strictly increase over the rounds and converge to $1/m$. The first phase ends when all buckets have contents $1/m - \varepsilon$ (where ε is an arbitrarily small positive real number). Without loss of generality GREEDY empties $I_{2m} = \{1, 3, 5, \ldots, 2m-1\}$, so that the $m+1$ buckets $2, 4, 6, \ldots, 2m-2, 2m$ and $2m+1$ take their contents of $1/m - \varepsilon$ into the second phase.

To keep the presentation simple, we let ε become infinitesimally small and will not further indicate the dependence of our bounds on ε from now on. On top of this, we will manipulate and control the tie-breaking behavior of GREEDY by introducing appropriate infinitesimally small fluctuations in the bucket loads. This is straightforward to do (but messy to describe in detail), and we will apply it without much further discussion. The second phase is built around the following sequence of real numbers.

$$\alpha_k = \frac{1}{2m}\left(k + 1 + 2^{-k}\right) \qquad \text{for } k \geq 0.$$

Note that $\alpha_0 = 1/m$ and that

$$4\,\alpha_k = 2\alpha_{k-1} + \frac{k+2}{m} \qquad \text{for } k \geq 1. \tag{10}$$

Now let us specify what is going on during the second phase. There are $m - 1$ additional rounds that we number by $k = 1, \ldots, m-1$. In the canonical situation at the beginning of round k

- the buckets $k + 2i - 1$ with $1 \leq i \leq m - k$ all have contents $1/m$;
- the two buckets $2m - k + 1$ and $2m - k + 2$ both have contents α_{k-1};
- all other buckets are empty.

Then the Stepmother moves. She fills the $m - k - 1$ empty buckets $k + 2i$ with $1 \leq i \leq m - k - 1$ to level $1/m$. Furthermore the Stepmother adds water to the four buckets $2m - k - 1$, $2m - k$, $2m - k + 1$, and $2m - k + 2$ so that their contents increases from respectively $1/m$, 0, α_{k-1}, α_{k-1} to respectively α_k, α_k, α_k, α_k. By using (10), we compute that the overall amount of water added by the Stepmother in this round is

$$(m - k - 1)\frac{1}{m} + 4\,\alpha_k - \frac{1}{m} - 2\,\alpha_{k-1} = 1.$$

After the Stepmother's move, the $2(m - k - 1)$ buckets in the interval $k + 1, \ldots, 2m - k - 2$ all have contents $1/m$ and the next four buckets $2m - k - 1, \ldots, 2m - k + 2$ all have contents α_k. The remaining buckets are empty.

Then GREEDY moves. Through infinitesimally small changes in the bucket loads the Stepmother tricks GREEDY into emptying the buckets $k + 2i - 1$ for $1 \leq i \leq m - k$ together with the bucket $2m - k + 2$. After GREEDY's move, the buckets $k + 2i$ with $1 \leq i \leq m - k - 1$ all have contents $1/m$, the two buckets $2m - k$ and $2m - k + 1$ both have contents α_k, and all other buckets are empty. Note that GREEDY has created the canonical situation for the beginning of round $k + 1$.

The third phase begins, as soon as round $m - 1$ of the second phase has been completed. The canonical situation for the beginning of the following round m does not have any buckets with contents $1/m$, but it does have the two buckets $m + 1$ and $m + 2$ with contents α_{m-1}. The Stepmother adds half a liter to both buckets and raises their contents to

$$\alpha_{m-1} + \frac{1}{2} = \frac{1}{2m}\left(m + 2^{-m+1}\right) + \frac{1}{2} = 1 + \frac{1}{m} \cdot 2^{-m}.$$

GREEDY empties bucket $m + 1$, so that bucket $m + 2$ takes its contents into the next round. This completes the proof of Theorem 10.

5   The Game on Anti-holes

In this section the considered game board will be the anti-hole $\overline{C_n}$ on $n \geq 4$ vertices. The vertices/buckets correspond to the integers to the integers in $B = \{1, 2, \ldots, n\}$, and two buckets i and j are adjacent, if and only if $|i - j| \not\equiv 1 \pmod{n}$. Bucket numbers are always taken modulo n, so that k and $k + n$ refer to the same bucket.

Since even anti-holes are perfect graphs, the results in Section 3 yield $\mathsf{bucket}(\overline{C_{2m}}) = \mathsf{g\text{-}bucket}(\overline{C_{2m}}) = \mathcal{H}\langle m - 1\rangle$. Since furthermore the anti-hole $\overline{C_5}$ is isomorphic to the hole C_5 (which has been treated in the preceding section), we will assume throughout the current section that the number of buckets is odd with $n = 2m + 1$ and $m \geq 3$. Since $\omega(\overline{C_{2m+1}}) = m$ and $\chi(\overline{C_{2m+1}}) = m + 1$, the inequalities in (6) turn into

$$\mathcal{H}\langle m - 1\rangle \ \leq \ \mathsf{bucket}(\overline{C_{2m+1}}) \ \leq \ \mathsf{g\text{-}bucket}(\overline{C_{2m+1}}) \ \leq \ \mathcal{H}\langle m\rangle \qquad (11)$$

If we imagine the buckets arranged along the circumference of a circle, then in each round Cinderella may empty two consecutive buckets from the circle; therefore the games on anti-holes also fall into the class of games investigated in [1]. The following conjecture from [1] states that in (11) the leftmost inequality should in fact hold with equality.

Conjecture 11. *(Bodlaender & al. [1])*
Every odd anti-hole $\overline{C_{2m+1}}$ satisfies $\mathsf{bucket}(\overline{C_{2m+1}}) = \mathcal{H}\langle m - 1\rangle$.

This conjecture has been settled in [1] for the odd anti-holes on $n = 5, 7, 9, 11$ vertices. The behavior of the general case, however, remains unclear and seems to be quite messy. In the remainder of this section, we will analyze the behavior of GREEDY on $\overline{C_{2m+1}}$ with $m \geq 3$, and we will strengthen (11) to the following bounds:

$$\mathcal{H}\langle m-1\rangle + \frac{m^2 - 3m + 1}{2m^2(m-1)} \leq \texttt{g-bucket}(\overline{C_{2m+1}}) \leq \mathcal{H}\langle m\rangle - \frac{1}{2m}.$$

Upper and lower bound are proved respectively in the following two theorems.

Theorem 12. *For $m \geq 3$ we have* $\texttt{g-bucket}(\overline{C_{2m+1}}) \leq \mathcal{H}\langle m\rangle - 1/(2m)$.

Proof. The proof can be found in the full version of the paper. □

Theorem 13. *For $m \geq 3$,* $\texttt{g-bucket}(\overline{C_{2m+1}}) \geq \mathcal{H}\langle m-1\rangle + \dfrac{m^2 - 3m + 1}{2m^2(m-1)}$.

Proof. We give an adversarial strategy for the Stepmother that works against GREEDY in two phases. During the first phase, the Stepmother always distributes her liter such that all $2m + 1$ buckets are filled to the same level and then GREEDY always empties two buckets. The filling level of all buckets increases and converges to $1/2$. The first phase ends when all buckets are sufficiently close to $1/2$ (similarly as in Section 4.2, we can make the difference to $1/2$ infinitesimally small). Then GREEDY empties buckets $2m$ and $2m + 1$, so that the remaining buckets $1, \ldots, 2m - 1$ take contents $1/2$ into the second phase.

The second phase goes through m further rounds. As in Section 4.2, we control the tie-breaking of GREEDY through infinitesimally small fluctuations in the bucket loads.

- In the first round, the Stepmother adds $1/2$ to bucket $2m$ and then adds $1/(2m)$ to the odd-numbered buckets $1, 3, 5, \ldots, 2m-5$ and to buckets $2m-2$ and $2m$. GREEDY empties $2m - 2$ and $2m - 1$.
- In the second round, the Stepmother adds $1/2$ to bucket $2m$; adds $1/(2m)$ to bucket $2m - 3$; and adds $(m - 1)/(2m^2)$ to the odd-numbered buckets $1, 3, 5, \ldots, 2m - 3$ and to bucket $2m$. GREEDY empties buckets 2 and 3.
- In the third round, the Stepmother adds $1/2$ to bucket 1, and then adds $1/(2m - 2)$ to the odd-numbered buckets $5, 7, \ldots, 2m - 3$ and to buckets 1 and $2m$. GREEDY empties buckets $2m - 4$ and $2m - 3$.
- In the fourth round, the Stepmother adds $1/(m - 2)$ to the odd-numbered buckets $5, 7, \ldots, 2m - 5$ and to buckets 1 and $2m$. GREEDY empties $2m - 6$ and $2m - 5$.

Let us summarize the situation after the fourth round. The six buckets $2m - 6, \ldots, 2m - 1$ and the three buckets $2, 3, 2m + 1$ are empty. Bucket 1 and bucket $2m$ each contain $\gamma = 1 + \frac{1}{m-2} + \frac{1}{2m-2} + \frac{1}{2m} + \frac{m-1}{2m^2}$. The $m - 5$ even-numbered buckets $4, 6, \ldots, 2m - 8$ each contain $1/2$, and the $m - 5$ odd-numbered buckets $5, 6, \ldots, 2m - 7$ each contain $\gamma - 1/2$.

- In round $k = 5, \ldots, m - 1$, the Stepmother adds $1/(m - k + 2)$ to buckets $5, 7, \ldots, 2m-2k+3$ and to buckets 1 and $2m$. GREEDY empties $2m-2k+2$ and $2m - 2k + 3$.

After round $m - 1$, all but two buckets are empty. The two non-empty buckets are 1 and $2m$, and the Stepmother adds $1/2$ to each of them, and thus increases their contents to $\gamma + \sum_{k=5}^{m} \frac{1}{m-k+2} = \mathcal{H}\langle m - 1\rangle + \frac{m^2 - 3m + 1}{2m^2(m-1)}$. GREEDY empties bucket $2m$, and bucket 1 takes its load into the next round.

6 Conclusions and Conjectures

We have analyzed the bucket number and the greedy bucket number for several families of graphs. For perfect graphs (Corollary 6), odd cycles (Theorem 9), and all graphs on $n \leq 6$ vertices (Theorem 8) the leftmost inequality in (6) is an equality. Under Conjecture 11, the same holds true for odd anti-holes. This perhaps suggests the following:

Wild Guess 14. *Every graph G satisfies* $\texttt{bucket}(G) = \mathcal{H}\langle\omega(G) - 1\rangle$.

We have shown that for every perfect graph the bucket number and the greedy bucket number coincide (Corollary 6), and we have seen that for every non-perfect hole and for every non-perfect anti-hole these two numbers are distinct (assuming the odd anti-hole Conjecture 11). One might speculate that perfect graphs are the only graphs on which the GREEDY strategy is optimal:

Wild Guess 15. *A graph G is perfect, if and only if* $\texttt{bucket}(G) = \texttt{g-bucket}(G)$.

The GREEDY strategy is primitive and simple, and performs extremely well on the graphs investigated by us. For instance on the odd cycles C_{2m+1} the difference between the greedy bucket number and the bucket number is $\frac{1}{m} \cdot 2^{-m}$, which rapidly tends to 0 as m increases. The largest gap known to us is $1/8$, and occurs for the cycle C_5.

Wild Guess 16. *The difference between* $\texttt{g-bucket}(G)$ *and* $\texttt{bucket}(G)$ *is bounded by an absolute constant (that does not depend on G).*

For getting a better understanding of these issues, we will have to analyze the bucket numbers of graphs whose clique number is far away from the chromatic number. Good candidates might be the graphs introduced by Mycielski [4]; recall that the Mycielski graph M_k (with $k \geq 3$) has $3 \cdot 2^{k-2} - 1$ vertices, is triangle-free and has chromatic number k.

References

1. Bodlaender, M., Hurkens, C.A.J., Kusters, V.J.J., Staals, F., Woeginger, G.J., Zantema, H.: Cinderella versus the wicked Stepmother. Working paper, TU Eindhoven (February 2011)
2. Chrobak, M., Csirik, J.A., Imreh, C., Noga, J., Sgall, J., Woeginger, G.J.: The Buffer Minimization Problem for Multiprocessor Scheduling with Conflicts. In: Yu, Y., Spirakis, P.G., van Leeuwen, J. (eds.) ICALP 2001. LNCS, vol. 2076, pp. 862–874. Springer, Heidelberg (2001)

3. Dietz, P.F., Sleator, D.D.: Two algorithms for maintaining order in a list. In: Proceedings of the 19th Annual ACM Symposium on Theory of Computing (STOC 1987), pp. 365–372 (1987)
4. Mycielski, J.: Sur le coloriage des graphs. Colloquium Mathematicum 3, 161–162 (1955)
5. Woeginger, G.J.: Combinatorics problem C5. In: Problem Shortlist of the 50th International Mathematical Olympiad, Bremen, Germany, pp. 33–35 (2009)

On the Complexity of Planar Covering
of Small Graphs[*]

Ondřej Bílka, Jozef Jirásek, Pavel Klavík, Martin Tancer, and Jan Volec

Department of Applied Mathematics, Faculty of Mathematics and Physics, Charles
University, Malostranské nám. 25, 118 00 Prague, Czech Republic
{ondra,klavik,tancer,janv}@kam.mff.cuni.cz,jirasekjozef@gmail.com

Abstract. The problem COVER(H) asks whether an input graph G covers a fixed graph H (i.e., whether there exists a homomorphism $G \rightarrow H$ which locally preserves the structure of the graphs). Complexity of this problem has been intensively studied. In this paper, we consider the problem PLANARCOVER(H) which restricts the input graph G to be planar.

PLANARCOVER(H) is polynomially solvable if COVER(H) belongs to P, and it is even trivially solvable if H has no planar cover. Thus the interesting cases are when H admits a planar cover, but COVER(H) is NP-complete. This also relates the problem to the long-standing Negami Conjecture which aims to describe all graphs having a planar cover. Kratochvíl asked whether there are non-trivial graphs for which COVER(H) is NP-complete but PLANARCOVER(H) belongs to P.

We examine the first nontrivial cases of graphs H for which COVER(H) is NP-complete and which admit a planar cover. We prove NP-completeness of PLANARCOVER(H) in these cases.

1   Introduction

Unless stated otherwise, we work with simple undirected finite graphs and we use standard notation from graph theory.

Graph Homomorphisms and Covers. Let G and H be graphs. A mapping $f : V(G) \rightarrow V(H)$ is a *homomorphism* from G to H if the edges of G are mapped to the edges H, i.e., for every edge $uv \in E(G)$, $f(u)f(v) \in E(H)$.

A homomorphism f is called *locally bijective* if for every $v \in V(G)$ the *closed neighborhood* $N_G[v] \subseteq V(G)$ is bijectively mapped to $N_H[f(v)] \subseteq V(G)$. Notice that $x, y \in N_G[v]$ and $f(x)f(y) \in E(H)$ may or may not imply $xy \in E(G)$. We say that G *covers* H (or G is a *cover* of H) if there exists a locally bijective homomorphism from G to H; see Figure 1. If G covers H, their local structures are somewhat similar. Note that if G covers H and H is connected, then $|V(G)|$ is a multiple of $|V(H)|$ and every vertex of H has the same number of preimages.

---

[*] The initial research was supported by DIMACS/DIMATIA REU program (grant number 0648985). The third and the fourth author were supported by Charles University as GAUK 95710. The fourth author is also affiliated with Institute for Theoretical Computer Science (supported by project 1M0545 of The Ministry of Education of the Czech Republic).

P. Kolman and J. Kratochvíl (Eds.): WG 2011, LNCS 6986, pp. 83–94, 2011.

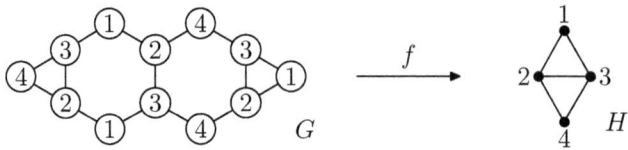

Fig. 1. An example of a cover of K_4^-. For every vertex v, its image $f(v)$ is written in the circle. Notice that for vertices mapped to 1, its neighbors 2 and 3 may or may not be adjacent.

Graph homomorphisms and covers provide a common language for various problems in graph theory. They are studied as generalized coloring. A graph is properly k-colorable if and only if it homomorphically maps to K_k. Similarly, covering is a generalization of coloring which puts additional restrictions on neighborhoods. From another point of view, G covers K_k if and only if G can be partitioned into k 1-perfect codes.

Computational Problems. For a fixed graph H, the problem $\text{HOM}(H)$ asks whether there exists a homomorphism from an input graph G to H. Hell and Nešetřil [HN90] proved a dichotomy for computational complexity: $\text{HOM}(H)$ is polynomially solvable if H is a bipartite graph, and it is NP-complete otherwise.

Similarly, the problem $\text{COVER}(H)$ asks whether an input graph G covers a fixed graph H. Study of this problem was pioneered by Bodlaender [Bod89]. First results depending on the graph H were proved by Abello et al. [AFS91]. Kratochvíl [Kra94] showed that $\text{COVER}(K_4)$ is NP-complete. Afterwards, Kratochvíl, Proskurowski and Telle [KPT97] and Fiala [Fia00] proved that $\text{COVER}(H)$ is NP-complete for every k-regular graph H with $k \geq 3$. Later, a dichotomy for all simple graphs with up to six vertices was proved, [KPT98].

We can restrict the input graph G and ask whether it changes $\text{COVER}(H)$ to be polynomially solvable. In this paper, we restrict G to be planar. We consider the following problem:

Problem: $\text{PLANARCOVER}(H)$.
Input: A planar graph G.
Output: Yes if G covers H, no otherwise.

Every $\text{PLANARCOVER}(H)$ problem trivially lies in NP. In the rest of the paper, we only question NP-hardness. Note that if $\text{COVER}(H)$ is polynomially solvable, then $\text{PLANARCOVER}(H)$ is also polynomially solvable.

Many NP-complete problems remain hard for planar inputs. Originally, the graph covering problems looked similar to problems such as Not-All-Equal Satisfiability or 3-Edge-Colorability of Cubic graphs. Both of them are polynomially solvable for planar graphs, the first was proved by Moret [Mor88], the latter is trivial to decide since Tait has showed that the 3-Edge-Colorability of Cubic planar graphs is equivalent to the Four Color Theorem. Indeed the NP-hardness reduction for $\text{COVER}(K_4)$ presented in [Kra94] is from 3-Edge-Colorability of Cubic graphs. This has led Kratochvíl to pose the problem of the complexity

of PLANARCOVER(K_4) at several occasions, including 6th Czech-Slovak Symposium on Graph Theory 2006, IWOCA 2007 and ATCAGC 2009. In this paper,
we prove that PLANARCOVER(K_4) is NP-complete.

Negami Conjecture. As a motivation to study the PLANARCOVER problems,
we describe a relation to a long-standing conjecture of Negami [Neg88]. For a
graph H, we can ask whether there exists a planar graph G that covers H. If the
answer is no, then the problem PLANARCOVER(H) is trivial—we output "no"
regardless of the input. Negami conjectured the following:

Conjecture 1. The connected graphs which admit a planar cover are exactly the
connected graphs embeddable in the projective plane.

If a connected graph is embeddable in the projective plane, it is straightforward
to construct one of its planar covers. The other implication is still open. The most
recent results can be found in Hliněný and Thomas [HT04]. For example, K_7
has no (finite) planar cover since there is no six regular planar graph. Therefore,
the problem PLANARCOVER(K_7) is polynomially solvable but COVER(K_7) is
NP-complete; see [KPT97].

It is natural to ask whether the restriction to planarity makes the problem
easier in a non-trivial case. Kratochvíl asked the following question in his talk
at Prague Midsummer Combinatorial Workshop 2009:

Question 1. Is it true that PLANARCOVER(H) is NP-complete if and only if
COVER(H) is NP-complete and the graph H admits a planar cover?

Our Results. In this paper, we show that PLANARCOVER(H) is NP-complete
for several small graphs H.

Theorem 1. *The problem* PLANARCOVER(K_6) *is* NP-*complete.*

The graph K_6 is a somewhat extremal case for the Negami Conjecture. If a
planar graph G covers K_6, it has to be 5-regular. The structure of 5-regular
planar graphs is very limited, but we show that the problem is still NP-complete.

Theorem 2. *The problems* PLANARCOVER(K_4) *and* PLANARCOVER(K_5) *are*
NP-*complete.*

Covering of these regular graphs is related to coloring squares of graphs. For
example, a cubic planar graph G covers K_4 if and only if its square G^2 is 4-
colorable. Coloring the squares of graphs (especially planar) is widely studied
as a special case of the channel assignment problem; see [RL93]. Dvořák et
al. [DŠT08, Theorem 25] prove that deciding whether the square of a given
subcubic planar graph is 4-colorable is NP-complete. Theorem 2 strengthens
this result.

We denote K_4 with a leaf attached to a vertex by K_4^+ and K_5 without an
edge by K_5^-; see Figure 6.

Theorem 3. *The problems* PLANARCOVER(K_4^+) *and* PLANARCOVER(K_5^-) *are* NP-*complete.*

Theorems 1, 2 and 3 together give an affirmative answer to Question 1 for all graphs with up to five vertices except for W_4; see details in Conclusions.

We also examine the smallest non-trivial multigraph case. The dumbbell graph D is a multigraph with two adjacent vertices with a loop on each vertex; see Figure 7 on the right. This graph is the smallest multigraph for which the problem COVER is NP-complete.

Theorem 4. *The problem* PLANARCOVER(D) *is* NP-*complete.*

This result strengthens a result of Janczewski et al. [JKM09, Proposition 5] which proves hardness for partial PLANARCOVER(D). By partial covers we mean locally *injective* homomorphisms. As described in Section 4, if G covers D, it has to be a cubic planar graph. For a partial cover of D, it has to be a subcubic planar graph. But if the input graph is cubic, then every partial cover of D is also a cover of D.[1] On the other hand, reductions for partial covers cannot be easily extended to covers while preserving planarity.

2   Hardness of Planar Covering of K_6

In this section, we prove Theorem 1: PLANARCOVER(K_6) is NP-complete. First, we describe a problem we reduce from.

An *intersection representation* of a graph is an assignment of sets to the vertices in such a way that two vertices are adjacent if and only if the corresponding sets intersect. A graph is called a *segment graph* if it has an intersection representations where the sets are segments in the plane. We consider only segment representations with all endpoints distinct and with no three segments crossing in one point.

> **Problem:** k-SEGMENTCOLORING
>   **Input:** A segment representation of a graph G.
> **Output:** Yes if G is k-colorable, no otherwise.

Ehrlich et al. [EET76] proved that k-SEGMENTCOLORING is NP-complete for $k \geq 3$. We note that there exist segment graphs which have every representation exponentially large in the number of vertices; see [KM94]. However this representation is a part of the input hence it does not pose a problem; see [EET76].

Overview of the Reduction. We reduce PLANARCOVER(K_6) from 6-SEGMENTCOLORING. For a graph G with a segment representation, we construct a plane graph G' which covers K_6 if and only if G is 6-colorable.

The reduction is sketched in Figure 2. Consider an arrangement of segments. Every segment is split by crossings into several *subsegments* which contain no crossings. We construct a graph G' with the same topology as the segment representation of G. Every subsegment is represented by two parallel edges.

---

[1] We note that even in general partial PLANARCOVER(H) problem is at least as hard as PLANARCOVER(H).

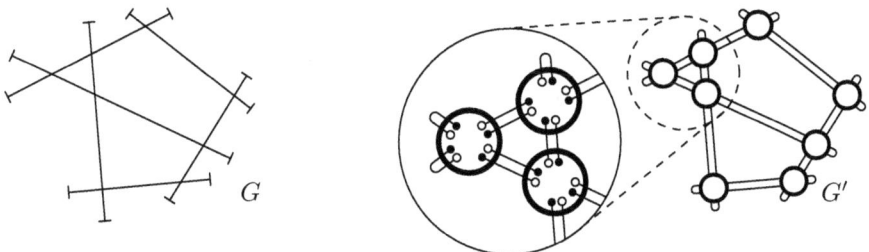

Fig. 2. We construct a planar graph G' having the same topology as the arrangement of the segments

We replace every crossing by a *crossing gadget*. Every crossing gadget has four pairs of outer vertices. These vertices are incident with the edges representing subsegments; see the detail in Figure 2. In other words, two crossing gadgets are connected by a pair of parallel edges if the crossings they represent lie on the same segment and there is no other crossing between them. A last subsegment of a segment is represented by one edge connecting both outer vertices of a crossing gadget. The obtained planar graph has the same topology as the arrangement of the segments.

Relation to Coloring. Every subsegment is represented by a pair of parallel edges. Mapping of vertices of these edges to K_6 gives a coloring of this subsegment in a way depicted in Figure 3a. On the other hand, crossing gadgets ensure that every covering of K_6 satisfy these properties. The vertices depicted in black are called *color vertices* and the vertices depicted in white are called *non-color vertices*.

Crossing Gadget. Every crossing gadget has four adjacent subsegments. Their color and non-color vertices alternate; see Figure 3b. Using the topology of the segment arrangement, the crossing gadgets can be connected in this way so that G' is planar.

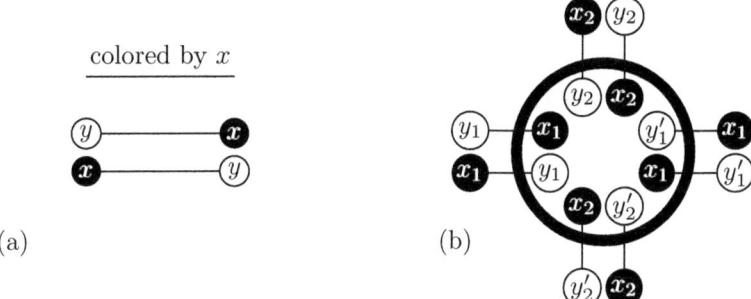

Fig. 3. (a) A subsegment colored by x is represented by two parallel edges, with two *color vertices* mapped to x (depicted in black) and the other two vertices mapped to an arbitrary y (depicted in white) where $y \neq x$. (b) A crossing gadget transfers color information between the opposite subsegments.

This gadget ensures three properties:

(1) *The subsegments are mapped in the way described in Figure 3a.*
(2) *The subsegments belonging to the same segment are colored by the same color.* Therefore, for every segment, all its subsegments are colored by the same color and the color of this segment is well-defined. The crossing gadget gives no additional restrictions on non-color vertices.
(3) *Every two intersecting segments are colored by different colors.* It is possible to map the crossing gadget only if $x_1 \neq x_2$.

The crossing gadget is built from several basic blocks, called *auxiliary gadgets*. The auxiliary gadget is a graph shown in Figure 4.

Lemma 1. *The auxiliary gadget can be mapped to K_6 in a unique way up to a permutation of the vertices of K_6.*

Proof. Observe that if we fix a mapping for any vertex and its neighbors, the rest of the mapping is uniquely determined.   □

In every covering f, the six outer vertices $u_1, \ldots, u_6$ of the auxiliary gadget are mapped to three distinct vertices of K_6 with $f(u_i) = f(u_{i+3})$, $i \in \{1, 2, 3\}$. The parallel edges adjacent to the auxiliary gadget are mapped in the way described in Figure 3a.

The crossing gadget consists of eight auxiliary gadgets; see Figure 5. We need to prove that the crossing gadget is correct.

Lemma 2. *The crossing gadget can be mapped to K_6 if and only if the properties (1) to (3) are satisfied.*

Proof. Let G' cover K_6. Consider one crossing gadget. Since all edges representing subsegments are connected to auxiliary gadgets, according to Lemma 1 these edges are mapped correctly as in Figure 3a. Colors are transfered between the opposite subsegments, as depicted in Figure 5. The central auxiliary gadgets force x_1 and x_2 to be distinct. Therefore, we know that every mapping satisfies properties (1) to (3).

We need to show that if the vertices of the edges adjacent to the crossing gadget are mapped correctly, we can extend this mapping to the rest of the gadget. By a straightforward case analysis we can see that an arbitrary correct mapping of non-color vertices can be extended.   □

We conclude this section with a proof of the main theorem:

Proof (Theorem 1). Let G' cover K_6. By Lemma 2, the mapping of every crossing gadget satisfies the properties (1) to (3). Using the properties (1) and (2), we can infer colors of the segments. By the property (3), this coloring is proper.

On the other hand, given a proper coloring, we map the color vertices according to this coloring. By Lemma 2, it is possible to extend the mapping to the entire graph G'.   □

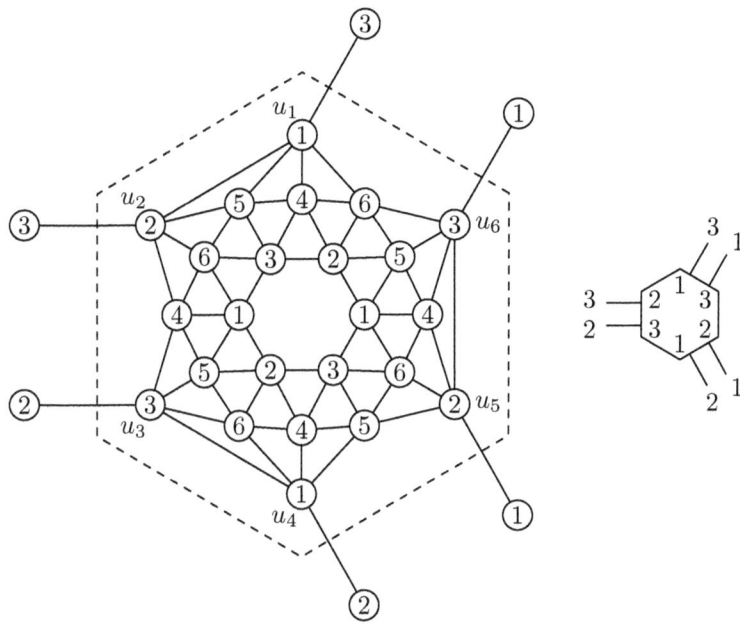

Fig. 4. The auxiliary gadget on the left, denoted by a hexagon on the right

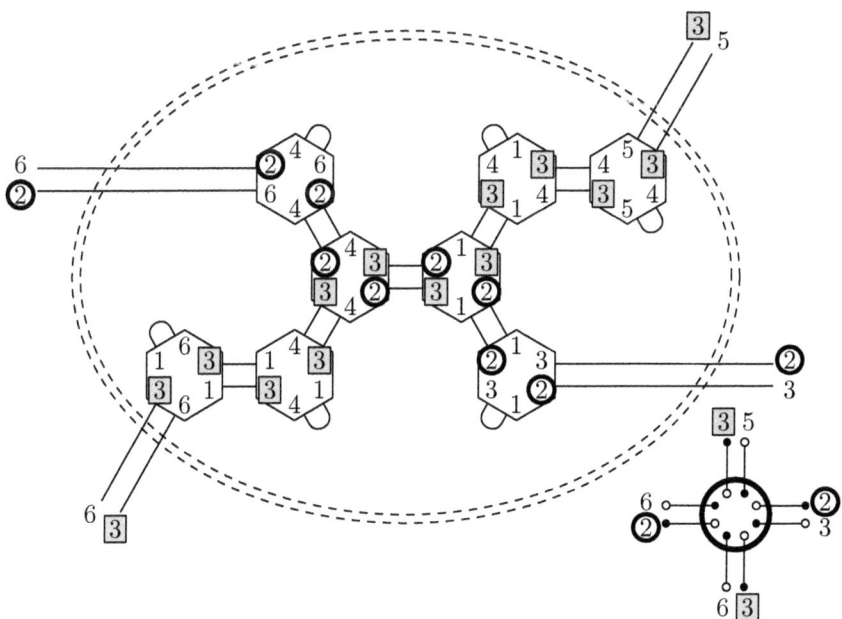

Fig. 5. The crossing gadget on the left, denoted by a circle on the right

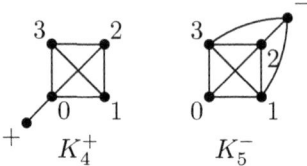

Fig. 6. Graphs K_4^+ and K_5^- with labeled vertices

3 Hardness of Planar Covering of K_4, K_5, K_4^+ and K_5^-

In this section, we sketch the proof of hardness of PLANARCOVER of K_4, K_5, K_4^+ and K_5^-. The reductions slightly modify the reduction described in Section 2.

In the case of K_4 and K_5, we just change the auxiliary gadget and reduce these problems from 4-SEGMENTCOLORING, resp. 5-SEGMENTCOLORING.

In the case of K_4^+ and K_5^-, we change both the auxiliary gadget and the crossing gadget and reduce these problems from 3-SEGMENTCOLORING. The color vertices are mapped to 1, 2 and 3, the non-color vertices are mapped to 0 (or $-$ in the case of K_5^-); see Figure 6.

Due to space limitations, details of these reductions are described in Appendix of pre-print version http://arxiv.org/abs/1108.0064.

4 Hardness of Planar Covering of the Dumbbell Graph

In this section, we prove hardness of PLANARCOVER(D) where D is the dumbbell graph (see Figure 7 on the right). It is a multigraph and the notion of covering can be extended to multigraphs, see [Kra94]. For the purpose of this paper, we need only the following: G is a planar cover of D if G is a cubic planar graph and can be colored by two colors (black and white) in such a way that every black vertex has two black neighbors and one white neighbor and every white vertex has two white neighbors and one black neighbor; see Figure 7. In the rest of the section, we use this coloring interpretation.

To prove the hardness of PLANARCOVER(D), we first describe the problem we reduce from. 2-IN-4-MONOTONEPLANARSAT is a satisfiability problem where:

- all clauses contain exactly four variables,
- the incidence graph of clauses and variables is planar, and
- all variables are in the positive form, i.e. there is no negation.

A clause is satisfied if exactly two variables are true. The entire formula is satisfied if all clauses are satisfied. For an example, see Figure 8a. Kára, Kratochvíl and Wood [KKW07] proved that this problem is still NP-complete.

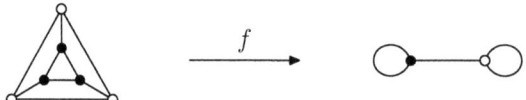

Fig. 7. An example of a cover of the dumbbell graph W

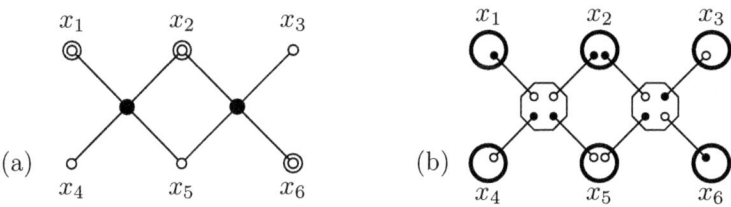

Fig. 8. (a) A graph G representation of the formula: $(x_1, x_2, x_4, x_5) \land (x_2, x_3, x_5, x_6)$. This formula can be satisfied by an assignment $x_1 = x_2 = x_6 = 1$ and $x_3 = x_4 = x_5 = 0$. (b) The constructed graph G' for this formula.

Overview of Reduction. Let G be a planar incidence graph of variables and clauses. We construct a graph G' such that G' covers D if and only if the formula is satisfiable. We replace every variable with a *variable gadget* and every clause with a *clause gadget*. If a variable is in a clause, we connect the variable gadget and the clause gadget by an edge. The variable gadgets and the clause gadgets are connected in the way that the overall topology of G is preserved in G'; see Figure 8b.

The variable gadget can be colored in two ways which encodes the assignment of the variable. Every clause gadget can be colored if and only if two of its variables gadget are true and the other two are false.

Variable Gadget. For a variable that appears in the formula k times, the variable gadget contains the cycle C_{4k}. Every fourth vertex of the cycle is connected to a clause gadget. The remaining vertices are connected to triangles; see Figure 9.

Lemma 3. *For every coloring of the variable gadget, $u_1, \ldots, u_k$ are colored by one color and $v_1, \ldots, v_k$ by the other one.*

Proof. Every triangle in the graph has to be monochromatic. The triangles force the cycle to be monochromatic as well. □

The color of $v_1, \ldots, v_k$ represents the value assigned to the variable. If the inner cycle is colored black, the variable is assigned true, otherwise the variable is false.

Clause Gadget. We start with *basic blocks* described in Figure 10. The *auxiliary gadget* consists of eight basic blocks; see Figure 11. The reader is encouraged to prove that there exists no other colorings of these gadgets.

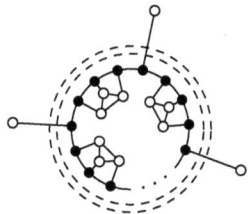

Fig. 9. The variable gadget with a unique coloring—up to swapping of the colors

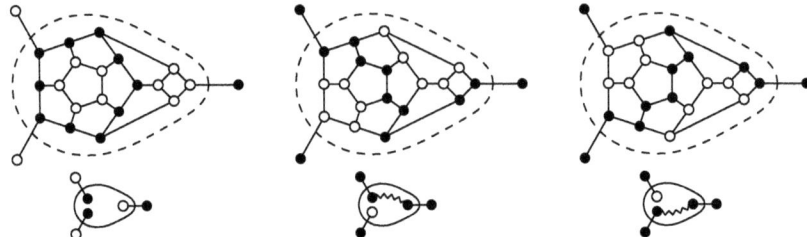

Fig. 10. The basic block has three different colorings—up to swapping of the colors

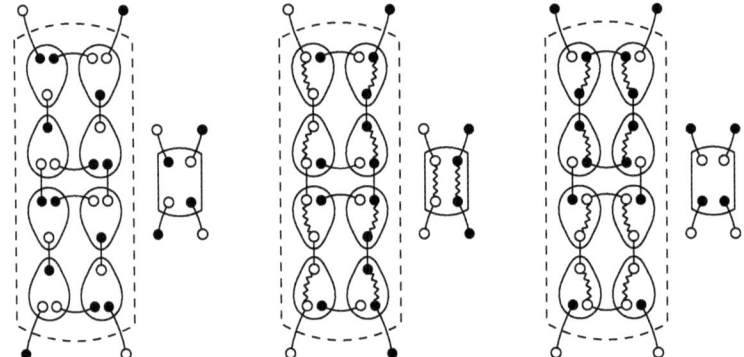

Fig. 11. The auxiliary gadget with all three colorings—up to swapping of the colors. Note that every coloring has two outer vertices black and the other two white.

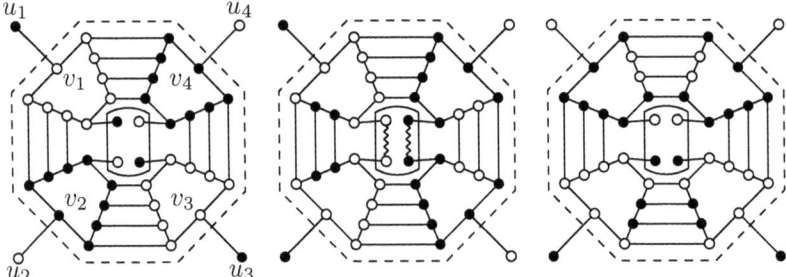

Fig. 12. The clause gadget with all three colorings—up to swapping of the colors

The clause gadget, described in Figure 12, contains an auxiliary gadget. Every clause gadget is connected by edges to the corresponding variable gadgets.

Lemma 4. *Let the vertices u_i and v_i have distinct colors for every $i \in \{1, 2, 3, 4\}$. The crossing gadget can be covered if and only if exactly two of v_i's are colored black and the other two are colored white.*

Proof. Observe that the coloring is forced by colors of u_i and v_i. The rest is ensured by the auxiliary gadget; see description in Figure 11.   □

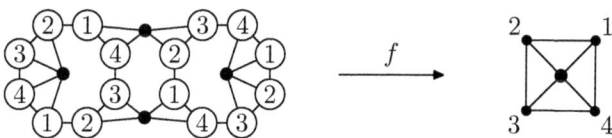

Fig. 13. An example planar cover of W_4. If G covers W_4, it has to consist of cycles of length divisible by four connected by black vertices of degree four. The cycles can be labeled in the cyclic order 1, 2, 3 and 4 (eight possible labellings for each cycle) in such a way that every vertex of degree four is adjacent to one vertex of each label.

Proof (Theorem 4). We need to show that there exists a correct assignment of the variables if and only if G' covers D. Let G' cover D. According to Lemma 3, every variable gadget has to be colored in one of two ways, one representing the true assignment and the other one the false assignment; see Figure 9. By Lemma 4, the clause gadget can be colored if and only if exactly two variables in the clause are true and the other two are false. Therefore, the colorings gives an assignment of the variables which satisfies the formula.

On the other hand, if there exists a correct assignment, we cover the variable gadgets according to it. Since every clause has two variables assigned true and the other two assigned false, the corresponding clause gadgets can be covered according to Lemma 4. We obtain a correct colorings of G'.

The reduction is clearly polynomial, which concludes the proof. □

5   Conclusions

In this paper, we prove hardness of PLANARCOVER(H) for several small graphs H. Our techniques can be generalized to prove hardness of other graphs. For example, if we replace the leaf in K_4^+ with any planar graph, the resulting PLANARCOVER problem is still NP-complete.

Our results give a positive answer to Question 1 for all graphs with at most five vertices except for W_4, the wheel graph with four outer vertices (COVER(W_4) is NP-complete; see [KPT98]). For an example, see Figure 13. Since the symmetries of W_4 are different from the symmetries of other graphs solved in this paper, a reduction would require a new technique. We note that we were able to prove hardness of partial PLANARCOVER(W_4).

Acknowledgment. We would like to thank Jan Kratochvíl for introducing us to the problem and kindly answering our questions. We would also like to thank Aaron D. Jaggard, Pavel Paták and Zuzana Safernová for fruitful discussions.

References

[AFS91]   Abello, J., Fellows, M.R., Stillwell, J.C.: On the complexity and combinatorics of covering finite complexes. Australian Journal of Combinatorics 4, 103–112 (1991)

[Bod89]   Bodlaender, H.L.: The classification of coverings of processor networks. Jour-
          nal of Parallel and Distributed Computing 6(1), 166–182 (1989)
[DŠT08]   Dvořák, Z., Škrekovski, R., Tancer, M.: List-coloring squares of sparse sub-
          cubic graphs. SIAM J. Discrete Math. 22(1), 139–159 (2008)
[EET76]   Ehrlich, G., Even, S., Tarjan, R.E.: Intersection graphs of curves in the
          plane. Journal of Combinatorial Theory, Series B 21(1), 8–20 (1976)
[Fia00]   Fiala, J.: Note on the computational complexity of covering regular graphs.
          In: 9th Annual Conference of Doctoral Students, WDS 2000, pp. 89–90.
          Matfyzpress (2000)
[HN90]   Hell, P., Nešetřil, J.: On the complexity of H-coloring. J. Combin. Theory
          Ser. B 48(1), 92–110 (1990)
[HT04]   Hliněný, P., Thomas, R.: On possible counterexamples to Negami's planar
          cover conjecture. J. Graph Theory 46(3), 183–206 (2004)
[JKM09]   Janczewski, R., Kosowski, A., Malafiejski, M.: The complexity of the l(p,q)-
          labeling problem for bipartite planar graphs of small degree. Discrete Math-
          ematics 309(10), 3270–3279 (2009)
[KKW07]   Kára, J., Kratochvíl, J., Wood, D.R.: On the complexity of the balanced
          vertex ordering problem. Discrete Mathematics & Theoretical Computer
          Science 9(1), 193–202 (2007)
[KM94]   Kratochvíl, J., Matoušek, J.: Intersection graphs of segments. Journal of
          Combinatorial Theory, Series B 62(2), 289–315 (1994)
[KPT97]   Kratochvíl, J., Proskurowski, A., Telle, J.A.: Covering regular graphs. J.
          Comb. Theory Ser. B 71(1), 1–16 (1997)
[KPT98]   Kratochvíl, J., Proskurowski, A., Telle, J.A.: Complexity of graph covering
          problems. Nordic J. of Computing 5(3), 173–195 (1998)
[Kra94]   Kratochvíl, J.: Regular codes in regular graphs are difficult. Discrete
          Math. 133(1-3), 191–205 (1994)
[Mor88]   Moret, B.M.E.: Planar NAE3SAT is in P. SIGACT News 19, 51–54 (1988)
[Neg88]   Negami, S.: The spherical genus and virtually planar graphs. Discrete
          Math. 70(2), 159–168 (1988)
[RL93]   Ramanathan, S., Lloyd, E.: Scheduling algorithms for multihop radio net-
          works. IEEE/ACM Transactions on Networking 1(2), 166–177 (1993)

Approximability of Economic Equilibrium for Housing Markets with Duplicate Houses

Katarína Cechlárová[1,*] and Eva Jelínková[2,**]

[1] Institute of Mathematics,
Faculty of Science, P. J. Šafárik University,
Jesenná 5, 040 01 Košice, Slovakia
[2] Department of Applied Mathematics
Faculty of Mathematics and Physics, Charles University
Malostranské nám. 25, 118 00 Praha 1, Czech Republic
katarina.cechlarova@upjs.sk, eva@kam.mff.cuni.cz

Abstract. In a modification of the classical model of housing market which includes duplicate houses, economic equilibrium might not exist. As a measure of approximation the value $sat(\mathcal{M})$ was proposed: the maximum number of satisfied agents in the market $\mathcal{M}$, where an agent is said to be satisfied if, given a set of prices, he gets a most preferred house in his budget set. Clearly, market $\mathcal{M}$ admits an economic equilibrium if $sat(\mathcal{M})$ is equal to the total number n of agents, but $sat(\mathcal{M})$ is NP-hard to compute.

In this paper we give a 2-approximation algorithm for $sat(\mathcal{M})$ in the case of trichotomic preferences. On the other hand, we prove that $sat(\mathcal{M})$ is hard to approximate within a factor smaller than 21/19, even if each house type is used for at most two houses. If the preferences are not required to be trichotomic, the problem is hard to approximate within a factor smaller than 1.2. We also prove that, provided the Unique Games Conjecture is true, approximation is hard within a factor 1.25 for trichotomic preferences, and within a factor 1.5 in the case of general preferences.

1 Introduction

A housing market consists of a finite set of agents and a finite set of houses. Each agent owns one house, considers some other houses acceptable and orders them according to their desirability. The aim of each agent is to get the house he finds to be the best possible. This model was introduced by Shapley and Scarf in [13], where the notion of the economic equilibrium in such markets was considered and its existence in all housing markets proved. A polynomial-time algorithm for finding an economic equilibrium, called the Top Trading Cycles (TTC for

---

* Supported by VEGA grants 1/0035/09 and 1/0325/10.
** Supported by project 1M0021620838 of the Czech Ministry of Education. Part of work was done while visiting ICE-TCS, Reykjavík University, Iceland.

short) algorithm was attributed to Gale (see [13]). An asymptotically optimal implementation of the TTC algorithm was proposed in [2].

The above results rely substantially on the assumption that each agent's house is unique. A modified version of the basic model, in which some houses are of the same type (and so must have equal price), was proposed by Fekete et al. [9]. In this model it may happen that the economic equilibrium does not exist. Fekete et al. even proved that the problem to decide its existence is NP-complete.

Cechlárová and Fleiner [4] further narrowed the dividing line between easy and difficult cases. They proved that if agents have strict preferences over house types, a polynomial-time algorithm decides the existence of an economic equilibrium. An efficient optimal implementation of their algorithm was proposed in [5]. On the other hand, Cechlárová and Fleiner showed that the problem remains NP-complete even if each agent distinguishes only three classes of house types: desired houses, houses of the same type as his original house and unacceptable houses. They call such preferences *trichotomic*.

In general markets with indivisible goods (i. e., each agent may own several units of each good) it has been known for many years that equilibrium might not exist. A recent result of Deng, Papadimitriou and Safra states that even in the case with linear utility functions the problem to decide its existence is NP-complete [7]. These authors studied the so-called ε-approximate equilibrium, i. e., such that the market clears approximately (at most ε units of each good remain unsold) and each agent obtains a commodity bundle such that its utility is within a factor of $(1 - \varepsilon)$ from the optimum in his budget set.

Cechlárová and Fleiner [4] proposed a different notion of approximate equilibrium for housing markets with duplicate houses. They studied the deficiency of housing markets, i. e., the minimum number of agents that cannot get a most preferred house in their budget set.

Cechlárová and Schlotter [6] examined deficiency from the parameterized complexity viewpoint. They proved that the deficiency problem is NP-hard even in the case when each agent prefers only one house type to his own, and the maximum number of houses of the same type is two. They further proved this problem to be W[1]-hard with the parameter α describing the desired value of deficiency, and fixed-parameter tractable when parameterized by the number of distinct house types.

In this paper, we focus on the approximability of the equilibrium in housing markets with duplicate houses when preferences contain ties, and in particular, in the case with trichotomic preferences. Technically, instead of minimizing the number of unsatisfied agents, we shall maximize $\mathrm{sat}(\mathcal{M})$, i. e., the number of satisfied agents in a housing market $\mathcal{M}$. In Section 3 we study bounds for $\mathrm{sat}(\mathcal{M})$ in trichotomic markets and present a 2-approximation algorithm. In Section 4 we prove that the problem is NP-hard to approximate within a factor smaller than $21/19$, even if the maximum number of houses of the same type is two, and agents endowed with a house of the same type have the same preference lists. We further prove that if the preferences are not required to be trichotomic, then $\mathrm{sat}(\mathcal{M})$ is NP-hard to approximate within a factor smaller than 1.2.

Assuming that the Unique Games Conjecture of Khot [11] is true, we obtain stronger innaproximability results, also in the case of trichotomic preference lists whose lengths are bounded by a constant.

2 Preliminaries

Let A be the set of n *agents*, H the set of m *house types*. The *endowment function* $\omega : A \to H$ assigns to each agent the type of house he originally owns. We shall denote by $A(h)$ the set of agents endowed with a house of type $h \in H$.

Each agent has preferences over house types in H in the form of a linearly ordered list $P(a)$, possibly with ties. Notation $h \succeq_a k$ means that agent a prefers houses of type h to houses of type k. The set of house types appearing in $P(a)$ is denoted by $H(a)$, and house types in $H(a)$ are said to be *acceptable* for a. We assume that $\omega(a)$ belongs to the least preferred acceptable house types for each agent. The remaining house types are called *unacceptable*.

In the special case of *trichotomic preferences*, each agent distinguishes only three kinds of house types: house types more preferred than the type of his own house, these are called *desired*; houses of the same type as his own house and *unacceptable* house types.

The n-tuple $\mathcal{P}$ of all preference lists is called the *preference profile*. The quadruple $\mathcal{M} = (A, H, \omega, \mathcal{P})$ is called a *housing market*. A market $\mathcal{M}' = (A', H', \omega', \mathcal{P}')$ is a *submarket* of a market $\mathcal{M} = (A, H, \omega, \mathcal{P})$ if $A' \subseteq A$, $\omega(A') \subseteq H' \subseteq H$, $H'(a) = H(a) \cap H'$ and the endowment function ω' and preference profile $\mathcal{P}'$ are restrictions of ω and $\mathcal{P}$ to A'.

In a housing market $\mathcal{M}$, we want to assign prices to house types and design trading consistent with prices so that each agent ends up with exactly one acceptable house. More formally, we say that a triple $\mathcal{T} = (x, \pi, p)$ is a *solution for* $\mathcal{M}$ if

(i) $x : A \to H$ is a function and $\pi : A \to A$ is a bijection such that $x(a) = \omega(\pi(a))$ and $x(a) \in H(a)$ for each $a \in A$,

(ii) $p : H \to \mathbb{R}$ is a *price function* such that $p(x(a)) \leq p(\omega(a))$ for each $a \in A$.

Condition (ii) ensures that each agent can afford a house of type $x(a)$. Function x and bijection π define who gets whose house. Moreover, they partition A into cycles of the form $(a_0, \ldots, a_{l-1})$, where $x(a_i) = \omega(a_{i+1})$ for all $i = 0, \ldots, l-1$ (modulo l), called *trading cycles*. We say that an agent a is *trading* in solution $\mathcal{T}$ if $a \neq \pi(a)$.

Given a price function $p : H \to \mathbb{R}^+$, the *budget set* of agent a with respect to p is the set of house types that a can afford, i. e., $\{h \in H : p(h) \leq p(\omega(a))\}$. An agent a is *satisfied* in a solution $\mathcal{T}$ of $\mathcal{M}$ if $x(a)$ is among the most preferred house types in the budget set of a. (Notice that in the case of trichotomic preferences, an agent a is satisfied if and only if a is either trading or $p(h) > p(\omega(a))$ for each $h \in H(a)$ such that $h \neq \omega(a)$; in other words, of all acceptable houses a can only afford houses of the same type as his endowment). Otherwise, we say that a is *dissatisfied*. By $\mathrm{Sat}(\mathcal{T})$ and $\mathrm{Dissat}(\mathcal{T})$ we denote the sets of satisfied and dissatisfied agents in $\mathcal{T}$, respectively, and by sat($\mathcal{T}$) and dissat($\mathcal{T}$) the cardinalities of these sets. The minimum of values dissat($\mathcal{T}$) over all solutions for a market $\mathcal{M}$ is called the *deficiency* of $\mathcal{M}$ [4]. As a dual notion, the maximum of sat($\mathcal{T}$) over all solutions $\mathcal{T}$ is denoted by sat($\mathcal{M}$); a solution achieving this value is called *optimal*. A solution $\mathcal{T}$ is called an *economic equilibrium for $\mathcal{M}$* if all agents are satisfied in $\mathcal{T}$.

A simple observation follows directly from the definitions (see also [4,9]).

Lemma 1. *If $\mathcal{T} = (x, \pi, p)$ is a solution for a market $\mathcal{M}$ then $p(x(a)) = p(\omega(a))$ for each agent $a \in A$.*

We study the following problems.

MAX-SHDTIES (Maximum Satisfied Housing with Duplicate houses and Ties): Given a market $\mathcal{M}$, find a solution $\mathcal{T}$ that maximizes sat($\mathcal{T}$).

MAX-SHDTRI (Maximum Satisfied Housing with Duplicate houses and Trichotomic preferences): Given a market $\mathcal{M}$ with trichotomic preferences, find a solution $\mathcal{T}$ that maximizes sat($\mathcal{T}$).

For our purpose it is very convenient to view a housing market $\mathcal{M}$ as a digraph $G_{\mathcal{M}} = (A, E)$. Vertices of $G_{\mathcal{M}}$ correspond to agents and each vertex is colored according to the type of house this agent is endowed with—hence the color classes correspond to the sets $A(h)$. An arc $ab \in E$ means that $\omega(b) \succ_a \omega(a)$. In case of a general housing market $\mathcal{M}$, the arcs may be labelled with numbers that express the preference order.

When there is no danger of confusion, we identify the digraph $G_{\mathcal{M}}$ with the market $\mathcal{M}$, and the vertices of $G_{\mathcal{M}}$ with agents of $\mathcal{M}$. We then speak of *in-neighbors* and *out-neighbors* of agents, of *directed cycles in $\mathcal{M}$*, and we say that $\mathcal{M}$ is *acyclic* if $G_{\mathcal{M}}$ does not contain any directed cycle (note that in $G_{\mathcal{M}}$ there are no loops and arcs between agents of the same type). Finally, as it will be clear from the context, we usually say simply *cycle* also for a directed cycle.

Recall that a set S of vertices of an undirected graph is called a *vertex cover* (VC for short) if each edge has at least one endpoint in S. A set of vertices in a directed graph is a *feedback vertex set* (FVS for short) if its removal leaves an acyclic graph. Both problems MIN-VC and MIN-FVS, asking for finding the size of a minimum vertex cover VC(G) and the size of a minimum feedback vertex set FVS(G), respectively, in a given (di)graph G are well-known NP-hard problems.

3   Bounds for sat($\mathcal{M}$)

In this section we deal with markets that exhibit special structure. For example, we show that each acyclic market admits an equilibrium and that sat($\mathcal{M}$) can be computed easily in markets with only two house types . Then we prove that in each trichotomic market at least half of the agents can always be satisfied. The proof of this theorem also provides a simple 2-approximation algorithm for sat($\mathcal{M}$). Finally, we show that the approximation guarantee 2 of this algorithm is tight.

Lemma 2. *Any acyclic market $\mathcal{M}$ admits an economic equilibrium, i. e., for any acyclic market $\mathcal{M}$ we have* $\mathrm{sat}(\mathcal{M}) = n$. *Moreover, an optimal solution can be found in time* $O(L)$, *where* $L = \sum_{a \in A} |H(a)|$.

Proof. Let us denote by $G_{\mathcal{M}}^*$ the digraph whose vertices correspond to house types and a pair hk is an arc if and only if there exists an agent a with $\omega(a) = h$ such that $k \succ_a \omega(a)$.

If $G_{\mathcal{M}}$ is acyclic then $G_{\mathcal{M}}^*$ is acyclic too, and the argument is as follows: Suppose that $(h_0, h_1, \ldots, h_{k-1})$ is a cycle in $G_{\mathcal{M}}^*$. This means that there are agents $a_0, a_1, \ldots, a_{k-1}$ such that $\omega(a_i) = h_i$ and a_i desires h_{i+1} (modulo k). It is easy to see that $(a_0, a_1, \ldots, a_{k-1})$ is a cycle in $G_{\mathcal{M}}$. Now assign prices to house types according to any topological ordering of vertices in $G_{\mathcal{M}}^*$ so that $hk \in E(G_{\mathcal{M}}^*)$ implies $p(h) < p(k)$. Such prices enable no trading, but all agents are satisfied.

Finally, the digraph $G_{\mathcal{M}}^*$ can easily be constructed from preference lists of agents in time $O(L)$, the number of its arcs is $O(L)$ too, so the topological ordering of its vertices can also be performed in time $O(L)$.

The main significance of the above lemma is in the possibility to extend a solution obtained for an acyclic submarket to a solution of the whole market while preserving the number of satisfied agents.

The proof of the following basic lemma is omitted due to space limits.

Lemma 3. *Let $\mathcal{M}'$ be a submarket of $\mathcal{M}$. Every solution $\mathcal{T}'$ of $\mathcal{M}'$ can be extended to a solution $\mathcal{T}$ of $\mathcal{M}$ such that* $\mathrm{sat}(\mathcal{T}) \geq \mathrm{sat}(\mathcal{T}')$. *Hence,* $\mathrm{sat}(\mathcal{M}) \geq \mathrm{sat}(\mathcal{M}')$.

The following lemma is an immediate corollary of Lemma 2 and Lemma 3.

Lemma 4. *Let $\mathcal{M}$ be a market. If a set $F \subseteq A$ is a FVS in $G_{\mathcal{M}}$, then there exists a solution $\mathcal{T}$ for $\mathcal{M}$ such that $A \setminus F \subseteq \mathrm{Sat}(\mathcal{T})$. Hence,* $\mathrm{sat}(\mathcal{M}) \geq |A \setminus F|$.

The following theorem provides a lower bound for the number of satisfied agents in each trichotomic market.

Theorem 1. *In each trichotomic market $\mathcal{M}$ with n agents at least $n/2$ agents can be satisfied.*

Proof. Let $\mathcal{C}$ be any maximum cycle packing of $G_{\mathcal{M}}$, i. e., a set of vertex-disjoint directed cycles that contains the maximum possible number of vertices. Let us denote by $A_{\mathcal{C}}$ the set of agents contained in $\mathcal{C}$.

If $|A_{\mathcal{C}}| \geq n/2$, a possible solution is as follows: all house types in $\mathcal{M}$ receive the same price, and the cycles of $\mathcal{C}$ are trading cycles. Then all trading agents are satisfied, hence $\mathrm{sat}(\mathcal{T}) \geq n/2$.

Now consider the case $|A_{\mathcal{C}}| < n/2$. As $\mathcal{C}$ is maximal, $A_{\mathcal{C}}$ is a FVS of $\mathcal{M}$ and so the submarket generated by $A \setminus A_{\mathcal{C}}$ is acyclic. Using Lemma 4 we obtain a solution $\mathcal{T}$ satisfying all the agents in $A \setminus A_{\mathcal{C}}$, i. e., $\mathrm{sat}(\mathcal{T}) \geq n/2$.

Notice that there is a folklore polynomial algorithm for finding a maximum cycle packing in a digraph, see e. g. [1]. So the proof of Theorem 1 immediately provides the following corollary.

Corollary 1. *There is a polynomial 2-approximation algorithm for the problem* MAX-SHDTRI.

Example 1. Consider the following market $\mathcal{M}$.

$$\mathcal{M}:\quad A = \{a_1, a_2, \ldots, a_q, b_1, b_2, \ldots, b_q, c\}$$
$$\omega(a_i) = \omega(b_i) = h_i;\ \omega(c) = h_{q+1}$$
$$P(a_i) = P(b_i) = h_{i+1} \succ h_i;\ P(c) = h_1 \succ h_{q+1}$$

Here, $|A| = 2q + 1$ and any maximum cycle packing contains a single cycle of length $q + 1$: $(c, a_1$ or b_1, a_2 or $b_2, \ldots, a_q$ or $b_q)$. Thus, the algorithm satisfies $q + 1$ agents only, and since the house types of the rest of the agents are already present in the trading cycle, q agents remain unsatisfied. On the other hand, if agent c is removed, an acyclic market with $2q$ satisfied agents is obtained.

If we let q grow indefinitely, this example shows that the bound of the above approximation algorithm cannot be tightened to $2 - \varepsilon$ for any $\varepsilon > 0$.

Example 2. Suppose that housing market $\mathcal{M}$ fulfills that $|H| = 2$ and $H(a) \setminus \{\omega(a)\} \neq \emptyset$ for each agent. We show that $\mathrm{sat}(\mathcal{M}) = \max\{2\min\{n_1, n_2\}, n_1, n_2\}$, where $|A(h_1)| = n_1$, $|A(h_2)| = n_2$.

In this case, $G_{\mathcal{M}}$ is a complete bipartite digraph. For any solution $\mathcal{T}$, there are three possibilities. If $p(h_1) = p(h_2)$ then each trading cycle is even and contains alternately agents from $A(h_1)$ and $A(h_2)$, hence the same number of agents of the two types. So $\mathrm{sat}(\mathcal{T}) = 2\min\{n_1, n_2\}$. If $p(h_1) < p(h_2)$ then there is no trading, but all the agents from $A(h_1)$ are satisfied, as they cannot afford a house of type h_2. So $\mathrm{sat}(\mathcal{T}) = n_1$. Finally, if $p(h_1) > p(h_2)$ then $\mathrm{sat}(\mathcal{T}) = n_2$ by a similar argument.

Example 2 gives an infinite number of housing markets with $\mathrm{sat}(\mathcal{M}) = \frac{2}{3}n$ if we set $n_2 = 2n_1$. We remark that it remains an open problem whether a housing market exists where the number of dissatisfied agents is more than one third of all agents.

4   Inapproximability

We derive our results from the hardness results for MIN-VC. Proposition 1, the main result of Subsection 4.1, provides a transformation of a graph G into a housing market for which the number of dissatisfied agents in any optimal solution is proportional to $\mathrm{VC}(G)$. In Subsection 4.2, we derive several consequences for the hardness of approximation of $\mathrm{sat}(\mathcal{M})$ in the case of trichotomic preferences.

Further, Proposition 1 contains a parameter k that influences the fraction of dissatisfied agents in any optimal solution. This is used to get stronger inapproximability bounds of $\mathrm{sat}(\mathcal{M})$ if preferences are not required to be trichotomic. These bounds are presented in Subsection 4.3.

4.1   The Transformation

Proposition 1. *For every integer $k \geq 1$, there is a polynomial-time transformation $\tilde{T}_k$ from MIN-VC to MAX-SHDTIES such that each graph G with $|V(G)|$ vertices is transformed into a housing market $\tilde{\mathcal{M}}_k = \tilde{T}_k(G)$ with $n = (2k+1)|V(G)|$ agents, such that $\mathrm{sat}(\tilde{\mathcal{M}}_k) = (2k+1)|V(G)| - k\,VC(G)$.*

Proof. Consider an instance G of MIN-VC. We construct a market $\tilde{\mathcal{M}}_k$.

For each vertex $v \in G$, there is a set of vertices A_v consisting of $k+1$ *incoming* agents $I_v = \{i_{v,1}, i_{v,2}, \ldots, i_{v,k+1}\}$, and k *outgoing* agents $O_v = \{o_{v,1}, o_{v,2}, \ldots, o_{v,k}\}$. Their endowments and preferences are as follows (the sets of house types in brackets in the preference lists of outgoing agents represent ties).

$$w(i_{v,1}) = w(i_{v,2}) = \cdots = w(i_{v,k+1}) = h_v$$
$$w(o_{v,1}) = h^*_{v,1}; \ w(o_{v,2}) = h^*_{v,2}; \ \ldots \ ; \ w(o_{v,k}) = h^*_{v,k}$$

$$P(i_{v,1}) = \cdots = P(i_{v,k+1}) = h^*_{v,1} \succ h^*_{v,2} \succ \cdots \succ h^*_{v,k} \succ h_v$$
$$P(o_{v,1}) = (h^*_{v,2}, \ldots, h^*_{v,k}) \succ (\text{all } h_w \text{ such that } \{vw\} \in E(G)) \succ h^*_{v,1}$$
$$P(o_{v,2}) = (h^*_{v,3}, \ldots, h^*_{v,k}) \succ (\text{all } h_w \text{ such that } \{vw\} \in E(G)) \succ h^*_{v,2}$$
$$\vdots$$
$$P(o_{v,k}) = (\text{all } h_w \text{ such that } \{vw\} \in E(G)) \succ h^*_{v,k}.$$

For illustration, see Figure 1. Incoming (white) agents have the same in-arcs and out-arcs, which is for simplicity illustrated by just one copy of the arcs coming into and going out of the oval shape. Numbers accompanying arcs express the preference ordering.

Clearly, this construction can be performed in polynomial time.

For a vertex subset B of G, we define the corresponding set O_B of outgoing agents: $O_B = \{o_{v,j} : v \in B, j = 1, \ldots, k\}$. We prove several properties of $\tilde{\mathcal{M}}_k$.

Lemma 5. *F is a vertex cover in G if and only if O_F is a FVS in $\tilde{\mathcal{M}}_k$.*

Due to space constraints, the simple proof of Lemma 5 is omitted.

Lemma 6. *There exists an optimal solution for $\tilde{\mathcal{M}}_k$ with no trading.*

Proof. Assume that every optimal solution has at least one trading cycle, and let $\mathcal{T}$ be an optimal solution with the minimum number of trading cycles. Let C be any trading cycle of $\mathcal{T}$. By Lemma 1, houses of all agents on C have the same price; we denote this price by p_C.

Now consider the set A_v for any v such that at least one agent $a \in I_v$ belongs to C. Then the price of the house of each agent from O_v who is trading must also be equal to p_C, as some of the agents preceding him on his trading cycle must belong to I_v. Thus, each agent of A_v is either trading with price p_C or not trading.

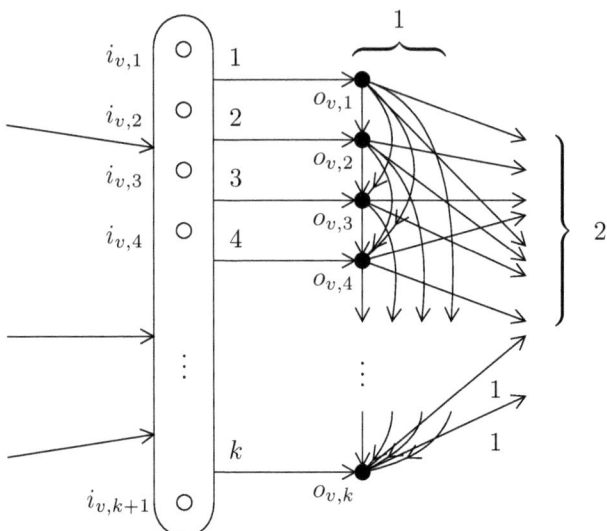

Fig. 1. Agents in $\tilde{\mathcal{M}}_k$ corresponding to a vertex v of G'

Let A_C^+ denote the union of all sets A_v such that at least one agent in A_v is trading with price p_C.

We construct another solution $\mathcal{T}'$ from $\mathcal{T}$ by modifying prices of houses according to their owner and their price in $\mathcal{T}$ as follows.

1. Owner in $A \setminus A_C^+$, price $\leq p_C$. These houses receive the same price in $\mathcal{T}'$ as in $\mathcal{T}$.
2. Owner in $A \setminus A_C^+$, price $> p_C$. These houses receive their price in $\mathcal{T}$ increased by $k + 1$.
3. Owner in A_C^+. For each v such that $A_v \subseteq A_C^+$, we set $p(h_v) = p_C$, $p(h_{v,1}^*) = p_C + 1$, $p(h_{v,2}^*) = p_C + 2$, $\ldots$, $p(h_{v,k}^*) = p_C + k$.

It is immediate that prices of the houses are defined consistently. Trading cycles from $\mathcal{T}$ with price different from p_C are in $\mathcal{T}'$ as well, while agents of A_C^+ are not trading in $\mathcal{T}'$. This concludes the construction of $\mathcal{T}'$.

Clearly, the number of trading cycles in $\mathcal{T}'$ is strictly smaller than the number of trading cycles in $\mathcal{T}$. Now we show that $\mathrm{sat}(\mathcal{T}') \geq \mathrm{sat}(\mathcal{T})$.

Let us first consider an agent outside A_C^+. His budget set in $\mathcal{T}'$ has not increased, and his trading/nontrading status has not changed compared to $\mathcal{T}$, so if this agent was satisfied in $\mathcal{T}$, he remains satisfied in $\mathcal{T}'$ as well.

Now take a set $A_v \subseteq A_C^+$. Notice that all agents from I_v have the same strict preferences and the same nonempty budget set in $\mathcal{T}$. Hence, since at most one of them can get their common unique most preferred affordable house from O_v, at most one agent from I_v is satisfied in $\mathcal{T}$. This implies that $|\mathrm{Sat}(\mathcal{T}) \cap A_v| \leq k+1$. In $\mathcal{T}'$ no agent from A_v is trading, but since agents from I_v cannot afford any house from O_v, they are all satisfied. This implies $|\mathrm{Sat}(\mathcal{T}') \cap A_v| \geq k+1$.

Summarizing, $\text{sat}(\mathcal{T}') \geq \text{sat}(\mathcal{T})$. Hence, $\mathcal{T}'$ is another optimal solution having less trading cycles than $\mathcal{T}$, which is a contradiction with the choice of $\mathcal{T}$.

Lemma 7. *Let $\mathcal{T}$ be any optimal solution of $\tilde{\mathcal{M}}_k$ with no trading and let $v \in G$ be arbitrary. Then $I_v \subseteq \text{Sat}(\mathcal{T})$ and either $O_v \subseteq \text{Sat}(\mathcal{T})$ or $O_v \subseteq \text{Dissat}(\mathcal{T})$.*

Proof. Let $\mathcal{T}$ be an optimal solution with no trading and v any vertex from $V(G)$. First we claim that

$$p(h_v) < p(h^*_{v,j}) \text{ for each } j \in \{1, \ldots, k\}. \tag{1}$$

Otherwise, if $p(h_v) \geq p(h^*_{v,j})$ for some j, then all the agents in I_v can afford the acceptable house $h^*_{v,j}$ and are dissatisfied. By increasing $p(h^*_{v,j})$ to $p(h_v) + 1$ for every such j, all $k+1$ agents in I_v become satisfied, while at most k agents from O_v may become dissatisfied. This is a contradiction with optimality of $\mathcal{T}$. Hence, (1) holds, and this implies that $I_v \subseteq \text{Sat}(\mathcal{T})$.

Now distinguish two cases.

Case 1. $p(h_v) < p(h_u)$ for each $u \in V$ such that $\{v, u\} \in E$. Let us denote $p^* = \min\{p(h_u); \{v, u\} \in E\}$ and define new prices p' fulfilling the following inequality:

$$p(h_v) < p'(h^*_{v,1}) < p'(h^*_{v,2}) < \cdots < p'(h^*_{v,k}) < p^*.$$

Thanks to (1), the satisfaction of any agent outside A_v w. r. t. new prices does not differ from his satisfaction according to old prices. Moreover, all agents in O_v are satisfied. So they all were satisfied in $\mathcal{T}$ too, or $\mathcal{T}$ was not optimal.

Case 2. There exists $u \in V$ such that $\{v, u\} \in E$ and $p(h_v) \geq p(h_u)$. Due to (1), all agents from O_v can afford the house h_u and so they are all dissatisfied, as there is no trading in $\mathcal{T}$.

To prove the Proposition, we show that $\text{VC}(G) = l$ if and only if $\text{sat}(\tilde{\mathcal{M}}_k) = n - kl$. Assume that F is a vertex cover of size l in G. Then by Lemma 5 the set O_F is a feedback vertex set in $\tilde{\mathcal{M}}_k$ and by Lemma 4, $\text{sat}(\tilde{\mathcal{M}}_k) \geq |A \setminus O_F| = n - kl$.

Now let $\mathcal{T}$ be an optimal solution in $\tilde{\mathcal{M}}_k$ with no trading (which exists by Lemma 6). Lemma 7 implies that $\text{Dissat}(\mathcal{T}) = \bigcup_{v \in B} O_v$ for some $B \subseteq V$, and hence $\text{dissat}(\mathcal{T}) = kl$ for some l. For any directed cycle C in $\tilde{\mathcal{M}}_k$, at least one agent will have maximum price among the agents of C, and because there is no trading, this agent is dissatisfied. Moreover, (1) implies that the dissatisfied agent is outgoing. Hence, the set O_B contains at least one agent out of every directed cycle in $\tilde{\mathcal{M}}_k$, and thus O_B is a FVS of $\tilde{\mathcal{M}}_k$. Lemma 5 directly implies that B is a vertex cover in G and hence $\text{VC}(G) \leq l$.

Therefore $\text{sat}(\tilde{\mathcal{M}}_k) = n - kl$ if and only if $\text{VC}(G) = l$.

In the case that $k = 1$, the constructed market $\tilde{\mathcal{M}}_k$ is trichotomic. Hence, we get the following proposition as a corollary.

Proposition 2. *There is a polynomial-time transformation T from MIN-VC to MAX-SHDTRI such that each graph G with $|V(G)|$ vertices is transformed into a housing market $\mathcal{M} = T(G)$ with $n = 3|V(G)|$ agents, such that $\text{sat}(\mathcal{M}) = 3|V(G)| - \text{VC}(G)$.*

4.2   Inapproximability for Max-SHDTri

Halldórsson et al. [10], when studying the approximability of the problem to find a stable matching of maximum size in the stable marriage problem with incomplete lists and ties (SMTI for short), presented a construction that assigns to each graph $G = (V, E)$ an SMTI instance I such that the number of men as well as the number of women in I is equal to $3|V(G)|$, and $|\mathrm{opt}(I)| = 3|V(G)| - \mathrm{VC}(G)$. Since the quantitative relations in our Proposition 2 and in the proof of their Theorem 3.2. are the same, we get by exactly the same argument the following analogies of Theorem 3.2., its Corollary 3.4. and Remark 3.6. of [10].

Proposition 3. *For any $\varepsilon > 0$ and $p < \frac{3-\sqrt{5}}{2}$, given an instance $\mathcal{M}$ of* MAX-SHDTRI *with n agents, it is NP-hard to distinguish between the following two cases:*

1. $\mathrm{sat}(\mathcal{M}) \geq \frac{2+p-\varepsilon}{3} n$, *and*
2. $\mathrm{sat}(\mathcal{M}) < \frac{2+\max\{p^2, 4p^3 - 3p^4\}+\varepsilon}{3} n$.

Theorem 2. *It is NP-hard to approximate* MAX-SHDTRI *within a factor smaller than $21/19$.*

Theorem 3. *If* MIN-VC *is NP-hard to approximate within a factor of $2 - \varepsilon$ then* MAX-SHDTRI *is NP-hard to approximate within a factor smaller than 1.25.*

The *Unique Games Conjecture* was introduced by Khot [11]. For the statement of the Conjecture and all necessary definitions, see [11] and [12]. Khot and Regev [12] proved that if UGC is true, then MIN-VC is NP-hard to approximate within a ratio smaller than 2. Thus, the assumption of Theorem 3 would be fulfilled. We get the following corollary.

Corollary 2. *If UGC is true, then* MAX-SHDTRI *is NP-hard to approximate within a factor smaller than 1.25.*

The results of Austrin et al. [3, page 3 and Theorem 4.1] may be stated in the following way.

Proposition 4. *If UGC is true, then for every sufficiently large integer d it is NP-hard to distinguish between the following two cases:*

1. $\mathrm{VC}(G) \leq (1/2 + \Theta(\frac{\log\log d}{\log d}))|V(G)|$, *and*
2. $\mathrm{VC}(G) > (1 - \frac{1}{\log d})|V(G)|$,

even on graphs of maximum degree d.

It is easy to see that in the transformation of Proposition 2, the resulting preference list lengths are bounded in terms of degrees of G. We therefore define a restricted variant of the MAX-SHDTRI problem with preference lists of length at most d; we call it MAX-SHDTRI$_d$. From Proposition 4 we derive the following result. Due to space constraints, the straightforward calculation is omitted.

Theorem 4. *If UGC is true, then for every sufficiently large integer d it is NP-hard to approximate* MAX-SHDTRI$_d$ *within a ratio $1.25 - \Theta(\frac{\log\log d}{\log d})$*

4.3 Inapproximability for Max-SHDTies

As the problem MAX-SHDTRI is a restricted version of MAX-SHDTIES, all inapproximability results for MAX-SHDTRI apply to MAX-SHDTIES as well. However, due to the general Proposition 1, we obtain even stronger inapproximability results here.

Theorem 5. *The problem* MAX-SHDTIES *is*

1. *NP-hard to approximate within a factor smaller than* 1.2, *and*
2. *NP-hard to approximate within a factor smaller than* 1.5, *if UGC is true.*

To prove the first part of the theorem, we use the following assertion for MIN-VC due to Dinur and Safra [8] (as in the reasoning that leads to Proposition 3).

Proposition 5. *For any* $\varepsilon > 0$ *and* $p < \frac{3-\sqrt{5}}{2}$, *given a graph* G, *it is NP-hard to distinguish between the following two cases:*

1. $VC(G) \leq (1 - p + \varepsilon)|V(G)|$, *and*
2. $VC(G) > (1 - \max\{p^2, 4p^3 - 3p^4\} - \varepsilon)|V(G)|$.

To prove the second part, we use the following special case of the results of Khot and Regev [12, Section 4].

Proposition 6. *If UGC is true, then for any* $\varepsilon > 0$, *given a graph* G, *it is NP-hard to distinguish between the following two cases:*

1. $VC(G) \leq (\frac{1}{2} + \varepsilon)|V(G)|$, *and*
2. $VC(G) > (1 - \varepsilon)|V(G)|$.

The proof of Theorem 5 then consists in combining Proposition 1 with Proposition 5 and Proposition 6. It is omitted due to space limits.

5 Conclusion and Open Problems

In this paper, we have presented a simple 2-approximation algorithm for MAX-SHDTRI. We have also shown that the number of agents satisfiable in any instance of MAX-SHDTRI is at least $n/2$.

Based on a reduction from MIN-VC, we have shown several inapproximability results for MAX-SHDTRI, MAX-SHDTRI$_d$ (where preference lists have length at most d), and MAX-SHDTIES (where preference lists are not required to be trichotomic).

All the markets constructed in this paper have a special property: all agents that own the same house type have the same preference lists. Markets with this property may be called *coherent*.

One could expect that for coherent markets, stronger results could be obtained. Further, no approximation algorithm is known for MAX-SHDTIES, nor for markets where preference lists are strictly ordered.

Acknowledgement. The authors would like to thank Magnús M. Halldórsson and Vít Jelínek for helpful suggestions.

References

1. Abraham, D., Blum, A., Sandholm, T.: Clearing algorithms for barter exchange markets: Enabling nationwide kidney exchanges. In: EC 2007, San Diego, California (2007)
2. Abraham, D.J., Cechlárová, K., Manlove, D.F., Mehlhorn, K.: Pareto Optimality in House Allocation Problems. In: Fleischer, R., Trippen, G. (eds.) ISAAC 2004. LNCS, vol. 3341, pp. 3–15. Springer, Heidelberg (2004)
3. Austrin, P., Khot, S., Safra, M.: Inapproximability of vertex cover and independent set in bounded degree graphs. In: CCC 2009: Proceedings of the 2009 24th Annual IEEE Conference on Computational Complexity, pp. 74–80. IEEE Computer Society, Washington, DC, USA (2009)
4. Cechlárová, K., Fleiner, T.: Housing markets through graphs. Algorithmica 58(1), 19–33 (2010)
5. Cechlárová, K., Jelínková, E.: An efficient implementation of the equilibrium algorithm for housing markets with duplicate houses. Information Processing Letters 111(13), 667–670 (2011)
6. Cechlárová, K., Schlotter, I.: Computing the Deficiency of Housing Markets with Duplicate Houses. In: Raman, V., Saurabh, S. (eds.) IPEC 2010. LNCS, vol. 6478, pp. 72–83. Springer, Heidelberg (2010)
7. Deng, X., Papadimitriou, C., Safra, S.: On the complexity of price equilibria. J. Computer and System Sciences 67, 311–324 (2003)
8. Dinur, I., Safra, S.: On the hardness of approximating minimum vertex cover. Annals of Mathematics 162, 439–485 (2005)
9. Fekete, S., Skutella, M., Woeginger, G.: The complexity of economic equilibria for house allocation markets. Information Processing Letters 88(5), 219–223 (2003)
10. Halldórsson, M.M., Iwama, K., Miyazaki, S., Yanagisawa, H.: Improved approximation results for the stable marriage problem. ACM Trans. Algorithms 3(3), 30 (2007)
11. Khot, S.: On the power of unique 2-prover 1-round games. In: Proc. 34th ACM Symp. on Theory of Computing, STOC 2002, pp. 767–775 (2002)
12. Khot, S., Regev, O.: Vertex cover might be hard to approximate to within $2 - \varepsilon$. Journal of Computer and System Sciences 74(3), 335–349 (2008); Computational Complexity 2003
13. Shapley, L., Scarf, H.: On cores and indivisibility. J. Math. Econ. 1, 23–37 (1974)

Planarization and Acyclic Colorings of Subcubic Claw-Free Graphs

Christine Cheng*, Eric McDermid**, and Ichiro Suzuki***

Department of Computer Science,
University of Wisconsin–Milwaukee, Milwaukee, WI 53211, USA
{ccheng,mcdermid,suzuki}@uwm.edu

Abstract. We study methods of planarizing and acyclically coloring claw-free subcubic graphs. We give a polynomial-time algorithm that, given such a graph G, produces an independent set Q of at most $n/6$ vertices whose removal from G leaves an induced planar subgraph P (in fact, P has treewidth at most four). We further show the stronger result that in polynomial-time a set of at most $n/6$ *edges* can be identified whose removal leaves a planar subgraph (of treewidth at most four). From an approximability point of view, we show that our results imply 6/5- and 9/8-approximation algorithms, respectively, for the (NP-hard) problems of finding a maximum induced planar subgraph and a maximum planar subgraph of a subcubic claw-free graph, respectively.

Regarding acyclic colorings, we give a polynomial-time algorithm that finds an optimal acyclic vertex coloring of a subcubic claw-free graph. To our knowledge, this represents the largest known subclass of subcubic graphs such that an optimal acyclic vertex coloring can be found in polynomial-time. We show that this bound is tight by proving that the problem is NP-hard for cubic line graphs (and therefore, claw-free graphs) of maximum degree $d \geq 4$. An interesting corollary to the algorithm that we present is that there are exactly three subcubic claw-free graphs that require four colors to be acyclically colored. For all other such graphs, three colors suffice.

1   Introduction

A simple, finite graph G is said to be *claw-free* if no vertex of G has three pairwise nonadjacent neighbors. It is *subcubic* if every vertex of G has degree at most three. Claw-free graphs are a well-studied and interesting class of graphs that generalize line graphs. Additionally, claw-free graphs are very well-understood, thanks to a complete structure theorem found by Chudnovsky and Seymour [5]. In this paper we explore methods of planarization and acyclic colorings of claw-free subcubic graphs.

Planarization. Planarization is a broad term used to refer to the process of modifying a graph in order to make it planar (see [13] for a survey). Interest in

---

* Supported by NSF award CCF-0830678.
** Supported by NSF award CCF-0830678 and UWM Research Growth Initiative.
*** Supported by UWM Research Growth Initiative.

P. Kolman and J. Kratochvíl (Eds.): WG 2011, LNCS 6986, pp. 107–118, 2011.

such techniques arises both for combinatorial and theoretical uses, but also for industrial applications (for example, VLSI and layout problems). In the *minimum nonplanar vertex deletion problem* (MINVD) we wish to compute a smallest set of vertices Q such that $G - Q$ is an *induced planar subgraph*. The complement problem of MINVD is to find a *maximum induced planar subgraph* (MAXIPS), where the size of the solution is the number of vertices of $G - Q$. Analogously, the *minimum nonplanar edge deletion problem* MINED asks for a smallest set of edges E' such that $G - E'$ is a *planar subgraph*. The complement of MINED is to find a *maximum planar subgraph* (MAXPS), where the size of the solution is the number of edges of $G - E'$.

From a complexity perspective, all of these optimization problems are hard, but some positive results are also known. On the negative side, Yannakakis [16,17] first showed that MINVD and MINED are NP-hard. Independently, Liu and Geldmacher [14] also proved that MINED is NP-hard. Călinescu et al. [4] improved upon this by showing that MINED and MAXPS are Max SNP-hard. Subsequently, Faria et al. [8,9,10] (note the two different sets of authors) showed that MINVD and MINED are both Max SNP-hard in cubic graphs, and that MAXIPS and MAXPS are NP-hard in cubic graphs. They also showed that MAXIPS in general graphs admits no polynomial-time approximation algorithm with a fixed ratio unless P = NP. We remark that, although Faria et al. [10] do not explicitly state it, the derived instance in their hardness proof for MAXIPS in cubic graphs has the property that every vertex is in a triangle (a necessary and sufficient condition for claw-freeness in a cubic graph). Hence MAXIPS is NP-hard in claw-free cubic graphs. It is also easy to see that their hardness result for MAXPS in cubic graphs implies that MAXPS is also NP-hard in claw-free cubic graphs: a cubic graph G contains a planar subgraph P of size $n - k$ (i.e., we remove k edges to obtain P) if and only if the claw-free graph G' obtained by replacing each vertex of G with a triangle has a planar subgraph of size $3n - k$.

On the positive side, Faria et al. [10] gave a 4/3-approximation algorithm for MAXIPS in subcubic graphs. Edwards and Farr [6] presented a polynomial-time algorithm that, given a graph of maximum degree d, finds an induced planar subgraph of size at least $3n/(d+1)$. They later generalized this result to graphs of average degree d (in fact, the induced subgraph found is series-parallel). This can be interpreted as a $(d+1)/3$-approximation algorithm, which matches the bound given by Faria et al. when the maximum degree is three. Finally, Călinescu et al. [4] gave a 9/4-approximation for MAXPS in general graphs. We remark that there is much other work on finding subgraphs (induced or otherwise) having a particular property; we have only mentioned those results most directly related to our own.

Acyclic Colorings. An *acyclic coloring* of a graph G is a proper vertex coloring of G with the property that no cycle is bicolored. The *acyclic coloring problem* is to compute an acyclic coloring for G using the fewest number of colors. Aside from its intrinsic interest, researchers are interested in this problem because it has applications in computing the Jacobian and Hessian of sparse matrices [11].

Using probabilistic methods, Alon et al. [1] proved that it is always possible to acyclically color the vertices of a graph of maximum degree d using $O(d^{4/3})$ colors, and showed that there exists graphs requiring $\Omega(d^{4/3}/(\log d)^{1/3})$ colors. For fixed values of d, they showed that a straightforward greedy algorithm uses at most $d^2 + 1$ colors. Several authors have presented polynomial-time algorithms [3,12,15] showing that graphs of maximum degree 3, 4, and 5, respectively, admit an acyclic 4-, 5-, and 7-coloring. Note that all of these results are approximation algorithms – they will not, in general, return an acyclic coloring with the fewest number of colors possible. Regarding special cases of graphs, Zhang and Bilka [18] (implicitly) give a polynomial-time algorithm showing that every cubic line graph admits an acyclic coloring using at most three colors – a result that will be useful to us later on.

We remark that much attention has also been devoted to the acyclic *edge* coloring of a graph G (defined analogously to acyclic vertex colorings). For our purposes, we mention only that Alon and Zaks [2] have shown that determining if a cubic graph G admits an acyclic coloring with at most three colors is NP-complete.

Our Contribution. On the surface, planarization and acyclic colorings seem to be quite different problems. A unifying approach to our results is our simple reduction technique, that, in linear time, yields a reduced graph with very desirable properties. In particular, we show that if the reduced graph admits a (induced) planar subgraph of a certain size, or is acyclically colorable, then the original graph also has this property. We believe that this reduction technique could be applicable to other problems restricted to subcubic claw-free graphs.

Regarding planarization, we give a polynomial-time algorithm that, given a claw-free subcubic graph, produces an independent set Q of at most $n/6$ vertices whose removal from G leaves an induced planar subgraph P (in fact, P has treewidth at most four). Hence, G always has an induced planar subgraph (of treewidth at most four) of size at least $5n/6$. Given this result, we further show the stronger result that in polynomial-time a set of at most $n/6$ *edges* can be identified whose removal leaves a planar subgraph (of treewidth at most four). From the perspective of the combinatorial bounds given by Edwards and Farr [7], our results show the existence of significantly larger induced planar subgraphs when restricted to subcubic claw-free graphs, rather than general subcubic graphs. From an approximability point of view, we show that our first two bounds give a 6/5- and a 9/8-approximation ratio, respectively, for MAX-IPS and MAXPS. This gives a better performance guarantee for the special case of subcubic claw-free graphs than that of the 4/3-approximation algorithm for MAXIPS in general subcubic graphs, and an improvement for this special case over the general 9/4-approximation algorithm for MAXPS.

Next, we give a polynomial-time algorithm that finds an optimal acyclic vertex coloring of a subcubic claw-free graph. To our knowledge, this represents the largest subclass of subcubic graphs (or any non-trivially degree bounded class of graphs) such that an optimal acyclic vertex coloring can be found in polynomial-time. We show that this bound is tight by proving that the problem is NP-hard

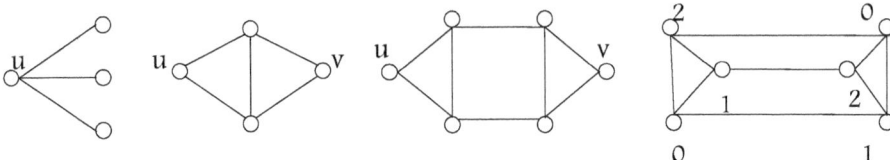

Fig. 1. A claw with base u, a diamond and double diamond with endpoints u and v, and an envelope with an acyclic coloring

for claw-free graphs of maximum degree $d \geq 4$. An interesting corollary to the algorithm that we present is that there are exactly three subcubic claw-free graphs that require four colors to be acyclically colored. For all other such graphs, three colors suffice.

2   Preliminaries and Definitions

We consider all graphs to be simple and finite, and use standard graph terminology and notation. A *claw, diamond, double-diamond*, and an *envelope*, respectively, are defined to be the graphs (from left to right) depicted in Figure 1. Let D be an induced diamond or double-diamond. The *endpoints* of D are defined to be the (only) two vertices in D with degree two in D. Let H be any induced subgraph of G. For a vertex $u \in H$ with degree two in H, but degree three in G, $t_H(u)$ is defined to be the (distinct) neighbor of u not in H. For example, if G is a cubic graph, and u is an endpoint of an induced diamond or double-diamond D, then $t_D(u)$ is the distinct neighbor of u that is not in D. A *tree decomposition* of a graph $G = (V, E)$ is a pair (X, T), where $X = \{X_1, \ldots, X_n\}$ is a family of subsets of V, and T is a tree whose nodes are the subsets X_i, satisfying that: (i) $\cup X_i = V$, (ii) every edge $uv \in E$ belongs to some $X_i \in X$, and (iii) for each vertex $u \in V$, the set of subsets of X containing u induce a connected subtree of T. The *treewidth* of a tree decomposition is $\max_i |X_i| - 1$. The *treewidth* of a graph is the minimum treewidth over all possible tree decompositions of a graph.

3   Simplifying the Graph

In this section we present our reduction algorithm, which takes an arbitrary subcubic claw-free graph $G = (V, E)$, and returns a reduced claw-free *cubic* graph $G_R = (V_R, E_R)$ with $V_R \subseteq V$, with no induced diamonds or double-diamonds.

   Reducing the graph is a crucial preprocessing step for each of the results that we present in the later sections of this paper. In particular, in each of the subsequent sections we will show that, roughly speaking, if we can find an (induced) planar subgraph of G_R of a certain size, or an acyclic coloring for G_R, then we can find such a solution for G as well.

Before presenting the reduction algorithm we need one lemma regarding the properties of induced diamonds and double-diamonds of cubic claw-free graphs, whose proof follows immediately from the claw-freeness of G.

Lemma 1. *Let G be a claw-free cubic graph, and D an induced diamond or double-diamond with endpoints u, v. Then, $t_D(u) \neq t_D(v)$.*

The reduction algorithm is as follows. We remark that it can be implemented to run in linear time; we omit the details.

Reduce(G)
1. Perform steps 1(a) and 1(b) until no further reduction is possible.
   (a) If G contains a pendant or isolated vertex u, delete u.
   (b) If G contains a vertex u of degree two, let v, w be the neighbors of u. Delete u, and add the edge vw if it does not already exist.
2. If G contains an induced diamond or an induced double-diamond D, with endpoints u, v, let $t_D(u) = w$ and $t_D(v) = x$. Delete D, and add the edge wx to G if it does not already exist. Return to Step 1. Else, return G.

In the next lemmas, we establish the properties of this procedure (namely, we will show that we can never introduce a claw into the graph), and characterize the properties of the final graph G_R returned by the reduction algorithm.

Lemma 2. *Let G be a subcubic claw-free graph, and G' the graph resulting from a single execution of one of the steps denoted by (1a), (1b), or (2) in the pseudocode. Then, G' is a subcubic claw-free graph.*

Proof. None of the operations described in the reduction algorithm can increase the degree of a vertex of G, so G' must also have maximum degree three.

It is clear that step (1a) cannot create a claw, so we need only consider the steps denoted by (1b) and (2) in the pseudocode, that involve potentially adding an edge to G. Suppose that a degree two vertex u, with neighbors v and w, is deleted from G, so that the edge vw is present in G', creating a claw C in G'. If C does not involve the edge vw, then C must be present in G, a contradiction – hence the edge vw is involved in C. Suppose without loss of generality that v is the base of C, and that the other two neighbors of v are x and y. If vw is present in G, v, w, x, y induces a claw in G, and otherwise, v, u, x, y induces a claw in G, a contradiction.

A very similar argument holds for the deletion of an induced diamond or double-diamond D from G as described in step 2. If u, v are the endpoints of D, and the edge $t_D(u)t_D(v)$ is involved in a claw in G', then $t_D(u)$ or $t_D(v)$ is involved in a claw in G.   □

The following definition allows us to characterize the graph G_R returned by the algorithm.

Definition 1. *A cubic disjoint triangle graph is a cubic graph with the property that every vertex is in exactly one triangle (hence any two triangles are vertex disjoint), and every edge of G is either a part of a triangle or is the unique edge joining two vertex-disjoint triangles.*

Lemma 3. *Let G be a subcubic claw-free graph, and $G_R = (V_R, V_E)$ the graph returned at the termination of the reduction algorithm. Then, either $V_R = \emptyset$, or each component of G_R is either: (i) isomorphic to K_4, or (ii) isomorphic to an envelope, or (iii) a cubic disjoint triangle graph.*

Proof. Consider a particular maximal component G'_R of G_R. If G'_R is not the empty set, then it must be a cubic graph, for otherwise the reduction algorithm cannot have terminated. In a claw-free cubic graph, every vertex must be in at least one triangle. Let $T = \{u, v, w\}$ be a triangle of G'_R, with $t_T(u) = x$, $t_T(v) = y$, and $t_T(w) = z$. If $x = y = z$, then G'_R consists entirely of the vertices u, v, w, x, and is isomorphic to K_4. If exactly two of x, y, z are identical, say, x and y, then G'_R contains an induced diamond, a contradiction. Suppose now that x, y, z are distinct. If x, y, z is a triangle, then G'_R is isomorphic to an envelope. If exactly two of x, y, z are in the same triangle, say, x and y, then, u, v, w and the triangle including x and y is an induced double-diamond, a contradiction. The only remaining possibility is that x, y, z are in vertex-disjoint triangles, implying that G'_R is a cubic disjoint triangle graph. $\square$

Finally, we bound the treewidth of the graph G in the case that G_R is not a cubic disjoint triangle graph. One finds a tree decomposition of width at most four for G as follows. Begin with a tree decomposition of width at most four for G_R. Next, iteratively 'undo' the operations of the reduction algorithm, one by one, extending the tree decomposition by adding an appropriate set of bags to the tree decomposition. The details are an easy exercise.

Proposition 1. *Let G be a subcubic claw-free graph, and G_R the graph returned at the termination of the reduction algorithm. If G_R is not a cubic disjoint triangle graph, then G has treewidth at most four.*

4   Finding Large Planar Subgraphs

Our main focus in this section is to present a polynomial-time algorithm that, given a subcubic, claw-free graph, finds an induced planar subgraph with at least $5n/6$ vertices. In particular, we construct a set of at most $n/6$ vertices Q such that $G - Q$ is planar. We then use the particular properties of this set Q to show the stronger result that, in fact, we need only remove $n/6$ *edges* from G to arrive at a planar graph.

As alluded to in the previous section, our algorithm begins by passing the input graph to the reduction algorithm. The usefulness of the reduction algorithm for finding large planar subgraphs emerges in the following lemma.

Lemma 4. *Let G be a subcubic claw-free graph, and G' the graph resulting from a single execution of one of the steps denoted by (1a), (1b), or (2) in the pseudocode. Then, for any subset V' of the vertices of G', $G' - V'$ is planar implies that $G - V'$ is planar.*

Proof. Fix a particular plane embedding Π' of $G' - V'$. We shall construct a plane embedding Π for $G - V'$. If G' is obtained from G by deleting a pendant or isolated vertex v, then v can obviously be added to Π' in a way that avoids edge crossings. Otherwise, G' is obtained from G by deleting a vertex of degree two, an induced diamond, or an induced double-diamond, along with the addition of at most one edge.

If a vertex u of degree two is deleted from G, then, the neighbors of u, say, v and w, are joined by an edge in G'. If both v and w are present in $G' - V'$, then u, along with the edges uw and uv can be drawn as close as necessary to the edge vw to avoid edge crossings. The edge vw may then be deleted. If at most one of v and w is in $G' - V'$, then u can be added as a pendant or isolated vertex to Π'.

The case for an induced diamond or double-diamond D is analogous to that of a single vertex of degree two. We omit the full details.   □

In light of Lemma 4, we may focus our attention on finding an induced planar subgraph of G_R, the graph returned by the reduction algorithm.

4.1   Induced Planar Subgraphs

The planarization algorithm is outlined in Figure 2. Let G denote the input graph, which is assumed to be subcubic and claw-free. Recall that the goal of the planarization algorithm is to compute a set Q of vertices such that $G - Q$ is planar. The algorithm begins by passing G to the reduction algorithm. Let G_R denote the graph returned. The main body of the algorithm is a while loop, which continues as long as there is some component G_i of G_R that is a cubic disjoint triangle graph. Inside the loop, an arbitrary vertex u of G_i is added to Q, and deleted from G_R. The reduction algorithm is then called on G_R, and the loop proceeds to its next iteration.

The following lemma establishes the correctness of the planarization algorithm, along with establishing the properties of Q.

Lemma 5. *Let G be a subcubic claw-free graph. When Planarize(G) terminates, (i) $G - Q$ is planar, (ii) $|Q| \leq n/6$, where n is the number of vertices in G, and (iii) Q is an independent set in G.*

Proof. (i) If there is any nonempty component of $G - Q$, then, by Lemma 3, it is isomorphic to K_4 or an envelope, which are both planar. Thus, by inductively applying Lemma 4, $G - Q$ is planar.

(ii) We will show that after the deletion of a vertex u as described in the while loop of the planarization algorithm, the reduction algorithm always deletes at least five additional vertices. Hence, the number of iterations cannot exceed $n/6$. Let $T = u, v, w$ be the triangle containing u in G, $t_T(u) = x$, and y, z the two vertices in a triangle with vertex x. Note that there is no edge with one endpoint in $\{v, w\}$ and one endpoint in $\{y, z\}$, for otherwise we either have an induced double-diamond, or these six vertices are isomorphic to an envelope, a contradiction. After removing the vertex u, it is easy to see that the reduction algorithm will delete, regardless of the exact order, the vertices v, w, x, y, and z (as these vertices will have degree two), and possibly additional vertices.

Planarize(G):
$G_R \leftarrow$ Reduce(G)
$Q \leftarrow \emptyset$
while G_R has a component G_i that is a cubic disjoint triangle graph:
        $u \leftarrow$ arbitrary vertex of G_i
        $Q \leftarrow Q \cup \{u\}$
        $G_R \leftarrow G_R - \{u\}$
        $G_R \leftarrow$ Reduce(G_R)
return Q

Fig. 2. The planarization algorithm

(iii) When a vertex u is chosen to be placed into Q, the reduction algorithm will subsequently delete all of u's neighbors, hence none of these vertices can ever be placed into Q at a later iteration. $\square$

Theorem 1. *Let G be a subcubic claw-free graph with n vertices. In $O(n^2)$ time, an independent set Q can be found containing at most $n/6$ vertices such that $G - Q$ is planar (in fact, $G - Q$ has treewidth at most four).*

Corollary 1. *There is a $O(n^2)$-time approximation algorithm for MAXIPS on subcubic claw-free graphs with a performance guarantee of $6/5$.*

4.2   Planar Subgraphs

Let us now bound the number of edge deletions required to planarize G. Let $P = G - Q$ be the planar graph resulting from the planarization algorithm. Consider any vertex $u \in Q$. Since Q is an independent set, u has neighbors v, w, and x in P. Since G is claw-free, some pair from the set $\{v, w, x\}$, say, v and w, are joined by an edge. If we delete the edge ux, then $P \cup \{u\}$ is planar: u, along with the edges uv and uw can be drawn as close as necessary to the edge vw in any plane embedding of P, as discussed in the proof of Lemma 2. If this operation is performed for every vertex in Q, the resulting graph is planar. The proof of the next theorem is very similar to that of Theorem 1.

Theorem 2. *Let $G = (V, E)$ be a subcubic claw-free graph. In $O(n^2)$-time, a set of at most $n/6$ edges E' can be found such that $G - E'$ is planar (in fact, $G - E'$ has treewidth at most four).*

We now show that the planar subgraph found by the edge deletions described in the above procedure is a 9/8-approximation algorithm for MAXPS. Suppose $G = (V, E)$, the original input graph, has n vertices and m edges. After the reduction algorithm is run once on G, the resulting graph, $G_R = (V_R, E_R)$, has, say, $n' \leq n$ vertices and $m' = 3n'/2$ edges (since G_R is cubic). The number of edges removed to make G_R planar, and thus, the number of edges removed to make G planar, is at most $n'/6 = m'/9$. Let E_A denote the set of edges in the planar subgraph found by our method. The set E_A contains every edge of the graph, except for one edge for each vertex in Q. Thus we have that $|E_A| \gtrsim$

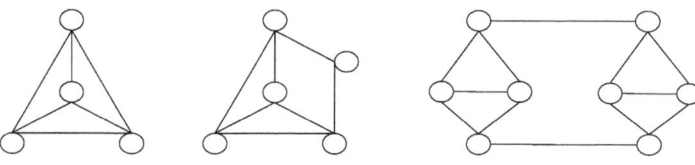

Fig. 3. The only three subcubic claw-free graphs requiring four colors for an acyclic coloring

$(m - |Q|) \geq m - \frac{n'}{6} \geq m - \frac{m'}{9} \geq \frac{8m}{9}$. Since the size of an optimal solution E_{OPT} is at most m, we have that $\frac{|E_A|}{|E_{OPT}|} \geq \frac{8}{9}$. This leads us to the following corollary.

Corollary 2. *There is a $O(n^2)$-time approximation algorithm for MAXPS on subcubic claw-free graphs with a performance guarantee of 9/8.*

5   Acyclic Colorings

In this section we give a polynomial-time algorithm for finding an optimal acyclic coloring of a subcubic claw-free graph G. An interesting consequence of the algorithm that we describe is that there are exactly three subcubic claw-free graphs that require four colors in order to be acyclically colored. Three colors suffice for all other subcubic claw-free graphs. Without loss of generality, we make a few assumptions to simplify our discussion. First, we assume that G is not a forest, for finding an optimal acyclic coloring in a forest is trivial. Secondly, we assume that G is connected, for if it is not, we can find an optimal coloring for each of the individual maximally connected components. Lastly, we assume that G is not one of the graphs in Figure 3 (for space reasons, we have moved the remaining figures to the Appendix), which, it turns out, are the only three graphs in this class that require four colors to be acyclically colored (it is easy to verify that four colors are necessary and sufficient for these graphs).

Similar to the approach used in the planarization algorithm, we first run the reduction algorithm on G. The following lemma demonstrates the usefulness of the reduction algorithm for finding acyclic colorings.

Lemma 6. *Let G be a subcubic claw-free graph, and G' the graph resulting from a single execution of one of the steps denoted by (1a), (1b), or (2) in the pseudocode. If G' is acyclically 3-colorable, then G is acyclically 3-colorable.*

Proof. Suppose that we are given an acyclic 3-coloring for G'. The graph G' is obtained from G by either (i) deleting an isolated vertex, or (ii) deleting a pendant vertex, or (iii) deleting a vertex u of degree two with neighbors v and w, and possibly adding a new edge vw, or (iv) deleting a diamond or a double-diamond, and possibly adding a new edge. We can extend the coloring of G' to an acyclic coloring of G if cases (i), (ii), or (iii) apply – simply color the new vertex with a different color than its neighbors (if it has any neighbors). Otherwise, suppose that D is a diamond or double-diamond deleted from G with endpoints

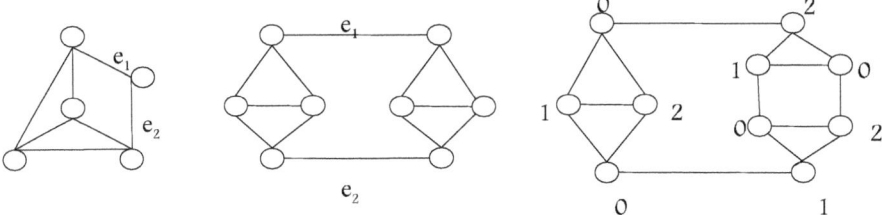

Fig. 4. The possible candidates for G_{s-1}, the last of which is acyclically colorable

u and v. Notice that, by the description of the reduction algorithm, G' is a cubic graph. By Lemma 1, $t_D(u) \neq t_D(v)$ in G, but $t_D(u)$ and $t_D(v)$ are joined by an edge in G', and are therefore colored different colors, say zero and one. It is a straightforward exercise to construct an acyclic coloring of D such that there is no bicolored path from $t_D(u)$ to $t_D(v)$ through D. We omit the full details. Therefore, G is acyclically 3-colorable.   □

Thus, by inductively applying Lemma 6, we have that if the reduced graph G_R is acyclically 3-colorable, then G is as well. We may therefore focus our attention on acyclically coloring G_R. Recall that, by Lemma 3, G_R is either an empty graph, or is isomorphic to K_4, or to an envelope, or is a cubic disjoint triangle graph. Trivially, we are already done if G_R is the empty graph. If G_R is isomorphic to an envelope, then G_R is acyclically 3-colored as described in Figure 1, and we are done. If, instead it is a cubic disjoint triangle graph, then G_R is acyclically 3-colored using the method of Zhang and Bylka [18]. In the next section, we describe how to color G_R if it is isomorphic to K_4, which is not immediately acyclically 3-colorable.

5.1 Acyclically Coloring $G_R \cong K_4$

Let us consider the sequence of operations performed by the reduction algorithm on G to arrive at G_R. This gives rise to a sequence of graphs $S = \langle G = G_0, G_1, \ldots, G_{s-1}, G_s = G_R \rangle$. Notice that this sequence of operations cannot be empty, for, by assumption, G is not isomorphic to K_4.

Consider the graph G_{s-1} – this graph must be obtained by replacing an edge of K_4 with a degree two vertex, diamond, or double-diamond, and is therefore isomorphic to one of the graphs shown in Figure 4. For, G_s cannot be obtained by deleting a pendant vertex of G_{s-1}, as this would force some vertex of G_{s-1} to have degree four. Also, since none of the operations of the reduction algorithm can disconnect the graph, the graph G_s cannot be obtained by deleting an isolated vertex from G_{s-1} either. If G_{s-1} is isomorphic to the rightmost graph in Figure 4, then it is acyclically 3-colored as shown in the figure, and we are done. Thus, we may restrict our attention to the case in which G_{s-1} is isomorphic to the first or second graphs (from left to right) in Figure 4. Notice that from this observation, we may conclude that $G_{s-1} \neq G_0$, for these are two of the graphs in Figure 3 that, by assumption, are not isomorphic to $G = G_0$.

Define an edge e of G_{s-k} to be *artificial* if it is not present in G_{s-k-1} (G_0 is defined to have no artificial edges). The following lemma begins to shed light on how to proceed.

Lemma 7. *Let* $\mathcal{S} = \langle G_0, G_1, \ldots, G_k \rangle$ *be the graphs arising from a non-empty sequence of reductions as described by the reduction algorithm on* G, *and* D *an induced diamond of* G_k *such that every vertex* $u \in D$ *has degree three in* G_k. *Then, in* G_k, *none of the edges with both endpoints in* D *are artificial.*

Proof. Suppose e is an artificial edge of an induced diamond D of G_k such that every vertex (crucially) has degree three in G_k. Then, G_k must have been obtained from G_{k-1} by replacing a vertex of degree two, a diamond, or a double-diamond of G_{k-1} with the edge e, as described by the reduction algorithm. It is easily verified that in all of the above cases, G_{k-1} must contain a claw, the base of which is one of the vertices of D, a contradiction to Lemma 2. □

In light of Lemma 7, we may immediately conclude that none of the edges of G_{s-1} that are a part of an induced diamond in either of the two left graphs depicted in Figure 4 are artificial. Thus, only two edges, denoted by e_1 and e_2 in Figure 4, can possibly be artificial. There are exactly five distinct graphs (up to isomorphism) that can therefore be candidates for G_{s-2}, which are those graphs that can be obtained by replacing e_1 or e_2 with a vertex of degree two, a diamond, or a double-diamond. There are five possible distinct graphs (up to isomorphism) that arise from this operation; it is an easy exercise to find an acyclic coloring for each of them (for space reasons, we omit the details, a fuller version of this paper demonstrating these coloring can be found on the first author's webpage).

Theorem 3. *Let* G *be a claw-free subcubic graph. Then, (i) there is a polynomial-time algorithm to compute an optimal acyclic coloring for* G, *and (ii)* G *admits an acyclic coloring with at most three colors, unless* G *is one of the three graphs shown in Figure 3.*

5.2   NP-Hardness for $d \geq 4$

Alon and Zaks [2] proved that the problem of deciding if a given cubic graph G can be acyclically edge colored using only k colors is NP-complete, even if $k = 3$. If we compute the line graph $L(G)$ of G, then $L(G)$ contains a vertex for every edge, and two vertices $\{e, e'\}$ are adjacent in $L(G)$ if and only if the corresponding edges are incident in G. Since G is cubic, $L(G)$ is 4-regular. Clearly, an acyclic vertex coloring of $L(G)$ corresponds to an acyclic edge coloring of G, and vice versa.

Theorem 4. *Deciding if the vertices of a line graph (and therefore, a claw-free graph) of maximum degree four can be acyclically colored using at most* k *colors is NP-complete even if* $k = 3$.

References

1. Alon, N., McDiarmid, C., Reed, B.: Acyclic colourings of graphs. Random Structures and Algorithms 2, 277–288 (1990)
2. Alon, N., Zaks, A.: Algorithmic aspects of acyclic edge colorings. Algorithmica 32, 611–614 (2002)
3. Burnstein, M.I.: Every 4-valent graph has an acyclic five coloring, Soobšč. Akad. Nauk Gruzin SSR 93, 21–24 (1979) (in Russian)
4. Călinescu, G., Fernandes, C.G., Finkler, U., Karloff, H.: A better approximation algorithm for finding planar subgraphs. Journal of Algorithms 27, 269–302 (1998)
5. Chudnovsky, M., Seymour, P.: The structure of claw-free graphs. In: Proceedings of the 20th British Combinatorial Conference, Surveys in Combinatorics 2005, Durham, pp. 153–171 (2005)
6. Edwards, K., Farr, G.: An Algorithm for Finding Large Induced Planar Subgraphs. In: Mutzel, P., Jünger, M., Leipert, S. (eds.) GD 2001. LNCS, vol. 2265, pp. 75–83. Springer, Heidelberg (2002)
7. Edwards, K., Farr, G.: Planarization and fragmentability of some classes of graphs. Discrete Mathematics 308, 2396–2406 (2008)
8. Faria, L., de Figueiredo, C.M.H., Mendonça, C.F.X.: Splitting number is NP-complete. In: Hromkovič, J., Sýkora, O. (eds.) WG 1998. LNCS, vol. 1517, pp. 285–297. Springer, Heidelberg (1998)
9. Faria, L., de Figueiredo, C.M.H., de Mendonça Neto, C.F.X.: On the complexity of the approximation of nonplanarity parameters for cubic graphs. Discrete Applied Mathematics 141(1-3), 119–134 (2004)
10. Faria, L., de Figueiredo, C.M.H., Gravier, S., de Mendonça Neto, C.F.X., Stolfi, J.: On maximum planar induced subgraphs. Discrete Applied Mathematics 154(13), 1774–1782 (2006)
11. Gebremedhin, A.H., Manne, F., Pothen, A.: What color is your Jacobian? Graph coloring for computing derivatives. SIAM Review 47, 629–705 (2005)
12. Kostochka, A., Stocker, C.: Graphs with maximum degree 5 are acyclically 7-colorable. Ars Mathematica Contemporanea 4, 153–164 (2011)
13. Liebers, A.: Planarizing graphs – a survey and annotated bibliography. Journal of Graph Algorithms and Applications 5(1), 1–74 (2001)
14. Liu, P.C., Geldmacher, R.C.: On the deletion of nonplanar edges of a graph. Cong. Numer. 24, 727–738 (1979)
15. Skulrattanakulchai, S.: Acyclic colorings of subcubic graphs. Information Processing Letters 92(4), 161–167 (2004)
16. Yannakakis, M.: Node and edge-deletion NP-complete problems. In: Proceedings of the 10th Annual ACM Symposium on Theory of Computing (STOC 1978), pp. 253–264 (1978)
17. Yannakakis, M.: Edge-deletion problems. SIAM J. Comput. 10, 297–309 (1981)
18. Zhang, X.-D., Bylka, S.: Disjoint triangles of a cubic line graph. Graphs and Combinatorics 20, 275–280 (2004)

List Coloring in the Absence of a Linear Forest[*]

Jean-François Couturier[1], Petr A. Golovach[2],
Dieter Kratsch[1], and Daniël Paulusma[2]

[1] Laboratoire d'Informatique Théorique et Appliquée,
Université Paul Verlaine - Metz, 57045 Metz Cedex 01, France
{couturier,kratsch}@univ-metz.fr
[2] School of Engineering and Computing Sciences, Durham University,
Science Laboratories, South Road, Durham DH1 3LE, United Kingdom
{petr.golovach,daniel.paulusma}@durham.ac.uk

Abstract. The k-COLORING problem is to decide whether a graph can be colored with at most k colors such that no two adjacent vertices receive the same color. The LIST k-COLORING problem requires in addition that every vertex u must receive a color from some given set $L(u) \subseteq \{1, \ldots, k\}$. Let P_n denote the path on n vertices, and $G + H$ and rH the disjoint union of two graphs G and H and r copies of H, respectively. For any two fixed integers k and r, we show that LIST k-COLORING can be solved in polynomial time for graphs with no induced $rP_1 + P_5$, hereby extending the result of Hoàng, Kamiński, Lozin, Sawada and Shu for graphs with no induced P_5. Our result is tight; we prove that for any graph H that is a supergraph of $P_1 + P_5$ with at least 5 edges, already LIST 5-COLORING is NP-complete for graphs with no induced H. We also show that LIST k-COLORING is fixed parameter tractable in $k + r$ on graphs with no induced $rP_1 + P_2$, and that k-COLORING restricted to such graphs allows a polynomial kernel when parameterized by k. Finally, we show that LIST k-COLORING is fixed parameter tractable in k for graphs with no induced $P_1 + P_3$.

1 Introduction

Graph coloring involves the labeling of the vertices of some given graph by integers called colors such that no two adjacent vertices receive the same color. The corresponding k-COLORING problem is to decide whether a graph can be colored with at most k colors. Due to the fact that k-COLORING is NP-complete for any fixed $k \geq 3$, there has been considerable interest in studying its complexity when restricted to certain graph classes. One of the most well-known results in this respect is due to Grötschel, Lovász, and Schrijver [11] who show that k-COLORING is polynomial-time solvable for perfect graphs. More information on this classic result and on the general motivation, background and related work on coloring problems restricted to special graph classes can be found in several surveys [24, 26] on this topic.

[*] This work has been supported by ANR Blanc AGAPE (ANR-09-BLAN-0159-03) and EPSRC (EP/G043434/1).

P. Kolman and J. Kratochvíl (Eds.): WG 2011, LNCS 6986, pp. 119–130, 2011.
© Springer-Verlag Berlin Heidelberg 2011

We continue the study of the computational complexity of the k-COLORING problem and related problems, in particular LIST k-COLORING when restricted to graph classes defined by one or more forbidden induced subgraphs. Such problems have been studied in many papers by different groups of researchers [3–7, 12, 15–19, 23, 27]. Before we summarize these results and explain our new results, we first state the necessary terminology and notations.

Terminology. We only consider finite undirected graphs $G = (V, E)$ without loops and multiple edges. We sometimes denote the vertex set of G by V_G. The subgraph of $G = (V, E)$ induced by $U \subseteq V$ is denoted by $G[U]$. We refer to Bondy and Murty [2] for any undefined graph terminology and to Downey and Fellows [8] and Niedermeier [22] for a discussion on parameterized complexity.

The graph P_n denotes the path on n vertices. The disjoint union of two graphs G and H is denoted $G + H$, and the disjoint union of r copies of G is denoted rG. A *linear forest* is the disjoint union of a collection of paths. Let $\{H_1, \ldots, H_p\}$ be a set of graphs. We say that a graph G is $(H_1, \ldots, H_p)$-*free* if G has no induced subgraph isomorphic to a graph in $\{H_1, \ldots, H_p\}$; if $p = 1$, we sometimes write H_1-free instead of (H_1)-free.

A *(vertex) coloring* of a graph $G = (V, E)$ is a mapping $\phi : V \to \{1, 2, \ldots\}$ such that $\phi(u) \neq \phi(v)$ whenever $uv \in E$. Here, $\phi(u)$ is referred to as the *color* of u. A k-*coloring* of G is a coloring ϕ of G with $\phi(V) \subseteq \{1, \ldots, k\}$. Here, we used the notation $\phi(U) = \{\phi(u) \mid u \in U\}$ for $U \subseteq V$. If G has a k-coloring, then G is called k-*colorable*. Recall that the problem k-COLORING is to decide whether a given graph admits a k-coloring. Here, k is *fixed*, i.e., not part of the input. If k is part of the input then we denote the problem as COLORING. The optimization version of this problem is to determine the *chromatic number* of a graph, i.e., the smallest k such that G has a k-coloring.

A *list assignment* of a graph $G = (V, E)$ is a function L that assigns a list $L(u)$ of so-called *admissible* colors to each $u \in V$. If $L(u) \subseteq \{1, \ldots, k\}$ for $u \in V$, then L is also called a k-*list assignment*. Equivalently, L is a k-list assignment if $|\bigcup_{u \in V} L(u)| \leq k$. We say that a coloring $\phi : V \to \{1, 2, \ldots\}$ *respects* L if $\phi(u) \in L(u)$ for all $u \in V$. For a fixed integer k, the LIST k-COLORING problem has as input a graph G with a k-list assignment L and asks whether G has a coloring that respects L. If $|L(u)| = 1$ for every vertex u of some subset $W \subseteq V$ and $L(u) = \{1, \ldots, k\}$ for $u \in V \setminus W$, then we obtain the k-PRECOLORING EXTENSION problem.

Related Work. Král', Kratochvíl, Tuza and Woeginger [17] completely determined the computational complexity of COLORING for graph classes characterized by a forbidden induced subgraph and achieved the following dichotomy.

Theorem 1 ([17]). *Let H be a fixed graph. If H is a (not necessarily proper) induced subgraph of P_4 or of $P_1 + P_3$ then COLORING can be solved in polynomial time for H-free graphs; otherwise it is NP-complete for H-free graphs.*

Theorem 1 can be extended in various ways. One way of doing this is to consider the computational complexity of COLORING for $\mathcal{H}$-free graphs where $\mathcal{H}$ is a family of two (or more) graphs. Some initial results have been obtained by Král' et al. [17], Schindl [25] and a number of authors studying the $\mathcal{H}$-free graphs, in which one of the two graphs in $\mathcal{H}$ is the triangle [5, 7, 15, 21]. Another way is to apply parameterized complexity to establish a more subtle classification of those problems being NP-complete. Finally, Theorem 1 can also be extended by classifying the computational complexity of k-COLORING and other variants of coloring for H-free graphs where k is a fixed integer and H is a fixed graph. The complexity classifications in all three directions are far from being finished. In this paper we consider the second and third direction. We focus on the case when H is a linear forest. Below we justify this.

Kamiński and Lozin [15] showed that for any $k \geq 3$, the k-COLORING problem is NP-complete for the class of graphs of girth (the length of a shortest induced cycle) at least p for any fixed $p \geq 3$. Their result implies that for any $k \geq 3$, the k-COLORING problem is NP-complete for the class of H-free graphs if H contains a cycle. Holyer [13] showed that 3-COLORING is NP-complete on line graphs. Later, Leven and Galil [20] extended this result by showing that k-COLORING is also NP-complete on line graphs for $k \geq 4$. Because line graphs are claw-free, i.e., they have no induced $K_{1,3}$, we find that for $k \geq 3$, the k-COLORING problem is NP-complete for the class of H-free graphs if H is a forest that contains a vertex with degree at least 3. Hence, only the case in which H is a linear forest remains.

It is known that 4-COLORING is NP-complete for P_8-free graphs [4] and that 6-COLORING is NP-complete for P_7-free graphs [3]. On the contrary, Randerath and Schiermeyer [23] showed that 3-COLORING can be solved in polynomial time for P_6-free graphs. A result which was generalized by Broersma et al. [3] who showed that 3-PRECOLORING EXTENSION can be solved in polynomial time for P_6-free graphs. Later, Broersma et al. [4] extended this result by showing that 3-PRECOLORING EXTENSION can be solved in polynomial time for H-free graphs if H is a linear forest on at most 6 vertices. The proof methods of both papers [3, 4] can directly be applied to show exactly the same results for LIST 3-COLORING. For P_5-free graphs, Hoàng et al. [12] could show a stronger result; note that COLORING is NP-complete for P_5-free graphs due to Theorem 1.

Theorem 2 ([12]). *For any fixed integer k, the* LIST k-COLORING *problem can be solved in polynomial time for P_5-free graphs.*

Our Results. The first aim of our paper is to generalize Theorem 2 as much as possible. We prove that for any fixed integers k and r, the LIST k-COLORING problem is polynomial-time solvable for $(rP_1 + P_5)$-free graphs. In order to prove our result, we show that our input graphs have a dominating set of small size should they be k-colorable. Hence, we search for such a dominating set. If we find it, then we color its vertices in every possible way. Afterwards, we use the technique of "separating the color lists of independent sets" of Hoàng et al. [12] on each resulting instance. They successfully applied this technique for coloring P_5-free graphs, and our result for $(rP_1 + P_5)$-free graphs can be seen as a second example of its usefulness. We present this technique in Section 2 in a more generic way. In order to obtain our result for $(rP_1 + P_5)$-free graphs we have to prove a number of additional structural results. This is done in Section 3. There, we also show that our result is *tight* by proving that already LIST 5-COLORING is

NP-complete for the class of H-free graphs whenever H has at least 5 edges and contains $P_1 + P_5$ as a subgraph.

The second aim of our paper is to initiate a parameterized complexity study for the k-COLORING and LIST k-COLORING problem restricted to H-free graphs, when H is some fixed linear forest. In Section 4 we prove the following three results: (i) LIST k-COLORING is fixed parameter tractable in $k + r$ for $(rP_1 + P_2)$-free graphs; (ii) k-COLORING restricted to $(rP_1 + P_2)$-free graphs allows a polynomial kernel when parameterized by k; and (iii) LIST k-COLORING is fixed parameter tractable in k for $(P_1 + P_3)$-free graphs.

2   A Generic Approach for Coloring H-Free Graphs

We generalize the technique Hoáng et al. [12] used to prove Theorem 2.

Given a graph $G = (V, E)$ with a k-list assignment L, we use the following terminology. Two adjacent vertices u and v are *essential* if $L(u) \cap L(v) \neq \emptyset$; otherwise u and v are *non-essential*. We observe that u is an essential neighbor of v if and only if v is an essential neighbor of u. Two disjoint sets of vertices are *separated* for L if no vertex in one of them has an essential neighbor in the other. Let $\mathcal{L}$ be a set of k-list assignments of G with $L'(u) \subseteq L(u)$ for all $L' \in \mathcal{L}$ and $u \in V$. Then L and $\mathcal{L}$ are *compatible* if the following holds: G has a coloring respecting L' for some $L' \in \mathcal{L}$ if G has a coloring respecting L. Note that the reverse implication holds by the definition of $\mathcal{L}$.

Assigning an admissible color to a vertex u does not influence the choice of admissible colors for its non-essential neighbors. Hence, in our coloring algorithm, we would like to branch in such a way that we obtain a compatible set of list assignments for which disjoint sets of vertices become separated. Then we can apply the algorithm recursively on smaller graphs induced by these disjoint sets. This idea has been applied more often but usually leads to a huge case analysis. However, Hoáng et al. [12] developed an elegant technique, which works well for P_5-free graphs. We present it in a more generic way below.

A subset $D \subseteq V$ is a *dominating* set of G if every vertex in G belongs to D or is adjacent to a vertex of D. In that case we also say that $G[D]$ is *dominating*. Suppose that we have ordered the vertices of D as $d_1, \ldots, d_p$. Then we can define (possibly empty) sets F_i for $i = 1, \ldots, p$ as follows. Let F_1 be the set of vertices in $V \setminus D$ adjacent to d_1, and for $i = 2, \ldots, p$, let F_i be the set of vertices in $V \setminus D$ adjacent to d_i but not to any d_h with $h \leq i - 1$. The sets $F_1, \ldots, F_p$ are called *fixed* sets for D. By this definition and because D is dominating, every vertex in $V \setminus D$ belongs to exactly one fixed set F_i. We note, however, that D can have several collections of fixed sets, depending on the ordering of the vertices of D. A subset $X \subseteq V$ is *independent* if there is no edge between any two vertices of X. We call a graph H a *dominator-separator* graph if every connected H-free graph $G = (V, E)$ satisfies the following two properties.

(i) If G is k-colorable for some integer $k \geq 1$, then G has a dominating set D of at most $f(k)$ vertices, where f is a function that only depends on k.

(ii) There exists a polynomial-time algorithm that on input G, two independent sets X and Y that are subsets of two different fixed sets of a dominating set of G and a k-list assignment L of G outputs a set $\mathcal{L}$ of k-list assignments of G with $L'(u) \subseteq L(u)$ for all $L' \in \mathcal{L}$ and all $u \in V$, such that

    1. $\mathcal{L}$ is compatible with L;
    2. $|\mathcal{L}| = O(h(k)n^{g(k)})$ for some functions $h(k)$ and $g(k)$ that only depend on k;
    3. X and Y are separated for every $L' \in \mathcal{L}$.

By a straightforward translation of the proof of Hoàng et al. [12] one finds that for P_5-free graphs, $f(k) = k$ satisfies property (i), whereas $h(k) = k^k$ and $g(k) = k$ satisfy property (ii). Hence, P_5 is a dominator-separator. The following theorem generalizes their approach. Its proof is a reformulation of their proof in terms of dominator-separator graphs. As such, we omit it from our paper.

Theorem 3. *Let H be a dominator-separator graph, and let k be a fixed integer. Then* LIST k-COLORING *can be solved in polynomial time for H-free graphs.*

3   Coloring $(rP_1 + P_5)$-Free Graphs

In order to apply Theorem 3 we must prove that $rP_1 + P_5$ is a dominator-separator graph for any fixed r. We start with the following more general lemma that we use in Section 4 as well. We omit its proof.

Lemma 1. *Let G be an $(rP_1 + P_\ell)$-free graph for integers r and ℓ. If G contains an induced P_ℓ, then G contains a dominating induced $sP_1 + P_\ell$ for some $s < r$.*

A vertex subset K in a graph G is called a *clique* of G if there is an edge between any two vertices of K. Just as Hoáng et al. [12], we need the following result of Bacsó and Tuza [1] for the class of connected P_5-free graphs.

Theorem 4 ([1]). *Every connected P_5-free graph G has a dominating P_3 or a dominating clique.*

We are now ready to show the following two lemmas which together show that $rP_1 + P_5$ is a dominator-separator for any fixed integer r.

Lemma 2. *Every connected $(rP_1 + P_5)$-free graph satisfies property (i).*

Proof. Let G be a connected $(rP_1 + P_5)$-free graph that is k-colorable for some integer $k \geq 1$. We show that G has a dominating set of size at most $\max\{3, k, r+4\}$. Then we may define $f(k) = \max\{3, k, r+4\}$ for all $k \geq 1$. This function only depends on k, because r is fixed. If G is P_5-free, then G has a dominating P_3 or a dominating clique due to Theorem 4. Because G is k-colorable, any clique in G has at most k vertices. Hence we find a dominating set of size 3 or of size at most k. If G is not P_5-free, then by Lemma 1, G has a dominating induced $sP_1 + P_5$ for some $s < r$. Hence, we find a dominating set of size $s + 5 \leq r + 4$. □

Lemma 3. *Every connected $(rP_1 + P_5)$-free graph satisfies property (ii).*

Proof sketch. Let $G = (V, E)$ be a connected $(rP_1 + P_5)$-free graph on n vertices with k-list assignment L. Let $D = \{d_1, \ldots, d_p\}$ be a dominating set of G, and let $F_1, \ldots, F_p$ be the collection of fixed sets for D. For some $1 \leq i < j \leq n$, let $X \subseteq F_i$ and $Y \subseteq F_j$ be two independent sets of G. Note that $i < j$ implies that d_i is not adjacent to any vertex in F_j, whereas d_j might be adjacent to one or more vertices of F_i.

Let the set C consist of every color c for which there exist two adjacent vertices $x \in X$ and $y \in Y$ such that $c \in L(x) \cap L(y)$. By definition, such x and y are essential neighbors of each other. If $C = \emptyset$, then X and Y are separated.

Suppose that $C \neq \emptyset$. We define a set X' as the set of all vertices in X that have an essential neighbor in Y, and a set Y' as the set of all vertices in Y that have an essential neighbor in X'. Because $C \neq \emptyset$, both X' and Y' are nonempty. Our goal is to reduce the size of X'. The reason is that when X' becomes empty, then C' will be empty, and consequently, X and Y will be separated.

We will use the following claim, the proof of which we omit. We say that $x \in X'$ is *maximal* if there is no vertex in X' that has more neighbors in Y' than x has. We say that a vertex $z \in X'$ is an *associate* of x if at least $|Y'| - r + 1$ vertices in Y' are adjacent to x or z.

Claim 1. Let $x \in X'$ be maximal. Then either x is adjacent to all vertices of Y', or every vertex in X' that is adjacent to a non-neighbor of x in Y' is an associate of x.

We are now ready to describe our algorithm that we use to prove property (ii). Recall that our goal is to reduce the size of X'. Hence, we branch on vertices of X'. Because X' may have a large size, we cannot branch by arbitrarily assigning colors to vertices of X'. Therefore, we do as follows as long as $X' \neq \emptyset$.

Determine a maximal vertex $x \in X'$ and start to branch on x.

Our algorithm either assigns to x a specific color c from C, creating a number of branches, or no color from C at all, yet another branch. In a branch of the first type we cannot only remove x from X' but we will also show that we may remove c from C; this is crucial for the running time analysis which we do afterwards. If x is not adjacent to every vertex in Y', then we may need to refine the branching by involving the associates of x. In a branch of the second type we remove every color in C from the list of x. Consequently, x can be removed from X' as desired (but we might not have decreased the size of C in this case).

The procedure `Reduce-to-empty-set` explains our approach in detail; see Pseudocode 2. Here, *updating* a list assignment after a vertex gets a color means removing this color from the list of every neighbor of that vertex. Further, for $x \in X'$, the set A_c^x denotes the set of associates of x that have color c in their list and that are adjacent to a vertex in Y' that is no neighbor of x. Finally, we note that at some places in this procedure we could also reduce the set Y'. However, for simplicity, we refrain from doing this, except in line 14 where it is necessary for the correctness.

We will use the `Reduce-to-empty-set` procedure as a subroutine inside our separation algorithm called `Separator`; see Pseudocode 1. The output of `Separator` is a set $\mathcal{L}$ of k-list assignments of G; at the start we set $\mathcal{L} = \emptyset$.

Separator

input   : sets X and Y
output : a set $\mathcal{L}$ of k-list assignments

1. determine the sets X', Y' and C
2. set $\mathcal{L} := \emptyset$
3. `Reduce-to-empty-set`$(X', Y', C, \mathcal{L})$
4. return $\mathcal{L}$

Pseudocode 1. Separating the two sets X and Y

`Reduce-to-empty-set`$(X', Y', C, \mathcal{L})$

1. **while** $X' \neq \emptyset$
2.   determine a maximal vertex $x \in X'$
3.   **for** every color $c \in C$ that is in the list of x **do**
4.   color x by c and update the list assignment
5.   determine the set A_c^x
6.   **if** $A_c^x = \emptyset$ **then**
7.   `Reduce-to-empty-set`$(X' \setminus \{x\}, Y', C \setminus \{c\}, \mathcal{L})$
8.   **else**
9.   **for** every $z \in A_c^x$ **do**
10.   color z by c and update the list assignment
11.   determine the set $Y'' \subseteq Y'$ of vertices that have c in their list
12.   **for** every coloring ϕ of Y'' that respects the lists **do**
13.   color Y'' according to ϕ and update the list assignment
14.   `Reduce-to-empty-set`$(X' \setminus \{x, z\}, Y' \setminus Y'', C \setminus \{c\}, \mathcal{L})$
15.   **end for**
16.   remove c from the lists of every vertex in A_c^x
17.   `Reduce-to-empty-set`$(X' \setminus \{x\}, Y', C' \setminus \{c\}, \mathcal{L})$
18.   **end for**
19.   **end if**
20.   remove every color in C from the list of x
21.   `Reduce-to-empty-set` $(X' \setminus \{x\}, Y', C, \mathcal{L})$
22.   **end for**
23. **end while**
24. put the obtained list assignment in $\mathcal{L}$

Pseudocode 2. Reducing the set X' to the empty set

Having completed the overall description of our branching algorithm we now prove that G satisfies property (ii).

From the description of the procedure Reduce-to-empty-set, we conclude that each time we process a maximal vertex $x \in X'$, the size of X' reduces by at least one vertex. Hence, this procedure will always terminate, and when it does X' will be empty. Consequently, our algorithm Separator will terminate as well. When it does, it will return as output a set $\mathcal{L}$ of k-list assignments of G. The sets X and Y are separated for each k-list assignment of $\mathcal{L}$, because X', and consequently, C are empty for each such list assignment. In other words, condition 3 of property (ii) is satisfied. Below we show that conditions 1 and 2 are also satisfied.

The procedure Reduce-to-empty-set only reduces lists of vertices of G. As a consequence, every list assignment $L' \in \mathcal{L}$ has the property that $L'(u) \subseteq L(u)$ for all $u \in V$. We will show that L and $\mathcal{L}$ are compatible.

In order to show this suppose that G has a coloring ϕ respecting L. Let $x \in X'$ be the maximal vertex that is under consideration. We show that in the search tree that represents our recursive procedure, there exist a branch that we can follow in order to prove the existence of a list assignment $L' \in \mathcal{L}$ that is respected by ϕ. The line numbers in our proof refer to lines in the Reduce-to-empty-set procedure.

If $\phi(x) \in C$, then we follow the branch that assigns color c to x in one of the executions of line 4. Afterwards, we may safely update the list assignment.

If $A_c^x = \emptyset$, then Claim 1 tells us that there is no vertex in X' left that has color c in its list and that is adjacent to a vertex in Y' with c in its list; if there were such vertices they would have been associates of x. Hence, we may remove c from C and x from X', as is done in line 7.

If $A_c^x \neq \emptyset$, then there are two cases to consider.

Case 1. At least one vertex $z \in A_c^x$ has color $\phi(z) = c$.
We will detect this case in one of the execution of line 10. If after updating the list assignment there is still a set Y'' of vertices in Y' left, then we will consider the coloring according to ϕ in one of the executions of line 13. We follow the corresponding branch that colors the vertices of Y'' according to ϕ. Afterwards, we may remove the vertices of Y'' from Y' as is done in line 14. Consequently, the lists of the remaining vertices of Y' do not contain c anymore. Hence, we may remove c from C in line 14. Because x and z received a color, we may remove x and z from X'; this is done in line 14 as well.

Case 2. None of the vertices in A_c^x has color c according to ϕ.
In this case we follow the branch that removes c from the lists of every vertices in A_c^x; see line 16. We claim that c is not in C anymore. This can be seen as follows. In order to obtain a contradiction suppose that $c \in C$. Then there are two adjacent vertices $x^* \in X'$ and $y^* \in Y'$ that each have c in their list. Because x received color c and we removed c from the lists of its neighbors, we find that y^* is no neighbor of x. However, then x^* must be in A_c^x by the definition of this set and Claim 1. This is not possible either, because we removed c from the list of every vertex in A_c^x. We conclude that $c \notin C$. Hence, we may remove c from C

in line 17, and as before, we may also remove x from X', which is done in line 17 as well.

Finally, we consider the case in which $\phi(x) \notin C$. In this case, we follow the branch that removes every color in C from the list of x; see line 20. Afterwards, we may remove x from X', as is done in line 21. We conclude that for every maximal vertex x, there exists a branch that assigns color $\phi(x)$ to x and that the adjustments in the sets X', Y' and C in lines 7, 14, 17 and 21 are permitted. Following these branches leads to a k-list assignment $L' \in \mathcal{L}$ that is respected by ϕ, as desired. This completes our proof of condition 1 of property (ii).

We are left to prove condition 2 of property (ii), namely that our algorithm Separator runs in polynomial time and that $|\mathcal{L}| = O(h(k)n^{g(k)})$ for some functions $h(k)$ and $g(k)$ that only depend on k. We note that the sets X', Y' and C can be computed in polynomial time. By the construction of the Reduce-to-empty procedure, each k-list assignment in $\mathcal{L}$ is the output of exactly one leaf of the search tree T. This means that the number of leaves of T is an upper bound for the number of the k-list assignments of $\mathcal{L}$. Also, finding a maximal vertex, assigning it a color and updating its list and the lists of its neighbors takes polynomial time. Hence our algorithm runs in polynomial-time if the number of leaves in T is $O(h(k)n^{g(k)})$ for some functions $h(k)$ and $g(k)$ that only depend on k.

Let ℓ be a leaf of T. Then there exists a sequence of vertices of X', on which we branched in order to arrive at ℓ. Each of these vertices was a maximal vertex at the moment it was considered. We call these vertices the ℓ-vertices. The procedure Reduce-to-Empty only assigns a color from C to a vertex in X' if it can remove this color from C afterwards. Maintaining this property has the following two consequences. First, the number of ℓ-vertices that received a color from C is at most $|C|$; all other ℓ-vertices got their list reduced by removing the colors of C. Second, no two ℓ-vertices received the same color from C. Recall that every vertex in every nonempty set A_c^x determined in line 5 is an associate of the minimal vertex x under consideration. Then, by definition, every set Y'' determined in line 11 has size at most $r - 1$. For a leaf ℓ of T, we let C_ℓ denote the set of colors from C used on the ℓ-vertices. Using the above observations, we can determine that the number of leaves of T and consequently the number of k-list assignments of $\mathcal{L}$ is at most $2^k \cdot n^k \cdot k! \cdot n^k \cdot r^k n^{rk} \cdot k^r$. Hence, we can set $h(k) = 2^k k! r^k k^r$ and $g(k) = 2k + rk$. This completes the proof of Lemma 3. □

Due to Lemmas 2 and 3, the graph $rP_1 + P_5$ is a dominator-separator for every fixed integer r. Hence we can apply Theorem 3 and obtain the main result of this section.

Theorem 5. *For any fixed integers k and r, the* LIST k-COLORING *problem can be solved in polynomial time for $(rP_1 + P_5)$-free graphs.*

This theorem is best possible in the sense that LIST k-COLORING becomes NP-complete for some integer k on H-free graphs, whenever H is a supergraph of $P_1 + P_5$ with at least 5 edges. Theorem 6 shows this for $k = 5$; we omit its proof.

Theorem 6. *Let H be a supergraph of $P_1 + P_5$ with at least 5 edges. Then* LIST *5-COLORING is* NP-*complete for H-free graphs.*

4   Parameterized Complexity Results

By Theorem 5, LIST k-COLORING is in XP for $(rP_1 + P_5)$-free graphs when k is the parameter and r is fixed. In this section we show that LIST k-COLORING is in FPT for graph classes defined by taking a smaller linear forest as the forbidden induced subgraph.

First we consider $(rP_1 + P_2)$-free graphs. Theorem 1 tells us that already COLORING is NP-complete for $(rP_1 + P_2)$-free graphs whenever $r \geq 2$. For a graph $G = (V, E)$, we let $N(u) = \{v \in V \mid uv \in E\}$ denote the set of neighbors of a vertex $u \in V$, $N(S) = \{v \in V \setminus S \mid uv \in E$ for some $u \in S\}$ denotes the set of neighbors of a set $S \subseteq V$, and $N[S] = N(S) \cup S$.

Let G be a graph with a k-list assignment L. Let $\mathcal{G} = \{G_1, \ldots, G_p\}$ be a set of graphs, where each G_i has a $(k-1)$-list assignment L_i. Then we say that G and $\mathcal{G}$ are $(k-1)$-*compatible* if the following holds: G has a coloring respecting L if and only if there exists a graph $G_i \in \mathcal{G}$ that has a coloring respecting L_i.

We need the following two lemmas; we omit their proofs.

Lemma 4. *Let $k \geq 2$ and $r \geq 1$. Let $G = (V, E)$ be an $(rP_1 + P_2)$-free graph on n vertices with a k-list assignment L. If G has a maximal independent set X with at least $(r-1)k + 1$ vertices, then it is possible to find in $O(k^2 n)$ time a $(k-1)$-compatible set $\mathcal{G}$ that consists of at most k induced subgraphs of G.*

Lemma 5. *Let $k \geq 2$ and $r \geq 1$. Let G be an $(rP_1 + P_2)$-free graph with $n \geq (r+1)^{k-1}((r-1)k + 1) + (r+1)\frac{(r+1)^{k-1}-1}{r}$ vertices and m edges. Then either G has a clique of size $k + 1$ or a maximal independent set X of size at least $(r-1)k + 1$. Moreover, it is possible to find such a clique or independent set in $O(k(n+m))$ time.*

Now we are ready to prove the following result.

Theorem 7. *The* LIST k-COLORING *problem is in* FPT *for $(rP_1 + P_2)$-free graphs when parameterized by k and r.*

Proof. Let G be an $(rP_1 + P_2)$-free graph on n vertices that has a k-list assignment L. If $k \leq 2$, then we can solve the problem in polynomial time. If $n < f(k, r) = (r+1)^{k-1}((r-1)k + 1) + (r+1)\frac{(r+1)^{k-1}-1}{r}$, then we can solve it in $O(f(k,r)^k)$ time by brute force. Otherwise, by Lemma 5, we either find a clique of size $k + 1$ or a maximal independent set of size at least $(r-1)k + 1$ in $O(k(n+m))$ time. In the first case, G has no coloring respecting L. In the second case, we construct in $O(k^2 n)$ time a $(k-1)$-compatible set $\mathcal{G}$ of at most k subgraphs of G by using Lemma 4. We branch on each of them and repeat the same steps. Since the depth of the search tree is bounded by k, the desired result follows.   □

If we only choose k as the parameter, then we can improve our result for the k-COLORING problem as shown in Theorem 8, the proof of which we omit. Here, we assume that $r \geq 2$ because COLORING can be solved in polynomial time for $(rP_1 + P_2)$-free graphs with $r \leq 1$, due to Theorem 1.

Theorem 8. *For any fixed integer $r \geq 2$, the k-COLORING problem restricted to $(rP_1 + P_2)$-free graphs has a kernel of size $k^2(r-1)$ when parameterized by k.*

We now consider $(P_1 + P_3)$-free graphs. Recall that COLORING is polynomial-time solvable for $(P_1 + P_3)$-free graphs due to Theorem 1. However, Jansen and Scheffler [14] showed that LIST k-COLORING is NP-complete when k is part of the input, already for complete bipartite graphs which form a subclass of the class of $(P_1 + P_3)$-free graphs. We show the following result, the proof of which we omit.

Theorem 9. *The LIST k-COLORING problem is in FPT for $(P_1 + P_3)$-free graphs when parameterized by k.*

5   Future Work

Theorem 5 implies that for any fixed integer k and any fixed graph H on at most 5 vertices, LIST k-COLORING is polynomially solvable, except when $H = P_2 + P_3$.

1. Is LIST k-COLORING polynomial-time solvable on $(P_2 + P_3)$-free graphs for any fixed k?

Due to the aforementioned polynomial-time result on LIST 3-COLORING for sP_3-free graphs [4], the first open case is $k = 4$. We note that the same question is also open with respect to k-COLORING. For this problem, the first open case is $k = 5$, as it is known that 4-COLORING is polynomial-time solvable on $(P_2 + P_3)$-free graphs [10]. A possible solution strategy would be to prove that $P_2 + P_3$ is a dominator-separator graph but this seems to be difficult.

Jansen and Scheffler [14] showed that LIST k-COLORING is in FPT for P_4-free graphs when parameterized by k. This result together with Theorems 7 and 9 implies that the two smallest open cases parameterized by k are the cases $H = 2P_2$ and $H = 2P_1 + P_3$.

2. Is LIST k-COLORING parameterized by k in FPT for $2P_2$-free graphs?
3. Is LIST k-COLORING parameterized by k in FPT for $(2P_1 + P_3)$-free graphs?

References

1. Bacsó, G., Tuza, Z.: Dominating cliques in P_5-free graphs. Periodica Mathematica Hungarica 21, 303–308 (1990)
2. Bondy, J.A., Murty, U.S.R.: Graph Theory. Springer Graduate Texts in Mathematics, vol. 244 (2008)
3. Broersma, H., Fomin, F.V., Golovach, P.A., Paulusma, D.: Three Complexity Results on Coloring P_k-Free Graphs. In: Fiala, J., Kratochvíl, J., Miller, M. (eds.) IWOCA 2009. LNCS, vol. 5874, pp. 95–104. Springer, Heidelberg (2009)

4. Broersma, H.J., Golovach, P.A., Paulusma, D., Song, J.: Updating the complexity status of coloring graphs without a fixed induced linear forest (manuscript)
5. Broersma, H.J., Golovach, P.A., Paulusma, D., Song, J.: Determining the chromatic number of triangle-free $2P_3$-free graphs in polynomial time (manuscript)
6. Bruce, D., Hoàng, C.T., Sawada, J.: A Certifying Algorithm for 3-Colorability of P_5-Free Graphs. In: Dong, Y., Du, D.-Z., Ibarra, O. (eds.) ISAAC 2009. LNCS, vol. 5878, pp. 594–604. Springer, Heidelberg (2009)
7. Dabrowski, K., Lozin, V., Raman, R., Ries, B.: Colouring Vertices of Triangle-Free Graphs. In: Thilikos, D.M. (ed.) WG 2010. LNCS, vol. 6410, pp. 184–195. Springer, Heidelberg (2010)
8. Downey, R.G., Fellows, M.R.: Parameterized Complexity. Springer, Heidelberg (1999)
9. Garey, M.R., Johnson, D.S.: Computers and Intractability: A Guide to the Theory of NP-Completeness. Freeman, San Francisco (1979)
10. Golovach, P.A., Paulusma, D., Song, J.: 4-Coloring H-free graphs when H is small (manuscript)
11. Grötschel, M., Lovász, L., Schrijver, A.: Polynomial algorithms for perfect graphs. Ann. Discrete Math., Topics on Perfect Graphs 21, 325–356 (1984)
12. Hoàng, C.T., Kamiński, M., Lozin, V., Sawada, J., Shu, X.: Deciding k-colorability of P_5-free graphs in polynomial time. Algorithmica 57, 74–81 (2010)
13. Holyer, I.: The NP-completeness of edge-coloring. SIAM J. Comput. 10, 718–720 (1981)
14. Jansen, K., Scheffler, P.: Generalized coloring for tree-like graphs. Discrete Appl. Math. 75, 135–155 (1997)
15. Kamiński, M., Lozin, V.V.: Coloring edges and vertices of graphs without short or long cycles. Contributions to Discrete Math. 2, 61–66 (2007)
16. Kamiński, M., Lozin, V.V.: Vertex 3-colorability of Claw-free Graphs. Algorithmic Operations Research 21 (2007)
17. Král', D., Kratochvíl, J., Tuza, Z., Woeginger, G.J.: Complexity of coloring graphs without forbidden induced subgraphs. In: Brandstädt, A., Le, V.B. (eds.) WG 2001. LNCS, vol. 2204, pp. 254–262. Springer, Heidelberg (2001)
18. Kratochvíl, J.: Precoloring extension with fixed color bound. Acta Math. Univ. Comen. 62, 139–153 (1993)
19. Le, V.B., Randerath, B., Schiermeyer, I.: On the complexity of 4-coloring graphs without long induced paths. Theoret. Comput. Sci. 389, 330–335 (2007)
20. Leven, D., Galil, Z.: NP completeness of finding the chromatic index of regular graphs. Journal of Algorithms 4, 35–44 (1983)
21. Maffray, F., Preissmann, M.: On the NP-completeness of the k-colorability problem for triangle-free graphs. Discrete Math. 162, 313–317 (1996)
22. Niedermeier, R.: Invitation to Fixed-Parameter Algorithms. Oxford Lecture Series in Mathematics and its Applications. Oxford University Press (2006)
23. Randerath, B., Schiermeyer, I.: 3-Colorability $\in$ P for P_6-free graphs. Discrete Appl. Math. 136, 299–313 (2004)
24. Randerath, B., Schiermeyer, I.: Vertex colouring and forbidden subgraphs - a survey. Graphs Combin. 20, 1–40 (2004)
25. Schindl, D.: Some new hereditary classes where graph coloring remains NP-hard. Discrete Math. 295, 197–202 (2005)
26. Tuza, Z.: Graph colorings with local restrictions - a survey. Discuss. Math. Graph Theory 17, 161–228 (1997)
27. Woeginger, G.J., Sgall, J.: The complexity of coloring graphs without long induced paths. Acta Cybernet. 15, 107–117 (2001)

Parameterized Complexity of Eulerian Deletion Problems

Marek Cygan[1], Dániel Marx[2], Marcin Pilipczuk[1],
Michał Pilipczuk[1], and Ildikó Schlotter[3,*]

[1] Institute of Informatics, University of Warsaw, Poland[**]
{cygan,malcin}@mimuw.edu.pl, mp248287@students.mimuw.edu.pl
[2] Institut für Informatik, Humboldt-Universität zu Berlin, Germany
dmarx@cs.bme.hu
[3] Department of Computer Science and Information Theory,
Budapest University of Technology and Economics, Hungary
ildi@cs.bme.hu

Abstract. We study a family of problems where the goal is to make a graph Eulerian by a minimum number of deletions. We completely classify the parameterized complexity of various versions: undirected or directed graphs, vertex or edge deletions, with or without the requirement of connectivity, etc. Of particular interest is a randomized FPT algorithm for making an undirected graph Eulerian by deleting the minimum number of edges.

1 Introduction

An undirected graph is Eulerian if it is connected and every vertex has even degree; a directed graph is Eulerian if it is strongly connected and every vertex is balanced (i.e., the indegree equals the outdegree). The class of Eulerian graphs is a well-studied and classical notion in the graph theory. We investigate several algorithmic problems related to the question of how to make a graph Eulerian. We focus on deletion problems, where either vertices or edges can be deleted from the input graph to make it Eulerian, using as few deletions as possible. What makes these problems interesting is the interplay of two different type of constraints: each vertex locally prescribes the constraint that it has to be even/balanced, while retaining connectivity is a global requirement. For comparison, we also investigate the variant of the problem where we have only the local constraints (i.e., the task is to delete the minimum number of edges or nodes to make every vertex even/balanced). As many of the studied problems turn out to be NP-hard, we apply the framework of parameterized complexity to get a more detailed insight.

The investigation of these problems was initiated by Cai and Yang [9] who presented parameterized results for some cases. We complement their work by answering here several open questions raised in [9]. Another motivation for our work comes from an

---

* Supported by the Hungarian National Research Fund (grant OTKA 67651), and by the European Union and the European Social Fund (grant TÁMOP 4.2.1./B-09/1/KMR-2010-0003).
** Authors from the University of Warsaw are partially supported by the Polish Ministry of Science grant N206 567140 and Foundation for Polish Science.

observation of Cechlárová and Schlotter [10]: computing the deficiency for a certain type of housing market is equivalent to finding the minimum number of arcs whose deletion makes every strongly connected component of the graph balanced. While we are not able to determine the parameterized complexity of this problem, our results shed light on the complexity of several related problems.

Related Work. Subgraph problems have been widely studied in the literature. To name a few examples, Lewis and Yannakakis [20] investigated the complexity of the vertex-deletion problem for hereditary properties, Alon et al. [2] examined edge-deletion problems for monotone properties, while Natanzon et al. [25] and Burzyn et al. [6] studied the classical complexity of edge modification problems for various graph classes.

Subgraph problems have also been looked at from the parameterized perspective. The most extensively studied variants are the vertex-deletion problems for hereditary properties: the results by Cai [8], and Khot and Raman [17], yield a complete characterization of the fixed-parameter tractable cases. Apart from hereditary properties, FPT algorithms are known for vertex-deletion problems where the task is to obtain a regular graph [24], a chordal graph [22], a grid [12], etc. Parameterized hardness results have been obtained in numerous cases as well [21,23]. Recently, researchers focused on the issue of kernelization, yielding both positive [4,16,26] and negative results [19].

There is much less known about directed graphs. Raman and Sikdar [29] investigated the parameterized complexity of hereditary vertex-deletion problems in digraphs, while Raman and Saurabh [28] examined feedback set problems in tournaments. The FPT algorithm by Chen et al. for finding a feedback vertex set in a directed graph [11] resolved a long-standing open question.

Work related to the class of Eulerian graphs mainly concentrated on the extension problem, where the task is to add a minimum number of edges or arcs in order to make the given graph Eulerian. FPT algorithms were given for various settings by Dorn et al. [13] and by Sorge [30]. Eulerian deletion problems were studied by Cai and Yang [9].

Our Contribution. To settle the classical complexity of the examined problems, first we observe (Thms. 1 and 2) that classical results imply polynomial-time algorithms for the edge-deletion problems where the task is to make the graph even/balanced: in the undirected case, this is essentially a T-join problem, while the directed case can be reduced to a flow problem. These observations answer a question raised by Cai and Yang [9], who observed that the analogous vertex-deletion problems are NP-hard. Moreover, the aforementioned algorithms are used as subroutines in our FPT results.

By contrast to the polynomial time algorithms, we show that the seemingly similar edge- (or arc-) deletion problems where we aim for an Eulerian graph are NP-hard, even in the extremely restricted case when the input is a cubic planar graph and the number of deletions can be arbitrary (Theorem 3). We investigate both the undirected and the directed cases of Eulerian edge-deletion problem thoroughly from the parameterized point of view: we present a fixed-parameter tractable algorithm for both cases where the parameter is the number of deletions allowed (Theorem 8), and prove that these problems do not admit a polynomial-size kernel unless NP $\subseteq$ coNP/poly (Theorem 9), which is known to imply a collapse of the polynomial hierarchy to its third level [31,7]. The FPT results use a novel argument that might be of independent interest. Intuitively,

Table 1. Summary of the main results. Parameterized results only appear when the corresponding problem is NP-hard; the parameter considered is the number of deletions allowed.

	Undirected even	Undirected Eulerian	Directed balanced	Directed Eulerian
Vertex deletion:	W[1]-hard [9]	W[1]-hard [9]	W[1]-hard Thm. 15	W[1]-hard Thm. 15
Edge deletion:	P Thm. 1	FPT, no poly kernel Thms. 3, 8, 9	P Thm. 2	FPT, no poly kernel Thms. 3, 8, 9

we need to find a solution S to a T-join problem and a witness (disjoint from S) certifying that the graph remains connected after the removal of S. Using a random colouring, we partition the edges into two types: each edge can contribute either to the solution or to the witness of the solution. This partition ensures that the solution and the witness are disjoint. While the use of random colourings is a standard technique for finding a solution consisting of disjoint objects [3], we use this technique to separate the solution from its proof of feasibility.

The undirected vertex-deletion problems, where the task is to obtain an Eulerian or an even graph, were already handled by Cai and Yang [9] who proved their W[1]-hardness. We complemented these results by showing W[1]-hardness for the directed cases as well in Theorem 15. Additionally, we also focus on a slight modification of the node-deletion problems where certain forbidden vertices are not allowed to be deleted. Theorem 16 shows that each of the four node-deletion problems remains W[1]-hard, even if we are only allowed to delete vertices of degree at most 4. This contrasts the easy FPT algorithm applicable if the parameter is not only the number of deletions but also the maximum degree of the graph (this algorithm will be included in the full version of the paper).

Table 1 shows a summary of our main results.

Organization of the Paper. Section 2 describes our notation, and provides basic concepts of parameterized complexity. Section 3 discusses polynomial-time solvable edge-deletion problems. We deal with the NP-hard Eulerian edge-deletion problems in Section 4, first covering the issue of NP-completeness, and then fixed-parameter tractability and kernelization in Sections 4.1 and 4.2. Node-deletion problems are discussed in Section 5. We summarize our results and draw conclusions in Section 6.

2 Notation and Preliminaries

Given a graph G, let $V(G)$ denote its vertex set and $E(G)$ denote its edge set (or, in the directed case, its arc set). The *degree* of a vertex v in an undirected graph G is denoted by $d_G(v)$; we say that v is *even*, if $d_G(v)$ is even. For a vertex v in a directed graph G, we denote by $d_G^{in}(v)$ and $d_G^{out}(v)$ its indegree and its outdegree, respectively. We say that v is *balanced*, if $d_G^{in}(v) = d_G^{out}(v)$. We define the *degree* of v in G (where G is directed), as $d_G(v) = d_G^{in}(v) + d_G^{out}(v)$. If G is clear from the context, we might omit the subscript. A directed graph is *weakly connected* if the underlying undirected graph is connected. An *even (balanced) graph* is an undirected (directed) graph where each vertex is even (balanced). An undirected Eulerian graph is a connected even graph, and a directed Eulerian graph is a strongly connected balanced graph.[1]

Given a path P in a (directed or undirected) graph, the *internal vertices* of P are the vertices lying on P except for the two end-vertices. If $d_G(v) = 2$ holds (meaning $d_G^{in}(v) = d_G^{out}(v) = 1$ in the directed case) for each internal vertex v of P, then we say that the path P is an *unattached path*. In a directed graph, a *pair of twin arcs* is two arcs (a, b) and (b, a).

Given a set X of vertices, edges, or arcs in a graph G, let $G \setminus X$ denote the graph obtained by deleting X from G. When X has only one element x, we might also write $G \setminus x$ instead of $G \setminus \{x\}$.

Parameterized Complexity. In the parameterized complexity setting, an instance comes with an integer parameter k — formally, a parameterized problem Q is a subset of $\Sigma^* \times \mathbb{N}$ for some finite alphabet Σ. We say that the problem is *fixed parameter tractable (FPT)* if there exists an algorithm solving any instance (x, k) in time $f(k)\mathrm{poly}(|x|)$ for some (usually exponential) computable function f. It is known that a problem is FPT iff it is kernelizable: a kernelization algorithm for a problem Q takes an instance (x, k) and in time polynomial in $|x| + k$ produces an equivalent instance (x', k') (i.e., $(x, k) \in Q$ iff $(x', k') \in Q$) such that $|x'| + k' \leq g(k)$ for some computable function g. The function g is the *size of the kernel*, and if it is polynomial, we say that Q admits a polynomial kernel.

3 Polynomial-Time Solvable Cases

First, we give a simple polynomial time algorithm for the following problem:

UNDIRECTED EVEN EDGE DELETION   **Parameter:** k
Input: An undirected graph G and an integer k.
Question: Does there exist a set S of at most k edges in G such that $G \setminus S$ is even?

It turns out that this problem is strongly connected to the concept of a *T-join*. If we define T to be the set of vertices having odd degree, then UNDIRECTED EVEN EDGE DELETION is equivalent with the following classical problem of finding a T-join of minimum size:

MINIMUM T-JOIN
Input: A graph $G = (V, E)$ and a set $T \subseteq V$ of even size.
Task: Find a minimum T-join, i.e., a set $S \subseteq E$ of minimum size such that T is exactly the set of vertices of odd degree in the graph $H = (V, S)$.

Since MINIMUM T-JOIN can be solved in cubic time by the algorithm of Edmonds and Johnson [14], we obtain the following consequence:

Theorem 1. UNDIRECTED EVEN EDGE DELETION *can be solved in $O(n^3)$ time for an n-vertex graph.*

---
[1] Strictly speaking, the usual definition of being Eulerian requires only that the graph is connected after removing the isolated vertices. However, we feel that requiring connectivity instead leads to more natural and fundamental problems.

Now we turn our attention to the directed version of the problem:

DIRECTED BALANCED EDGE DELETION   **Parameter:** k
Input: A directed graph G and an integer k.
Question: Does there exist a set S of at most k arcs in G such that $G \backslash S$ is balanced?

This problem can be formulated as a minimum cost flow problem with unit costs as follows. We create a digraph G' by taking G and adding two vertices s,t (source and sink). Each edge of $E(G)$ has unit capacity and unit cost. For each vertex $v \in V(G)$ such that $d^{in}(v) < d^{out}(v)$ we add to G' an arc (s, v) of capacity $d^{out}(v) - d^{in}(v)$ and cost zero. Similarly, for each vertex $v \in V(G)$ such that $d^{in}(v) > d^{out}(v)$ we add to G' an arc (v, t) of capacity $d^{in}(v) - d^{out}(v)$ and cost zero. Let f^* denote the total capacity of the added arcs (s, v). In a solvable instance we know that $f^* \leq k$.

It is straightforward to see that a flow of size f^* and cost k corresponds to a set S of k arcs for which $G \setminus S$ balanced, and vice versa. Thus, in order to find a solution of minimum size it suffices to find a minimum cost flow of size f^*. As $f^* \leq k$ and each arc has unit cost, this can be done in $O(nm \log n \log \log k)$ time [1], where $n = |V(G)|$ and $m = |E(G)|$. Note that the above argument also handles an annotated case, where we require that $S \subseteq E_a$ for a set $E_a \subseteq E$ given in the input, as we can put zero capacities on $E \setminus E_a$. This yields the following:

Theorem 2. DIRECTED BALANCED EDGE DELETION *can be solved in* $O(nm \log n \log \log k)$ *time for an input graph with* n *vertices and* m *edges, even in an annotated case where some edges are forbidden to delete.*

4   Eulerian Edge-Deletion Problems

In this section we examine the following problems:

UNDIRECTED EULERIAN EDGE DELETION   **Parameter:** k
Input: A connected undirected graph G and an integer k.
Question: Does there exist a set S of at most k edges of G such that $G \setminus S$ is Eulerian, i.e., even and connected?

DIRECTED EULERIAN EDGE DELETION   **Parameter:** k
Input: A strongly connected directed graph G and an integer k.
Question: Does there exist a set S of at most k arcs of G such that $G \backslash S$ is Eulerian, i.e., balanced and strongly connected?

The undirected problem can be easily seen to be NP-hard by observing that a cubic graph contains a Hamiltonian cycle if and only if it can be made Eulerian by edge deletions. Indeed, if deleting a set of edges from a cubic graph G results in an Eulerian graph G', then each vertex in G' must have degree 2, so G' must be a Hamiltonian cycle of G. Since the HAMILTONIAN CYCLE problem restricted to cubic planar graphs is NP-hard [15] the result follows. The directed version can be treated in a similar way using NP-hardness from [27].

Theorem 3. *The* UNDIRECTED *and* DIRECTED EULERIAN EDGE DELETION *problems are NP-hard, even when restricted to inputs* (G, k) *where G is a planar (directed) graph with maximum degree at most 3, and* $k = |E(G)|$.

In Section 4.1, we show that both versions of the problem are FPT and can be solved in time $2^{O(k \log k)} n^{O(1)}$. The algorithm is based on a novel randomized selection argument. In Section 4.2, we sharpen Theorem 3 by showing that the problems do not admit a polynomial kernel. In some sense, the nonexistence of polynomial kernels suggests that randomized selection or a similar technique is inherently required for the problems, as they cannot be solved by simple reduction rules.

4.1 FPT Algorithms

We have seen in Section 3 that removing edges to make all the vertices even can be expressed as a T-join problem, where T is the set of odd vertices. Thus UNDIRECTED EULERIAN EDGE DELETION requires us to find a T-join S such that $G \backslash S$ is connected. Observe that if G is connected, and $G \setminus S$ has a connected subgraph W containing the endpoints of every edge in S, then $G \setminus S$ is connected as well. We will call such a subgraph W a *witness* of S. Therefore, the right way to look at the problem is that we need to find a pair (S, W), where is S is a T-join and W is the witness of S. It is clear that the problem has a solution if and only if such a pair exists.

Our approach for finding a pair (S, W) is the following. We randomly colour the edges of the graph red and blue, and try to find a pair (S, W) where S uses only red edges and the subgraph W uses only blue edges. We would like to ensure that if a suitable pair (S, W) exists, then it is correctly coloured red and blue with probability at least $2^{-O(k \log k)}$. However, in general the size of W can be very large (unbounded in k) and therefore the probability of a correct colouring can be very small. We get around this problem by observing that edges "far" from T can be always coloured blue, and there is a witness W that uses only a bounded number of edges "close" to T. Formally, we say that an edge e is *close* if at least one endpoint of e is at distance at most k from T; otherwise, e is *far*. The following two lemmas contain the crucial combinatorial ideas of the algorithm:

Lemma 4. *If S is an optimum solution of size at most k, then each edge of S is close.*

Proof. As removing a cycle from S would still yield a solution, $H = (V, S)$ has to be a forest for an optimum solution S. Each connected component of H that is not an isolated vertex contains a vertex from T, as each tree contains vertices of odd degree (for example, leaves). Since $|S| \leq k$, each vertex in such a connected component is at distance at most k from T, and thus each edge in S is close.   □

Lemma 5. *If S is an optimum solution of size at most k, then S has a witness W having at most $(2k - 1)(2k + 2)$ close edges.*

Proof. Let X be the set of endpoints of the edges in S. Note that $T \subseteq X$ and $|X| \leq 2|S| \leq 2k$. Let i be the smallest integer such that $G \setminus S$ has a subgraph W containing X, having exactly i connected components and at most $(|X| - i)(2k + 2)$ close edges (such i and W always exist as for $i = |X|$ we can take $W = (X, \emptyset)$). If $i = 1$, then we are done. Otherwise, we can assume that each component of W contains a vertex of X; let P be a shortest path in $G \setminus S$ that connects two different components of W. Denote these components K_1 and K_2.

We claim that only the first $k + 1$ and the last $k + 1$ edges of P may be close. If this is true, then adding P to W decreases the number of components and increases the number of close edges by at most $2k + 2$, contradicting the minimality of i.

Suppose that an edge e is close, but it is not among the first or last $k + 1$ edges, i.e., both of its endpoints are at distance greater than k from both K_1 and K_2 on P. As e is close, it has an endpoint v such that there is a path P' of length at most k connecting v and T. As $T \subseteq X$, the path P' connects v to a component K' of W. Assuming without loss of generality that $K' \neq K_1$, the concatenation of P' and the subpath of P from K_1 to v is a walk P'' connecting two different components of W. As the distance of v from K_2 on P is more than k, the walk P'' is shorter than P, contradicting the minimality of P. $\qquad\square$

Now, we are ready to state our algorithm, working as follows:

1. Determine which edges are close and which are far.
2. Make each close edge independently with probability $1/k^2$ red; every edge that is not red becomes blue.
3. If there is more than one connected component of the blue edges containing a vertex from T, return NO; otherwise let K_B be this unique component.
4. Solve MINIMUM T-JOIN instance (G_R, T), where G_R is the graph induced by the red edges with both endpoints in K_B. If the solution is of size at most k, return it, otherwise return NO.

Lemma 6. *If the algorithm returns a solution S, then S is a proper solution to* UNDIRECTED EULERIAN EDGE DELETION.

Proof. By the definition of MINIMUM T-JOIN, $G \setminus S$ is even. The component K_B of blue edges ensures that the endpoints of S are in the same component of $G \setminus S$, i.e., $G \setminus S$ is connected. $\qquad\square$

Lemma 7. *If the* UNDIRECTED EULERIAN EDGE DELETION *instance (G, k) was a YES-instance, the algorithm returns a solution with probability at least $1/2^{O(k \log k)}$.*

Proof. Let S be an optimum solution to (G, k), and let W be a witness having at most $(2k - 1)(2k + 2)$ close edges, guaranteed by Lemma 5. In the algorithm:

1. With probability at least $(1/k^2)^k = 1/2^{2k \log k}$ each edge of S becomes red.
2. With probability at least $(1 - 1/k^2)^{(2k-1)(2k+2)} = \Omega(1)$ each close edge of W becomes blue (and hence every edge of W is blue).

The above events are independent, since S and W do not share edges. Furthermore, if both events happen, then W will connect all the endpoints of the edges from S. Therefore, all of these endpoints will be contained in one connected component K_B of the graph induced by blue edges, which in particular connects all the vertices from T. Thus, with probability $1/2^{O(k \log k)}$, every edge of S appears in G_R in the last step of the algorithm and the MINIMUM T-JOIN instance has a solution of size at most k. $\qquad\square$

Theorem 8. *Both the* UNDIRECTED *and* DIRECTED EULERIAN EDGE DELETION *problems are fixed-parameter tractable with parameter k.*

Proof. By Lemmas 6 and 7, the presented algorithm for UNDIRECTED EULERIAN EDGE DELETION finds a solution with probability $1/2^{O(k \log k)}$, and never produces a wrong output, that is removal of the returned set of edges always makes the graph Eulerian. Since the algorithm runs in $O(n^3)$ time for an n-vertex graph, we immediately obtain a randomized FPT Monte-Carlo algorithm, running in $2^{O(k \log k)} n^3$ time.

We can derandomize the above algorithm using the standard technique of splitters, which will be described in the full version.

Regarding DIRECTED EULERIAN EDGE DELETION, we can use a slightly modified version of our randomized algorithm (which then can be derandomized). After defining the set T of terminals to contain the unbalanced vertices, we forget about the orientation of the arcs, and perform Steps $1 - 3$ of the algorithm. We adjust Step 4 by solving an annotated DIRECTED BALANCED EDGE DELETION instance (G, k) where only red arcs can be deleted. Observe that this algorithm in fact looks for a set of edges S of size at most k such that $G \setminus S$ is balanced and weakly connected. However, every graph that is weakly connected and balanced is Eulerian, thus the algorithm returns the solution to DIRECTED BALANCED EDGE DELETION with high probability, if one exists. □

4.2 Non-existence of a Polynomial Kernel for UNDIRECTED and DIRECTED EULERIAN EDGE DELETION

The aim of this subsection is to prove the following theorem.

Theorem 9. *If NP $\not\subseteq$ coNP/poly, then there is no polynomial kernel for the* UNDIRECTED *and* DIRECTED EULERIAN EDGE DELETION *problems with parameter k, even if the input graph has maximum degree at most 4.*

We use the cross-composition technique introduced by Bodlaender et al. [5]. Let us recall the crucial definitions.

Definition 10 (Polynomial equivalence relation [5]). *An equivalence relation $\mathcal{R}$ on Σ^* is called a* polynomial equivalence relation *if (1) there is an algorithm that given two strings $x, y \in \Sigma^*$ decides whether $\mathcal{R}(x, y)$ in $(|x| + |y|)^{O(1)}$ time; (2) for any finite set $S \subseteq \Sigma^*$ the equivalence relation $\mathcal{R}$ partitions the elements of S into at most $(\max_{x \in S} |x|)^{O(1)}$ classes.*

Definition 11 (Cross-composition [5]). *Let $L \subseteq \Sigma^*$ and let $Q \subseteq \Sigma^* \times \mathbb{N}$ be a parameterized problem. We say that L* cross-composes *into Q if there is a polynomial equivalence relation $\mathcal{R}$ and an algorithm which, given t strings $x_1, x_2, \ldots x_t$ belonging to the same equivalence class of $\mathcal{R}$, computes an instance $(x^*, k^*) \in \Sigma^* \times \mathbb{N}$ in time polynomial in $\sum_{i=1}^{t} |x_i|$ such that (1) $(x^*, k^*) \in Q$ iff $x_i \in L$ for some $1 \leq i \leq t$; (2) k^* is bounded polynomially in $\max_{i=1}^{t} |x_i| + \log t$.*

Theorem 12 ([5], Theorem 9). *If $L \subseteq \Sigma^*$ is NP-hard under Karp reductions and L cross-composes into the parameterized problem Q that has a polynomial kernel, then NP $\subseteq$ coNP/poly.*

We apply Theorem 12 on the following language L:

UNDIRECTED or DIRECTED $s - t$ PATH WITH FORBIDDEN PAIRS OF EDGES
Input: An undirected or directed graph $G = (V, E)$, two vertices $s, t \in V$, and a set $\mathcal{C} \subseteq E \times E$ called *the constraints*.
Task: Does there exist an $s - t$ path $\mathcal{P}$ in G such that from each constraint $(e_1, e_2) \in \mathcal{C}$ at least one edge (arc) does not lie on $\mathcal{P}$?

The undirected version of this problem with forbidden pairs of vertices was proven to be NP-hard by Kolman and Pangrác [18] and their proof can be easily modified to handle our case as well.

Lemma 13. UNDIRECTED *and* DIRECTED $s - t$ PATH WITH FORBIDDEN PAIRS OF EDGES *are NP-hard under Karp reductions, even in the case where each vertex has maximum degree three, s and t have degree one, and, in the directed case, each vertex has maximum in- and outdegree two.*

To finish the proof of Theorem 9 we need to show a cross-composition algorithm. This is done in the following lemma.

Lemma 14. UNDIRECTED (DIRECTED) $s - t$ PATH WITH FORBIDDEN PAIRS OF EDGES *cross-composes to* UNDIRECTED (DIRECTED) EULERIAN EDGE DELETION. *If the input instances have degrees bounded as in Lemma 13 then the output instance can be made to have maximum degree 4.*

Proof. For the equivalence relation $\mathcal{R}$ we take an almost trivial relation that sorts all malformed instances into one equivalence class and all well-formed into another one. If we are given malformed instances, we simply output a trivial NO-instance. Thus in the rest of the proof we assume we are given a sequence $(G_i, s_i, t_i, \mathcal{C}_i)_{i=1}^t$ of UNDIRECTED or DIRECTED $s - t$ PATH WITH FORBIDDEN PAIRS OF EDGES instances.

We now construct an UNDIRECTED or DIRECTED EULERIAN EDGE DELETION instance (G, k). We start by obtaining a graph G'_i for each $1 \le i \le t$ as follows. First we subdivide each edge $e \in E(G_i)$ with new vertices x_e^C, one for each constraint $C \in \mathcal{C}_i$ that contains e. Then for each constraint $C = (e_1, e_2) \in \mathcal{C}_i$ we introduce vertices z_1^C and z_2^C and create a (directed) cycle $x_{e_1}^C, z_1^C, x_{e_2}^C, z_2^C$. By $V(G_i)$ we denote the subset of $V(G'_i)$ containing vertices different than x_e^C and z_α^C. To construct the graph G, we first take the union of all graphs G'_i and identify all vertices s_i into one vertex s^* and all vertices t_i into one vertex t^*. Let $V^0 = \{s^*, t^*\} \cup \bigcup_{i=1}^t V(G_i) \setminus \{s_i, t_i\}$. Second, we introduce a new vertex r and connect it to the rest of the graph as follows. In the undirected case for each $v \in V^0 \setminus \{s^*, t^*\}$ we connect r and v with one or two unattached paths of length 2, so that in G the vertex v is even. In the directed case, we connect r and v with some positive number of unattached directed paths of length 2, so that in G the vertex v is balanced. We do almost the same construction to connect s^* and t^* to r, but we ensure that the degrees of s^* and t^* are odd (in the undirected case) or that $d_G^{in}(s^*) + 1 = d_G^{out}(s^*)$ and $d_G^{in}(t^*) = d_G^{out}(t^*) + 1$ (in the directed case). Note that r is even (balanced). Finally, we set $k = \max_{i=1}^t |V(G'_i)| - 1 = O(\max_{i=1}^t |V(G_i)| + |\mathcal{C}_i|)$.

It is clear that the above construction can be done in polynomial time and that the parameter k is bounded polynomially in the maximum size of the input instances. We verify its correctness and explain the degree reduction in the full version.   □

5   Node-Deletion Problems

We first consider the following two node-deletion problems:

DIRECTED BALANCED (or EULERIAN) NODE DELETION   **Parameter:** k
Input: A directed graph G and an integer k
Question: Does there exist a set of at most k vertices $S \subseteq V(G)$ such that $G \setminus S$ is balanced (or Eulerian)?

The undirected versions of these problems, namely UNDIRECTED EVEN and UNDI-RECTED EULERIAN NODE DELETION, are defined analogously. While these undirected variants were already shown to be W[1]-hard with parameter k by Cai and Yang [9], the complexity of the directed versions has not been studied yet. In the full version by the following theorem we show that they are intractable as well.

Theorem 15. DIRECTED BALANCED NODE DELETION *and* DIRECTED EULERIAN NODE DELETION *are NP-hard and W[1]-hard with parameter* k.

As Table 1 shows, the vertex-deletion variant is W[1]-hard in all four cases, while the edge-deletion version is FPT or even polynomial-time solvable. What makes the vertex-deletion versions harder? One obvious difference is that in the edge-deletion problem the answer is trivially no if there are more than $2k$ odd/unbalanced vertices, but the vertex-deletion versions can have a solution even if the number of such nodes is unbounded. This suggests that the higher complexity comes from the ability of affecting the degree of many vertices by a single vertex deletion. Indeed, if every vertex has degree bounded by Δ, then we can solve all of the above defined node-deletion problems in $O((\Delta + 1)^k(|V(G)| + |E(G)|))$ time by a simple branching algorithm which will be described in the full version. However, this interpretation is not fully correct: as we shall show, the vertex-deletion problems are hard even if we are allowed to delete only vertices of constant degree.

To this end, we define the following variation of the four different node-deletion problems, where α can be UNDIRECTED EVEN, UNDIRECTED EULERIAN, DIRECTED BALANCED, or DIRECTED EULERIAN:

α NODE DELETION WITH FORBIDDEN NODES   **Parameter:** k
Input: A graph G, a set $F \subseteq V(G)$ of *forbidden nodes*, and an integer k.
Question: Does there exist a solution $S \subseteq V(G)$ for (G, k) with respect to the corresponding α NODE DELETION problem such that $S \cap F = \emptyset$ and $|S| \leq k$?

In other words, we require the solution to be disjoint from a set of *forbidden vertices*. A vertex is *allowed*, if it is not forbidden. For each of the four node-deletion problems, the above variant is at least as hard as the original problem, and in fact has the same complexity: this variant can easily be reduced to the original version, by attaching long unattached cycles to every forbidden vertex. Furthermore, we show that allowing only the deletion of bounded-degree vertices does not make the problem easier:

Theorem 16. *Each of the problems α NODE DELETION WITH FORBIDDEN NODES where α is* UNDIRECTED EVEN, UNDIRECTED EULERIAN, DIRECTED BALANCED, *or* DIRECTED EULERIAN *remains W[1]-hard with parameter k, even if each allowed vertex has degree at most 4.*

6   Conclusion

We completed the analysis of the complexity of making a graph Eulerian via edge or vertex deletions. There are two open problems that we would like to emphasise here.

First, do there exist FPT algorithms for the edge-deletions problems running in time $c^k n^{O(1)}$? It seems hard to obtain such algorithms using our techniques, mainly due to the fact that the witness subgraph W may contain $\Omega(k^2)$ close edges.

Second, Cechlárová and Schlotter in [10] asked for the parameterized complexity of a related problem, where the task is to delete at most k arcs from a directed graph to obtain a graph where each strongly connected component is Eulerian. This problem seems to be significantly different than the problems considered in this paper, as for example it includes DIRECTED FEEDBACK VERTEX SET [10], and, to the best of our knowledge, the question of its parameterized complexity still remains open.

References

1. Ahuja, R.K., Goldberg, A.V., Orlin, J.B., Tarjan, R.E.: Finding minimum-cost flows by double scaling. Math. Program. 53, 243–266 (1992)
2. Alon, N., Shapira, A., Sudakov, B.: Additive approximation for edge-deletion problems. In: FOCS, pp. 419–428 (2005)
3. Alon, N., Yuster, R., Zwick, U.: Color-coding. J. ACM 42(4), 844–856 (1995)
4. van Bevern, R., Moser, H., Niedermeier, R.: Approximation and tidying—a problem kernel for s-plex cluster vertex deletion. Algorithmica (February 2011)
5. Bodlaender, H.L., Jansen, B.M.P., Kratsch, S.: Cross-composition: A new technique for kernelization lower bounds. CoRR abs/1011.4224 (2010)
6. Burzyn, P., Bonomo, F., Durán, G.: NP-completeness results for edge modification problems. Discrete Appl. Math. 154, 1824–1844 (2006)
7. Cai, J., Chakaravarthy, V.T., Hemaspaandra, L.A., Ogihara, M.: Competing provers yield improved Karp-Lipton collapse results. Inf. Comput. 198(1), 1–23 (2005)
8. Cai, L.: Fixed-parameter tractability of graph modification problems for hereditary properties. Inf. Process. Lett. 58(4), 171–176 (1996)
9. Cai, L., Yang, B.: Parameterized Complexity of Even/Odd Subgraph Problems. In: Calamoneri, T., Diaz, J. (eds.) CIAC 2010. LNCS, vol. 6078, pp. 85–96. Springer, Heidelberg (2010)
10. Cechlárová, K., Schlotter, I.: Computing the Deficiency of Housing Markets with Duplicate Houses. In: Raman, V., Saurabh, S. (eds.) IPEC 2010. LNCS, vol. 6478, pp. 72–83. Springer, Heidelberg (2010)
11. Chen, J., Liu, Y., Lu, S., O'Sullivan, B., Razgon, I.: A fixed-parameter algorithm for the directed feedback vertex set problem. J. ACM 55(5), 1–19 (2008)
12. Díaz, J., Thilikos, D.M.: Fast FPT-Algorithms for Cleaning Grids. In: Durand, B., Thomas, W. (eds.) STACS 2006. LNCS, vol. 3884, pp. 361–371. Springer, Heidelberg (2006)

13. Dorn, F., Moser, H., Niedermeier, R., Weller, M.: Efficient Algorithms for Eulerian Extension. In: Thilikos, D.M. (ed.) WG 2010. LNCS, vol. 6410, pp. 100–111. Springer, Heidelberg (2010)

14. Edmonds, J., Johnson, E.: Matching, Euler tours and the Chinese postman problem. Math. Program. 5, 88–124 (1973)

15. Garey, M.R., Johnson, D.S., Tarjan, R.E.: The planar Hamiltonian circuit problem is NP-complete. SIAM J. on Computing 5, 704–714 (1976)

16. Guo, J.: Problem Kernels for NP-Complete Edge Deletion Problems: Split and Related Graphs. In: Tokuyama, T. (ed.) ISAAC 2007. LNCS, vol. 4835, pp. 915–926. Springer, Heidelberg (2007)

17. Khot, S., Raman, V.: Parameterized complexity of finding subgraphs with hereditary properties. Theor. Comput. Sci. 289, 997–1008 (2002)

18. Kolman, P., Pangrác, O.: On the complexity of paths avoiding forbidden pairs. Discrete Applied Mathematics 157(13), 2871–2876 (2009)

19. Kratsch, S., Wahlström, M.: Two Edge Modification Problems Without Polynomial Kernels. In: Chen, J., Fomin, F.V. (eds.) IWPEC 2009. LNCS, vol. 5917, pp. 264–275. Springer, Heidelberg (2009)

20. Lewis, J.M., Yannakakis, M.: The node-deletion problem for hereditary properties is NP-complete. J. Comput. Syst. Sci. 20(2), 219–230 (1980)

21. Lokshtanov, D.: Wheel-Free Deletion is W[2]-Hard. In: Grohe, M., Niedermeier, R. (eds.) IWPEC 2008. LNCS, vol. 5018, pp. 141–147. Springer, Heidelberg (2008)

22. Marx, D.: Chordal deletion is fixed-parameter tractable. Algorithmica 57(4), 747–768 (2010)

23. Mathieson, L., Szeider, S.: The parameterized complexity of regular subgraph problems and generalizations. In: CATS, pp. 79–86 (2008)

24. Moser, H., Thilikos, D.M.: Parameterized complexity of finding regular induced subgraphs. Journal of Discrete Algorithms 7, 181–190 (2009)

25. Natanzon, A., Shamir, R., Sharan, R.: Complexity classification of some edge modification problems. Discrete Appl. Math. 113, 109–128 (2001)

26. Philip, G., Raman, V., Villanger, Y.: A Quartic Kernel for Pathwidth-one Vertex Deletion. In: Thilikos, D.M. (ed.) WG 2010. LNCS, vol. 6410, pp. 196–207. Springer, Heidelberg (2010)

27. Plesník, J.: The NP-completeness of the Hamiltonian cycle problem in planar digraphs with degree bound two. Inf. Process. Lett. 8(4), 199–201 (1979)

28. Raman, V., Saurabh, S.: Parameterized algorithms for feedback set problems and their duals in tournaments. Theor. Comput. Sci. 351, 446–458 (2006)

29. Raman, V., Sikdar, S.: Parameterized complexity of the induced subgraph problem in directed graphs. Inf. Process. Lett. 104, 79–85 (2007)

30. Sorge, M.: On making directed graphs Eulerian. CoRR abs/1101.4283 (2011)

31. Yap, C.K.: Some consequences of non-uniform conditions on uniform classes. Theor. Comput. Sci. 26, 287–300 (1983)

Restricted Cuts for Bisections in Solid Grids: A Proof via Polygons*

Andreas Emil Feldmann[1], Shantanu Das[2], and Peter Widmayer[1]

[1] Institute of Theoretical Computer Science, ETH Zürich, Switzerland
{feldmann,widmayer}@inf.ethz.ch
[2] Laboratoire d'Informatique Fondamentale, Aix-Marseille University, France
shantanu.das@lif.univ-mrs.fr

Abstract. The *graph bisection problem* asks to partition the n vertices of a graph into two sets of equal size so that the number of edges across the cut is minimum. We study finite, connected subgraphs of the infinite two-dimensional grid that do not have holes. Since bisection is an intricate problem, our interest is in the tradeoff between runtime and solution quality that we get by limiting ourselves to a special type of cut, namely cuts with at most one bend each (corner cuts). We prove that optimum corner cuts get us arbitrarily close to equal sized parts, and that this limitation makes us lose only a constant factor in the quality of the solution. We obtain our result by a thorough study of cuts in polygons and the effect of limiting these to corner cuts.

1   Comparing Optimal with Restricted Cuts

We consider the *bisection problem*: partition the vertex set of a given graph into two (almost) equal sized subsets such that the number of edges with an endpoint in each partition is minimised. The problem has been studied extensively, due to its utility in divide-and-conquer algorithms. It is NP-hard in general [5] and the best approximation algorithm known [9] guarantees an approximation ratio of $\mathcal{O}(\log n)$. For planar graphs a PTAS [2] has been found, while for trees an optimum solution can be computed in $\mathcal{O}(n^2)$ time [6,7]. Our motivation to study the bisection problem comes from the need to parallelise a finite element computation of a human bone structure model in order to diagnose osteoporosis [1]. In such an application the aim is to distribute the data, modelled by the vertices (of a graph G), evenly onto a given number p of processors to achieve a balanced computation load. At the same time the communication between data points, modelled by the edges (of G), needs to be kept at a minimum between processors since this constitutes a bottleneck in parallel computing. One way to distribute the data is to recursively solve the bisection problem. The finite element models corresponding to the bones in our application constitute a porous 3D grid. In a

* We gratefully acknowledge discussions with Peter Arbenz who introduced the human bone simulation problem to us, and the support of this work through the Swiss National Science Foundation under Grant No. 200021_125201/1.

P. Kolman and J. Kratochvíl (Eds.): WG 2011, LNCS 6986, pp. 143–154, 2011.
© Springer-Verlag Berlin Heidelberg 2011

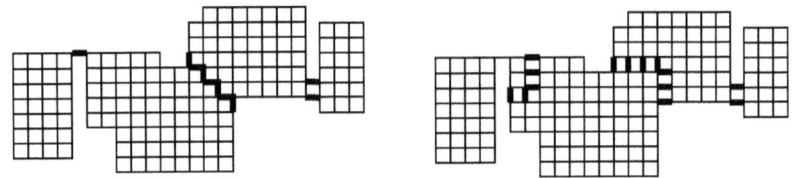

Fig. 1. An optimal (left) and a corner cut (right) in a solid grid, each cutting out $k = 110$ many vertices. The bold edges indicate the segments.

first step towards understanding the problem for 3D grids, in this paper we limit ourselves to 2D *grid graphs*, which are finite, connected subgraphs of the infinite two-dimensional grid. A grid graph is called *solid* if it has no hole, i.e. if it has no interior face surrounded by more than four edges. For solid grid graphs the fastest algorithm known [8] computes an optimum bisection in time $\mathcal{O}(n^5)$.

We aim at understanding the intricacies of bisection from a novel point of view: we study particular classes of cuts, for which it is possible to find optimum solutions faster. It is well known that any cut in a planar graph $G = (V, E)$ corresponds to a set of cycles in its dual graph (i.e. the graph whose vertices are the faces of G and whose edges represent a shared boundary between faces). We call a set of edges in a planar graph a *segment* if it corresponds to a simple cycle in the dual graph. Hence a *cut* can be defined as a set $S \subseteq 2^E$ of segments. In this paper we consider cuts that contain segments with at most one bend, so-called *corner cuts* (Figure 1). We prove that optimum corner cuts do not need a lot more cut-edges than arbitrary cuts. Using only segments without any bend (straight cuts) does not achieve this high quality: these cuts can be a $\sqrt{n}$ factor away from optimum. We achieve our result by proving a number of theorems for polygons that we relate to the case of grid graphs. The main part of this paper will therefore be concerned with thoroughly analysing corner cuts in polygons.

We call the number of edges $\sum_{s \in S} |s|$ in a cut S its *cut-size*. The cut minimising the cut-size is *optimal*. Notice that some edges may be counted several times in the sum. However in the non-restricted case, edges that appear more than once can be removed. Hence this generalisation does not change the optimal solution. A simple cycle in the dual of a planar graph corresponds to a closed curve in its embedding in the plane. Hence the cycle divides the plane into an interior and an exterior area. We say that a pair of cycles *cross* if the corresponding closed curve of one of them both contains points belonging to the interior and to the exterior area into which the other cycle divides the plane. Note that any pair of crossing simple cycles can be seen as a (different) pair of simple cycles that do not cross. Hence for the non-restricted optimal cut we may consider only *non-crossing* cuts in which no corresponding cycles cross. In our application we can assume that the grid graph G is given together with its natural embedding in the plane, i.e. G is a plane graph in which the vertices are coordinates in $\mathbb{N}^2$ and all edges have unit length. We propose to use only segments that correspond to orthogonal lines with at most one right-angled bend in the dual graph, when disregarding the part of the cycle that connects to the

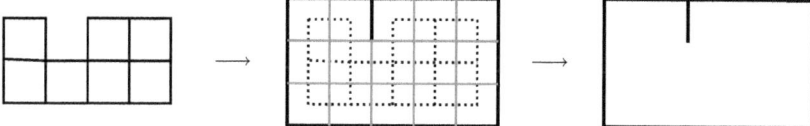

Fig. 2. Converting a grid to a polygon

exterior face (Figure 1). If the corresponding line of a segment contains no bend we call it a *straight segment* and if it contains exactly one right-angled bend a *corner segment*.

If removing the edges included in a cut leaves a set of connected components together containing exactly k vertices, we call it a k-*cut*. We say that it *cuts out* k vertices. We call the set containing k vertices the $\mathcal{A}$-*part*, and the other set the $\mathcal{B}$-*part* of the k-cut. A natural generalisation of the bisection problem is to find the optimal k-cut for a given integer $k \in \{0, ..., n\}$. The main result of this paper is summarised in the following theorem.

Theorem 1. *Let l be the cut-size of an optimal k-cut in G and $\varepsilon \in \,]0, 1]$. Then there exists a non-crossing corner k'-cut, for some $k' \in [(1 - \varepsilon)k, (1 + \varepsilon)k]$, which has a cut-size of $\mathcal{O}(1/\sqrt{\varepsilon}) \cdot l$.*

This result was already used in [3] where we showed that the optimal corner k-cuts for all $k \in \{0, ..., n\}$ can be computed in time $\mathcal{O}(n^4)$ for a given solid grid graph G. Thus the above theorem implies that one can find a good approximation to the optimal bisection of G faster than by using the $\mathcal{O}(n^5)$ time algorithm from [8]. Since the set sizes of the resulting k-cut correspond to the load on each machine in our application, the deviation by a factor of ε on the load balance is acceptable as it does not significantly affect the resulting running time of the parallel computation. However the speed-up of the bisection algorithm is a considerable improvement since a typical data set will have billions of vertices.

We will prove Theorem 1 by going through several steps, each of which is an interesting problem in itself. We start by comparing cuts in grid graphs to cuts in polygons. For this we convert a given solid grid graph into a simple orthogonal polygon, and hence all polygons considered in this paper are orthogonal and simple (but can be degenerate in the sense that the polygonal chain corresponding to its border can have overlapping edges; see Figure 2). Given a solid grid graph G the conversion is done by replacing each vertex $(x, y) \in V$ by a unit square that has its centre at the coordinate (x, y). Notice that the squares of two neighbouring vertices of V will share a boundary, but the converse is not necessarily true. Ignoring those boundaries that correspond to an edge in G leaves a connected line that is the boundary of the polygon. The region enclosed by the boundary is of size exactly n, the number of vertices in G. The k-cut in the polygon that corresponds to the optimal k-cut in the grid obviously has a cut-size that is at least the cut-size of the optimal k-cut in the polygon.

All the notions used for cuts in grids carry over naturally to the case of polygons. For a polygon $\mathcal{P}$ with fixed orientation in the plane we call an orthogonal line[1] within $\mathcal{P}$ that has no bend and starts and ends at the boundary of $\mathcal{P}$, a *straight line*. Accordingly an orthogonal line within $\mathcal{P}$ that has exactly one right-angled bend and starts and ends at the boundary of $\mathcal{P}$ is called a *corner line*. The cut-size of a set of lines L in a polygon is the sum of the lengths of the lines in L, which are measured using the Manhattan distance. We will first show the existence of corner cuts in simple polygons that cut out almost the required area and have small cut-size (close to optimal). We will then convert such a cut in a polygon derived from a grid graph to a corresponding cut in the grid having the properties described in Theorem 1. More precisely we prove the following results for polygons which together imply the theorem:

1. We show that there is an optimal k-cut in a polygon that is almost a corner cut, in the sense that the cut consists of only straight and corner lines except at most one other line. This line may be shaped like a staircase (a so called *staircase line*), or it may be a *rectangular line*, which is defined as a continuous part of the boundary of an orthogonal rectangle (Figure 3).
2. We show how to remove a rectangular line from a cut containing only straight and corner lines otherwise. We replace the rectangular line by a set of straight and corner lines, and at most one staircase line. Together these cut out the same area as the rectangular line. While doing this we need to take other lines from the cut into consideration so that the newly introduced lines do not interfere with these. The new cut will also be a k-cut but its cut-size may not be optimal. However, we show that the cut-size of the new cut is only a constant factor away from the optimal.
3. Given a k-cut of the polygon consisting of straight and corner lines, and one staircase line, we next show how to replace the staircase line with a set of corner and straight lines, such that the new area that is cut out is close to k. To be more precise, the new cut is a k'-cut where $k' \in [(1-\varepsilon)k, (1+\varepsilon)k]$ for any desired constant $\varepsilon \in]0,1]$. Further, the cut-size of the new cut is only a constant factor (depending on ε) times the cut-size of the original cut.
4. At last we show how to convert a cut containing only straight and corner lines in a polygon corresponding to a grid graph G into a cut in G. Note that this step would be straightforward if all the lines in the cut were passing through exactly the midpoints of the edges of the grid. We call such lines *grid lines*. We show that all lines in the cut obtained in the previous steps can be moved to grid lines in such a way that the cut-size remains the same, but we lose a small area a from the cut out area. Since a is small we can cut this area from the polygon using a recursive method using only grid lines so that the cut-size grows by only a small factor.

The next sections explain these techniques in more detail.

2   Cuts in Polygons

We will now show that in an optimal k-cut of a polygon all but at most one line are corner and straight lines. Lines with more bends include *staircase lines* and

---

[1] All lines considered in this paper have finite length unless otherwise stated.

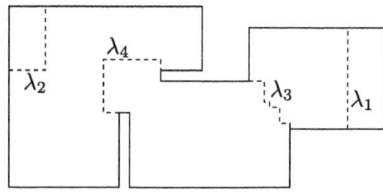

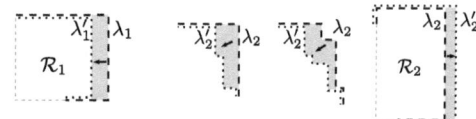

Fig. 3. A straight, corner, staircase, and rectangular line in a polygon denoted by λ_1 through λ_4 respectively

Fig. 4. A rectangular line λ_1 with its defining rectangle $\mathcal{R}_1$ is replaced by the rectangular line λ_1'. To compensate for the area a (shaded grey), another line λ_2 is replaced by λ_2'. It can be a corner, staircase, or a rectangular line (with defining rectangle $\mathcal{R}_2$).

rectangular lines. The former have at least two bends and are monotonic in x- and in y-direction. The latter have two or three bends and lie on the boundary of an orthogonal rectangle (Figure 3). In a first step, we convince ourselves (Lemma 5)[2] that in any simple polygon there is an optimal k-cut that contains only straight, corner, staircase, and rectangular lines. Furthermore, none of these lines cross or overlap. These results are analogous to those attained in [8] for grid graphs.

In a next step we show that if an optimal k-cut contains a rectangular line, then all other lines are straight or corner lines (Lemma 9). Generally speaking the reason is that cuts can be modified so that the cut out area remains the same. This is easy to see for two rectangular lines where the $\mathcal{A}$-part of the cut out area is on the inside of one of the rectangles and on the outside of the other: we can simply make both rectangular lines smaller by the same area, thereby decreasing the length of the cut (Figure 4)—a contradiction to optimality. More generally, we call a corner or rectangular line *convex* w.r.t. the area next to its 90 degree angles and *concave* w.r.t. the area next to its 270 degree angles (Figure 5). Similar area exchange arguments show that for an optimal k-cut with a rectangular line, the area on its concave side will belong to the same part of the cut as the area on the concave sides of all corner lines. This fact will become important later when a rectangular line is replaced by a staircase line.

For staircase lines, area exchange works by changing the staircase line while still keeping it monotonic between its end points. The potential area exchanged is the *deficit* or the *surplus*, which are the areas of maximal size with monotone borders contained in the $\mathcal{B}$- and $\mathcal{A}$-part respectively (Figure 6). These areas are used to prove that an optimal cut requires at most one staircase line (Lemma 10): for more than one staircase line we trade the smaller deficit or surplus of one staircase with the larger of another one, turning the former into only straight and corner lines. Putting all these observations together (Corollary 11), we obtain the following result.

Theorem 2. *For any simple polygon $\mathcal{P}$ there is an optimal k-cut L such that L is non-crossing and all lines in L are corner or straight lines except at most one which is either a staircase line or a rectangular line. If there is a rectangular line in L that is concave with respect to the $\mathcal{A}$-part (resp. $\mathcal{B}$-part) then all corner lines in L are concave with respect to the same area $\mathcal{A}$ (resp. $\mathcal{B}$).*

---

[2] All lemmas and theorems, along with their proofs, can be found in the full version of the paper [4].

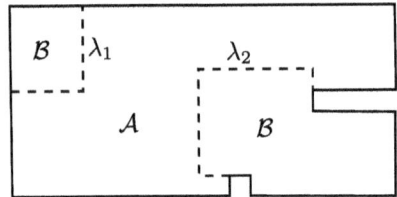
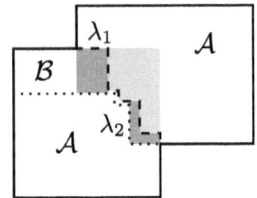

Fig. 5. A corner, and rectangular line in a polygon denoted by λ_1, λ_2, respectively. Both are concave with respect to the $\mathcal{A}$-part and convex with respect to the $\mathcal{B}$-part.

Fig. 6. A staircase line λ_1 together with its surplus (in light grey shading) and its deficit (in dark grey shading)

3   Removing Rectangular Lines

We now show how to convert an optimal k-cut containing straight and corner lines and one rectangular line into a k-cut containing only straight and corner lines except at most one which is a staircase line. Consider the area inside the defining rectangle of the rectangular line (Figure 7). This region may contain a part of the boundary of the polygon (and possibly some other lines of the cut). We can replace the rectangular line with a set Ξ of straight and corner lines lying within the defining rectangle such that these lines have total length less than the length of the rectangular line. By doing this we do not increase the cut size, but we now have to cut out an additional area of size a equal to the difference in sizes of the parts cut out by the original and the new cut. We show how to find a set of lines that cut out an area of size a and has total length not too large (compared to the optimal cut-size l). Note that the length of the rectangular line (and thus l) is at least $\sqrt{a}$. So, it is sufficient to show that the area of size a can be cut out using a set of lines of total length not much larger than $\sqrt{a}$.

Consider any corner line of infinite length in the plane. Since the line can be rotated to any particular orientation, let us assume w.l.o.g. the corner line has a vertical section going up and a horizontal section going right from the corner point (x, y). We call this line the corner line *at* (x, y). Given any specific polygon, the parts of this infinite line that are inside this polygon are said to form a *virtual corner line* (Figure 8). Notice that a virtual corner line is a set of straight and corner lines lying inside the polygon. If a virtual corner line cuts out an area of size a on the upper right side of its corner, we say that it is a virtual corner line *for* a. We can analogously define a *virtual staircase line* by considering any staircase line of infinite length in the plane and taking the parts of the line that lie inside some specific polygon. The easy case is when the required area a can be cut out from the polygon using a single virtual corner line of short length (say, of length at most $c\sqrt{a}$ for some fixed constant c). However, depending on the shape of the polygon, it is not always possible to find such a virtual corner line. For example, in the polygon shown in Figure 9, any virtual corner line cutting out the required area contains a long vertical section or a long horizontal section.

Given any simple polygon, we can search along the x-axis between the two extremities of the polygon, and for each value of x try to find a y such that the

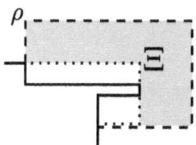

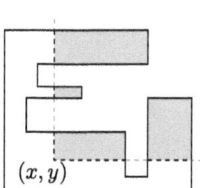

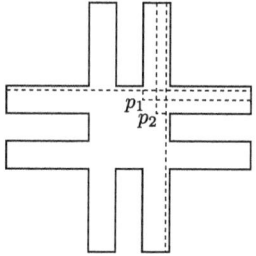

Fig. 7. A rectangular line ρ (dashed) together with the set of lines Ξ (dotted) with which it is replaced. The area of size a is shaded in grey.

Fig. 8. A virtual corner line (black dashed) at (x, y). The cut out area is shaded in grey.

Fig. 9. A polygon in which every virtual corner line for a is too long. At p_1 the vertical section switches from short to long and at p_2 the horizontal section switches from long to short.

virtual corner line at (x, y) cuts out exactly an area of size a (Figure 10). We can show that if there does not exist any single virtual corner line for a having sufficiently small length, then there exist virtual corner lines for a at two points (x_1, y_1) and (x_2, y_2) such that the former has a short (i.e. at most $c\sqrt{a}$) vertical section, the latter has a short horizontal section, and for all virtual corner lines in between both sections are long (Lemma 13). Using these properties we can show that the intervals $[x_1, x_2]$ and $[y_2, y_1]$ are short (Lemma 14). With these results we find a virtual staircase line which cuts out exactly the required area a and has a short total length (Lemma 16). The corresponding staircase line goes along the vertical section of the first virtual corner line, to some y^* and then turns to the right and goes to some x^*, turns again and then finally follows the horizontal part of the second virtual corner line (Figure 11).

Let L be the set of straight and corner lines in the cut after replacing the rectangular line, such that L cuts out an area $k - a$ or $k + a$. We now know that there exists a virtual staircase line Λ that can be used to cut out the remaining area of size a from the $\mathcal{A}$- or $\mathcal{B}$-part. Notice that the underlying staircase line (of infinite length in the plane) may be intersecting with other lines in the cut (Figure 12). So the parts of the line included in Λ may not have endpoints on the boundary of the polygon. Thus, we need to convert Λ into a set M of staircase, corner, and straight lines, none of which ends at any other line in L (however, the lines may partially overlap). This is done by adding those parts of lines in L to the lines in Λ that are monotonic extensions of the latter in x- and in y-direction (Lemma 17). This is possible since the corner lines in L are all concave w.r.t. the same cut out part, as pointed in the previous section. Thus the set M may contain several staircase lines, but its total length is at most that of L.

The next step is to convert the staircase lines from the set $M \cup L$ so that at most one of them remains but the cut-size does not increase. Similar to the techniques seen before, we will use the lines contained in the boundary of the surplus or deficit of a staircase line for the transformation. Unfortunately some of the previous arguments can not be used here since $M \cup L$ is not an optimal

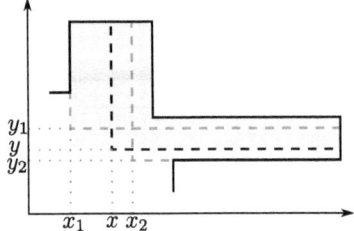

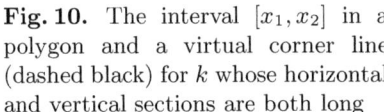

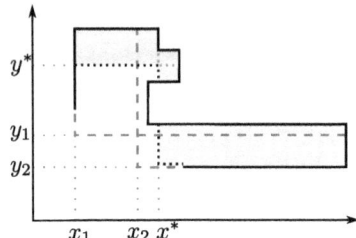

Fig. 10. The interval $[x_1, x_2]$ in a polygon and a virtual corner line (dashed black) for k whose horizontal and vertical sections are both long

Fig. 11. A virtual staircase line (dotted black) cutting out the area of size a shaded in grey. It is constructed using the the two virtual corner lines at (x^*, y_2) and (x_1, y^*).

cut. Instead we need some observations on the nature of the boundary of the deficit and surplus of a staircase line $\lambda \in M$: it turns out that any staircase line λ' different from λ at the boundary of the deficit or surplus of λ overlaps with exactly one corner line $\mu \in L$ (Figure 13). This corner line μ together with the staircase line λ' can be used to construct a pair of corner lines which can be replaced with μ and λ' so that the same area is cut out by the new set of lines. The cut-size decreases during this process (Lemma 18). Hence an area exchange between two staircase lines is possible even in this case (Lemma 19).

Using the above techniques we can find a k-cut containing at most one staircase line for any optimal k-cut containing a rectangular line, such that the cut-size of the former k-cut is at most a constant times the cut-size of the latter. The following theorem (Theorem 20 in the full paper) summarizes these results.

Theorem 3. *For any simple polygon $\mathcal{P}$ with an optimal k-cut L of $\mathcal{P}$ containing a rectangular line, there exists a non-crossing k-cut M which contains only corner and straight lines except at most one which is a staircase line and M has a cut-size of at most $9l$, where l is the cut-size of L.*

4   Removing Staircase Lines

We now turn to the task of converting a (not necessarily optimal) cut L containing only straight and corner lines except one which is a staircase line, into a cut containing only straight and corner lines. Similar to the case of rectangular lines we will replace the staircase line with a set of appropriate corner and straight lines having a short cut-size. It is easy to see that if the deficit (or surplus) area of the staircase line λ has size a, then $\sqrt{a} < l$, where l is the cut-size of L. Thus, if we can cut out the excess area a using straight and corner lines of total length in $\mathcal{O}(\sqrt{a})$, then our cut-size will still be close to optimal. Given any simple polygon $\mathcal{P}$ of area n, $a \in [0, n]$, and constant $\varepsilon \in]0, 1]$ we can find a set of at most three virtual corner lines that cut out an area whose size is in the interval $[(1 - \varepsilon)a, (1 + \varepsilon)a]$, and has a cut-size that is a constant (depending on ε) times $\sqrt{a}$. Furthermore the corners of these virtual corner lines all have either

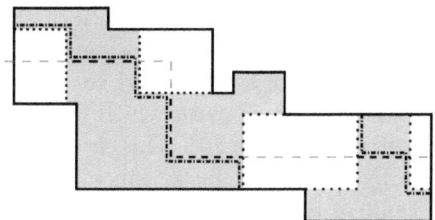

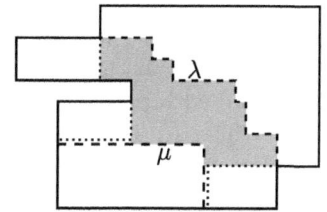

Fig. 12. A virtual staircase lined (dashed) that is converted to a set of staircase, corner, and straight lines (dashed and dotted). For this parts of the corner lines (dotted) from the original cut are used. These are all concave w.r.t. the same part of the cut shaded in grey.

Fig. 13. A staircase line λ with its surplus shaded in grey. The lines on the border of the surplus can be replaced by a set of straight and corner lines (dotted). The corner line μ is also removed.

the same x-coordinate or the same y-coordinate (Lemma 21). They can be found using the short interval $[x_1, x_2]$ that was identified before (Figure 10). We use the virtual corner line with corner (x_1, y_2) which has short length but cuts out an area that is too large. To correct for the area we additionally find two virtual corner lines (of short length) with corners at either points (x', y_2) and (x'', y_2), for some $x', x'' \geq x_2$, or points (x_1, y') and (x_1, y''), for some $y', y'' \geq y_1$.

To apply the above result we need to find a region of the polygon from which to cut out the excess area. We want this area to contain no lines of the cut so that we do not interfere with these. For this we define the concept of a *tail*: for any cut L in a polygon $\mathcal{P}$, consider all the connected pieces of $\mathcal{P}$ cut out by it. We call a piece $\mathcal{T}$ that is cut out by a single line $\tau \in L$ a tail of the polygon, and we refer to τ as the line of $\mathcal{T}$ (Figure 14).

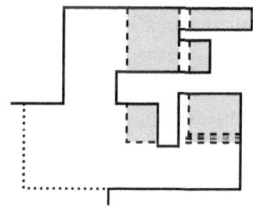

Fig. 14. A tail (dotted line) and three virtual corner lines (dashed) cut out the area shaded in grey (The lines overlap on the bottom right)

Note that we can shift the staircase line λ in either direction, i.e. going into either the $\mathcal{A}$- or the $\mathcal{B}$-part. However all the tails in the polygon may belong to only one part. We need to consider two cases, one of which is when L contains only λ. This means that there are exactly two tails, one on each side of λ. If we assume w.l.o.g. that the size a of λ's deficit is at most that of its surplus, we can replace the staircase line with the set of straight and corner lines on the border of its deficit. We then cut out the area $a' \in [(1-\varepsilon)a, (1+\varepsilon)a]$ from the original $\mathcal{A}$-part (containing the surplus) using the at most three virtual corner lines which were shown to exist above. The other case is when there is a tail contained in, say, the $\mathcal{A}$-part whose line μ is not the staircase line λ. We can safely assume that the size of the tail is larger than the size a of the deficit of λ (Lemma 23). If this was not the case then we could remove μ from the cut by using an area exchange with the staircase line λ, without increasing the cut-size. Hence we can replace λ by the corner and straight lines on the border of its deficit and cut out the area a' from the tail, again using the virtual corner lines of short length. In both cases it may be that some of the virtual corner lines end at the line μ of the tail. If this happens we can find a set of straight and corner lines that overlap with parts of the virtual corner lines and μ with which to replace the latter lines (in the same way as suggested by Figure 13). The cut out area is the same while the cut-size only grows by a constant factor since there are at most three virtual corner lines. The result of the above described method is summarized in the following theorem (Theorem 24 in the full paper).

Theorem 4. *Given a non-crossing k-cut L of a simple polygon $\mathcal{P}$ with cut-size l containing only straight and corner lines except one which is a staircase line, for any desired $\varepsilon \in \,]0,1]$ there exists a non-crossing corner k'-cut L', where $k' \in [(1-\varepsilon)k, (1+\varepsilon)k]$, having a cut-size of at most $(6\sqrt{7/\varepsilon} + 7) \cdot l$.*

5   Converting Lines in Polygons to Segments in Grids

We now face the task of finding a cut in the grid G given a cut in the polygon $\mathcal{P}_G$ constructed from G. Our transformation from a grid to a polygon implies that an optimal k-cut in G transforms into a k-cut in $\mathcal{P}_G$, but not necessarily into an optimal one, since the cut lines in the polygon are not limited to integer positions (these are integer positions in the dual of the grid, and thus halfway positions between grid points). In other words, a cut in the polygon does not in general translate directly into a cut in the grid (note that if we would just cut grid edges with polygon cut lines, that is, not cut the edges in the middle, this would not translate the cut out area into the same number of grid vertices). Whenever a line in $\mathcal{P}_G$ happens to lie in integer position however, we will just take the corresponding segment to cut the grid G (Figure 15).

For non-grid lines, we start with a clean-up phase that modifies a pair of these lines so that one of them becomes a grid line, and the other compensates for the area difference that this creates. We start the clean-up phase by first focussing on the unit length open intervals on the polygon boundary between adjacent integer positions, as shown in Figure 15. Because a grid line does not hit any such open unit interval, we are concerned only with cut lines that do. For any open unit interval hit by more than one cut line, we can shift one of these cut

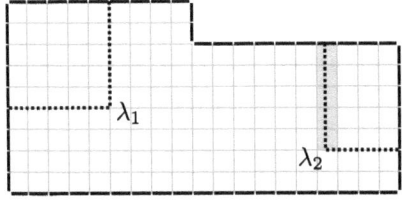

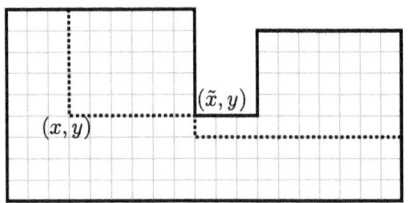

Fig. 15. A grid line λ_1 and a non-grid line λ_2. The corridor of λ_2 is shaded in grey. The boundary of the polygon is divided into the unit length lines.

Fig. 16. A virtual pseudo corner line with its corner at (x, y) and its unit sized step at $\tilde{x}$

lines to the boundary and compensate for the area difference by also shifting one other of these cut lines accordingly. Repeating this leaves us with at most one cut line per open unit interval on the boundary (Lemma 27). Similarly, as long as there is more than one such non-grid line (now in different open unit intervals on the boundary), we can shift one of them to become a grid line, and shift the other accordingly, to compensate for the area difference. This ends in a situation with at most one non-grid line in the cut (Lemma 28). During the whole process, the cut length does not increase (it might even decrease, since the cut was not necessarily optimal).

From now on, we can limit ourselves to the situation with only one non-grid line in the polygon cut. We shift this line to the nearest integer position (Figure 15), creating the need to compensate for the area difference (which will be the more involved part of the argument). We do this by introducing more grid lines. But since this increases the cut length, we need to prove that suitable short extra grid lines exist. In the end, this will preserve the property that the cut out area lies in the interval defined by k and ε, but will increase the cut-size by a constant factor. Next, we will look at a way to cut out for compensation, and then argue that there is a place from which to cut out in this way.

We manage to compensate in a recursive manner. We compensate for an area difference a by first finding a particular way to cut out an area guaranteed to be between a and $3a/2$, with the exact value not under our control (Lemma 31). This leaves us with the problem to compensate for at most half the previous area (since we are at most $a/2$ away from a). A recursive repetition of this compensation step ends after at most $\lfloor \log_2(\lceil a \rceil) \rfloor$ steps. The particular way to cut out the area between a and $3a/2$ makes use of a staircase grid line of three consecutive bends, with a step of unit height at the middle bend (Figure 16). Furthermore, the middle bend is guaranteed to lie outside or on the boundary of the polygon, so that the intersection of the staircase with the polygon results in a set of corner and straight lines (Figure 16) in the cut. We call this a *virtual pseudo corner line*. The analysis of the recursion reveals that the total length of the additional lines to cut out area a is limited to $3a$ (Lemma 30).

We still need to convince ourselves that such a line exists. In case there is a virtual corner line that cuts out the required area and contains only grid lines we are done. This is because such a set of lines can be seen as a virtual pseudo corner line, as the unit step of the underlying staircase line of the latter can entirely lie outside of the polygon. In the other case a suitable set of lines can be constructed using three virtual corner lines at some integer points (x^*, y^*), $(x^* + 1, y^*)$, and $(x^*, y^* + 1)$. These three virtual corner lines are chosen such that the first one cuts out an area larger than $3a/2$, while the other two each cut out at most $a - 1$. Using these properties it is then possible to show that there must be a unit sized step between the virtual corner lines at (x^*, y^*) and $(x^*, y^* + 1)$ with which a suitable virtual pseudo corner line can be constructed. That is, the corresponding set of lines cuts out an area between a and $3a/2$, and the upper most point of the unit sized step is on the boundary or outside of the polygon.

It remains to be shown that there is a place in the polygon to cut out from using the recursive method above. For this we use a tail of the cut (Figure 17), as for the staircase line argument in the previous section. However, we need to make sure that no additional lines are produced while cutting out the area of size a which would increase the cut-size by some non-constant factor. For this we break the tail into four sectors using four virtual corner lines having the same corner as the line of the tail. We then greedily assign these virtual corner lines to the

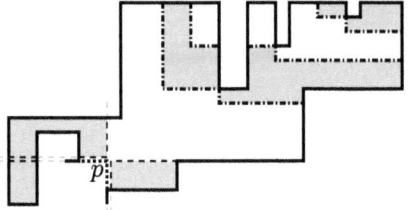

Fig. 17. A tail with its corner line at p (black dotted). The excess area in grey is cut out using the four virtual corner lines at p (thin dashed) together with the recursive method that uses virtual pseudo-corner lines (dashed and dotted).

cut as long as the cut out area does not exceed a. The difference still needed to reach the desired area a is finally cut out using the recursive method presented above from one of the four sectors that was not yet used (Lemma 33).

Thus the main result of this paper, as stated in Theorem 1, follows from the theorem below (Theorem 34 in the full paper) which summarizes this section.

Theorem 5. *Let l be the cut-size of an optimal k-cut L, for some $k \in \{0, ..., n\}$, in the polygon $\mathcal{P}_G$ of a grid G. For any $\varepsilon \in \,]0, 1]$ there exists a non-crossing corner k'-cut L' for some $k' \in [(1 - \varepsilon)k, (1 + \varepsilon)k]$, such that all lines in L' are grid lines and the cut-size is at most $(216\sqrt{7}/\varepsilon + 260) \cdot l$.*

References

1. Arbenz, P., Müller, R.: Microstructural finite element analysis of human bone structures. ERCIM News 74, 31–32 (2008)
2. Díaz, J., Serna, M.J., Torán, J.: Parallel approximation schemes for problems on planar graphs. Acta Informatica 33(4), 387–408 (1996)
3. Feldmann, A.E., Das, S., Widmayer, P.: Simple Cuts are Fast and Good: Optimum Right-Angled Cuts in Solid Grids. In: Wu, W., Daescu, O. (eds.) COCOA 2010, Part I. LNCS, vol. 6508, pp. 11–20. Springer, Heidelberg (2010)
4. Feldmann, A.E., Das, S., Widmayer, P.: Restricted cuts for bisections in solid grids: A proof via polygons. Technical Report 731, Institute of Theoretical Computer Science, ETH Zürich (July 2011)
5. Garey, M.R., Johnson, D.S.: Computers and Intractability: A guide to the Theory of NP-completeness. W.H. Freeman and Co., San Fransisco (1979)
6. Goldberg, M., Miller, Z.: A parallel algorithm for bisection width in trees. Computers & Mathematics with Applications 15(4), 259–266 (1988)
7. MacGregor, R.M.: On partitioning a graph: a theoretical and empirical study. PhD thesis, University of California, Berkeley (1978)
8. Papadimitriou, C., Sideri, M.: The bisection width of grid graphs. Theory of Computing Systems 29, 97–110 (1996)
9. Räcke, H.: Optimal hierarchical decompositions for congestion minimization in networks. In: Proceedings of the 40th Annual ACM Symposium on Theory of Computing (2008)

Maximum Independent Set in 2-Direction Outersegment Graphs*

Holger Flier, Matúš Mihalák, Peter Widmayer, and Anna Zych

Institute of Theoretical Computer Science, ETH Zürich, Switzerland
{firstname.lastname}@inf.ethz.ch,
http://www.pw.inf.ethz.ch/

Abstract. An outersegment graph is the intersection graph of line-segments lying inside a disk and having one end-point on the boundary of the disk. We present a polynomial-time algorithm for the problem of computing a maximum independent set in outersegment graphs where every segment is either horizontally or vertically aligned. We assume that a geometric representation of the graph is given as input.

1 Introduction

In this paper we study the problem of computing a maximum independent set (MIS) in intersection graphs of segments lying inside a disk and having one endpoint attached to the boundary of the disk. The problem of computing a MIS in various classes of intersection graphs has been intensively studied in the literature. For an extensive survey on many graph classes, refer to [1]. Despite the numerous efforts, the problem is by far not solved or fully understood. This paper adds to these efforts by presenting a polynomial-time algorithm for computing a MIS in a specific class of intersection graphs.

Motivated by general interest in the field of computational geometry and graph theory, intersection graphs of curves in the plane have received considerable attention in the literature, e.g., see [3,4,5,8,9,10,13,14]. A graph is a *string graph* if each vertex can be represented by a *string*, i.e., a curve in the Euclidean plane, such that there is an edge connecting two vertices if and only if the corresponding strings intersect. A set of strings representing the vertices is called a *representation* of the graph.

Most of the classical NP-hard optimization problems on graphs (such as finding a maximum clique, a maximum independent set, a minimum vertex cover, a minimum dominating set, or a minimum coloring) remain NP-hard for string graphs even if the representation is given [7,9,11,16]. The problem of recognizing string graphs, i.e., deciding whether a given graph is a string graph, is NP-hard, too [9]. Finding a MIS remains NP-hard even for the yet narrower class of *segment graphs*, which are the intersection graphs of straight line segments in the plane [11]. If every segment of the representation of a segment graph follows one

---

* This work was partially funded by the Swiss National Science Foundation (SNF grant no. 200021-125033/1).

P. Kolman and J. Kratochvíl (Eds.): WG 2011, LNCS 6986, pp. 155–166, 2011.
© Springer-Verlag Berlin Heidelberg 2011

of k directions, we say that the graph is a *k-direction segment graph*. It has been shown that the problem of computing a MIS in k-direction segment graphs is NP-hard for every $k \geq 2$ [11], whereas it is solvable in polynomial time for $k = 1$, since a 1-direction segment graph is an interval graph.

Considerably less is known about the problem of computing a MIS in string graphs if we restrict the strings to lie entirely inside a disk and to have one endpoint on the boundary of the disk. Such a string graph is called an *outerstring graph*. While finding a maximum clique is NP-hard in outerstring graphs [12], the complexity of finding a MIS in outerstring graphs is, to the best of our knowledge, an open problem. Our original motivation to study the problems considered in this paper stems from searching for a polynomial time algorithm for the latter problem, which would have interesting applications in railways [2].

We call an outerstring graph an *outersegment* graph if it has a representation where every string is a straight line segment. If further every segment of the representation follows one of k fixed directions, we say that the graph is a *k-direction outersegment graph*. Up to now, the complexity of computing a MIS in 2-direction outersegment graphs has been an open problem, too. We refer to this computational problem as MIS-2-DIR-OUTER-SEG and to the class of graphs as 2-DIR-OUTER-SEG.

Let us remark that the restriction to 2-direction outersegment graphs still allows for chordless cycles of length 5 (refer to [1] for terminology). Hence, outersegment graphs are not perfect. An interesting subclass of outerstring graphs that is neither a sub- nor a superclass of outersegment graphs is the class of interval filament graphs, for which the problem of computing a MIS is solvable in polynomial time [6].

In this paper we present a polynomial-time algorithm for MIS-2-DIR-OUTER-SEG if a representation of the outersegment graph is given. For the ease of presentation, we assume that each segment is either horizontally or vertically aligned. The presented techniques can be generalized to the case where the segments are aligned to two arbitrary directions (i.e., not only horizontally and vertically). A sketch of how to generalize our results is given at the end of this paper. The main ingredient of our solution is a dynamic-programming algorithm that solves the problem on restricted instances where no vertical segment attached to the upper half of the disk appears. Then, by a careful guessing of few segments of an optimal solution, we can decompose the original problem into four restricted subproblems where we can apply the dynamic programming algorithm.

We leave the complexity of computing a MIS in k-direction outersegment graphs open for $k \geq 3$. To the best of our knowledge, the complexity of recognizing 2-DIR-OUTER-SEG graphs is open, too.

Notation and Definitions. An instance of MIS-2-DIR-OUTER-SEG is a set $\mathcal{I}$ of straight line segments in the plane lying in a disk $\mathcal{D}$ such that each segment s has at least one endpoint on the boundary of $\mathcal{D}$. We call this endpoint the *disk-endpoint* of s. We assume w.l.o.g. that the other endpoint of s does not lie on the boundary of $\mathcal{D}$. We call this endpoint the *free-endpoint* of s.

To facilitate our discussion, we assume w.l.o.g. that the center of $\mathcal{D}$ is aligned with the origin of a Cartesian coordinate system. Furthermore, every segment $s \in \mathcal{I}$ is either *horizontal* (i.e., parallel with the x-axis) or *vertical* (i.e., parallel with the y-axis). We assume w.l.o.g. that no segment lies on the x-axis or on the y-axis. Thus, each segment is either a *left-, right-, top-* or *bottom-* segment, depending on the location of its disk-endpoint: a horizontal segment is a left- (right-) segment, if its disk-endpoint has a negative (positive) x-coordinate; a vertical segment is a top- (bottom-) segment, if its disk-endpoint has a positive (negative) y-coordinate. We denote the set of left-, right-, top-, and bottom-segments as $\mathcal{L}$, $\mathcal{R}$, $\mathcal{T}$, and $\mathcal{B}$, respectively. These sets form a partition of $\mathcal{I}$.

As our goal is to compute a MIS in the intersection graph of $\mathcal{I}$, we assume w.l.o.g. that no two segments in $\mathcal{L}$ have the same disk-endpoint: observe that no MIS can contain more than one such segment; thus we can preprocess the input by keeping in $\mathcal{L}$ the shortest segment of all segments with the same disk-endpoint. Similarly, we assume the same about segments in $\mathcal{R}$, $\mathcal{T}$, and $\mathcal{B}$. Thus, the segments within one set of the partition do not intersect (and so form an independent set in the underlying intersection graph). Note, however, that a horizontal segment from $\mathcal{L}$ may intersect with one from $\mathcal{R}$. Similarly, a vertical segment from $\mathcal{T}$ may intersect with one from $\mathcal{B}$. We call an instance *bipartite* if at least two of the sets of the partition $\mathcal{L}$, $\mathcal{R}$, $\mathcal{T}$, $\mathcal{B}$ are empty. Clearly, the intersection graph of a bipartite instance is a bipartite graph, for which the problem of finding a MIS can be solved in polynomial time [15]. We call an instance *tripartite* if one of the sets of the partition $\mathcal{L}$, $\mathcal{R}$, $\mathcal{T}$, $\mathcal{B}$ is empty. We refer to the version of MIS-2-DIR-OUTER-SEG that is restricted to tripartite instances as *tripartite* MIS-2-DIR-OUTER-SEG.

We distinguish between two different locations of the vertical segments: a *western vertical segment* lies to the left of the y-axis, and an *eastern vertical segment* lies to the right of the y-axis. Similarly, we distinguish two different locations of the horizontal segments: a *northern horizontal segment* lies above the x-axis, and a *southern horizontal segment* lies below the x-axis.

Finally, for a region $X \subset \mathcal{D}$ and a set of segments S, we denote by $S[X]$ the segments of S contained entirely in X.

Outline. In Section 2 we present a polynomial-time algorithm for tripartite MIS-2-DIR-OUTER-SEG, i.e., the restricted version of MIS-2-DIR-OUTER-SEG where one set of the partition $\mathcal{L}$, $\mathcal{R}$, $\mathcal{T}$, $\mathcal{B}$ is empty. Based on this, we present a polynomial-time algorithm for MIS-2-DIR-OUTER-SEG in Section 3.

2 Solving the Tripartite MIS-2-Dir-Outer-SEG

In this section we consider tripartite instances of MIS-2-DIR-OUTER-SEG, i.e., instances for which one of the sets $\mathcal{L}$, $\mathcal{R}$, $\mathcal{T}$, and $\mathcal{B}$ is empty. Without loss of generality, we will assume that $\mathcal{T} = \emptyset$, i.e., that there is no top segment. In the following we present a polynomial-time algorithm that finds a maximum independent set in any such restricted instance. We will observe the existence of a certain *decomposition of every solution* into solutions of independent subproblems, where every subproblem is a bipartite instance of MIS-2-DIR-OUTER-SEG. We describe this decomposition in the next section, before presenting the actual algorithm which is based on the dynamic programming technique.

2.1   Structure of an Optimal Solution

In the following we show that an optimal solution for tripartite MIS-2-DIR-OUTER-SEG is the disjoint union of optimal solutions for a certain set of subproblems. Let $\mathcal{I}$ be an instance of tripartite MIS-2-DIR-OUTER-SEG. Assume that we are given an optimal solution OPT for $\mathcal{I}$ together with a partition of the disk $\mathcal{D}$ into regions $\mathcal{D}_1, \ldots, \mathcal{D}_r$ such that each segment of OPT lies entirely inside one of these regions. For $i = 1, \ldots, r$, recall that $\mathcal{I}[\mathcal{D}_i]$ and $\mathrm{OPT}[\mathcal{D}_i]$ denote the set of segments of the instance $\mathcal{I}$ and OPT, respectively, that lie completely within region $\mathcal{D}_i$. Note that while $\mathrm{OPT} = \bigcup_{i=1}^{r} \mathrm{OPT}[\mathcal{D}_i]$, it holds that $\mathcal{I} \supsetneq \bigcup_{i=1}^{r} \mathcal{I}[\mathcal{D}_i]$ if there is a segment of $\mathcal{I}$ that lies in more than one of the regions. Denote by $\mathrm{MIS}[S]$ a maximum independent set for a set S of segments. For a region $\mathcal{D}_i$, we abbreviate $\mathrm{MIS}[\mathcal{I}[\mathcal{D}_i]]$ to $\mathrm{MIS}[\mathcal{D}_i]$. Clearly, it follows that $|\mathrm{OPT}[\mathcal{D}_i]| = |\mathrm{MIS}[\mathcal{D}_i]|$ for all $i = 1, \ldots, r$.

Hence, if we can find such a partition in polynomial time such that for all $i = 1, \ldots, r$, a $\mathrm{MIS}[\mathcal{D}_i]$ can be computed in polynomial time, then we can solve tripartite MIS-2-DIR-OUTER-SEG in polynomial time. We show that such a partition always exists before presenting our algorithm in the next section. The argument is based on the structure of an (unknown) optimal solution OPT. We show that a particular traversal of the bottom segments of OPT yields the desired partition of $\mathcal{D}$. We say that a bottom segment s_i *towers above* a bottom segment s_j if the y-coordinate of the free-endpoint of s_i is greater than that of s_j. W.l.o.g. we assume that the free-endpoints lie in general position and hence that for each region, there is at most one segment towering above all other segments.

Lemma 1. *Given an instance $\mathcal{I}$ of tripartite* MIS-2-DIR-OUTER-SEG *and an optimum solution OPT, there exists a partition $\mathcal{D}_1, \ldots, \mathcal{D}_r$ of the disk $\mathcal{D}$ into regions such that $|OPT| = \sum_{i=1}^{r} |\mathrm{MIS}[\mathcal{D}_i]|$ and a $\mathrm{MIS}[\mathcal{D}_i]$ can be computed in polynomial time for each $i = 1, \ldots, r$.*

Proof. Given an optimal solution OPT, we partition the disk recursively as follows, see Figure 1 for an example. At each step i of the recursion, there is an *unprocessed* region U_i of the disk for which $\mathcal{I}[U_i]$ is tripartite. Let $U_0 := \mathcal{D}$ be the initial region. At each step $i \geq 1$, let s_i be the bottom segment in U_{i-1} that towers above all other bottom segments in U_{i-1}. Segment s_i naturally divides U_{i-1} into three regions, namely A_i, B_i, and U_i, as follows.

A_i is the region (strictly) above s_i. Note that $\mathcal{I}[A_i]$ may contain bottom segments if it does not lie completely in the northern half of $\mathcal{D}$, and thus be tripartite. In this case, however, it must hold that a MIS for $\mathcal{I}[A_i] \setminus \mathcal{B}$ is also a MIS for $\mathcal{I}[A_i]$, since by choice of s_i, $\mathrm{OPT}[A_i]$ does not contain a bottom segment. Hence, a $\mathrm{MIS}[A_i]$ can be computed in polynomial time, as $\mathcal{I}[A_i] \setminus \mathcal{B}$ is bipartite.

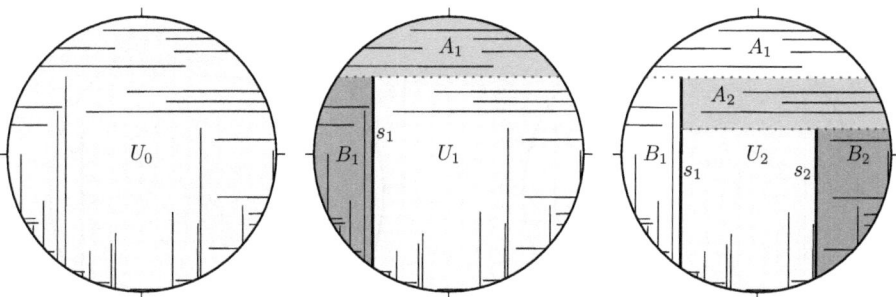

Fig. 1. All three pictures depict the same set of independent segments of an optimal solution to a tripartite instance of MIS-2-DIR-OUTER-SEG. Left: Initially, the unprocessed region U_0 is the whole disk $\mathcal{D}$. Middle: The bottom-segment s_1 of an optimal solution OPT partitions U_0 into three regions A_1, B_1, and U_1. Right: Recursively, s_2, which towers above all bottom-segments in U_1, partitions U_1 into three regions A_2, B_2, and U_2.

B_i and U_i are the regions to either side of s_i (and below A_i), where B_i includes s_i and U_i contains the lowest point of the disk. (For example, if s_i is a western bottom segment, B_i is to the left of s_i.) Because B_i does not contain the lowest point of the disk, it either lies completely in the left or completely in the right half of the disk. Hence, no two horizontal segments of $\mathcal{I}[B_i]$ can intersect, as they are all either left or right segments. Thus, $\mathcal{I}[B_i]$ is bipartite as well.

The recursion continues by partitioning U_i if OPT$[U_i]$ contains a bottom segment, and stops otherwise. Let U_ℓ denote the region at which the recursion stops. Similar to the argument for A_i above, it suffices to compute a MIS$[\mathcal{I}[U_\ell] \setminus \mathcal{B}]$, as OPT$[U_\ell]$ consists of horizontal segments only. As $\mathcal{I}[U_\ell] \setminus \mathcal{B}$ is bipartite, a MIS can be computed in polynomial time.

By the choice of s_i, the regions A_i, B_i, $i = 1, \ldots, \ell$ and U_ℓ are a partition of $\mathcal{D}$ such that each segment of OPT lies completely within one of these regions. Hence,

$$|\text{OPT}| = \sum_{i=1}^{\ell} |\text{MIS}[A_i]| + |\text{MIS}[B_i]| + |\text{MIS}[U_\ell]|$$

must hold. This completes the proof.   □

In the following, we call A_i the part *above* s_i in U_{i-1}, B_i the part *behind* s_i in U_{i-1}, and U_i the *unprocessed* part of U_{i-1} by s_i.

2.2   Algorithm for Tripartite MIS-2-Dir-Outer-SEG

From the above discussion we know that there exists a sequence of bottom segments $s_1, \ldots, s_\ell$ that yields a partition of $\mathcal{D}$ into regions such that a MIS for $\mathcal{I}$ can be computed in polynomial time by independently computing a MIS for a set of bipartite subproblems induced by these regions. The partition of $\mathcal{D}$ is based on the structure of an optimal solution which, of course, is unknown.

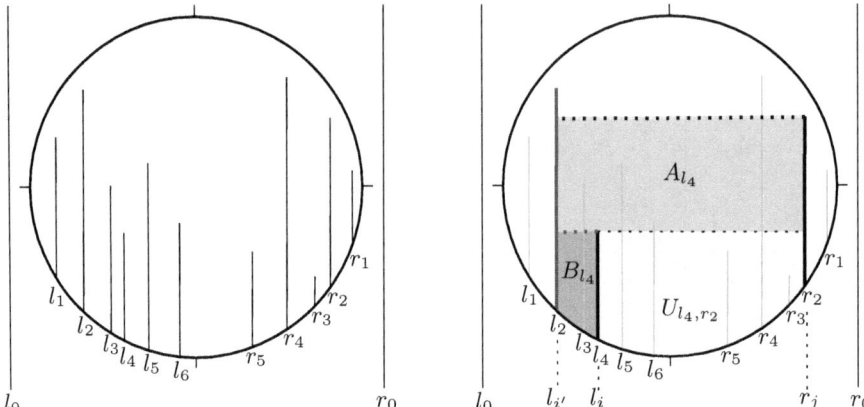

Fig. 2. Both figures depict the bottom segments of a (tripartite) instance of MIS-2-DIR-OUTER-SEG. Left: the western bottom-segments $l_0, \ldots, l_{n_l}$ are ordered from left to right (here, $n_l = 6$), and the eastern bottom-segments $r_0, \ldots, r_{n_r}$ are ordered from right to left (here, $n_r = 5$). Right: Computing $\mathrm{T}[l_i, r_j]$ as $\mathrm{T}[l_{i'}, r_j] + |\mathrm{MIS}[A_{l_i}]| + |\mathrm{MIS}[B_{l_i}]|$ for $i = 4$, $j = 2$ and $i' = 2$.

Next, we develop a dynamic programming approach that allows us to find such a partition.

To develop the algorithm, we need a few more definitions. Let $s_L(i)$ and $s_R(i)$ denote the last visited western and eastern bottom-segment, respectively, after step $i = 1, \ldots, \ell$, of the recursion in the proof of Lemma 1. In the following we will use two *phantom boundary* segments $s_{-\infty}$ and s_{∞} to allow $s_L(i)$ and $s_R(i)$ to be always defined: we denote by $s_{-\infty}$ the infinite vertical segment defined by the equation $x = -\infty$ (i.e, a line), and by ℓ_{∞} the infinite vertical segment $x = +\infty$. We set $s_L(0) = s_{-\infty}$ and $s_R(0) = s_{\infty}$.

Observe that with these definitions, the region U_i, $i = 0, \ldots, \ell$, is defined by $s_L(i)$ and $s_R(i)$: U_i is the region of $\mathcal{D}$ to the right of $s_L(i)$ and to the left of $s_R(i)$ and below the free-endpoints of both $s_L(i)$ and $s_R(i)$. The regions A_{i+1} and B_{i+1} can thus be defined by U_i and the $(i+1)$-th visited segment, i.e., by $s_L(i)$, $s_R(i)$ and by either $s_L(i+1)$ or $s_R(i+1)$.

Let $l_1, \ldots, l_{n_l}$ be the western bottom-segments sorted by x-coordinate in increasing order, and let $r_1, \ldots, r_{n_r}$ be the eastern bottom-segments sorted by x-coordinate in decreasing order. Further, let l_0 be the phantom segment $s_{-\infty}$ and let r_0 be the phantom segment s_{∞}. See Figure 2 for illustration.

For a segment l_i and a segment r_j we define by U_{l_i, r_j} the unprocessed region that we would obtain by the recursion if l_i and r_j were the last visited western and eastern bottom-segment, respectively. Thus, this is the region between segments l_i and r_j and below the free-endpoints of l_i and r_j.

We will compute the table $\mathrm{T}[l_i, r_j]$ for every $i = 0, \ldots, n_l$ and every $j = 0, \ldots, n_r$, where $\mathrm{T}[l_i, r_j]$ is the maximum number of non-intersecting segments in the subproblem defined by the segments lying completely inside $\mathcal{D} \setminus U_{l_i, r_j}$ where segments l_i and r_j are required to be part of the solution (of the subproblem).

Clearly, if we have such a table at hand, we can compute the optimal number of non-intersecting segments of the whole instance: consider the recursion in the proof of Lemma 1 on an (unknown) optimal solution OPT. Let $l_i \in \text{OPT}$ be the last western and $r_j \in \text{OPT}$ be the last eastern bottom-segment encountered in the sequence $s_1, \ldots, s_\ell$ of the recursion. By Lemma 1, a $\text{MIS}[U_{l_i,r_j}]$ can be computed in polynomial time (because it suffices to compute a MIS for the bipartite instance $\mathcal{I}[U_{l_i,r_j}] \setminus \mathcal{B}$). Hence, $|\text{OPT}| = \text{T}[l_i, r_j] + |\text{MIS}[U_{l_i,r_j}]|$ can be computed in polynomial time.

As we do not know OPT, our algorithm tries all pairs $l_i, r_j \in \mathcal{B}$, and outputs the maximum of the computed values $\text{T}[l_i, r_j] + |\text{MIS}[U_{l_i,r_j}]|$ over all i, j. The solution itself can be computed using standard book-keeping techniques.

We now show how to compute the entries of the table $\text{T}[\cdot, \cdot]$. Again, for a pair $s, t \in \mathcal{B}$, we say that s *towers above* t if the free-endpoint of s has a greater y-coordinate than that of t.

We set $\text{T}[l_0, r_0] = 0$. Then, for every $i = 0, \ldots, n_l$ and every $j = 0, \ldots, n_r$, we need to distinguish the following cases in order to compute the value of entry $\text{T}[l_i, r_j]$. Namely, for the case that r_j towers above l_i, we compute

$$\text{T}[l_i, r_j] = \max_{\substack{i' < i \\ l_{i'} \text{ towers above } l_i \text{ and } r_j}} \{\text{T}[l_{i'}, r_j] + |\text{MIS}[A_{l_i}]| + |\text{MIS}[B_{l_i}]|\}, \quad (1)$$

and otherwise (if l_i towers above r_j), we compute

$$\text{T}[l_i, r_j] = \max_{\substack{j' < j \\ r_{j'} \text{ towers above } l_i \text{ and } r_j}} \{\text{T}[l_i, r_{j'}] + |\text{MIS}[A_{r_j}]| + |\text{MIS}[B_{r_j}]|\}. \quad (2)$$

As in the proof of Lemma 1, A_{l_i} is the region of $\mathcal{D}$ between segments $l_{i'}$ and r_j and above the free-endpoint of l_i and not above the free-endpoints of $l_{i'}$ and r_j; B_{l_i} is the region of $\mathcal{D}$ below A_{l_i} and between $l_{i'}$ and l_i, including l_i but not $l_{i'}$. The regions A_{r_j} and B_{r_j} are defined symmetrically. By Lemma 1, the cardinalities of $|\text{MIS}[A_s]|$ and $|\text{MIS}[B_s]|$, $s \in \{l_i, r_j\}$ can be computed in polynomial time.

Theorem 1. *The table entry* $\text{T}[l_i, r_j]$, $i = 0, \ldots, n_l$, $j = 0, \ldots, n_r$, *contains the size of an optimal solution of an instance* $\mathcal{I}[\mathcal{D} \setminus U_{l_i,r_j}]$ *further restricted to contain the segments* l_i *and* r_j.

Proof. To prove the theorem we need to show that $\text{T}[\cdot, \cdot]$ indeed has the recursive property of Equations (1) and (2). This, however, follows directly from the existence of a decomposition as described in Section 2.1: The recursive computation of $\text{T}[l_i, r_j]$ takes the last visited segment x (x is the "smaller" segment of l_i or r_j, i.e., the one that is not towering above the other) and finds the segment that is the predecessor of x in the sequence $s_1, \ldots, s_\ell$ of bottom segments of an unknown optimal solution OPT. This predecessor of x naturally defines, together with x, the region A_x above x and the region B_x behind x, just as in the recursion of Lemma 1. The correctness of the recursive definition of T then directly follows from Lemma 1. □

Corollary 1. *There is a polynomial-time algorithm for tripartite instances of* MIS-2-DIR-OUTER-SEG.

3   Decomposing MIS-2-Dir-Outer-SEG

In this section we provide a polynomial time algorithm for the general setting. We show how to decompose an arbitrary instance of MIS-2-DIR-OUTER-SEG into few tripartite instances of MIS-2-DIR-OUTER-SEG. The decomposition we describe can be computed in polynomial time. Therefore, combined with the polynomial time algorithm for tripartite MIS-2-DIR-OUTER-SEG presented in the previous section, it yields a polynomial time algorithm for MIS-2-DIR-OUTER-SEG. The decomposition is determined by a constant number of segments in an optimal solution. Since we do not know these segments, we have to perform an exhaustive search, namely by enumerating through all sets of segments whose cardinality is bounded by a constant.

We will use the following notation. A vertical *overlap* is a pair of vertical segments that cannot be separated by a horizontal line. If a vertical overlap consists of western segments only, i.e., if it lies entirely to the left of the y-axis, we call it a *left overlap*. If an overlap lies entirely to the right of the y-axis, we call it a *right overlap*.

Observe that if a left overlap is part of an optimal solution OPT then there is no right segment of OPT that intersects with the region to the left of the overlap. This region thus induces a tripartite instance of MIS-2-DIR-OUTER-SEG. Similarly, the region to the right of a right overlap of OPT induces a tripartite instance of MIS-2-DIR-OUTER-SEG. In the following, we show that the region between the two overlaps can be decomposed into two tripartite instances of MIS-2-DIR-OUTER-SEG. For this, we will consider special (left and right) overlaps.

Lemma 2. *Let $\mathcal{I}$ be an instance of* MIS-2-DIR-OUTER-SEG *and let OPT be an optimal solution for it. If OPT contains a left overlap, then $\mathcal{D}$ can be partitioned into regions R_1, R_2, and R_3 such that*

- *$\mathcal{I}[R_1]$ is a tripartite instance of* MIS-2-DIR-OUTER-SEG
- *$OPT[R_2]$ does not contain a left overlap*
- *there are segments $f, c \in OPT$ s.t. $OPT = OPT[R_1] \cup OPT[R_2] \cup \{f, c\}$*

Proof. Assume that OPT contains a left overlap. Each overlap consists of two segments: the one further from the y-axis, which we call the *far* segment, and the one closer to the y-axis, which we call the *close* segment. Let $\{f, c\}$ be the left overlap in OPT where the far segment f is the rightmost far segment occurring in a left overlap of OPT, and where further the close segment c is the segment that is closest to f among all $c' \in OPT$ that form a left overlap $\{f, c'\}$.

Let E be the rectangle from the free-endpoint of c to the free-endpoint of f. Due to the choice of f and c, no segments in OPT lie within E. Let R_1 and R_2 be the region to the left and right of $f \cup c \cup E$, respectively, such that R_1, R_2, and $f \cup c \cup E$ are a partition of $\mathcal{D}$. See Figure 3 for illustration.

Clearly, OPT = $OPT[R_1] \cup OPT[R_2] \cup \{f, c\}$. Since $\{f, c\}$ is a left overlap, $\mathcal{I}[R_1]$ does not contain a right segment, and thus it is a tripartite instance of MIS-2-DIR-OUTER-SEG.

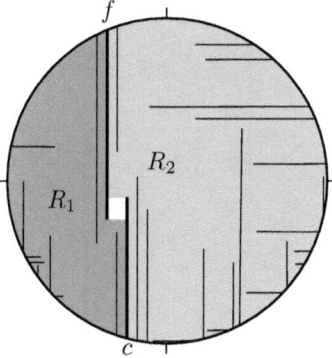

Fig. 3. Illustration for Lemma 2. A left-overlap $\{f, c\}$, yielding a tripartite subproblem $\mathcal{I}[R_1]$. Due to the choice of f and c, no segment can cross the boundary of the white rectangle.

Now let g be the chord of $\mathcal{D}$ containing f. Note that by the choice of f, for each left overlap in OPT, its far segment lies to the left of or on g. Thus, no pair of segments of OPT$[R_2]$ can form a left overlap. This completes the proof. □

Theorem 2. *Let $\mathcal{I}$ be an instance of* MIS-2-DIR-OUTER-SEG *and let OPT be an optimal solution for it. There is a set of segments $S \subseteq$ OPT that allows to determine in polynomial time pairwise disjoint instances $\mathcal{I}_1, \ldots, \mathcal{I}_h \subseteq \mathcal{I}$ of tripartite* MIS-2-DIR-OUTER-SEG, *such that*

$$|OPT| = |S \cup \bigcup_{i=1}^{h} OPT(\mathcal{I}_i)|$$

Moreover, $|S|$ is bounded from above by a constant.

Proof. The set S that we will construct in the following separates $\mathcal{D}$ into a constant number of regions. The i'th region determines $\mathcal{I}_i$ as a subset of $\mathcal{I}$ contained in that region. We proceed with the construction of S.

Lemma 2 shows that we may focus on the case when OPT contains neither a left nor a right overlap: If OPT contains a left overlap then $\mathcal{I}$ can be decomposed into two independent subproblems, namely a tripartite MIS-2-DIR-OUTER-SEG instance $\mathcal{I}_1$ and a MIS-2-DIR-OUTER-SEG instance $\mathcal{I}'$, each induced by segments lying completely inside the region to the left and, respectively, right of the overlap. Thus, it suffices to consider $\mathcal{I}'$. Lemma 2 also states that $\mathcal{I}'$ admits an optimum not containing a left overlap. Symmetrically, we can use Lemma 2 to eliminate right overlaps in the optimal solution. Hence, we further consider only the case where OPT contains neither a left nor a right overlap.

We distinguish two cases: first, we consider the case when OPT contains a vertical overlap, i.e., consisting of both a western and eastern vertical segment, and then the case when OPT does not contain a vertical overlap.

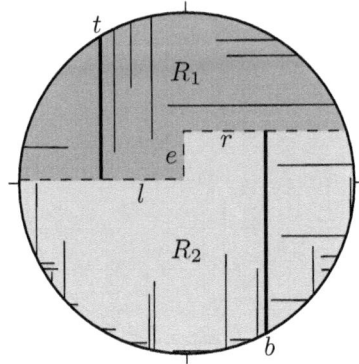

Fig. 4. Ilustration to Theorem 2. Segments t and b yield a separation line.

Case 1. We assume that OPT contains a vertical overlap. Let t be a top segment in OPT with the lowest y coordinate of its free-endpoint. Let b be a bottom segment in OPT with the highest y coordinate of its free-endpoint. Due to our assumption that OPT contains an overlap, t and b overlap in particular. Since there is no left and no right overlap, t and b lie on different sides of the y-axis. We assume w.l.o.g. that t lies to the left of the y-axis and b lies to the right of the y-axis. Let l be the horizontal line connecting the y-axis with the boundary of $\mathcal{D}$ passing through the free-endpoint of t and let r be the horizontal line connecting the y-axis with the boundary of $\mathcal{D}$ passing through the free-endpoint of b (see Figure 4). Let e be the line on y axis connecting the endpoints of l and r.

Observe that there are no top segments in OPT below l or crossing l, because a top segment in OPT below or crossing l would have the y-coordinate of its free-endpoint lower than the y-coordinate of the free-endpoint of t, a contradiction to the choice of t. Also, there is no top segment crossing or below r: any such segment would form a right overlap with r, a contradiction to our assumption. Similarly, there are no bottom segments in OPT above or crossing l or r. Since l and r are horizontal, only vertical segments could possibly cross them. Therefore no segments in OPT cross l or r.

Now observe, that no (horizontal) segment in OPT crosses e, as it would have to cross either t or b. The curve consisting of l, e and r divides $\mathcal{D}$ into two regions R_1 and R_2 that lie above and below the curve, respectively. These regions separate OPT into two independent parts. The part of OPT contained in R_1 is an optimal solution for instance $\mathcal{I}_1 = \mathcal{I}[R_1] \setminus \mathcal{B}$. The part of OPT contained in R_2 is an optimal solution for $\mathcal{I}_2 = \mathcal{I}[R_2] \setminus \mathcal{T}$. Both $\mathcal{I}_1$ and $\mathcal{I}_2$ are tripartite instances of MIS-2-DIR-OUTER-SEG. This completes the proof of the theorem for Case 1.

Case 2. We assume that OPT does not contain a vertical overlap. Consider the horizontal line l passing through the free-endpoint of the bottom segment in OPT with the highest y-coordinate of its free-endpoint. Clearly, l separates the top segments in OPT from the bottom segments in OPT. Thus, l divides $\mathcal{D}$

into two regions R_1 and R_2 lying above and below l, respectively. These regions separate OPT into two independent parts. Again, the part of OPT contained in R_1 is an optimal solution for instance $\mathcal{I}_1 = \mathcal{I}[R_1] \setminus \mathcal{B}$. The part of OPT contained in R_2 is an optimal solution for $\mathcal{I}_2$ defined as $\mathcal{I}[R_2] \setminus \mathcal{T}$. Both $\mathcal{I}_1$ and $\mathcal{I}_2$ are tripartite instances of MIS-2-DIR-OUTER-SEG. This completes the proof of the theorem for Case 2 (and thus of the whole theorem). $\square$

Corollary 2. MIS-2-DIR-OUTER-SEG *can be solved in polynomial time given a polynomial time algorithm for tripartite* MIS-2-DIR-OUTER-SEG.

Combining the results of the previous sections, we state our main result:

Theorem 3. MIS-2-DIR-OUTER-SEG *can be solved in polynomial time.*

We remark that the algorithms developed in this paper require a geometric representation of the graph, even if it is known that the graph under consideration is a 2-DIR-OUTER-SEG.

As mentioned earlier, the techniques of this paper can be generalized to the case where the segments are aligned to two arbitrary directions. Lets us sketch some steps of such a generalization. First, the disk can be rotated such that the segments of one direction are horizontally aligned. Then, there is an axis, say y', parallel to the other direction, and an axis x' that is orthogonal to y'. The definitions of top and bottom segments have to be adapted to x'. The definitions of eastern and western bottom segments, left and right segments, left and right half of the disk, as well as the notion of a bottom segment towering above another remain based on the y-axis, however. With these definitions, the results of Section 2 carry over easily. Further, it is not too difficult to adapt the results of Section 3. Some care has to be taken, however, to ensure that the choice of segments of an overlap indeed yields a partition such that each segment of OPT lies within exactly one region.

References

1. Brandstädt, A., Le, V.B., Spinrad, J.P.: Graph classes: a survey. SIAM Monographs on Discrete Mathematics and Applications. Society for Industrial and Applied Mathematics Mathematics (1999)
2. Flier, H., Mihalák, M., Schöbel, A., Widmayer, P., Zych, A.: Vertex Disjoint Paths for Dispatching in Railways. In: Proceedings of the 10th Workshop on Algorithmic Approaches for Transportation Modelling, Optimization, and Systems (ATMOS), Schloss Dagstuhl–Leibniz-Zentrum für Informatik, vol. 14, pp. 61–73 (2010)
3. Fox, J., Pach, J.: Coloring K_k-free intersection graphs of geometric objects in the plane. In: Proceedings of the 24th ACM Symposium on Computational Geometry (SoCG), pp. 346–354. ACM (2008)
4. Fox, J., Pach, J.: Erdős-Hajnal-type results on intersection patterns of geometric objects. In: Győri, E., Katona, G.O.H., Lovász, L. (eds.) Horizons of Combinatorics, vol. 17, pp. 79–103. Springer, Heidelberg (2008)
5. Fox, J., Pach, J.: A separator theorem for string graphs and its applications. Combinatorics, Probability and Computing 19(03), 371–390 (2010)

6. Gavril, F.: Maximum weight independent sets and cliques in intersection graphs of filaments. Information Processing Letters 73(5-6), 181–188 (2000)
7. Mark Keil, J.: The complexity of domination problems in circle graphs. Discrete Applied Mathematics 42(1), 51–63 (1993)
8. Kratochvíl, J.: String graphs. I. The number of critical nonstring graphs is infinite. Journal of Combinatorial Theory, Series B 52(1), 53–66 (1991)
9. Kratochvíl, J.: String graphs. II. Recognizing string graphs is NP-hard. Journal of Combinatorial Theory, Series B 52(1), 67–78 (1991)
10. Kratochvíl, J., Matoušek, J.: String graphs requiring exponential representations. Journal of Combinatorial Theory, Series B 53(1), 1–4 (1991)
11. Kratochvíl, J., Nešetřil, J.: INDEPENDENT SET and CLIQUE problems in intersection-defined classes of graphs. Commentationes Mathematicae Universitatis Carolinae 31(1), 85–93 (1990)
12. Middendorf, M., Pfeiffer, F.: The max clique problem in classes of string-graphs. Discrete Mathematics 108(1-3), 365–372 (1992)
13. Pach, J., Tóth, G.: Recognizing string graphs is decidable. Discrete & Computational Geometry 28, 593–606 (2002)
14. Schaefer, M., Sedgwick, E., Štefankovič, D.: Recognizing string graphs in NP. Journal of Computer and System Sciences 67(2), 365–380 (2003); Special Issue on STOC 2002
15. Schrijver, A.: Combinatorial Optimization: Polyhedra and Efficiency. Springer, Heidelberg (2003)
16. Unger, W.: On the k-colouring of circle-graphs. In: Cori, R., Wirsing, M. (eds.) STACS 1988. LNCS, vol. 294, pp. 61–72. Springer, Heidelberg (1988)

Complexity of Splits Reconstruction
for Low-Degree Trees[*]

Serge Gaspers[1], Mathieu Liedloff[2], Maya Stein[3], and Karol Suchan[4,5]

[1] Institute of Information Systems, Vienna University of Technology, Vienna, Austria
gaspers@kr.tuwien.ac.at
[2] LIFO, Université d'Orléans, Orléans, France
mathieu.liedloff@univ-orleans.fr
[3] CMM, Universidad de Chile, Santiago, Chile
mstein@dim.uchile.cl
[4] FIC, Universidad Adolfo Ibáñez, Santiago, Chile
karol.suchan@uai.cl
[5] WMS, AGH - University of Science and Technology, Krakow, Poland

Abstract. Given a vertex-weighted tree T, the split of an edge xy in T is $\min\{s_x, s_y\}$ where s_x (respectively, s_y) is the sum of all weights of vertices that are closer to x than to y (respectively, closer to y than to x) in T. Given a set of weighted vertices V and a multiset of splits $\mathcal{S}$, we consider the problem of constructing a tree on V whose splits correspond to $\mathcal{S}$. The problem is known to be NP-complete, even when all vertices have unit weight and the maximum vertex degree of T is required to be no more than 4. We show that

- the problem is strongly NP-complete when T is required to be a path. For this variant we exhibit an algorithm that runs in polynomial time when the number of distinct vertex weights is constant. We also show that
- the problem is NP-complete when all vertices have unit weight and the maximum degree of T is required to be no more than 3, and
- it remains NP-complete when all vertices have unit weight and T is required to be a caterpillar with unbounded hair length and maximum degree at most 3.

Finally, we shortly discuss the problem when the vertex weights are not given but can be freely chosen by an algorithm.

The considered problem is related to building libraries of chemical compounds used for drug design and discovery. In these inverse problems, the goal is to generate chemical compounds having desired structural properties, as there is a strong correlation between structural properties, such as the Wiener index, which is closely connected to the considered problem, and biological activity.

[*] The authors acknowledge the support of Conicyt Chile via projects Fondecyt 11090390 (M.L., K.S.), Fondecyt 11090141 (M.S.), Anillo ACT88 (K.S.), and Basal-CMM (S.G., M.S., K.S.). The first author acknowledges partial support from the European Research Council (COMPLEX REASON, 239962). The second and fourth authors acknowledge the support of the French Agence Nationale de la Recherche (ANR AGAPE ANR-09-BLAN-0159-03).

P. Kolman and J. Kratochvíl (Eds.): WG 2011, LNCS 6986, pp. 167–178, 2011.
© Springer-Verlag Berlin Heidelberg 2011

1   Introduction

In this paper, we consider trees $T = (V, E)$ where integer weights are associated to vertices by a function $\omega : V \to \mathbb{N}$, where $\mathbb{N}$ denotes the set of natural numbers excluding 0.

Definition 1. *Let T be a tree and $\omega : V \to \mathbb{N}$ be a function. The split of an edge e in T is the minimum of $\Omega(T_1)$ and $\Omega(T_2)$, where T_1 and T_2 are the two trees obtained by deleting e from T, and $\Omega(T_i) = \sum_{v \in T_i} \omega(v)$. We use $\mathcal{S}(T)$ to denote the multiset of splits of T.*

We consider the problem of reconstructing a tree with a given multiset of splits and a given set of weighted vertices.

> WEIGHTED SPLITS RECONSTRUCTION (WSR): Given a set V of n vertices, a weight function $\omega : V \to \mathbb{N}$, and a multiset $\mathcal{S}$ of integers, is there a tree T whose multiset of splits is $\mathcal{S}$ (i.e. $\mathcal{S}(T) = \mathcal{S}$)?

The WEIGHTED SPLITS RECONSTRUCTION FOR TREES OF MAXIMUM DEGREE k problem (WSR$_k$) is defined in the same way, except that we restrict T to have maximum degree at most k. When we require T to belong to a class of trees $\mathcal{T}$, the problem is called WEIGHTED SPLITS RECONSTRUCTION FOR $\mathcal{T}$. When ω assigns unit weights to the vertices, the problem is simply called SPLITS RECONSTRUCTION (SR). The SPLITS RECONSTRUCTION FOR TREES OF MAXIMUM DEGREE k problem (SR$_k$) and the SPLITS RECONSTRUCTION FOR $\mathcal{T}$ are the obvious unweighted counterparts of the weighted variants defined above.

Related Work. In the field of Chemical Graph Theory [2,3,18], molecules are modeled by graphs in order to study the physical properties of chemical compounds. A chemical graph is a graph, where vertices represent atoms of a chemical compound and edges the chemical bonds between them. Within the area of quantitative structure-activity relationship (QSAR), several structural measures of chemical graphs were identified that quantitatively correlate with a well defined process, such as biological activity or chemical reactivity. Probably the most widely known example is the *Wiener index* (see [12]): the sum of the distances in a graph between each pair of vertices, where the distance between two vertices is the number of edges on a shortest path from one to the other. Wiener [19] found a strong correlation between the boiling points of paraffins and the Wiener index. From then on, many other topological (using the information of the chemical graph) and topographical (using the information of the chemical graph and the location of its vertices in space) indices were introduced and their correlation with various other biological activities was investigated.

In Combinatorial Chemistry, drug design is facilitated by building libraries of molecules that are structurally related (via the Wiener index or any of the other numerous indices). We face inverse problems where the goal is to design new compounds that have a prescribed structural information (see also [6]).

Goldman et al. [11] study problems related to the design of combinatorial libraries for drug design from an algorithmic and complexity-theoretic point of view, following the heuristic approaches of [17] and [10]. They show that for every positive integer W, except 2 and 5, there exists a graph with Wiener index W. They also show that every integer, except a finite set, is the Wiener index of some tree. For constructing a tree (of unbounded or bounded maximum degree) with a given Wiener index, they devise pseudo-polynomial dynamic programming algorithms. Goldman et al. also introduce the SPLITS RECONSTRUCTION problem and recall a result due to Wiener [19]: the Wiener index of a tree T on n vertices with unit weights is $\sum_{s \in \mathcal{S}(T)} s \cdot (n - s)$. They show that SR is NP-complete and give an exponential-time algorithm without running time analysis.

As it is not reasonable to construct chemical trees with arbitrarily high vertex degrees, Li and Zhang [15] studied SR_4 and showed that it is also NP-complete. Their algorithm to construct a tree with maximum degree at most 4 to solve SR_4 runs in exponential time (no running time analysis is provided) and creates weighted vertices in intermediate steps.

In order to reconstruct glycans or carbohydrate sugar chains, Aoki-Kinoshita et al. [1] study the reconstruction of a node-labeled supertree from a set of node-labeled subtrees. They give a 6-approximation algorithm for this problem, which generalizes the smallest superstring problem. We refer to [4] surveying results on the Wiener index for trees.

Our Results. By the result of Li and Zhang [15], SR_4 is NP-complete, while SR_2 is trivially in P. We close this gap by showing that SR_3 is NP-complete by a reduction from NUMERICAL MATCHING WITH TARGET SUMS (defined below). It is even NP-complete for caterpillars with unbounded hair length. Identifying small classes of trees for which the problem is NP-complete may be important for future investigations in the spirit of the deconstruction of hardness proofs [14] which aim at identifying parameters for which the problem becomes tractable if these parameters are small.

Our main result proves that WSR_2 is strongly NP-complete by a reduction from a variant of NUMERICAL MATCHING WITH TARGET SUMS in which all integers of the input are distinct. For the case where the weights of the vertices are chosen from a small set of values, our dynamic-programming algorithm solves WSR_2 in time $O(n^{k+3} \cdot k)$, where k is the number of distinct vertex weights.

Definitions. A *caterpillar* is a tree consisting of a path, called its *backbone*, and paths attached with one end to the backbone. Its *hair length* is the maximum distance from a leaf to the closest vertex of the backbone. A *star* $K_{1,k}$ is a tree with k leaves and one internal vertex, called the *center*. In our hardness proofs, we reduce from the following problem (problem [SP17] in [9]).

> NUMERICAL MATCHING WITH TARGET SUMS (NMTS): Given three disjoint multisets A, B, and $S = \{s_1, \ldots, s_m\}$, each containing m elements from $\mathbb{N}$, can $A \cup B$ be partitioned into m disjoint sets $C_1, C_2, \ldots, C_m$, each containing exactly one element from each of A and B, such that, for $1 \le i \le m$, $\sum_{c \in C_i} c = s_i$?

Due to space constraints, the proofs of some statements are omitted in this extended abstract.

2 WSR$_2$ is Strongly NP-Complete

In this section, we show that WSR$_2$ is strongly NP-complete. First we introduce a new problem that is polynomial-time-reducible to WSR$_2$, and then show that this new problem is strongly NP-hard.

> SCHEDULING WITH COMMON DEADLINES (SCD): Given n jobs with positive integer lengths $j_1, \ldots, j_n$ and n deadlines $d_1 \leq \ldots \leq d_n$, can the jobs be scheduled on two processors P_1 and P_2 such that at each deadline there is a processor that finishes a job exactly at this time, and processors are never idle between the execution of two jobs?

To reinforce the intuition on this problem one may imagine that we want to satisfy delivery deadlines and avoid using any warehouse space to store a product between its fabrication and the delivery date. There is no restriction as to which product should be delivered at a given time. (Another possibility is imagining computer scientists scheduling paper production to fit conference deadlines.)

Given an instance $(j_1, \ldots, j_n, d_1, \ldots, d_n)$ for SCD, we construct an instance for WSR$_2$ as follows. For each job j_i, $1 \leq i \leq n$, create a vertex v_i with weight $\omega(v_i) = j_i$. For each deadline d_i, $1 \leq i \leq n-1$, create a split d_i. We may assume that $\sum_{i=1}^{n} j_i = d_{n-1} + d_n$, otherwise we add a deadline $d_{n+1} = d_n$ and a job of length $2d_n - \sum_{i=1}^{n} j_i$.

Suppose the path $P = (v_{\pi(1)}, v_{\pi(2)}, \ldots, v_{\pi(n)})$ is a solution to WSR$_2$. Say $\{v_{\pi(\ell)}, v_{\pi(\ell+1)}\}$ is the edge associated to the split d_{n-1}. We construct a solution for SCD by assigning the jobs $j_{\pi(1)}, j_{\pi(2)}, \ldots, j_{\pi(\ell)}$ to processor P_1, and the jobs $j_{\pi(n)}, j_{\pi(n-1)}, \ldots, j_{\pi(\ell+2)}, j_{\pi(\ell+1)}$ to processor P_2, in this order. Note that then, one of the jobs $j_{\pi(\ell)}, j_{\pi(\ell+1)}$ ends at d_{n-1}, and the other at $-d_{n-1} + \sum_{i=1}^{n} j_i = d_n$, which is as desired.

On the other hand, if SCD has a solution, then WSR$_2$ has a solution as well, because the previous construction is easily inverted. Visually, the list of jobs of P_2 is reversed and appended to the list of jobs of P_1. Job lengths correspond to vertex weights and deadlines correspond to splits (the last deadline where a job from P_1 finishes is merged with the last deadline where a job from P_2 finishes). Thus, SCD is polynomial-time-reducible to WSR$_2$.

Lemma 1. SCD $\leq_p$ WSR$_2$.

In the remainder of this section, we show that dNMTS is polynomial-time-reducible to SCD. The dNMTS problem is equal to the NMTS problem, except that all integers in $A \cup B \cup S$ are pairwise distinct. This variant has been shown to be strongly NP-hard by Hulett et al. (see corollary 8 in [13]). As the proof becomes somewhat simpler, we use dNMTS instead of NMTS for our reduction.

Let us first give a high level description of the main ideas of the reduction. For a dNMTS instance (A, B, S), the elements of $A \cup B$ will be encoded as jobs, and the elements of S will be encoded as deadlines. A convenient way to represent an element $s \in S$ is by introducing segments which are delimited to the left and the right by double deadlines, and whose distance is equivalent to s. The elements of $A \cup B \cup S$ are blown up by well-chosen additive factors that preserve solutions and make sure that the length of each segment can only be met by the sum of exactly two job-lengths, one corresponding to an element of A and the other to an element of B.

Our reduction will create an instance whose solution assigns, in each segment, one x-job (a job corresponding to an A-element) and one y-job (a job corresponding to a B-element) to the same processor, such that these two jobs are the only jobs executed on this processor in this segment, thus providing a solution to dNMTS. W.l.o.g., the x-job is scheduled first. As we must not introduce any restriction which x-jobs can be assigned to which segments, we introduce a deadline for each length of an x-job; these are the real deadlines. We refer to the x- and y-jobs as green jobs. The job lengths were blown up such that in each segment, exactly one processor starts with a green x-job, and in each segment, exactly one processor ends by executing a green y-job. In each segment, the green jobs must not overlap; this is achieved by multiplying all deadlines created so far and the corresponding job lengths by a factor 2, and introducing fake deadlines at odd positions one unit before the real deadlines. If an x-job and a y-job overlapped, there would be no job ending at the fake deadline preceding the real deadline at which the x-job ends, as all green jobs have even length and all real deadlines and double deadlines are even. Blue, red, and black jobs are created to meet all deadlines on the processor that is not currently executing green jobs. The blow-up of the elements of $A \cup B \cup S$ ensures that these jobs cannot equate the green jobs (except the black jobs whose lengths might equal the lengths of green y-jobs, but, w.l.o.g., one can assign them to the last part of each segment of the processor not executing a green job). That none of these jobs is executed between two green jobs within a segment is ensured as the sum of all green job lengths equals the sum of the lengths of the segments. This summarizes the reduction and gives the reasons for the different elements of the construction. Let us now turn to the formal reduction.

Let (A, B, S) be an instance for dNMTS. We suppose, w.l.o.g., that $\sum_{i=1}^{m} s_i = \sum_{x \in A \cup B} x$, otherwise (A, B, S) is trivially a No-instance for dNMTS. Let $A = \{a_1, \ldots, a_m\}$ and $B = \{b_1, \ldots, b_m\}$. We also assume, w.l.o.g., that $a_i < a_{i+1}$, $b_i < b_{i+1}$, $s_i < s_{i+1}$, for all $i \in \{1, \ldots, m-1\}$, that $a_m < b_m$, and that $s_m \leq a_m + b_m$.

First, we construct an equivalent instance (X, Y, Z) for dNMTS. Each of $X := \{x_1, \ldots, x_n\}$, $Y := \{y_1, \ldots, y_n\}$, and $Z := \{z_1, \ldots, z_n\}$ has $n := m + 1$ elements:

for $i \in \{1, \ldots, n-1\}$,

$$x_i := 2 \cdot (a_i + (b_m + 2)), \qquad x_n := 2 \cdot (a_m + 1 + (b_m + 2)),$$
$$y_i := 2 \cdot (b_i + 3 \cdot (b_m + 2)), \qquad y_n := 2 \cdot (b_m + 1 + 3 \cdot (b_m + 2)),$$
$$z_i := 2 \cdot (s_i + 4 \cdot (b_m + 2)), \text{ and} \qquad z_n := 2 \cdot (a_m + b_m + 2 + 4 \cdot (b_m + 2)).$$

The elements of X, Y, and Z have the following properties.

Property 1. *Each element of $X \cup Y \cup Z$ is an even positive integer.*

Property 2. *For every* $i \in \{1, \ldots, n-1\}$, *we have that* $x_i < x_{i+1}$, *that* $y_i < y_{i+1}$, *and that* $z_i < z_{i+1}$.

Property 3. *For every* $i \in \{1, \ldots, n\}$, *we have*

$$2 \cdot b_m + 4 \leq x_i \leq 4 \cdot b_m + 4,$$
$$6 \cdot b_m + 12 \leq y_i \leq 8 \cdot b_m + 14, \quad and$$
$$8 \cdot b_m + 16 \leq z_i \leq 12 \cdot b_m + 18.$$

In particular, Property 3 implies that $y_1 > x_n$, that $z_1 > y_n$, and that $2 \cdot y_1 > z_n$. Properties 1–3 easily follow by construction of X, Y, and Z.

Property 4. *If* k *and* ℓ *are integers such that* $x_k + y_\ell = z_n$, *then* $k = \ell = n$.

Property 4 holds because x_n and y_n are the only elements of X and Y, resp., that are large enough to sum to z_n.

Property 5. *Let* $p, q \in X \cup Y$, $p \leq q$, *and* $z \in Z$. *If* $p + q = z$, *then* $p \in X$ *and* $q \in Y$.

By Property 3, the sum of any two X-elements is smaller and the sum of any two Y-elements is larger than any element of Z.
For our SCD instance, we create the following deadlines:

- *real* deadlines: $r_{i,j} := x_i + \sum_{k=1}^{j} z_k$, for each $j \in \{0, \ldots, n-1\}$ and each $i \in \{1, \ldots, n\}$,
- *fake* deadlines: $f_{i,j} := r_{i,j} - 1$, for each $j \in \{0, \ldots, n-1\}$ and each $i \in \{1, \ldots, n\}$, and
- *sum* deadlines: two deadlines $ds_{1,j} := ds_{2,j} := \sum_{k=1}^{j} z_k$, for each $j \in \{1, \ldots, n\}$.

The sum deadlines we just defined partition the interval $[0, ds_{1,n}]$ into n segments $I_j := [ds_{1,j-1}, ds_{1,j}]$, $j = 1, \ldots n$, where for convenience, we let $ds_{1,0} = 0$. We create jobs with the following lengths, where $x_0 = 0$:

- green x-jobs: x_i, for each $i \in \{1, \ldots, n\}$,
- green y-jobs: y_i, for each $i \in \{1, \ldots, n\}$,
- blue jobs: $n \cdot (n-1)$ times a job of length 1,
- red fill jobs: $n-1$ times a job of length $x_i - 1 - x_{i-1}$, for each $i \in \{1, \ldots, n\}$,
- red overlap jobs: $x_i - x_{i-1}$, for each $i \in \{1, \ldots, n\}$,
- black fill jobs: $z_i - x_n$ for $i \in \{1, \ldots, n-1\}$, and
- a black overlap job: $z_n - x_n + 1$.

To illustrate these definitions, we start by showing that if we have a YES-instance (X, Y, Z) for dNMTS, then we have an SCD YES-instance as well. Let $C_1, C_2, \ldots, C_n$ be n couples such that $C_j = \{x_{\pi_1(j)}, y_{\pi_2(j)}\}$ and $x_{\pi_1(j)} + y_{\pi_2(j)} = z_j$, $j \in \{1, \ldots, n\}$, for two permutations π_1 and π_2 of the set $\{1, \ldots, n\}$. We construct a solution for SCD. Let us construct the schedules for P_1 and P_2. For each $j \in \{1, \ldots, n-1\}$,

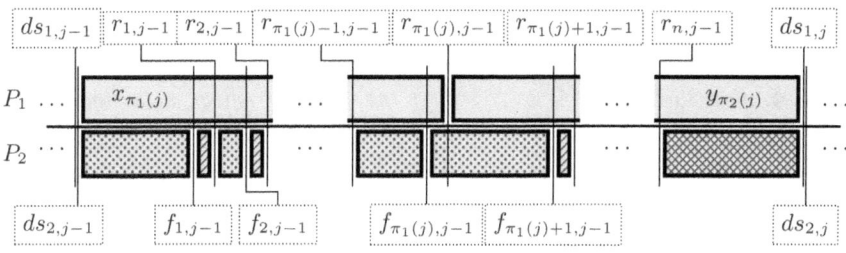

Fig. 1. How jobs are assigned to processors in the SCD instance in segment $j < n$. (The patterns of the jobs are the following: Red jobs are dotted, blue jobs are hatched, black jobs are cross-hatched, and green jobs have no pattern.)

- assign the green x-job $x_{\pi_1(j)}$ to the interval $[ds_{1,j-1}, r_{\pi_1(j),j-1}]$ of P_1,
- assign the green y-job $y_{\pi_2(j)}$ to the interval $[r_{\pi_1(j),j-1}, ds_{1,j}]$ of P_1,
- assign a red fill job of length $x_1 - 1$ to the interval $[ds_{1,j-1}, f_{1,j-1}]$ of P_2,
- for every $i \in \{1, \ldots, n-1\} \setminus \pi_1(j)$, assign a red fill job of length $x_{i+1} - 1 - x_i$ to the interval $[r_{i,j-1}, f_{i+1,j-1}]$ of P_2,
- for every $i \in \{1, \ldots, n\} \setminus \pi_1(j)$, assign a blue job to the interval $[f_{i,j-1}, r_{i,j-1}]$ of P_2,
- assign a red overlap job of length $x_{\pi_1(j)+1} - x_{\pi_1(j)}$ to the interval $[f_{\pi_1(j),j-1}, f_{\pi_1(j)+1,j-1}]$ of P_2, and
- assign a black fill job of length $z_j - x_n$ to the interval $[r_{n,j-1}, ds_{1,j}]$ of P_2.

It only remains to assign jobs to the last segment. The last segment of P_1 contains the green x-job x_n and the green y-job y_n, in this order. The last segment of P_2 contains a red fill job of length $x_1 - 1$, a blue job, a red fill job of length $x_2 - 1 - x_1$, a blue job, ..., a red fill job of length $x_n - 1 - x_{n-1}$, and the black overlap job, in this order. See Fig. 1 for an illustration.

Now suppose the SCD instance is a YES-instance. We will show some structural properties of any valid assignment of jobs to the processors, which will help to extract a solution for our original dNMTS instance. We will show that in each segment I_j, any valid solution for the SCD instance has exactly one green x-job x_k and exactly one green y-job y_ℓ, and x_k and y_ℓ sum to z_j.

Consider a valid assignment of the jobs to the processors P_1 and P_2. As two jobs with the same length are interchangeable, when we encounter a job whose length belongs to more than one category (for example "black fill" and "green y") we may choose in this case, w.l.o.g., to which category the job belongs.

Claim 1. *A black fill job is assigned to each interval* $[r_{n,j}, ds_{1,j+1}]$, $j \in \{0, \ldots, n-2\}$.

This uses up all black fill jobs.

Claim 2. *The green y-job y_n is assigned to the interval* $[r_{n,n-1}, ds_{1,n}]$.

Claim 3. *The black overlap job is assigned to the interval* $[f_{n,n-1}, ds_{1,n}]$.

This uses up all black jobs. Now, the only jobs left whose length is between $6b_m + 12$ and $8b_m + 14$ are the green y-jobs $y_1, \ldots, y_{n-1}$.

Claim 4. *For each $\ell \in \{1, \ldots, n - 1\}$, the green y-job y_ℓ is assigned to an interval $[r_{i,j-1}, ds_{1,j}]$ for some $i \in \{1, \ldots, n - 1\}$ and $j \in \{1, \ldots, n - 1\}$.*

Proof. Each job is assigned to an interval inside some segment, as the double deadlines prevent jobs to span more than one segment. Suppose the green y-job y_ℓ is assigned to segment p. As $ds_{1,p} + y_\ell > ds_{1,p} + x_n$, by Properties 2 and 3, and the deadline following $r_{n,p} = ds_{1,p} + x_n$ is $ds_{1,p+1}$, it must be that the green y-job y_ℓ finishes at $ds_{1,p+1}$. Moreover, $ds_{1,p+1} - y_\ell$ is equal to a real deadline as $ds_{1,p+1} - y_\ell$ is even. □

Each of the $2n$ jobs that have been assigned so far finish at a double deadline $ds_{1,j}, ds_{2,j}$. Thus, no other jobs may end at a double deadline.

Claim 5. *A red fill job of length $x_1 - 1$ is assigned to each interval $[ds_{1,j}, f_{1,j}]$, $0 \leq j \leq n - 1$.*

This uses up all red fill jobs of length $x_1 - 1$.

Claim 6. *For each $\ell \in \{1, \ldots, n\}$, the green x-job x_ℓ is assigned to an interval $[ds_{1,j}, r_{i,j}]$ for some $i \in \{1, \ldots, n\}$ and $j \in \{0, \ldots, n - 1\}$.*

By Claims 2, 4, and 6, and since we have the same amount of segments as green x-jobs, resp. green y-jobs, we obtain that each segment I_j, $1 \leq j \leq n,$, contains exactly one green x-job and exactly one green y-job.

Claim 7. *For $j \in \{1, \ldots, n\}$, the green x-job and the green y-job in the segment I_j do not overlap.*

Proof. Suppose otherwise, that is, suppose there is a $j \in \{1, \ldots, n\}$ such that I_j contains a green x-job, say x_ℓ, and a green y-job, say y_k, that overlap (i.e. the intervals they are assigned to overlap). Since x_ℓ ends at a real deadline by Claim 6 and y_k starts at a real deadline by Claim 4, no job ends at the fake deadline situated at $ds_{1,j-1} + x_\ell - 1$, which contradicts the validity of the SCD solution. □

The last claim implies that in each segment I_j, $1 \leq j \leq n$, there is a green x-job x_{ℓ_j} and a green y-job y_{k_j} which together have the same size as the interval. Hence the couples $C_j = \{a_{\ell_j}, b_{k_j}\}, 1 \leq j \leq n$, form a solution of dNMTS. Thus, we have the following lemma.

Lemma 2. *dNMTS $\leq_p$ SCD.*

Our main theorem follows from the strong NP-hardness of dNMTS, Lemmata 1 and 2, and the membership of WSR$_2$ in NP, which is easily verified as the certificate is a path and an assignment of the splits to its edges, all of which can be encoded in polynomial space.

Theorem 1. *WSR$_2$ is strongly NP-complete.*

Corollary 1. SPLITS RECONSTRUCTION FOR CATERPILLARS OF UNBOUNDED HAIR-LENGTH AND MAXIMUM DEGREE 3 *is NP-complete.*

3   Algorithm for WSR$_2$ with Few Distinct Vertex Weights

Let $k = |\{\omega(v) : v \in V\}|$ denote the number of distinct vertex weights in an instance $(V, \omega, \mathcal{S})$ for WSR$_2$. We exhibit a dynamic programming algorithm for WSR$_2$ that works in polynomial time when k is a constant. Moreover, standard backtracking can be used to actually construct a solution, if one exists.

Suppose $|V| = n$ and the multiset of splits, $\mathcal{S}$, contains the splits $s_1 \leq s_2 \ldots \leq s_{n-1}$. Let $w_1 < w_2 \ldots < w_k$ denote the distinct vertex weights and $m_1, m_2, \ldots, m_k$ denote their respective multiplicities, i.e. $m_i = |\{v \in V : \omega(v) = w_i\}|$ for all $i \in \{1, 2, \ldots, k\}$.

Our dynamic programming algorithm computes the entries of a boolean table A. The table A has an entry $A[p, W_L, W_R, v_1, v_2, \ldots, v_k]$ for each integer p with $1 \leq p \leq n - 1$, each two integers $W_L, W_R \in \mathcal{S}$, and each $v_i \in \{0, 1, \ldots, m_i\}$, where $i \in \{1, 2, \ldots, k\}$. The entry $A[p, W_L, W_R, v_1, v_2, \ldots, v_k]$ is set to **true** iff there is an assignment of the splits $s_1, s_2, \ldots, s_p$ to the ℓ leftmost edges and the r rightmost edges of the path P_n on n vertices, such that

- $p = \ell + r$;
- v_1 weights w_1, v_2 weights w_2, $\ldots$, and v_k weights w_k are assigned to the ℓ leftmost and the r rightmost vertices of P_n such that each split assigned to the left (respectively to the right) part of the path corresponds to the sum of the vertex weights assigned to vertices to the left (respectively to the right) of this split; and
- W_L is equal to the value of the ℓth split from the left and W_R is equal to the rth split from the right.

Intuitively, our algorithm assigns splits and weights by starting from both endpoints of the path and trying to meet these two sub-solutions.

For the base case, set $A[0, W_L, W_R, v_1, v_2, \ldots, v_k]$ to **true** if $W_L = W_R = v_1 = v_2 = \ldots = v_k = 0$ and to **false** otherwise. We compute the remaining entries of A by increasing values of p using the following recurrence.

$$A[p, W_L, W_R, v_1, v_2, \ldots, v_k] = \bigvee_{i=1}^{k} \begin{cases} A[p-1, W_L - w_i, W_R, v_1, v_2, \ldots, v_{i-1}, \\ \quad v_i - 1, v_{i+1}, v_{i+2}, \ldots, v_k] \\ \vee A[p-1, W_L, W_R - w_i, v_1, v_2, \ldots, v_{i-1}, \\ \quad v_i - 1, v_{i+1}, v_{i+2}, \ldots, v_k] \end{cases}$$

In the previous recurrence, each table entry that does not exist is set to **false**. The final result of the algorithm is computed by evaluating the expression

$$\bigvee_{\substack{W_L, W_R \in \mathcal{S} \\ i \in \{1, 2, \ldots, k\} \\ (W_L \leq w_i + W_R) \wedge (W_R \leq w_i + W_L)}} A[|\mathcal{S}|, W_L, W_R, m_1, m_2, \ldots, m_{i-1}, m_i - 1, m_{i+1}, m_{i+2}, \ldots, m_k].$$

Finally, we establish the following theorem.

Theorem 2. WSR$_2$ can be solved in time $O(n^{k+3} \cdot k)$, where k is the number of distinct vertex weights of any input instance $(V, \omega, \mathcal{S})$ and $n = |V|$.

4   SR$_3$ is NP-Complete

In this section we show that SPLITS RECONSTRUCTION with unit weights is NP-complete for trees with maximum degree 3. Our polynomial-time reduction is done from the strongly NP-complete NMTS problem recalled in Section 1. This problem remains NP-complete even if each integer of the NMTS instance is at most $p(m)$, where p is a polynomial and m is the length of the description of the instance. Let us just mention that the next theorem does not immediately follow from Corollary 1.

Theorem 3. SR$_3$ *is NP-complete.*

Proof. Let $\tilde{A} = \{\tilde{a}_1, \tilde{a}_2, \ldots, \tilde{a}_m\}$, $\tilde{B} = \{\tilde{b}_1, \tilde{b}_2, \ldots, \tilde{b}_m\}$ and $\tilde{S} = \{\tilde{s}_1, \tilde{s}_2, \ldots, \tilde{s}_m\}$ be an instance of NMTS. Let $C = \max\{x \ : \ x \in \tilde{A} \cup \tilde{B}\}$ be the maximum over $\tilde{A} \cup \tilde{B}$. W.l.o.g., we construct the following equivalent NMTS instance:

$$a_i := \tilde{a}_i + 2 + 3C, \quad 1 \le i \le m,$$
$$b_i := \tilde{b}_i + 3 + 5C, \quad 1 \le i \le m, \text{ and}$$
$$s_i := \tilde{s}_i + 5 + 8C, \quad 1 \le i \le m.$$

Let $A = \bigcup_{1 \le i \le m}\{a_i\}$, $B = \bigcup_{1 \le i \le m}\{b_i\}$, and $S = \bigcup_{1 \le i \le m}\{s_i\}$. Clearly, the instance $(\tilde{A}, \tilde{B}, \tilde{S})$ has a solution iff the instance (A, B, S) has a solution.

Now we describe an instance $(V, \mathcal{S})$ of SR$_3$, which is a YES-instance iff the previous instance (A, B, S) of NMTS is a YES-instance (see also Figure 2).

Let $n = 2m - 2 + \sum_{i=1}^{m} a_i + \sum_{i=1}^{m} b_i$ be the number of vertices in V; we recall that they have unit weight. The multiset $\mathcal{S}$ of splits is defined as follows.

- For each value s_i, $1 \le i \le m$, the value $1 + s_i$ is added to $\mathcal{S}$ and we refer to these splits as *red* splits.
- For each value s_i, $2 \le i \le m - 2$, the value $(i - 1) + \sum_{j=1}^{i}(1 + s_j)$ is added to $\mathcal{S}$ and we refer to these splits as *black* splits.

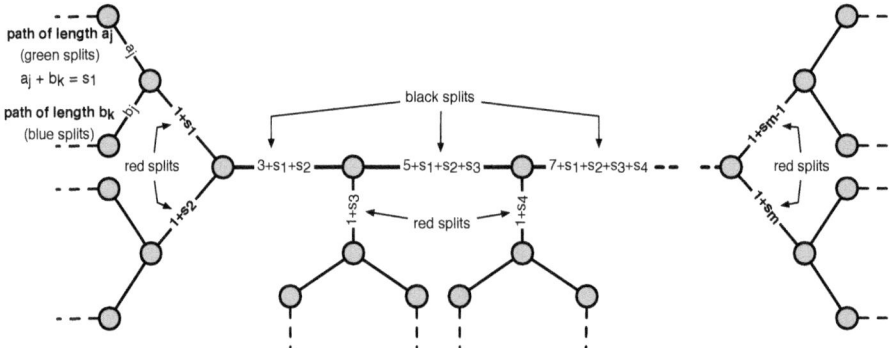

Fig. 2. A tree with maximum degree 3 representing a solution to a SR$_3$ instance constructed as described in the proof of Theorem 3

- For each value a_i, $1 \leq i \leq m$, the values $\{1, 2, \dots, a_i\}$ are added to $\mathcal{S}$ and we refer to these splits as *green* splits.
- For each value b_i, $1 \leq i \leq m$, the values $\{1, 2, \dots, b_i\}$ are added to $\mathcal{S}$ and we refer to these splits as *blue* splits.

Finally each value x of $\mathcal{S}$ is replaced by $\min(x, n - x)$.

Lemma 3. *(A, B, S) is a* YES-*instance for* NMTS *if and only if $(V, \omega : V \rightarrow \{1\}, \mathcal{S})$ is a* YES-*instance for* SR$_3$.

As the certificate is a tree on n vertices, the membership in NP is obvious.   □

5   Freely Choosable Weights

We remark that the following modification of WSR makes any set of splits realizable in some tree. Suppose the weight function ω is not given, but freely choosable, that is, we ask whether, given a multiset $\mathcal{S}$ of integers, there exists a tree $T = (V, E)$ and a weight function $\omega : V \rightarrow \mathbb{N}$, such that $\mathcal{S}$ is the multiset of splits of T. We call this problem CHWSR.

Theorem 4. CHWSR *always admits a solution.*

6   Conclusion

In Section 3, we have shown that, in the framework of parameterized complexity [5,8,16], WSR$_2$ is in XP when parameterized by the number of distinct vertex weights. A generalization of this problem is W[1]-hard [7], but it remains open whether this problem is fixed parameter tractable. For practical purposes, it would further be important to identify other quantities that are small in practice (e.g. the number of leaves, the diameter of the tree, or topological indices), and investigate the multivariate complexity of the considered problems parameterized by combinations of these quantities.

There is a large contrast between the complexities of WSR, where we are given n vertex weights, and CHWSR, where we can freely choose the vertex weights, or, alternatively, we can choose the vertex weights from an infinite multiset containing n times each element of $\mathbb{N}$. It would be interesting to know some restrictions on the multiset of vertex weights such that the problem becomes tractable with respect to interesting parameterizations. Ideally, these restrictions should be consistent with the applications in drug design and discovery.

Acknowledgment. We thank Ming-Yang Kao for communicating this problem.

References

1. Aoki-Kinoshita, K.F., Kanehisa, M., Kao, M.-Y., Li, X.-Y., Wang, W.: A 6-Approximation Algorithm for Computing Smallest Common Aon-Supertree with Application to the Reconstruction of Glycan Trees. In: Asano, T. (ed.) ISAAC 2006. LNCS, vol. 4288, pp. 100–110. Springer, Heidelberg (2006)

2. Balaban, A.T.: Chemical Applications of Graph Theory. Academic Press, Inc. (1976)
3. Bonchev, D., Rouvray, D.H.: Chemical Graph Theory: Introduction and Fundamentals. Taylor & Francis (1991)
4. Dobrynin, A.A., Entringer, R., Gutman, I.: Wiener index of trees: Theory and applications. Acta Applicandae Mathematicae 66(3), 211–249 (2001)
5. Downey, R.G., Fellows, M.R.: Parameterized complexity. Springer, Heidelberg (1999)
6. Faulon, J.-L., Bender, A.: Handbook of Chemoinformatics Algorithms, 1st edn. Chapman and Hall/CRC (2010)
7. Fellows, M.R., Gaspers, S., Rosamond, F.A.: Parameterizing by the Number of Numbers. In: Raman, V., Saurabh, S. (eds.) IPEC 2010. LNCS, vol. 6478, pp. 123–134. Springer, Heidelberg (2010)
8. Flum, J., Grohe, M.: Parameterized Complexity Theory. Texts in Theoretical Computer Science. An EATCS Series. Springer, Berlin (2006)
9. Garey, M.R., Johnson, D.S.: Computers and Intractability, A Guide to the Theory of NP-Completeness. W.H. Freeman and Company, New York (1979)
10. Gillet, V.J., Willett, P., Bradshawand, J., Green, D.V.S.: Selecting combinatorial libraries to optimize diversity and physical properties. Journal of Chemical Information and Computer Sciences 39(1), 169–177 (1999)
11. Goldman, D., Istrail, S., Lancia, G., Piccolboni, A., Walenz, B.: Algorithmic strategies in combinatorial chemistry. In: SODA, pp. 275–284 (2000)
12. Hammer, P.L. (ed.): Special issue on the 50th anniversary of the Wiener index. Discrete Applied Mathematics, vol. 80. Elsevier (1997)
13. Hulett, H., Will, T.G., Woeginger, G.J.: Multigraph realizations of degree sequences: Maximization is easy, minimization is hard. Operations Research Letters 36(5), 594–596 (2008)
14. Komusiewicz, C., Niedermeier, R., Uhlmann, J.: Deconstructing Intractability: A Case Study for Interval Constrained Coloring. In: Kucherov, G., Ukkonen, E. (eds.) CPM 2009 Lille. LNCS, vol. 5577, pp. 207–220. Springer, Heidelberg (2009)
15. Li, X., Zhang, X.: The edge split reconstruction problem for chemical trees is NP-complete. MATCH Communications in Mathematical and in Computer Chemistry 51, 205–210 (2004)
16. Niedermeier, R.: Invitation to Fixed-Parameter Algorithms. Oxford Lecture Series in Mathematics and Its Applications. Oxford University Press, Oxford (2006)
17. Sheridan, R.P., Kearsley, S.K.: Using a genetic algorithm to suggest combinatorial libraries. Journal of Chemical Information and Computer Sciences 35(2), 310–320 (1995)
18. Trinajstić, N.: Chemical Graph Theory, 2nd edn. CRC Press (1992)
19. Wiener, H.: Structural determination of paraffin boiling points. Journal of the American Chemical Society 69(1), 17–20 (1947)

Empires Make Cartography Hard: The Complexity of the Empire Colouring Problem

Andrew R.A. McGrae and Michele Zito

Department of Computer Science, University of Liverpool, Liverpool, L69 3BX, UK
{A.McGrae,M.Zito}@liverpool.ac.uk

Abstract. We study the empire colouring problem (as defined by Percy Heawood in 1890) for maps containing empires formed by exactly $r > 1$ countries each. We prove that the problem can be solved in polynomial time using s colours on maps whose underlying adjacency graph has no induced subgraph of average degree larger than s/r. However, if $s \geq 3$, the problem is NP-hard for forests of paths of arbitrary lengths (if $s < r$) for trees (if $r \geq 2$ and $s < 2r$) and arbitrary planar graphs (if $s < 7$ for $r = 2$, and $s < 6r - 3$, for $r \geq 3$). The result for trees shows a perfect dichotomy (the problem is NP-hard if $3 \leq s \leq 2r-1$ and polynomial time solvable otherwise). The one for planar graphs proves the NP-hardness of colouring with less than 7 colours graphs of thickness two and less than $6r - 3$ colours graphs of thickness $r \geq 3$.

1 Introduction

Let r and s be fixed positive integers. Assume that the n vertices of a planar graph G are partitioned into blocks (or *empires*) each containing exactly r vertices. The (s,r)-*colouring* problem (s-COL$_r$) asks for a vertex colouring of G that uses at most s colours, never assigns the same colour to adjacent vertices in different empires and, conversely, assigns the same colour to all vertices in the same empire, disregarding adjacencies. s-COL$_1$ coincides with the classical vertex colouring problem on planar graphs. The generalization for $r \geq 2$ was defined by Heawood [10] in the same paper in which he refuted a previous "proof" of the famous Four Colour Theorem. It has since been shown that $6r$ colours are always sufficient and in some cases necessary to solve this problem [12].

In [17] (also see [16]), we proved that $2r$ colours suffice and are sometimes needed to colour a collection of empires defined in an arbitrary tree. We also looked at the proportion of (s, r)-colourable trees on n vertices. We showed that, as n tends to infinity, for each r there exists a value s_r such that almost no tree can be coloured with at most s_r colours and, conversely, for s sufficiently larger than s_r, s colours are sufficient with (at least) constant positive probability. Later on [5] we improved on this showing that, as n tends to infinity, the minimum value s for which a random tree is (s, r)-colourable is concentrated in a very short interval with high probability. Although our investigation considerably expanded

P. Kolman and J. Kratochvíl (Eds.): WG 2011, LNCS 6986, pp. 179–190, 2011.

the state of knowledge on s-COL_r, it failed to shed light on its computational complexity. Heawood [10] was the first to argue that there is a simple algorithm that can find a $(6r, r)$-colouring in any planar graph G in polynomial time. The same process uses at most $2r$ colours if G is a tree. But what if we only have r available colours? How difficult is it to decide whether G has an (r, r)-colouring? In this paper we show that s-COL_r can be solved in polynomial time on planar graphs containing no induced subgraph of average degree greater than s/r. This implies that, for instance, $(2r - 1)$-COL_r (resp. $(6r - 1)$-COL_r) can be solved in polynomial time on forests consisting of paths of length at most $2r - 1$ (resp. planar graphs with components of size at most $12r$). Unfortunately, the outcome of our investigation seems to indicate that such algorithmic results cannot be extended much further. If $s \geq 3$, we prove that, if no bound is known on the length of the paths, s-COL_r on paths is NP-hard if $s < r$. Furthermore, the hardness extends to $s < 6r - 3$ (resp. $s < 7$) when $r \geq 3$ (resp. for $r = 2$) on arbitrary planar graphs. Finally, for trees, our argument entails a nice dichotomy: s-COL_r is NP-hard for $s \in \{3, \ldots, 2r - 1\}$ and solvable in polynomial time for any other positive value of s.

The hardness proofs mentioned above hinge on the fact that the connectivity within empires has no effect on the graph colourability. Essentially, to find an (s, r)-colouring in a planar graph G, it suffices to be able to colour with at most s distinct colours (in such a way that no two distinct vertices connected by an edge receive the same colour) its *reduced graph* $R_r(G)$. This is a (multi)graph obtained by contracting each empire to a distinct pseudo-vertex and adding an edge between a pair of pseudo-vertices u and v for each edge connecting two vertices in the original graph, one belonging to the empire represented by u, the other one to that represented by v. The algorithmic results are based on the use of simple minimum degree greedy colouring strategies [10] or more refined heuristics providing algorithmic proofs (see [9, Theorem 7.9] or [14, Exercises 9.12, 9.13]) of the well-known Brooks theorem [3] on such reduced graphs.

The reader at this point may question the reasons for studying this type of colourings. Our main interest in the problem comes from its relationship with other important colouring problems. Each instance of s-COL_r can be translated to an instance of the classical colouring problem, but it is not clear to what extent the two problems are equivalent. The empire colouring problem is related to the problem of colouring graphs of given thickness (a graph has *thickness* t [11], if t is the minimum integer such that its edges can be partitioned into at least t planar graphs). Bipartite graphs can have high thickness but only need two colours, and on the other hand a graph of thickness t may have chromatic number as large as $6t$. Theorem 9 in this paper implies that deciding whether a graph of thickness $t \geq 3$ can be coloured with $s < 6t - 3$ colours is NP-hard.

The rest of the paper is organized as follows. In Section 2 we present our positive results concerning sparse planar graphs. We then move on (Section 3) to describe a new reduction from the well-known satisfiability problem to the problem of colouring a particular type of graph. Hardness results for the colourability of these graphs will be instrumental to our main results. Section

4 deals with the hardness result for forests of paths. A substantial part of this section is devoted to the definition of a number of gadgets that will be used in subsequent proofs. The last two sections deal with the hardness results about trees and arbitrary planar graphs.

Let k and s be positive integers greater than two. In what follows k-SAT (resp. s-COL) denotes the well known [8,13] NP-complete problem of checking the satisfiability of a k-CNF boolean formula (resp. deciding whether the vertices of a graph G can be coloured using at most s distinct colours in such a way that no edge of G is monochromatic). Also, if Π is a decision problem and $\mathcal{I}$ is a particular set of instances for it, then $\Pi(\mathcal{I})$ will denote the restriction of Π to instances belonging to $\mathcal{I}$. If Π_1 and Π_2 are decision problems, then $\Pi_1 \leq_p \Pi_2$ will denote the fact that Π_1 is polynomial-time reducible to Π_2. Unless otherwise stated we follow [7] for all our graph-theoretic notations.

2 Algorithms

The main outcome of our work is that the empire colouring problem is much harder than the problem of colouring planar graphs in the classical sense. However there are cases where things are easy. Let σ be a positive real number. In the following result SPARSE(σ) denotes the class of planar graphs G containing no induced subgraph of average degree larger than σ.

Theorem 1. *Let σ be a positive rational number and r be a positive integer such that $r\sigma$ is a whole number. The decision problem $r\sigma$-COL$_r$(SPARSE(σ)) can be solved in polynomial time.*

Proof. Let r and σ be two positive numbers satisfying the assumptions above, and assume that $G \in$ SPARSE(σ), and its vertex set is partitioned into empires of size r. If $R_r(G)$ contains a copy of $K_{r\sigma+1}$ then there can be no $(r\sigma, r)$-colouring of G. We now argue that if $R_r(G)$ does not contain a copy of $K_{r\sigma+1}$ then it is $r\sigma$-colourable (and therefore G admits an $(r\sigma, r)$-colouring).

Let S be a connected component of $R_r(G)$. In what follows we denote by G^S the subgraph of G such that $R_r(G^S) \equiv S$. Because all edges of S are edges in G^S, the average degree of this graph satisfies

$$|E(S)| = |E(G^S)| = d(G^S) \cdot |V(G^S)|/2.$$

Note that $|V(S)| = |V(G^S)|/r$. Thus, using the definition of SPARSE(σ), we have

$$|E(S)| \leq \frac{r\sigma}{2 \cdot r} \cdot r \cdot |V(S)| = \frac{r\sigma}{2} \cdot |V(S)|.$$

This implies that the average degree of S is at most $r\sigma$. It follows that S is either a regular graph of degree $r\sigma$ or it must contain at least a vertex of degree less than $r\sigma$. In the former case S is can be coloured with $r\sigma$ colours using, say, the algorithm in the proof of Brooks' Theorem described in [9]. If S contains a vertex of degree less than $r\sigma$ we argue that, in fact, the assumptions about the average degree of all subgraphs of G imply that any induced subgraph of S is either $r\sigma$-regular or, in turn, contains a vertex of degree at most $r\sigma - 1$. Assume that some induced subgraph of S, S' is not $r\sigma$-regular and its minimum degree is at least $r\sigma$. This implies that in particular $d(S') \geq r\sigma$. But, by the assumptions on G the average degree of S' cannot exceed $r\sigma$. Therefore $d(S') = r\sigma$ and this implies S' must contain a vertex of degree less than $r\sigma$. ∎

The result above has a number of interesting consequences. Let k be a positive integer. Any induced subgraph on n vertices of a forest of paths of length at most k cannot span more than $kn/(k + 1)$ edges. Hence Theorem 1 implies, for instance, that $\left\lceil \frac{2kr}{k+1} \right\rceil$-COL$_r$ can be decided in polynomial time for forests of paths of length at most k. Similarly $(6r - 1)$-COL$_r$ can be decided in polynomial time for graphs G formed by arbitrary planar components of size at most $12r$.

3   A Useful Reduction

Let s and k be positive integers with $s > \max(2, k)$. An (s, k)-*formula graph* is an undirected graph Φ such that $V(\Phi) = \mathcal{T} \cup \mathcal{C} \cup \mathcal{A}$ where $\mathcal{T} = \{T, F, X^1, \ldots, X^{s-2}\}$, $\mathcal{C}$ contains m groups of vertices, denoted by $\{c^{1,1}, \ldots, c^{1,s-1}\}$, $\{c^{2,1}, \ldots c^{2,s-1}\}$, $\ldots, \{c^{m,1}, \ldots, c^{m,s-1}\}$ and $\mathcal{A}$ is a set of $2n$ vertices paired up in some recognizable way. In particular, in what follows we will denote the elements of $\mathcal{A}$ by $a_1, \ldots, a_n, \overline{a_1}, \ldots, \overline{a_n}$, and we will say that for each $i \in \{1, \ldots, n\}$, a_i and $\overline{a_i}$ are a *pair of complementary vertices*. Set $\mathcal{T}$ spans a complete graph; for each pair of complementary vertices a and $\overline{a}$, $\{a, \overline{a}, X^j\}$ spans a complete graph for each $j \in \{1, \ldots, s-2\}$; for each $i \in \{1, \ldots, m\}$, $\{T, c^{i,1}, \ldots, c^{i,s-1}\}$ spans a complete graph and if $j \in \{1, \ldots, k\}$ then there is a single edge connecting $c^{i,j}$ to some vertex in $\mathcal{A}$, else if $j \geq k+1$ then $\{c^{i,j}, F\} \in E(\Phi)$. Fig. 1 gives a simple example of a $(5, 3)$-formula graph.

Let FG(s, k) denote the class of all (s, k)-formula graphs. We will now describe a reduction from k-SAT to the problem of colouring using at most s distinct colours the vertices of a given (s, k)-formula graph. The reduction shows the NP-hardness of s-COL(FG(s, k)) for any $k \geq 3$ and $s > k$. This in turn will be used repeatedly to prove our hardness results on s-COL$_r$.

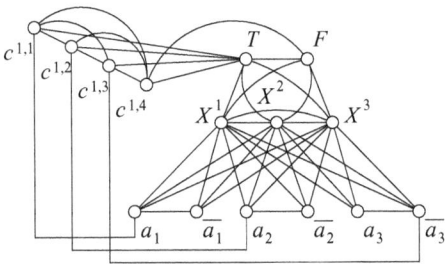

Fig. 1. A small formula graph

Theorem 2. *Let s be an integer with $s \geq 3$. Then k-SAT $\leq_p s$-COL(FG(s,k)) for any integer $k < s$.*

Proof. Given a k-CNF formula $\phi \equiv C_1 \wedge \ldots \wedge C_m$ where C_i is the disjunction of k literals $\mathsf{c}^{i,1}, \ldots, \mathsf{c}^{i,k}$ for each $i \in \{1, \ldots, m\}$, we devise an (s,k)-formula graph Φ that admits an s-colouring if and only if ϕ is satisfiable. The graph Φ will consist of one *truth gadget*, one *variable gadget* for each variable in ϕ, and one *clause gadget* for each clause in ϕ. The truth gadget is a complete graph on s vertices labelled T, F, and $X^1, \ldots, X^{s-2}$. Note that every vertex in this gadget must be given a different colour in any s-colouring. Hence w.l.o.g. we call these colours "TRUE", "FALSE", "OTHER^1", ..., "OTHER^{s-2}" respectively. For each variable a of ϕ the variable gadget consists of two complementary vertices labelled a, and $\bar{a}$, connected by an edge and also adjacent to $X^1, \ldots, X^{s-2}$. There are therefore only two ways to colour a and $\bar{a}$: either a is TRUE and $\bar{a}$ is FALSE or a is FALSE and $\bar{a}$ is TRUE. Thus the two colourings of a and $\bar{a}$ encode the two truth-assignments of the variable a. Each clause $\mathsf{c}^{i,1} \vee \ldots \vee \mathsf{c}^{i,k}$ will be represented by $s + k + 1$ vertices of Φ. Of these, k will correspond to the clause literals and will be labelled $c^{i,1}, \ldots, c^{i,k}$, $s - 1 - k$ will be labelled $c^{i,k+1}, \ldots, c^{i,s-1}$, and the remaining $k+2$ will be k vertices from variable gadgets and the vertices T and F from the truth gadget. Vertices $T, c^{i,1}, \ldots, c^{i,s-1}$ form a clique and, furthermore, for each $j \in \{1, \ldots, k\}$, the vertex $c^{i,j}$ is connected to the corresponding literal in a variable gadget. For $k \leq s - 2$ vertices $c^{i,j}$, for $j \in \{k+1, \ldots, s-1\}$, are adjacent to F. Note that, in any colouring of a clause gadget, vertices $c^{i,j}$, for $j \leq k$, cannot have the same colour of vertex T, and vertices $c^{i,j}$ for $j \geq k$ cannot be coloured like F either. The reader can readily verify that $\Phi \in$ FG(s,k). The graph in Fig. 1 is the $(5,3)$-formula graph corresponding to the formula ϕ consisting of the single clause $\mathsf{a}_1 \vee \mathsf{a}_2 \vee \overline{\mathsf{a}_3}$.

If ϕ is satisfiable, the elements of $\mathcal{A}$ in Φ can be assigned a colour in {TRUE, FALSE} so that, for each $i \in \{1, \ldots, m\}$ at least one of the $c^{i,j}$ (say for $j = j^*$) is adjacent to some literal coloured TRUE. This implies that c^{i,j^*} can be coloured FALSE, while all other $c^{i,j}$ for $j \in \{1, \ldots, s-1\} \setminus \{j^*\}$ can be assigned a distinct colour in {OTHER^1, OTHER^2, ..., OTHER^{s-2}}. Conversely if there is no way to colour $\mathcal{A}$ so that for each $i \in \{1, \ldots, m\}$ at least one of the $c^{i,j}$ is adjacent to some literal coloured TRUE, then the clause gadget will need $s + 1$ colours as the $s - 1$ vertices $c^{i,j}$ only have $s - 2$ colours available (as TRUE and FALSE are used up by T, F, and the corresponding literals). Thus Φ admits an s-colouring if and only if there is some way to assign the variables of ϕ as TRUE or FALSE in such a way that every clause contains at least one TRUE literal. ∎

4   Forests of Paths

In Section 2 we showed that there are specific values for s such that s-COL$_r$ becomes easy if the input graph is a collection of short paths. Here we argue that if the paths are allowed to have arbitrary length (let PATH denote the set of all forests of this form) then the problem becomes NP-hard. We will prove the following result.

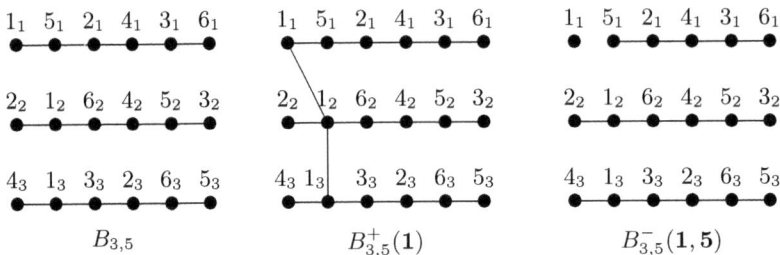

Fig. 2. $B_{3,5}$ and two related constrained clique gadgets

Theorem 3. *Let r and s be positive integers with $r > s \geq 3$. Then the s-COL$_r$(PATH) problem is NP-hard.*

The proof is split in two parts which are covered at the end of this section. The argument for $s = 3$ is based on a direct construction which is reminiscent of a well-known hardness proof for 3-COL [6, p.1103]. For $s > 3$, given an $(s, s-1)$-formula graph Φ, we will argue that there is a number of gadgets that can be used to define a forest of paths P and a partition of $V(P)$ into empires of size r in such a way that Φ is s-colourable if and only if P admits an (s, r)-colouring. The hardness of s-COL$_r$(PATH) will then follow from that of s-COL(FG$(s, s-1)$).

Gadgets

Before moving to the proof of Theorem 3 we introduce a number of useful gadgets.

Clique Gadgets. Let r and s be positive integers with $s < 2r$. In what follows the *clique gadget* $B_{r,s}$ is a graph satisfying the following properties.

B0 It has $r(s+1)$ vertices partitioned into $s+1$ empires of size r.
B1 It is a forest consisting of r paths.
B2 No path in the graph contains two vertices from the same empire.
B3 Its reduced graph contains a copy of K_{s+1}. Hence the graph admits an $(s+1, r)$-colouring and cannot be coloured with fewer colours.

Also, if $r > 1$ and $\mathbf{v} \equiv \{v_1, \ldots, v_r\}$ is some set of r vertices, the *connected clique gadget rooted at* $\mathbf{v}$, $B_{r,s}^+(\mathbf{v})$, is formed by adding edges $\{v_i, v_{i+1}\}$ for all i such that $1 \leq i \leq r - 1$ to $B_{r,s}$. Note that the resulting graph is a tree. However $B_{r,s}^+(\mathbf{v})$ still satisfies **B0**, and **B3**. Finally, if $\mathbf{u}$ and $\mathbf{v}$ are two sets of r vertices, the $(\mathbf{u}, \mathbf{v})$-*colour constraining gadget* $B_{r,s}^-(\mathbf{u}, \mathbf{v})$ is a graph obtained from $B_{r,s}$, without loss of generality, by removing a single edge connecting the end-point u_1 of a path to its neighbour v_1. Thus u_1 becomes isolated in $B_{r,s}^-(\mathbf{u}, \mathbf{v})$. The graph $R_r(B_{r,s}^-(\mathbf{u}, \mathbf{v}))$ contains a copy of K_{s-1} in which every vertex is also adjacent to the vertices corresponding to $\mathbf{u}$ and $\mathbf{v}$. Thus any (s, r)-colouring of $B_{r,s}^-(\mathbf{u}, \mathbf{v})$ must give $\mathbf{u}$ and $\mathbf{v}$ the same colour. Examples are given in Fig. 2. The clique gadgets $B_{r,s}$ can be easily constructed from the Hamiltonian decomposition of K_{2r+1} (see, for instance [4, p. 71]).

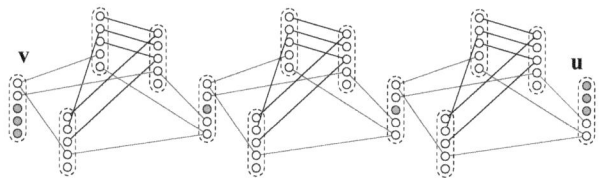

Fig. 3. The graph $A_{5,4,8}(\mathbf{v})$, dark vertices are in $Z(\mathbf{v})$

Connectivity gadgets. Let $\mathbf{v}$ be a given empire (i.e. a set of r vertices). For positive integers r, s and m with $r > s \geq 3$, an *m-connector for* $\mathbf{v}$, denoted by $A_{r,s,m}(\mathbf{v})$, is a graph satisfying the following properties.

A0 It contains no more than $r(m \cdot s + 1)$ vertices split into empires of size r.
A1 The graph is a forest of paths.
A2 The empire $\mathbf{v}$ contains at least $r - \lceil \frac{s-1}{2} \rceil$ isolated vertices.
A3 The graph has at least m isolated vertices which belong to empires different from $\mathbf{v}$ that must be given the same colour as $\mathbf{v}$ in any (s, r)-colouring. The set of such vertices, denoted by $Z(\mathbf{v})$ is called $\mathbf{v}$'s *monochromatic set*.

Gadgets $A_{r,s,m}(\mathbf{v})$ will be used to connect $\mathbf{v}$ to other parts of a bigger graph where there are degree constraints on the vertices. The gadget $A_{r,s,m}(\mathbf{v})$ will be linked to other parts of this graph through some of the isolated vertices in $\mathbf{v}$ and the elements belonging to its monochromatic set. The elements of $Z(\mathbf{v})$ (generically denoted by $z_{\mathbf{v}}$) are useful to "pass" the colour constraints on $\mathbf{v}$ to other parts of the bigger graph. Gadget $A_{r,sm}$ can be constructed recursively by adding a copy of $B_{r-2,s-2}$ along with a new set of r isolated vertices u to $A_{r,s,m'}(\mathbf{v})$ for some $m' < m$. Fig. 3 gives an example.

Proof of Theorem 3

We are now ready to tackle the proof of Theorem 3. We start from the case $s = 3$.

Theorem 4. *Let r be an integer with $r \geq 4$. Then* 3-SAT $\leq_p$ 3-COL$_r$(PATH).

Proof. Given an instance ϕ of 3-SAT we can produce a forest of paths $P(\phi)$ and a partition of $V(P(\phi))$ into empires of size r such that $P(\phi)$ admits a $(3, r)$-colouring if and only if ϕ is satisfiable. $P(\phi)$ consists of one *truth gadget*, one *variable gadget* for each variable used in ϕ, and one *clause gadget* for each clause in ϕ. To define the truth gadget, we start by adding $r - 2$ distinct isolated vertices to each empire in $B_{2,2}$. The empires in the resulting graph will be labelled $\mathbf{T}$, $\mathbf{F}$ and $\mathbf{X}$. Then, if ϕ uses n different variables and m clauses, one copy of $A_{r,3,2m}(\mathbf{T})$, and one copy of $A_{r,3,n}(\mathbf{X})$ are attached to $B_{2,2}$ via $\mathbf{T}$ and $\mathbf{X}$ respectively. Since $\mathbf{T}$, $\mathbf{F}$ and $\mathbf{X}$ are all adjacent (in the gadget's reduced graph) they must have different colours which, again, we call TRUE, FALSE and OTHER respectively. For each variable a in ϕ, $P(\phi)$ contains a variable gadget (Fig. 4),

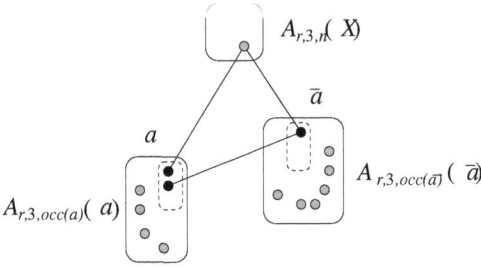

Fig. 4. The shape of a variable gadget for $s = 3$. The dashed shapes represent the empires **a** and **ā**.

including an empire named **a**, and another one named **ā**. A path connects two vertices in **a** with an element of **ā** and an element of $Z(\mathbf{X})$. Also, a copy of $A_{r,3,\text{occ}(a)}(\mathbf{a})$ (resp. $A_{r,3,\text{occ}(\bar{a})}(\mathbf{\bar{a}})))$ is attached to empire **a** (resp. **ā**). Here $\text{occ}(\cdot)$ is a function taking as input a literal and returning the number of occurrences of its argument in the given formula. Since $\mathbf{X}$ has colour OTHER, there are only two possible colourings for **a** and **ā** — either **a** is TRUE and **ā** is FALSE, or **a** is FALSE and **ā** is TRUE. Finally, for each clause in ϕ, $P(\phi)$ contains a gadget like the one depicted in Fig. 5. This is connected to the rest of the graph via four connectivity gadgets: $A_{r,3,2m}(\mathbf{T})$, and three of the form $A_{r,3,\text{occ}(\ell)}(\ell)$ where ℓ is a literal of ϕ. Since $\mathbf{T}$ will always be coloured TRUE, it can be shown that the clause gadget admits a proper $(3,r)$-colouring if and only if at least one of the empires corresponding to a literal is coloured TRUE.

Note that $P(\phi)$ is $(3,r)$-colourable if and only if ϕ is satisfiable. This follows from the properties of the well known [6, p.1103] reduction 3-SAT $\leq_p$ 3-COL, as the graph obtained from $P(\phi)$ by shrinking each connectivity gadget or empire to a distinct (pseudo-)vertex (removing loops or parallel edges created in the process) coincides with that created from ϕ using the classical 3-COL reduction. ∎

For $s > 3$ the NP-hardness of $s\text{-COL}_r(\text{PATH})$ follows from that of $s\text{-COL}(\text{FG}(s, s-1))$.

Theorem 5. *Let r and s be fixed positive integers with $3 < s < r$. Then $s\text{-COL}(\text{FG}(s, s-1)) \leq_p s\text{-COL}_r(\text{PATH})$.*

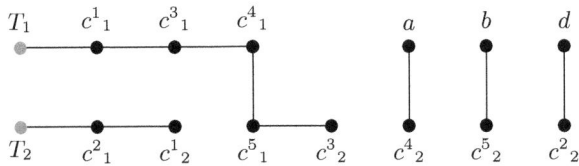

Fig. 5. The clause gadget for the clause $(a \lor b \lor d)$. Only at most two vertices from each empire are shown. In particular vertices labelled T_1 and T_2 are in $Z(\mathbf{T})$, while vertices labelled a, b and d are in $Z(\mathbf{a})$, $Z(\mathbf{b})$ and $Z(\mathbf{d})$ respectively.

Proof. Let Φ be an $(s, s-1)$-formula graph. Given $r > s$, a few simple replacement rules enable us to define a forest of paths $P(\Phi)$ and a partition of $V(P(\Phi))$ into empires of size r such that Φ is s-colourable if and only if $P(\Phi)$ is (s, r)-colourable. More specifically, the complete graph on $\{T, F, X^1, \ldots, X^{s-2}\}$ is replaced by s empires of size r labelled $\mathbf{T}$, $\mathbf{F}$, and $\mathbf{X}^1, \ldots, \mathbf{X}^{s-2}$ so that $\lceil \frac{s}{2} \rceil$ vertices from each empire induce a copy of $B_{\lceil \frac{s}{2} \rceil, s-1}$. Moreover we attach a copy of $A_{r,s,\lceil \frac{s}{2} \rceil m}(\mathbf{T})$ (resp. $A_{r,s,2n}(\mathbf{X}^1), \ldots, A_{r,s,2n}(\mathbf{X}^{s-2})$) to these empires. Note that, by $\mathbf{A2}$, the resulting graph is a collection of paths and isolated vertices. Next, for each $a, \bar{a} \in \mathcal{A}$, we define two empires on r vertices, and for each positive integer i such that $2i \leq s-2$ we replace the cycle $\{X^{2i-1}, a, X^{2i}, \bar{a}\}$ in Φ with a path $z_{\mathbf{X}^{2i-1}}, a_i, z_{\mathbf{X}^{2i}}, \bar{a}_i, z'_{\mathbf{X}^{2i-1}}$ (distinct cycles replaced by paths using distinct elements of $Z(\mathbf{X}^{2i-1})$ and $Z(\mathbf{X}^{2i})$). We also replace the edge $\{a, \bar{a}\}$ with $\{a_{\lceil \frac{s-1}{2} \rceil}, \bar{a}_{\lceil \frac{s-1}{2} \rceil}\}$, and if s is odd we replace the path $a, X^{s-2}, \bar{a}$ with the path $a_{\lceil \frac{s-1}{2} \rceil}, z_{\mathbf{X}^{s-2}}, \bar{a}_{\lceil \frac{s-1}{2} \rceil}$. As a result of these replacements and the properties of $A_{r,s,2n}(\mathbf{X}^i)$, empires $\mathbf{a}$ and $\bar{\mathbf{a}}$ in $P(\Phi)$ are adjacent to empires that must be given the same colour as the neighbours X^i of a and $\bar{a}$ in Φ. To finish with the variable gadgets we attach a copy of $A_{r,s,\mathrm{occ}(a)}(\mathbf{a})$ (resp. $A_{r,s,\mathrm{occ}(\bar{a})}(\bar{\mathbf{a}})$) to the empires $\mathbf{a}$ and $\bar{\mathbf{a}}$.

Finally, the clique on $\{T, c^{i,1}, \ldots, c^{i,s-1}\}$ is replaced by a copy of $B_{\lceil \frac{s}{2} \rceil, s-1}$ on the empires $\mathbf{c}^{i,1}, \ldots, \mathbf{c}^{i,s-1}$ and $\lceil \frac{s}{2} \rceil$ vertices from $Z(\mathbf{T})$. For each i, j and vertex $\ell \in \mathcal{A}$ such that $\{c^{i,j}, \ell\} \in E(\Phi)$ we add an edge connecting $c_s^{i,j}$ and a vertex from $Z(\ell)$.

The graph obtained from $P(\Phi)$ by shrinking each connectivity gadget or empire to a distinct (pseudo-)vertex (removing loops or parallel edges created in the process) coincides with the initial formula graph. The correctness of the reduction follows. ∎

5   Trees

The result on forests of paths of Section 4 already proves that s-COL_r is NP-hard on planar graphs if s is sufficiently small. In this section we investigate the effect of connectedness on the computational complexity of the s-COL_r. The outcome of our investigation is the following dichotomy result (in the next theorem TREE is the class of all trees).

Theorem 6. *Let r and s be fixed positive integers with $r \geq 2$, then the s-COL_r(TREE) problem is NP-hard if $2 < s < 2r$, and polynomial time solvable otherwise.*

The argument for $s = 3$ is very similar to the one we used for forests of paths, but simpler. We present the proof in some details only for the case $r = 2$ (see Theorem 7 below). For $r > 2$ note that a tree T_1 with empires of size r_1 can be translated into a tree T_2 with empires of size $r_2 > r_1$ by simply attaching $r_2 - r_1$ new leaves to a fixed element in each empire of T_1. For $s > 3$ we argue as in Section 4, translating formula graphs into pairs formed by a tree and a

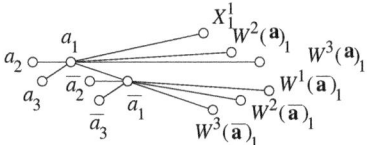

Fig. 6. The gadget for the complementary pair a and $\bar{a}$ when $r = 3$, $s = 5$

partition of its vertices into empires. The hardness of $s\text{-COL}_r(\text{TREE})$ follows from Theorem 2. Details in Theorem 8 below.

Theorem 7. $3\text{-SAT} \leq_p 3\text{-COL}_2(\text{TREE})$.

Proof. (Sketch) Given an instance ϕ of 3-SAT we define a tree $T(\phi)$ and a partition of its vertices into empires such that $T(\phi)$ admits a $(3,2)$-colouring if and only if ϕ is satisfiable. $T(\phi)$ will consist of one *truth gadget*, one *variable gadget* for each variable used in ϕ, and one *clause gadget* for each clause in ϕ. The truth gadget is a copy of $B_{2,2}^+(\mathbf{T})$. Since empires $\mathbf{T}$, $\mathbf{F}$ and $\mathbf{X}$ are adjacent to each other (in the gadget's reduced graph) w.l.o.g. we assume they are coloured TRUE, FALSE and OTHER respectively. For each variable $\mathbf{a}$ in ϕ, $T(\phi)$ contains a copy of $B_{2,2}$ spanned by empires labelled $\mathbf{a}$, $\bar{\mathbf{a}}$, and $\mathbf{X}$. The construction forces empires $\mathbf{a}$, $\bar{\mathbf{a}}$ to be coloured differently from $\mathbf{X}$ (and each other). Finally, for each clause in ϕ, we use a clause gadget like the one in Fig. 5.

Arguing like in the proof of Theorem 4 it is easy to see that $T(\phi)$ is $(3,2)$-colourable if and only if there is some way to assign the variables of ϕ as TRUE or FALSE so that every clause contains at least one TRUE literal. ∎

Theorem 8. $s\text{-COL}(\text{FG}(s, s-1)) \leq_p s\text{-COL}_r(\text{TREE})$, *for any $r \geq 3$ and $3 < s < 2r$.*

Proof. As in the proof of Theorem 5 we give a set of replacement rules that translate an $(s, s-1)$-formula graph Φ into a tree $T(\Phi)$ and a partition of $V(T(\Phi))$ into empires of size r such that $T(\Phi)$ is (s, r)-colourable if and only if the formula graph is s-colourable.

The complete graph on vertices $\{T, F, X^1, \ldots, X^{s-2}\}$ is replaced by a copy of $B_{r,s-1}^+(\mathbf{T})$ with empires labelled $\mathbf{T}$, $\mathbf{F}$, and $\mathbf{X}^1, \mathbf{X}^2, \ldots$ Note that this graph is in fact a tree. Also, because of constraint **B3** in the definition of $B_{r,s}$, w.l.o.g. we may assume that colours "TRUE", "FALSE", "OTHER^1", …, "OTHER^{s-2}" are assigned to empires $\mathbf{T}$, $\mathbf{F}$, $\mathbf{X}^1$ … respectively. For each complementary pair $a, \bar{a}$ of $V(\Phi)$ we create $2s - 5$ empires $\mathbf{W}^2(a), \ldots, \mathbf{W}^{s-2}(a)$ and $\mathbf{W}^1(\bar{a}), \ldots,$ $\mathbf{W}^{s-2}(\bar{a})$. For each $a \in \mathcal{A}$, the subgraph spanned by $\bigcup_i \{a, \bar{a}, X^i\}$ is replaced by a graph including the one given in Fig. 6 consisting of empires $\mathbf{a}$, $\bar{\mathbf{a}}$, $\mathbf{X}^1$, $\mathbf{W}^2(a), \ldots, \mathbf{W}^{s-2}(a)$ and $\mathbf{W}^1(\bar{a}), \ldots, \mathbf{W}^{s-2}(\bar{a})$, connected to $B_{r,s-1}^+(\mathbf{T})$ using the graphs $B_{r,s}^-(\mathbf{W}^i(a), \mathbf{X}^i)$ and $B_{r,s}^-(\mathbf{W}^i(\bar{a}), \mathbf{X}^i)$ for all $i \in \{1, \ldots, s-2\}$. Note that vertices of each empire corresponding to an element of $\mathcal{A}$ are connected to the tree in $B_{r,s-1}^+(\mathbf{T})$ via X_1^1. Also vertices in $\mathbf{a}$ (resp. $\bar{\mathbf{a}}$) are connected to the isolated vertices $W^i(a)_1$ (resp. $W^i(\bar{a})_1$). This prevents $\mathbf{a}$ and $\bar{\mathbf{a}}$ from being able to use the colours of the $\mathbf{X}^i$ in any colouring of $T(\Phi)$. Each group $\{c^1, \ldots, c^{s-1}\}$ in $\mathcal{C}$ is replaced by empires $\mathbf{c}^1, \ldots, \mathbf{c}^{s-1}$ (different groups replaced by different sets of empires). The complete graph on $\{T, c^1, \ldots, c^{s-1}\}$ is replaced by a copy of $B_{r,s-1}$ on the corresponding empires. We then attach to this graph $s-1$ graphs $B_{r,s}^-(\mathbf{b}^j, \mathbf{c}^j)$, for $j \in \{1, \ldots, s-1\}$. Empire $\mathbf{b}^j$ must have the same colour as $\mathbf{c}^j$ and it has, in $B_{r,s}^-(\mathbf{b}^j, \mathbf{c}^j)$, an isolated vertex, b_1^j. If ℓ is an element of $\mathcal{A}$ adjacent to c^j then $\{b_1^j, \ell_1\}$ is an edge of $T(\Phi)$.

The overall construction is such that for each vertex in $V(\Phi)$ there is an equivalent empire in $V(T(\Phi))$, and for each edge in $E(\Phi)$ there is an edge $\{u, v\} \in E(T(\Phi))$ that either connects the corresponding empires $\mathbf{u}$ and $\mathbf{v}$ or connects $\mathbf{u}$ to an empire that must be given the same colour as $\mathbf{v}$ in any (s, r)-colouring of $T(\Phi)$. From this we can see that $T(\Phi)$ admits an (s, r)-colouring if and only if Φ admits an s-colouring. ∎

6 General Planar Graphs

Theorem 6 of last section does not exclude the possibility that $s\text{-COL}_r$ be solvable in polynomial time for arbitrary planar graphs provided $s \geq 2r$. Here we show that in fact this is not the case. Let $\delta_{x,y}$ is equal to one if and only if $x = y$ (and equal to zero otherwise). The main result of this section is the following:

Theorem 9. *Let r and s be fixed positive integers with $r \geq 2$, then the $s\text{-COL}_r$ problem is NP-hard if $3 \leq s < 6r - 3 - 2\delta_{r,2}$, and solvable in polynomial time if $s = 2$ or $s \geq 6r$.*

Note that $s\text{-COL}_r$ can be solved in polynomial time for $s = 2$ (as checking if the reduced graph of a planar graph is bipartite is easy) and for $s \geq 6r$ (because of Heawood's result). Also, Theorem 6 proves the case $s < 2r$. Therefore only the case $s \geq 2r$ needs further discussion. The argument is similar to that of Theorems 8 with a couple of differences. First, this time we only need the graph resulting from the transformation of the initial formula graph to be planar (note that the formula graph in general is NOT planar). On the other hand, we want the transformation to work for much larger values of s. Our solution hinges on proving that all complete subgraphs of the starting formula graph and a number of other gadgets (see below) attached to them have sufficiently large thickness. For the complete graphs we may use well-known results [1], whereas for the specific gadgets we need a bespoke construction. In particular, if $r = 2$ for $s \leq 6$ and if $r \geq 3$ for $s \leq 6r - 4$, it is possible to define a family of planar graphs $D_{r,s}(\mathbf{u}, \mathbf{v})$ satisfying the following properties.

D0 It has $r(s + 1)$ vertices partitioned into $s + 1$ empires all of size r.
D1 It contains an isolated vertex v_1.
D2 No connected component of the given graph contains two vertices from the same empire.
D3 The graph K_{s+1} minus the edge $\{\mathbf{u}, \mathbf{v}\}$ is a subgraph of $R_r(D_{r,s}(\mathbf{u}, \mathbf{v}))$.

$D_{r,s}(\mathbf{u}, \mathbf{v})$ serves a similar purpose to $B^-_{r,s}(\mathbf{u}, \mathbf{v})$ in Theorem 8. Our construction of such graphs is based on a result by Beineke [2] showing that the thickness of K_{6r-3} is r.

A careful reader will realize that a proof of Theorem 9, using more direct reductions from 3-SAT for $s = 3$ and one from s-COL(FG$(s, s - 1)$) for $4 \leq s < 6r - 3 - 2\delta_{r,2}$ can be used to prove the NP-hardness of colouring, in the traditional sense, graphs of thickness r. Thus it is NP-hard to decide whether a graph of thickness $r > 1$ can be coloured with $s < 6r - 3 - 2\delta_{r,2}$ colours.

References

1. Beineke, L.W.: Biplanar graphs: a survey. Computers and Mathematical Applications 34(11), 1–8 (1997)
2. Beineke, L.W., Harary, F.: The thickness of the complete graph. Canadian Journal of Mathematics 17, 850–859 (1965)
3. Brooks, R.L.: On colouring the nodes of a network. Proc. Cambridge Phil. Soc. 37, 194–197 (1941)
4. Bryant, D.E.: Cycle decompositions of the complete graphs. In: Hilton, A.J.W., Talbot, J.M. (eds.) Surveys in Combinatorics. London Mathematical Society Lecture Notes Series, vol. 346, pp. 67–97. Cambridge University Press (2007)
5. Cooper, C., McGrae, A.R.A., Zito, M.: Martingales on Trees and the Empire Chromatic Number of Random Trees. In: Kutyłowski, M., Gebala, M., Charatonik, W. (eds.) FCT 2009. LNCS, vol. 5699, pp. 74–83. Springer, Heidelberg (2009)
6. Cormen, T.H., Leiserson, C.E., Rivest, R.L., Stein, C.: Introduction to Algorithms, 3rd edn. M.I.T. Press (2009)
7. Diestel, R.: Graph Theory. Graduate Texts in Mathematics, vol. 173. Springer, Heidelberg (1999)
8. Garey, M.R., Johnson, D.S.: Computer and Intractability, a Guide to the Theory of NP-Completeness. Freeman and Company (1979)
9. Gibbons, A.M.: Algorithmic Graph Theory. Cambridge University Press (1985)
10. Heawood, P.J.: Map colour theorem. Quarterly Journal of Pure and Applied Mathematics 24, 332–338 (1890)
11. Hutchinson, J.P.: Coloring ordinary maps, maps of empires, and maps of the moon. Mathematics Magazine 66(4), 211–226 (1993)
12. Jackson, B., Ringel, G.: Solution of Heawood's empire problem in the plane. Journal für die Reine und Angewandte Mathematik 347, 146–153 (1983)
13. Karp, R.M.: Reducibility among combinatorial problems. In: Miller, R.E., Thatcher, J.W. (eds.) Complexity of Computer Computations, pp. 85–103. Plenum Press, New York (1972)
14. Lovász, L.: Combinatorial Problems and Exercises, 2nd edn. North-Holland (1993)
15. Lucas, E.: Récreations Mathématiqués, vol. II. Gauthier-Villars (1892)
16. McGrae, A.R., Zito, M.: Colouring Random Empire Trees. In: Ochmański, E., Tyszkiewicz, J. (eds.) MFCS 2008. LNCS, vol. 5162, pp. 515–526. Springer, Heidelberg (2008)
17. McGrae, A.R.A.: Colouring Empires in Random Trees. PhD thesis, Department of Computer Science, University of Liverpool as technical report ULCS-10-007 (2010), http://www.csc.liv.ac.uk/research/techreports/techreports.html

Alternation Graphs

Magnús M. Halldórsson[1,*], Sergey Kitaev[1,2,**], and Artem Pyatkin[3,***]

[1] School of Computer Science, Reykjavik University, 101 Reykjavik, Iceland
mmh@ru.is
[2] Department of Computer and Information Sciences, University of Strathclyde,
Glasgow, G1 1XH, UK
sergey.kitaev@gmail.com
[3] School of Engineering and Computing Sciences, Durham University, Science
Laboratories, South Road, Durham DH1 3LE, UK
artempyatkin@gmail.com

Abstract. A graph $G = (V, E)$ is an *alternation graph* if there exists a word W over the alphabet V such that letters x and y alternate in W if and only if $(x, y) \in E$ for each $x \neq y$.

In this paper we give an effective characterization of alternation graphs in terms of orientations. Namely, we show that a graph is an alternation graph if and only if it admits a *semi-transitive orientation* defined in the paper. This allows us to prove a number of results about alternation graphs, in particular showing that the recognition problem is in NP, and that alternation graphs include all 3-colorable graphs.

We also explore bounds on the size of the word representation of the graph. A graph G is a *k-alternation* graph if it is represented by a word in which each letter occurs exactly k times; the alternation number of G is the minimum k for which G is a k-alternation graph. We show that the alternation number is always at most n, while there exist graphs for which it is $n/2$.

1 Introduction

Consider a scenario with n recurring tasks with requirements on the alternation of certain pairs of tasks. This captures typical situations in periodic scheduling, where there are recurring *precedence* requirements.

When tasks occur only once, the pairwise requirements form precedence constraints, which are modeled by partial orders. When the orientation of the constraints is omitted, the resulting pairwise constraints form comparability graphs. The focus of this paper is to study the class of undirected graphs induced by the alternation relationship of recurring tasks.

Consider, e.g., the following five tasks that may be involved in the operation of a given machine: 1) Initialize controller, 2) Drain excess fluid, 3) Obtain

---

* Partially supported by grant no. 090032021 from the Iceland Research Fund.
** Partially supported by grant no. 090038011 from the Icelandic Research Fund.
*** Partially supported by EPSRC, Grant EP/F064551/1 and by the Ministry of education and science of the Russian Federation (contract number 14.740.11.0868).

P. Kolman and J. Kratochvíl (Eds.): WG 2011, LNCS 6986, pp. 191–202, 2011.

permission from supervisor, 4) Ignite motor, 5) Check oil level. Tasks 1 & 2, 2 & 3, 3 &4, 4 & 5, and 5 & 1 are expected to alternate between all repetitions of the events. This is shown in Fig. 1(b). One possible task execution sequence that obeys these recurrence constraints – and no other – is shown in Fig. 1(a). We introduce later an orientation of such graphs that will be called semi-transitive.

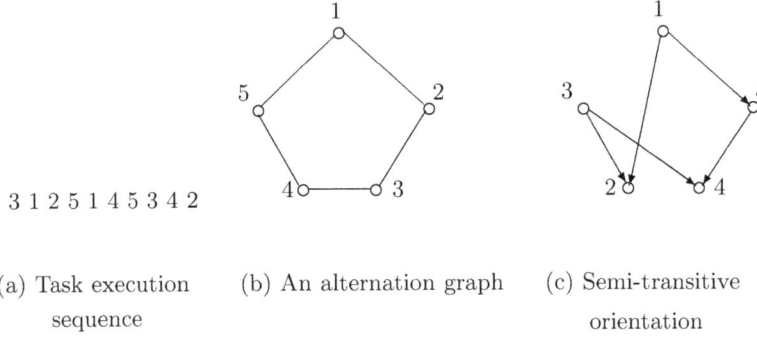

3 1 2 5 1 4 5 3 4 2

(a) Task execution   (b) An alternation graph   (c) Semi-transitive
    sequence   orientation

Fig. 1. The word in (a) corresponds to the alternation graph in (b). A semi-transitive orientation of the graph is given in (c).

Execution sequences of recurring tasks can be viewed as words over an alphabet V, where V is the set of tasks. A graph $G = (V, E)$ is an *alternation graph* if there exists a word W over the alphabet V such that letters x and y alternate in W if and only if $(x, y) \in E$ for each $x \neq y$. If each letter appears exactly k times in the word, the graph is said to be a *k-alternation graph*. It is known that any alternation graph is a k-alternation graph for some k [8]. Alternation graphs are also known as representable graphs [9,8,5].

Our results. We introduce the following notion. A directed graph (digraph) $G = (V, E)$ is *semi-transitive* if it is acyclic and for any directed path $v_1 v_2 ... v_k$ either $v_1 v_k \notin E$ or $v_i v_j \in E$ for all $1 \leq i < j \leq k$. Clearly, all transitive (i.e., comparability) graphs are semi-transitive.

The main result of this paper is that the graph is an alternation graph if and only if it admits a semi-transitive orientation. This result allows us to make progress on the three most fundamental issues about alternation graphs:

– Which types of graphs are alternation graphs and which ones are not?
– How large words can be needed to represent alternation graphs?
– Are there alternative representations of these graphs that aid in reasoning about their properties?

We show that the class of alternation graphs captures non-trivial graph properties. In particular, all 3-colorable graphs are alternation graphs, whereas various types of 4-chromatic graphs cannot all be represented in this way. This resolves a conjecture of [8] regarding the Petersen graph, showing that it is an alternation graph. The result also properly captures all the previously known classes of alternation graphs: outerplanar, prisms, and comparability graphs.

Finally, we show that any alternation graph on n vertices is an n-alternation graph, again utilizing the semi-transitive orientability. This result implies that the problem of deciding whether a given graph is an alternation graph is contained in NP. Previously, no polynomial upper bound was known on the alternation number, which is the smallest value k such that the given graph is k-alternation. This bound on the alternation number is tight up to a constant factor, as we construct graphs with alternation number $n/2$. We also show that deciding if an alternation graph is k-alternation is NP-complete for $3 \leq k \leq n/2$, while the polynomially decidable class of circle graphs coincides with the class of graphs with alternation number at most 2.

Related work. Several graph classes are defined in terms of interrelationships between letters in words, where the vertices represent the letters. *Circle graphs* are those whose vertices can be represented as chords on a circle in such a way that two nodes in the graph are adjacent if and only if the corresponding chords overlap. By viewing each chord as a letter and listing the chords in order of appearance on the circle we find that these graphs correspond to words where each letter appears twice and two nodes are adjacent if and only if the letter occurrences alternate [2]. They therefore correspond to 2-alternation graphs in our vocabulary.

This has been generalized to polygon-circle graphs (see [12]), which are the intersection graphs of polygons inscribed in a circle. If we view each polygon as a letter and read the incidences of the polygons on the circle in order, we see that two polygons intersect if and only if there *exists* a pair of occurrences of the two polygons that alternate. This compares with alternation graphs where *all* occurrences of the two letters must alternate in order for the nodes to be adjacent.

The notion of directed alternation graphs was introduced in [9] to obtain asymptotic bounds on the free spectrum of the widely-studied Perkins semigroup which has played central role in semigroup theory since 1960, particularly as a source of examples and counterexamples. The class of alternation graphs is known to contain comparability graphs [9]; in fact, the comparability graphs are precisely the permutational alternation graphs (see Sec. 2). In [8] numerous properties of alternation graphs were derived and several types of alternation and non-alternation graphs pinpointed. In particular, outerplanar graphs, prisms and 3-subdivision graphs are all alternation graphs. Also, the neighborhood of each vertex in an alternation graph induces a comparability graph. Some open questions from [8] were resolved recently in [5], including the representability of the Petersen graph. These works however do not give alternative representations or essential structural characteristics of alternation graphs.

Cyclic (or periodic) scheduling problems have been studied extensively in the operations research literature [6,11], as well as in the AI literature [3]. These are typically formulated with more general constraints, where, e.g., the 10th occurrence of task A must be preceded by the 5th occurrence of task B. The focus of this work is then on obtaining effective periodic schedules, while maintaining a small cycle time. We are, however, not aware of work on characterizing the graphs formed by the cyclic precedence constraints.

A different periodic scheduling application related to alternation graphs was considered by Graham and Zang [4], involving a counting problem related to the cyclic movements of a robot arm. More generally, given a set of jobs to be performed periodically, certain pairs (a, b) must be done alternately, e.g. since the product of job a is used as a resource for job b. Any valid execution sequence corresponds to a word over the alphabet formed by the jobs. The alternation graph given by the word must then contain the constraint pairs as a subgraph.

Organization. The paper is organized as follows. In Section 2 we give definitions of objects of interest and review some of the known results. In Section 3 we give a characterization of alternation graphs in terms of orientations and discuss some important corollaries of this fact. In Section 4 we examine the alternation number, and show that it is always at most n but can be as much as $n/2$. We explore in Section 5 which classes of graphs are alternation graphs, showing, in particular, that 3-colorable graphs are alternation graphs, but numerous other properties are orthogonal to the alternation property. The construction for triangle-free non-alternation graphs is also presented there. Finally, we conclude with a discussion of algorithmic complexity and some open problems in Section 6.

2   Definitions, Notation, and Known Results

In this section we follow [8] to define the objects of interest.

Let W be a finite word. If W involves the letters $x_1, x_2, \ldots, x_n$ then we write $Var(W) = \{x_1, \ldots, x_n\}$. A word is *k-uniform* if each letter appears in it exactly k times. A 1-uniform word is also called a *permutation*. Denote by $W_1 W_2$ the concatenation of words W_1 and W_2. We say that the letters x_i and x_j *alternate* in W if the word induced by these two letters contains neither $x_i x_i$ nor $x_j x_j$ as a factor. If a word W contains k copies of a letter x then we denote these k appearances of x by $x^1, x^2, \ldots, x^k$. We write $x_i^j < x_k^l$ if x_i^j occurs in W before x_k^l, i. e., x_i^j is to the left of x_k^l in W.

We say that a word W *represents* the graph $G = (V, E)$ if there is a bijection $\phi : Var(W) \to V$ such that $(\phi(x_i), \phi(x_j)) \in E$ if and only if x_i and x_j alternate in W. We call a graph G an *alternation graph* if there exists a word W that represents G. It is convenient to identify the vertices of an alternation graph and the corresponding letters of a word representing it. If G can be represented by a *k-uniform* word, then we say that G is a *k-alternation* graph. The *alternation number* of an alternation graph G is the minimum k such that G is a k-alternation graph. We call a graph a *permutational alternation* graph if it can be represented by a word of the form $P_1 P_2 \ldots P_k$ where all P_i are permutations.

A digraph is *transitive* if the adjacency relation is transitive, i. e. for every vertices $x, y, z \in V$, the existence of the arcs $xy, yz \in E$ yields that $xz \in E$. A *comparability graph* is an undirected graph having an orientation of the edges that yields a transitive digraph.

The following properties of alternation graphs are useful [8]. A graph G is an alternation graph if and only if it is k-alternation for some k. If $W = AB$ is k-uniform word representing a graph G, then the word $W' = BA$ also k-represents G.

The wheel W_5 is the smallest non-alternation graph. The non-alternation graphs on 6 and 7 vertices (from [8]) are given in Fig. 2.

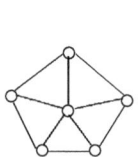 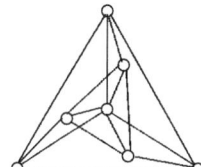 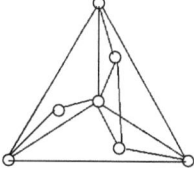 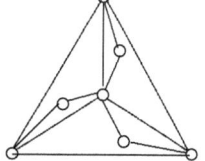

Fig. 2. Small non-alternation graphs

3   Characterization of Alternation Graphs by Orientability

The word representation of alternation graphs is simple and natural. Yet it does not lend itself to easy arguments for the characteristic of alternation graphs. Non-alternation is even harder to argue. The main result of this section is a new characterization of alternation graphs that is effective algorithmically.

We give a characterization in terms of orientability, which implies that alternation corresponds to a property of a digraph obtained by directing the edges in certain way. It is known that a graph is a permutational alternation graph if and only if it has a transitive orientation (i.e., is a comparability graph) [9]. We prove a similar fact on alternation graphs, namely, that a graph is an alternation graph if and only if it has a certain *semi-transitive orientation* that we shall define. Our definition, in fact, generalizes that of a transitive orientation.

Other orientations have been defined in order to capture generalizations of comparability graphs. As transitive orientations form constraints on the orderings of induced P_3, these generalizations form constraints on the orderings of induced P_4. These include *perfectly orderable graphs* (and its subclasses) and *opposition graphs* [1]. None of these properties captures our definition below, nor does our characterization subsume any of them.

We turn to the characterization and start with definitions of certain directed graphs. A *semi-cycle* is the directed acyclic graph obtained by reversing the direction of one arc of a directed cycle. An acyclic digraph is a *shortcut* if it is induced by the vertices of a semi-cycle and contains a pair of non-adjacent vertices. Thus, a digraph on the vertex set $\{v_0, v_1, \ldots, v_t\}$, is a shortcut if it contains a directed path $v_0 v_1 \ldots v_t$, the arc $v_0 v_t$ and it is missing an arc $v_i v_j$, $0 \le i < j \le t$ (in particular, $t \ge 3$).

A digraph is *semi-transitive* if it is acyclic and contains no shortcuts. A graph is *semi-transitively orientable* if there exists an orientation of the edges that results in a semi-transitive graph. Our main result in this paper is the following.

Theorem 1. *A graph is an alternation graph if and only if it is semi-transitively orientable.*

We first need some additional definitions and lemmas. A *linear extension* (a.k.a. topological order) of an acyclic digraph is a permutation of the vertices that obeys the arcs, i. e. for each arc uv, u precedes v in the permutation. For a node-labeled digraph, let the linear extension also refer to the word obtained by visiting the nodes in that order. Let $D = (V, E)$ be a digraph. The *t-string* digraph D^t of D is defined as follows. The vertices of D^t are v^i, for $v \in V$ and $i = 1, 2, \ldots, t$, and $v^i u^j$ is an arc in D^t if and only if either $i = j$ and $vu \in E$ or $i < j$ and $uv \in E$. Intuitively, the t-string digraph of D has t copies of D strung together. Given a word S, let G_S denote the graph represented by S. If S is a linear extension of D^t then we also denote by G_S the graph represented by the word S' obtained from S by omitting the superindices of the vertices (i. e. the copies of the same vertex in S are considered as the same letters in S').

Given a digraph D, let G_D be the graph obtained by ignoring orientation.

We argue that the word representing a semi-transitive digraph comes from a special linear extension of the t-string digraph D^t for some t. We first observe that any linear extension of D^t preserves arcs.

Lemma 1. *Let D be a digraph with distinct node-labels. Let S be a linear extension of a D^t. Then G_D is a subgraph of G_S.*

Proof. Consider an edge uv in G_D, and suppose without loss of generality that it is directed as uv in D. Then, in D^t, there is a directed path $u^1 v^1 u^2 v^2 \ldots u^t v^t$. Thus, occurrences of u and v in a linear extension of D^t are alternating. Hence, $uv \in G_S$. ∎

To prove equivalence, we now give a method to produce a linear extension of D^t that generates all non-arcs. We say that an induced subgraph H *covers* a set A of non-arcs if each non-arc in A is also a non-arc in H. A word covers the non-arc if the digraph that it represents covers them.

Lemma 2. *The non-arcs incident with a (directed) path P in a semi-transitive digraph D can be covered with a 2-uniform word.*

Proof. Consider the 2-string digraph D^2 and and let P^1 (P^2) be the first (second) copy of P in D^2. We say that a node x of D^2 *depends on* node y, and denote it by $y \rightsquigarrow x$, if there is a directed path from y to x in D^2, i. e. y must appear before x in a linear extension of D^2. Thus, $(D^2, \rightsquigarrow)$ is a partial order formed by the transitive closure of D^2.

We inductively form a linear extension S of D^2 as follows. Set $Q = D^2$. Select a source node (i.e., of indegree 0) v in Q, giving first priority to nodes in P^2, next nodes neither in P^1 or P^2, and finally nodes in P^1. Order v first in S, remove it from Q, and inductively form the linear extension of Q as the tail of S. Intuitively speaking, the nodes in P^1 are listed as late as possible, while the nodes in P^2 are listed as early as possible.

We claim that this word S covers all non-arcs involving nodes in P. Consider a pair u, v, where $uv \notin G_D$ and $u \in P$. Note that v may also belong to P, in which case we may assume that the path goes from u to v. Observe that u may depend on v, or vice versa, but not both. Let u^1, v^1, u^2, v^2 be the corresponding vertices of D^2. There are three cases to consider.

Case (i): There is a path from u to v in D. We claim that u^2 does not depend on v^1. Suppose it does, i. e. $v^1 \rightsquigarrow u^2$. Then, there is an arc $x^1 y^2 \in D^2$ such that $v^1 \rightsquigarrow x^1$ and $y^2 \rightsquigarrow u^2$. By the assumptions and the symmetry of the two copies of D in D^2, it follows that $y^1 \rightsquigarrow u^1 \rightsquigarrow v^1 \rightsquigarrow x^1$. By the definition of 2-string graphs, yx is an arc in D, so $y^1 x^1 \in E(D^2)$. Then, by semi-transitivity, $u^1 v^1 \in E(D^2)$, which implies that $uv \in E(G_D)$, which is a contradiction. Thus, $u^2 \not\rightsquigarrow v^1$. From the priority given by the algorithm, u^2 will then be listed before v^1, resulting in the order $u^1 u^2 v^1 v^2$ in S. Thus, $uv \notin E(G_S)$.

Case (ii): There is a path from v to u in D. This is symmetric to case (i), with u replaced by v. Thus, the nodes will occur as $v^1 v^2 u^1 u^2$ in S.

Case (iii): The nodes u and v are incomparable in D. In particular, v is not in P. Then, u^1 and v^1 do not depend on each other, nor do u^2 and v^2. If v^2 depends on u^1 then the nodes occur as $v^1 u^1 u^2 v^2$ in S. Otherwise, their order is $v^1 v^2 u^1 u^2$.

We now return to the proof of Theorem 1, starting with the forward direction. Given a word S, we direct an edge of G_S from x to y if the first occurrence of x is before that of y in the word. Let us show that such an orientation D of G_S is semi-transitive. Indeed, assume that $x_0 x_t \in E(D)$ and there is a directed path $x_0 x_1 \ldots x_t$ in D. Then in the word S we have $x_0^i < x_1^i < \ldots < x_t^i$ for every i. Since $x_0 x_t \in E(D)$ we have $x_t^i < x_0^{i+1}$. But then for every $j < k$ and i there must be $x_j^i < x_k^i < x_j^{i+1}$, i. e. $x_i x_j \subset E(D)$. So, D is semi-transitive.

For the other direction, denote by G the graph and by D its semi-transitive orientation. Let $P_1, P_2, \ldots, P_\tau$ be the set of directed paths covering all vertices of D. For every $i = 1, 2, \ldots, \tau$ denote by S_i the linear extension of the digraph D^2 satisfying the conditions of Lemma 2 for the path P_i. Put $S = S_1 S_2 \ldots S_\tau$. Clearly, S is a 2τ-uniform word; it can be treated as a linear extension of a 2τ-string $D^{2\tau}$. Then $G = G_S$. Indeed, by Lemma 1 we have $E(G) \subset E(G_S)$. On the other hand, if $uv \notin E(G)$ then $u \in P_i$ for some i, and thus by Lemma 2 the letters u and v are not alternating in the subword S_i. Therefore, $uv \notin E(S)$. Theorem 1 is proved. □

Theorem 1 makes clear the relationship to comparability graphs, which are those that have transitive orientations. Since transitive digraphs are also semi-transitive, this immediately implies that comparability graphs are alternation graphs.

The construction in Lemma 2 shows that all alternation graphs can be represented "almost" permutationally. This is made more precise as follows.

Observation 2. *Let G be an alternation graph. Then there is a word W representing G such that for any prefix P of W and any pair a, b of letters, the number of occurrences of a and b in P differ by at most two.*

4   The Alternation Number of Graphs

We focus now on the following question: Given an alternation graph, how large is its alternation number? In [8], certain classes of graphs were proved to be 2- or 3-alternation, and an example was given of a graph (the triangular prism) with the alternation number of 3. On the other hand, no examples were known of graphs with alternation numbers larger than 3, nor were there any non-trivial upper bounds known. We show here that the maximum alternation number of alternation graphs is linear in the number of vertices.

For the upper bound, we use the results of the preceding section. We have the following directly from the proof of Theorem 1.

Corollary 1. *An alternation graph G is a $2\tau(G)$-alternation graph, where $\tau(G)$ is the minimum number of paths covering all nodes in some semi-transitive orientation of G.*

This immediately gives an upper bound of $2n$ on the alternation number. We can improve this somewhat with an effective procedure.

Theorem 3. *Given a semi-transitive digraph D on n vertices, there is a polynomial time algorithm that generates an n-uniform word representing G_D. Thus, each alternation graph is an n-alternation graph.*

Proof. The algorithm works as follows.

Step 0. Start with $A = \emptyset$ and $i = 1$.

Step i. If D contains a path P_i covering at least two vertices from $V \setminus A$ then let $A := A \cup V(P_i)$ and $i := i + 1$. Otherwise, let $B = V \setminus A$ and go to the Final Step.

Final Step. Let S_i be the linear extension of the digraph D^2 satisfying the conditions of Lemma 2 for the path P_i and put $S' = S_1 S_2 \ldots S_t$ where t is the number of paths found at previous steps. If $|B| \leq 1$ then let $S = S'$. Otherwise, consider a linear extension S_0 of D where the vertices of B are listed in a row (since the vertices of B form an antichain, i.e. are mutually incomparable, such a linear extension must exist) and in particular in the reverse order of their appearance in S_1. Let $S = S'S_0$.

Clearly, $G_D = G_S$ (the proof is the same as in Theorem 1). It is easy to verify that each letter appears in S at most n times.

Theorem 3 implies that the graph property of alternation is polynomially verifiable, answering an open question in [8]. Indeed, having an alternation graph G, we may ask for a word representing it and verify this fact in time bounded by the polynomial in n.

Corollary 2. *The recognition problem for alternation graphs is in NP.*

We now show that there are graphs with alternation number of $n/2$, matching the upper bound within a factor of 2.

The *crown graph* $H_{k,k}$ is the graph obtained from the complete bipartite graph $K_{k,k}$ by removing a perfect matching. Denote by G_k the graph obtained from a crown graph $H_{k,k}$ by adding a universal vertex (adjacent to all vertices in $H_{k,k}$).

Theorem 4. *The graph G_k has alternation number $k = \lfloor n/2 \rfloor$.*

The proof is based on three statements.

Lemma 3. *Let H be a graph and G be the graph obtained from H by adding an all-adjacent vertex. Then G is a k-alternation graph if and only if H is a permutational k-alternation graph.*

Proof. Let 0 be the letter corresponding to the all-adjacent vertex. Then every other letter of the word W representing G must appear exactly once between two consecutive zeroes. We may assume also that W starts with 0. Then the word $W \setminus \{0\}$, formed by deleting all occurrences of 0 from W, is a permutational k-representation of H. Conversely, if W' is a word permutationally k-representing H, then we insert 0 in front of each permutation to get a (permutational) k-representation of G.

Lemma 4. *A comparability graph is permutational k-alternation graph if and only if the poset induced by this graph has dimension at most k.*

Proof. Let H be a comparability graph and W be a word permutationally k-representing it. Each permutation in W can be considered as a linear order where $a < b$ if a meets before b in the permutation (and vice versa). We want to show that the comparability graph of the poset induced by the intersection of these linear orders coincides with H.

Two vertices a and b are adjacent in H if and only if their letters alternate in the word. So, they must be in the same order in each permutation, i. e. either $a < b$ in every linear order or $b < a$ in every linear order. But this means that a and b are comparable in the poset induced by the intersection of the linear orders, i. e. a and b are adjacent in its comparability graph.

Lemma 5 ([13]). *The poset P over $2k$ elements $\{a_1, a_2, \ldots, a_k, b_1, b_2, \ldots, b_k\}$ such that $a_i < b_j$ for every $i \neq j$ and all other elements are not comparable has dimension k.*

Now we can prove Theorem 4.

Proof. Since the crown graph $H_{k,k}$ is a comparability graph of the poset P, we deduce from Lemmas 5 and 4 that $H_{k,k}$ is permutational k-alternation graph but not a permutational $(k-1)$-alternation graph. Then by Lemma 3 we have that G_k is a k-alternation graph but not a $(k-1)$-alternation graph. Theorem 4 is proved. □

The above arguments help us also in deciding the complexity of determining the alternation number. From Lemmas 3 and 4, we see that it is as hard as determining the dimension k of a poset. Yannakakis [14] showed that the latter is NP-hard, for any $3 \leq k \leq \lceil n/2 \rceil$. We therefore obtain the following result, which matches the situation for the related by different k-polygon circle graphs [10].

Proposition 1. *Deciding whether a given graph is a k-alternation graph, for any given $3 \leq k \leq \lceil n/2 \rceil$, is NP-complete.*

It was further shown by Hegde and Jain [7] that it is NP-hard to approximate the dimension of a poset within almost a square root factor. We therefore obtain the same hardness for the alternation number.

Proposition 2. *Approximating the alternation number within $n^{1/2-\epsilon}$-factor is NP-hard, for any $\epsilon > 0$.*

5   Characteristics of Alternation Graphs

When faced with a new graph class, the most basic questions involve the kind of properties it satisfies: which known classes are properly contained (and which not), which graphs are otherwise contained (and which not), what operations preserve alternation (or non-alternation), and which properties hold for these graphs.

Previously, it was known that the class of alternation graphs includes comparability graphs, outerplanar graphs, subdivision graphs, and prisms. The purpose of this section is to clarify this situation significantly, including resolving some conjectures. We start with exploring the relation of colorability and alternation.

Theorem 5. *3-colorable graphs are semi-transitively orientable, and thus alternation graphs.*

Proof. Given a 3-coloring of a graph, direct its edges from the first color class through the second to the third class. It is easy to see that we obtain a semi-transitive digraph.

This implies a number of earlier results on alternation, including that of outerplanar graphs, subdivision graphs, and prisms. The theorem also shows that 2-degenerate graphs, graphs of maximum degree 3 (via Brooks theorem), and triangle-free planar graphs (via Grötzch's theorem) are all alternation graphs.

This result does not extend to higher chromatic numbers. The examples in Fig. 2 show that 4-colorable graphs can be non-alternation. We can, however, obtain a result in terms of the *girth* of the graph, which is the length of its shortest cycle.

Proposition 3. *Let G be a graph whose girth is greater than its chromatic number. Then, G is an alternation graph.*

Proof. Suppose the graph is colored with $\chi(G)$ natural numbers. Orient the edges of the graph from small to large colors. There is no directed path with more than $\chi(G) - 1$ arcs, but since G contains no cycle of $\chi(G)$ or fewer edges, there can be no shortcut. Hence, the digraph is semi-transitive.

The next theorem shows us how to construct an infinite series of triangle-free non-alternation graphs. This answers an open question in [8].

Theorem 6. *There exist triangle-free non-alternation graphs.*

Proof. Let H be a 4-chromatic graph with girth at least 10 (such graphs exist by Erdös theorem). For every path P of length 3 in H add to H the edge e_P connecting its ends. Denote the obtained graph by G. Let us show that G is a triangle-free non-alternation graph.

If G contains a triangle on the vertices u, v, w then H contains three paths P_{uv}, P_{uw}, and P_{vw} of lengths 1 or 3 connecting these vertices. Let T be a graph spanned by these three paths. Since T has at most 9 edges and the girth of H is at least 10, T is a tree. Clearly, it cannot be a path. So, it is a subdivision of $K_{1,3}$ with the leafs u, v, w. But then at least one of the paths P_{uv}, P_{uw}, P_{vw} must have an even length, a contradiction.

So, G is triangle-free. Assume that G has a semi-transitive orientation. Then it induces a semi-transitive orientation on H. Since H is 4-chromatic, each of its acyclic orientation must contain a directed path P of length at least 3. But then the orientation of the edge e_P in G produces either a 4-cycle or a shortcut, contradicting the semi-transitivity. So, G is a triangle-free non-alternation graph.

6   Concluding Remarks and Open Questions

It is natural to ask about optimization problems on alternation graphs. The known NP-hardness of many classical optimization problems on 3-colorable graphs implies NP-hardness on alternation graphs, due to Theorem 5.

Observation 7. *The optimization problems Independent Set, Dominating Set, Graph Coloring, Clique Partition, Clique Covering are NP-hard on alternation graphs.*

Note that it may be relevant whether the representation of the graph as a semi-transitive digraph is given; solvability under these conditions is open. However, some problems remain polynomially solvable:

Observation 8. *The Clique problem is polynomially solvable on alternation graphs.*

Indeed, we can simply use the fact that the neighborhood of any node is a comparability graph. The clique problem is easily solvable on comparability graphs. Thus, it suffices to search for the largest clique within all induced neighborhoods.

We conclude with several open questions about alternation graphs:

1. Is it NP-hard to decide whether a graph is an alternation graph?
2. What is the maximum alternation number of a graph? We know that it lies between $n/2$ and n.
3. Are all graphs of maximum degree 4 alternation graphs?
4. Is there an algorithm that forms an $f(k)$-representation of a k-alternation graph, for some function f? Namely, can the alternation number be approximated as a function of itself? The same question holds also for the partial order (or poset) dimension [7].

References

1. Brandstädt, A., Bang Lee, V., Spinrad, J.P.: Graph Classes: A Survey. Monographs on Discrete Mathematics and Applications. SIAM (1987)
2. Courcelle, B.: Circle graphs and Monadic Second-order logic. J. Applied Logic 6(3), 416–442 (2008)
3. Draper, D.L., Jonsson, A.K., Clements, D.P., Joslin, D.E.: Cyclic Scheduling. In: Proc. IJCAI (1999)
4. Graham, R., Zang, N.: Enumerating split-pair arrangements. J. Combin. Theory, Series A 115(2), 293–303 (2008)
5. Halldórsson, M.M., Kitaev, S., Pyatkin, A.: Graphs Capturing Alternations in Words. In: Gao, Y., Lu, H., Seki, S., Yu, S. (eds.) DLT 2010. LNCS, vol. 6224, pp. 436–437. Springer, Heidelberg (2010)
6. Hanen, C., Munier, A.: Cyclic scheduling on parallel processors: An overview. In: Chretienne, P., Coffman Jr., E.G., Lenstra, J.K., Liu, Z. (eds.) Scheduling Theory and its Applications, ch. 9, John Wiley & Sons (1995)
7. Hegde, R., Jain, K.: The hardness of approximating poset dimension. Electronic Notes in Discrete Mathematics 29, 435–443 (2007)
8. Kitaev, S., Pyatkin, A.: On representable graphs. Automata, Languages and Combinatorics 13, 45–54 (2008)
9. Kitaev, S., Seif, S.: Word problem of the Perkins semigroup via directed acyclic graphs. Order (2008), doi: 10.1007/s11083-008-9083-7
10. Kratochvíl, J., Pergel, M.: Two Results on Intersection Graphs of Polygons. In: Liotta, G. (ed.) GD 2003. LNCS, vol. 2912, pp. 59–70. Springer, Heidelberg (2004)
11. Middendorf, M., Timkovsky, V.G.: On scheduling cycle shops: Classification, complexity and approximation. Journal of Scheduling 5, 135–169 (2002)
12. Pergel, M.: Recognition of Polygon-Circle Graphs And Graphs of Interval Filaments Is NP-Complete. In: Brandstädt, A., Kratsch, D., Müller, H. (eds.) WG 2007. LNCS, vol. 4769, pp. 238–247. Springer, Heidelberg (2007)
13. Trotter, W.T.: Combinatorics and partially ordered sets: Dimension theory. Johns Hopkins Univ. Press (2001)
14. Yannakakis, M.: The complexity of the partial order dimension problem. SIAM J. Algebraic Discrete Methods 3(3), 351–358 (1982)

Improved Bounds for Minimum Fault-Tolerant Gossip Graphs*

Toru Hasunuma[1] and Hiroshi Nagamochi[2]

[1] Institute of Socio-Arts and Sciences, The University of Tokushima,
Tokushima 770–8502 Japan
`hasunuma@ias.tokushima-u.ac.jp`
[2] Department of Applied Mathematics and Physics, Kyoto University,
Kyoto 606-8501, Japan
`nag@amp.i.kyoto-u.ac.jp`

Abstract. A k-fault-tolerant gossip graph is a (multiple) graph whose edges are linearly ordered such that for any ordered pair of vertices u and v, there are $k+1$ edge-disjoint ascending paths from u to v. Let $\tau(n,k)$ denote the minimum number of edges in a k-fault-tolerant gossip graph with n vertices. In this paper, we present upper and lower bounds on $\tau(n,k)$ which improve the previously known bounds. In particular, from our upper bounds, it follows that $\tau(n,k) \le \frac{nk}{2} + O(n \log n)$. Previously, it has been shown that this upper bound holds only for the case that n is a power of two.

1 Introduction

Throughout the paper, a graph may have multiple edges, but not self loops. Let $G = (V, E)$ be a graph. An *edge-ordering* of G is a bijection from $E(G)$ to $\{1, 2, \ldots, |E(G)|\}$. A graph G with an edge-ordering ρ is an *ordered graph* (G, ρ). Let $P = (v_0, e_1, v_1, e_2, v_2, \ldots, e_k, v_k)$ be a path from a vertex v_0 to a vertex v_k in G, where $v_i \in V(G)$ for $0 \le i \le k$ and $e_i \in E(G)$ for $1 \le i \le k$ such that all v_i's are distinct and e_i joins v_{i-1} and v_i for $1 \le i \le k$. If $\rho(e_i) < \rho(e_j)$ for $1 \le i < j \le k$, then P is an *ascending path* from v_0 to v_k in (G, ρ). An ordered graph (G, ρ) is a *k-fault-tolerant gossip graph* if for any ordered pair of vertices u and v in (G, ρ), there are $k+1$ edge-disjoint ascending paths from u to v. A 0-fault-tolerant gossip graph is simply called a *gossip graph*. Let $\tau(n, k)$ be the minimum number of edges in a k-fault-tolerant gossip graph with n vertices.

The term of a gossip graph comes from the gossiping problem, first proposed by Boyd. Suppose that there are n persons such that each person has a unique message, and all the n persons want to know all the n messages by telephone. In each telephone call, the two persons exchange every message which they have at the time of the call. The gossiping problem is to find the minimum number of calls. A process that the n persons communicate by telephone can be modeled by an ordered graph (G, ρ), where each vertex (respectively, edge) corresponds

---

* This work was supported by JSPS KAKENHI 20500012, 21500017.

P. Kolman and J. Kratochvíl (Eds.): WG 2011, LNCS 6986, pp. 203–214, 2011.

to each person (respectively, telephone call) such that the edge-ordering ρ indicates the ordering of telephone calls. A person (vertex) v receives the message originated from a person (vertex) u if and only if there is an ascending path from u to v in the ordered graph (G, ρ). Thus, a gossiping for n persons can be modeled by a gossip graph with n vertices. The minimum number of calls in the gossiping problem on n persons was determined to be $2n - 4$ for $n \geq 4$ by several researchers independently (see [1], [4], [6], [11]). Besides, several variations of the problem have been studied in [10], [12]. In some situation, a telephone call may fail in the sense that the messages in the failed call are not exchanged. Berman and Hawrylycz [2] first proposed the gossiping problem with at most k failed calls which can be modeled by a k-fault-tolerant gossip graph (G, ρ). Gossiping is now a fundamental problem in computer networks. When we study gossiping problems in computer networks, we need to specify many assumptions. Our gossiping model corresponds to gossiping in computer networks under the assumptions that communication mode is full-duplex and whispering, packet size is unbounded, fault type is transient link-fault, fault model is bounded, and algorithm is nonadaptive. For each terminology in these assumptions, the reader is referred to the survey [9] by Pelc.

Berman and Hawrylycz [2] showed that

$$\left\lceil \left(\tfrac{k+4}{2}\right)(n-1)\right\rceil - 2\lceil\sqrt{n}\,\rceil + 1 \leq \tau(n,k) \leq \left\lfloor \left(k+\tfrac{3}{2}\right)(n-1)\right\rfloor \quad \text{for } k \leq n-2,$$

$$\left\lceil \left(\tfrac{k+3}{2}\right)(n-1)\right\rceil - 2\lceil\sqrt{n}\,\rceil \qquad \leq \tau(n,k) \leq \left\lfloor \left(k+\tfrac{3}{2}\right)(n-1)\right\rfloor \quad \text{for } k \geq n-2.$$

Haddad, Roy, and Schäffer [5] proved that

$$\tau(n,k) \leq \left(\frac{k}{2} + 2p\right)\left((n-1) + \frac{n-1}{2^p-1} + 2^p\right),$$

where p is any integer between 1 and $\log_2 n$ inclusive. By choosing p appropriately, this upper bound improves the Berman and Hawrylycz's upper bounds for almost all k. In particular, by choosing $p = \lceil\frac{\log_2 n}{2}\rceil$, the following bound is obtained: $\tau(n,k) \leq \frac{nk}{2} + O(k\sqrt{n} + n\log n)$. For the special case of $n = 2^p$ for some p, Haddad, Roy, and Schäffer also showed that

$$\tau(n,k) \leq \min\left\{\left(\left\lceil\frac{k+1}{\log_2 n}\right\rceil + 1\right)\frac{n\log_2 n}{2},\right.$$
$$\left.\left(\left\lfloor\frac{k+1}{\log_2 n}\right\rfloor + 1\right)\frac{n\log_2 n}{2} + (k+1 \bmod \log_2 n)(2n-4)\right\}.$$

Thus, $\tau(n,k) \leq \frac{nk}{2} + O(n\log n)$ when n is a power of two.

Later on Berman and Paul [3] improved Berman and Hawrylycz's lower bounds by showing that

$$2n - 2 + \left\lceil\frac{k(n-1)}{2}\right\rceil - \lfloor\log_2 n\rfloor \leq \tau(n,k).$$

Recently, Hou and Shigeno [8] showed that

$$\left\lfloor\frac{n(k+2)}{2}\right\rfloor \leq \tau(n,k) \leq \frac{n(n-1)}{2} + \left\lceil\frac{nk}{2}\right\rceil.$$

Thus, it holds that $\frac{nk}{2}+\Omega(n) \leq \tau(n,k) \leq \frac{nk}{2}+O(n^2)$. Hou and Shigeno's bounds improve the previous bounds for small n and sufficiently large k.

In this paper, we show that

$$
\tau(n,k) \leq \begin{cases} \dfrac{n \log_2 n}{2} + \dfrac{nk}{2} & \text{if } n \text{ is a power of two,} \\[2ex] 2n \lfloor \log_2 n \rfloor + n \left\lceil \dfrac{k-1}{2} \right\rceil & \text{otherwise.} \end{cases}
$$

From our results, it holds that $\tau(n,k) \leq \frac{nk}{2}+O(n \log n)$. In particular, our upper bound improves Hou and Shigeno's upper bound for all $n \geq 13$. We also improve the upper bound by Haddad et al. by showing that the factor $(k/2 + 2p)$ in their upper bound can be replaced with a smaller factor $(k/2 + p)$:

$$
\tau(n,k) \leq \left(\frac{k}{2} + p \right) \left((n-1) + \frac{n-1}{2^p - 1} + 2^p \right),
$$

where p is any integer between 1 and $\log_2 n$ inclusive.

Besides, we show that

$$
\left\lceil \frac{3n-5}{2} \right\rceil + \left\lceil \frac{1}{2} \left(nk + \left\lfloor \frac{n+1}{2} \right\rfloor - \lfloor \log_2 n \rfloor \right) \right\rceil \leq \tau(n,k).
$$

Our lower bound improves Berman and Paul's lower bound when $k > n/2$ and Hou and Shigeno's lower bound when $n \geq 5$.

This paper is organized as follows. Section 2 presents our general method for constructing fault-tolerant gossip graphs. Upper bounds on the minimum number of edges in fault-tolerant gossiping graphs based on the hypercubes and circulant graphs are given in Sections 3 and 4, respectively. Section 5 gives our lower bound.

2   Construction of Fault-Tolerant Gossip Graphs

In order to simplify the discussion for edge-disjoint paths, we often omit the vertices (or edges) in the description of a path if there is no confusion. Let $P = (e_0, e_1, \ldots, e_k)$ be a path in an ordered graph (G, ρ), where $e_i \in E(G)$ for $0 \leq i \leq k$. If P is divided into $s+1$ subpaths $P^{(0)} = (e_0, \ldots, e_{p_0})$, $P^{(1)} = (e_{p_0+1}, \ldots, e_{p_1})$, $\ldots$, $P^{(s)} = (e_{p_{s-1}+1}, \ldots, e_k)$, then we write $P = P^{(0)} \odot P^{(1)} \odot \cdots \odot P^{(s)}$, where $\odot$ is the *concatenation* operation on two paths for which the last vertex of one path is the first vertex of the other. If $P = P^{(0)} \odot P^{(1)} \odot \cdots \odot P^{(s)}$ such that $P^{(j)}$ is an ascending path for $0 \leq j \leq s$ and $P^{(j)} \odot P^{(j+1)}$ is not an ascending path for $0 \leq j < s$, then P is an *s-folded ascending path*. For an s-folded ascending path P, the *folded number* of P is defined to be s.

Based on an ordered graph (G, ρ), we define $h \cdot (G, \rho)$ to be the ordered graph obtained from (G, ρ) by adding $h-1$ copies $E_1, E_2, \ldots, E_{h-1}$ of $E_0 = E(G)$ and setting the order of each edge $e_{t,i} \in E_i$ as $\rho(e_t) + i \cdot |E(G)|$, where e_t is the original edge in $E(G)$ corresponding to $e_{t,i}$. For a path $P = (e_0, e_1, \ldots, e_k)$

in (G, ρ), let $P_0 = P$ and $P_i(= (e_{0,i}, e_{1,i}, \ldots, e_{k,i}))$ be the corresponding path using edges in E_i for $1 \leq i < h$. Let $P = P^{(0)} \odot P^{(1)} \odot \cdots \odot P^{(s)}$ be an s-folded ascending path from a vertex u to a vertex v in (G, ρ), where $P^{(j)}$ is an ascending subpath for $0 \leq j \leq s$. Then, P_i is also an s-folded ascending path and $P_i = P_i^{(0)} \odot P_i^{(1)} \odot \cdots \odot P_i^{(s)}$. Now consider the path $P(k) = P_k^{(0)} \odot P_{k+1}^{(1)} \odot \cdots \odot P_{k+s}^{(s)}$ in $h \cdot (G, \rho)$. Then, $P(k)$ is an ascending path from u to v for $0 \leq k < h - s$ such that $P(k)$ and $P(k')$ are edge-disjoint if $k \neq k'$. Thus, based on an s-folded ascending path P, we can construct $(h - s)$ edge-disjoint ascending paths from u to v in $h \cdot (G, \rho)$. Similarly, based on another s-folded ascending path P' from u to v, we can construct $(h - s)$ edge-disjoint ascending paths $P'(k)$ from u to v for $0 \leq k < h - s$. If P and P' are edge-disjoint, then $P(0), \ldots, P(h - s - 1)$ and $P'(0), \ldots, P'(h - s - 1)$ are clearly edge-disjoint each other. Therefore, the following lemma holds. This lemma was shown by Haddad et. al. [5] in a slightly different form[1].

Lemma 1. *Let u and v be vertices in an ordered graph (G, ρ). If there are p edge-disjoint s-folded ascending paths from u to v in (G, ρ). then there are $p(h - s)$ edge-disjoint ascending paths from u to v in $h \cdot (G, \rho)$ for any integer $h \geq s$.*

From this lemma, if there are p edge-disjoint s-folded ascending paths from u to v in (G, ρ), then there are $k + 1$ edge-disjoint ascending paths from u to v in $(s + \lceil \frac{k+1}{p} \rceil) \cdot (G, \rho)$. Thus, the following corollary is obtained.

Corollary 1. *Let G be a graph with n vertices and m edges. If there are p edge-disjoint s-folded ascending paths from any vertex to any other vertex in an ordered graph (G, ρ), then $\tau(n, k) \leq (s + \lceil \frac{k+1}{p} \rceil)m$.*

Based on this corollary, Haddad et al. derived upper bounds. In order to improve their upper bounds, we need a proposition stronger than Corollary 1.

Theorem 1. *Let (G, ρ) be an ordered graph with n vertices. Suppose that*

- *$E(G)$ can be decomposed into ℓ subsets $F_0, F_1, \ldots, F_{\ell-1}$ such that for any two edges $e \in F_i$ and $e' \in F_j$, $\rho(e) < \rho(e')$ if $i < j$,*
- *for any two vertices u and v, there are p edge-disjoint paths from u to v such that the sum of their folded numbers is at most q, and the last edges of r_i paths are in F_i for $0 \leq i < \ell$.*

Then, $\tau(n, k) \leq \sum_{0 \leq i \leq w} |F_{i \bmod \ell}|$, where w is an integer satisfying $\sum_{0 \leq i \leq w} r_{i \bmod \ell} \geq k + q + 1$.

Proof. Suppose that there are p_i edge-disjoint i-folded ascending paths from a vertex u to a vertex v for $0 \leq i \leq s$ such that $p = \sum_{i=0}^{s} p_i$ and $q \geq \sum_{0 \leq i \leq s} i p_i$. Let $S(u, v)$ be the set of such p edge-disjoint paths from u to v in (G, ρ). Also, let $h = \lfloor \frac{w}{\ell} \rfloor$.

---

[1] Haddad et al. define an s-folded ascending path as a path which can be decomposed into *at most* s ascending subpaths.

- Case 1: $h \geq s$. From Lemma 1, we can see that based on p_i edge-disjoint i-folded ascending paths from u to v in (G, ρ), there are $p_i(h - i)$ edge-disjoint ascending paths from u to v in $h \cdot (G, \rho)$ for each i. Thus, there are $\sum_{0 \leq i \leq s} p_i(h - i) \geq hp - q$ edge-disjoint ascending paths from u to v in $h \cdot (G, \rho)$.

  Let F_i' be the subset of $E((h+1) \cdot (G, \rho)) - E(h \cdot (G, \rho))$ corresponding to $F_i \subseteq E(G)$ for $0 \leq i < \ell$. Define the ordered graph H_i $(-1 \leq i < \ell)$ as follows. Let $H_{-1} = h \cdot (G, \rho)$. For $0 \leq i < \ell$, let H_i be the ordered graph obtained from H_{i-1} by adding the edges in F_i'. Note that $H_{\ell-1} = (h+1) \cdot (G, \rho)$.

  Let $P = P^{(0)} \odot P^{(1)} \odot \cdots \odot P^{(t)}$ $(0 \leq t \leq s)$ be a t-folded ascending path in $S(u, v)$ such that each $P^{(j)}$ is an ascending subpath. Now consider the path $P(h - t + 1) = P^{(0)}_{h-t+1} \odot P^{(1)}_{h-t+2} \odot \cdots \odot P^{(t)}_{h+1}$ in $(h+1) \cdot (G, \rho)$. If the last edge of $P(h - t + 1)$ is in F_i', then $P(h - t + 1)$ exists in H_i. Note that all the edges in $P^{(t)}_{h+1}$ are in $\cup_{0 \leq j \leq i} F_j'$, since $P^{(t)}_{h+1}$ is an ascending subpath. There are r_i paths in $S(u, v)$ whose last edge is in F_i. Therefore, compared to H_{i-1}, r_i ascending paths are newly constructed in H_i while preserving edge-disjointness. Consequently, in the ordered graph with $h|E(G)| + \sum_{h\ell \leq i \leq w} |F_{i \bmod \ell}| = \sum_{0 \leq i \leq w} |F_{i \bmod \ell}|$ edges, there are $hp - q + \sum_{h\ell \leq i \leq w} r_{i \bmod \ell} = \sum_{0 \leq i \leq w} r_{i \bmod \ell} - q \geq k + 1$ edge-disjoint ascending paths from u to v.

- Case 2: $h < s$. In this case, there are $\sum_{0 \leq i \leq h} p_i(h-i) \geq hp - q + \sum_{h < i \leq s} p_i(i - h)$ edge-disjoint ascending paths from u to v in $h \cdot (G, \rho)$. The discussion in Case 1 can be similarly applied to t-folded ascending paths in $S(u, v)$ for $0 \leq t \leq h$ while for $h < t \leq s$, any t-folded ascending path is not newly appeared in H_i compared to H_{i-1}. Hence, in the ordered graph with $\sum_{0 \leq i \leq w} |F_{i \bmod \ell}|$ edges, there are at least $hp - q + \sum_{h < i \leq s} p_i(i - h) + (\sum_{h\ell \leq i \leq w} r_{i \bmod \ell} - \sum_{h < i \leq s} p_i) \geq hp - q + \sum_{h\ell \leq i \leq w} r_{i \bmod \ell} = \sum_{0 \leq i \leq w} r_{i \bmod \ell} - q \geq k + 1$ edge-disjoint ascending paths from u to v.

□

3   Fault-Tolerant Gossip Graphs Based on Hypercubes

The p-dimensional hypercube Q_p is the graph whose vertex set is the set of all 0-1 vectors of length p and two vertices are adjacent if and only if their coordinates differ in exactly one place. For an edge $e = \{u, v\}$ in Q_p, where $u = (u_1, u_2, \ldots, u_p)$ and $v = (v_1, v_2, \ldots, v_p)$, and an integer $i \in \{1, 2, \ldots, p\}$, if $u_i \neq v_i$ and $u_j = v_j$ for all $j \neq i$, then the *dimension* $dim(e)$ of the edge e is defined to be i. Haddad et al. [5] showed that there are p inner vertex-disjoint paths from any vertex to any other vertex in an ordered graph based on Q_p such that the folded number of each path is at most one, which implies that $\tau(n, k) \leq (\lceil \frac{k+1}{p} \rceil + 1) \frac{np}{2}$ by Corollary 1.

Applying Theorem 1 to the vertex-disjoint paths in the hypercube shown in [5], we can improve their upper bound result.

Theorem 2. $\tau(n, k) \leq \frac{n \log_2 n}{2} + \frac{nk}{2}$ *if n is a power of two.*

Proof. Let $n = 2^p$. Let F_{i-1} be the set of edges with dimension i in Q_p for $1 \leq i \leq p$. For each vertex $v = (v_1, v_2, \ldots, v_p)$ in Q_p, let $bin(v) = \sum_{1 \leq i \leq p} 2^{p-i} v_i$ and $bin_j(v) = \sum_{1 \leq i < j} 2^{p-1-i} v_i + \sum_{j < i \leq p} 2^{p-i} v_i$ for $1 \leq j \leq p$. Define the edge-ordering ρ of Q_p as follows: for each edge $e = \{u, v\}$, $\rho(e) = 2^{p-1}(dim(e) - 1) + bin_{dim(e)}(u) + 1$. By definition, it holds that for any two edges e_1 and e_2, if $dim(e_1) < dim(e_2)$, then $\rho(e_1) < \rho(e_2)$. Let $u = (u_1, u_2, \ldots, u_p), v = (v_1, v_2, \ldots, v_p) \in V(Q_p)$. Also, let $A = \{i \mid u_i = v_i, 1 \leq i \leq p\} = \{a_1, a_2, \ldots, a_{|A|}\}$ and $B = \{i \mid u_i \neq v_i, 1 \leq i \leq p\} = \{b_1, b_2, \ldots, b_{|B|}\}$ such that $a_1 < a_2 < \cdots < a_{|A|}$ and $b_1 < b_2 < \cdots < b_{|B|}$. A path from u to v in Q_p can be represented by a sequence of a_i's and b_i's, i.e., a sequence of the dimensions of edges. For $1 \leq j \leq |A|$, define i_j to be the integer satisfying $b_{i_j} < a_j < b_{i_j + 1}$ if $b_1 < a_j < b_{|B|}$. If $a_j < b_1$ or $b_{|B|} < a_j$, then let $i_j = 0$. Now consider the following paths which were previously defined in [5]: $P_{b_{|B|}} = (b_1, b_2, \ldots, b_{|B|})$, $P_{b_1} = (b_2, \ldots, b_{|B|}, b_1)$, $\ldots$, $P_{b_{|B|-1}} = (b_{|B|}, b_1, \ldots, b_{|B|-1})$, $P_{a_1} = (a_1, b_{i_1+1}, \ldots, b_{|B|}, b_1, \ldots, b_{i_1}, a_1)$, $P_{a_2} = (a_2, b_{i_2+1}, \ldots, b_{|B|}, b_1, \ldots, b_{i_2}, a_2)$, $\ldots$, $P_{a_{|A|}} = (a_{|A|}, b_{i_{|A|}+1}, \ldots, b_{|B|}, b_1, \ldots, b_{i_{|A|}}, a_{|A|})$. As shown in [5], these paths are inner vertex-disjoint, thus edge-disjoint, each other. The first path is 0-folded and the others are 1-folded. Also, the dimensions of the last edges of the paths are distinct, i.e., there is exactly one path whose last edge is in F_i, where $|F_i| = n/2$, for $0 \leq i < p$. Therefore, by setting $q = p - 1$ and $r_i = 1$ for $0 \leq i < p$ in Theorem 1, we obtain $\tau(n, k) \leq \frac{n(p+k)}{2}$.   □

Haddad et al. defined the (h, p)-*hypercube* $Q_{h,p}$ to be the graph obtained from h copies of Q_p by selecting one vertex from each Q_p and identifying such h vertices as a single vertex called the *center vertex*. Let $h = \lceil (n-1)/(2^p - 1) \rceil$. The number $\tilde{n}$ of vertices in the (h, p)-hypercube is $h(2^p - 1) + 1 \geq n$. It is not difficult to see that $\tau(n, k) \leq \tau(\tilde{n}, k)$. Then, they showed that $\tau(\tilde{n}, k) \leq \left(\frac{k}{2} + 2p \right) \left((n-1) + \frac{n-1}{2^p - 1} + 2^p \right)$. In what follows, we show that the factor $(\frac{k}{2} + 2p)$ in the upper bound by Haddad et al. can be replaced with a smaller factor $(\frac{k}{2} + p)$.

Theorem 3. $\tau(n, k) \leq \left(\frac{k}{2} + p \right) \left((n-1) + \frac{n-1}{2^p - 1} + 2^p \right)$, *where p is any integer between 1 and $\log_2 n$ inclusive.*

Proof. For the (h, p)-hypercube $Q_{h,p}$, let $Q_p^1, Q_p^2, \ldots, Q_p^h$ be the h hypercubes by which $Q_{h,p}$ is constructed. Define the edge-ordering ρ of $Q_{h,p}$ as follows: for each edge $e = \{u, v\} \in E(Q_p^i)$, $1 \leq i \leq h$, $\rho(e) = h2^{p-1}(dim(e) - 1) + bin_{dim(e)}(u) + (i-1)2^{p-1} + 1$. By definition, it holds that for any two edges e_1 and e_2, if $dim(e_1) < dim(e_2)$, then $\rho(e_1) < \rho(e_2)$. Let u and v be any two vertices of $Q_{h,p}$ and x the center vertex of $Q_{h,p}$. Suppose that $u \in V(Q_p^i)$ and $v \in V(Q_p^j)$. We construct p edge-disjoint paths from u to v in $Q_{h,p}$ by concatenating edge-disjoint paths from u to x in Q_p^i and edge-disjoint paths from x to v in Q_p^j. The set S_i (S_j) of such edge-disjoint paths in Q_p^i (Q_p^j) is defined similarly to that shown in the proof of Theorem 2. For $1 \leq i \leq p$, P_i (respectively, P_i') denotes the path in S_i in which the dimension of the last edge is i (respectively, the path in S_j in which the dimension of the first edge is i). Then, we define p edge-disjoint paths from u to v as follows: $R_1 = P_1 \odot P_2'$, $R_2 = P_2 \odot P_3'$, $\ldots$, $R_{p-1} = P_{p-1} \odot P_p'$,

$R_p = P_p \odot P_1'$. In the set $S_i = \{P_i \mid 1 \le i \le p\}$ $(S_j = \{P_i' \mid 1 \le i \le p\})$, there is exactly one ascending path and the remaining paths are 1-folded ascending paths. The folded number of R_i is the sum of the folded numbers of P_i and P_{i+1}' for $1 \le i < p$, while the folded number of R_p is just one more than the sum of the folded numbers of P_p and P_1'. Thus, the sum of the folded numbers of R_i's is $2p - 1$.

In Theorem 1, by letting F_i be the set of edges with the dimension i in $\cup_{1 \le j \le h} E(Q_p^j)$, i.e., $|F_i| = h|V(Q_p)|/2$, and $r_i = 1$ for $1 \le i \le p$, it holds that
$$\tau(\tilde{n}, k) \le (k + 2p)h|V(Q_p)|/2 = (k + 2p)\lceil\tfrac{n-1}{2^p-1}\rceil 2^{p-1} \le (k + 2p)(\tfrac{n-1}{2^p-1} + 1)2^{p-1} = (\tfrac{k}{2} + p)((n-1) + \tfrac{n-1}{2^p-1} + 2^p).$$
$\square$

4   Fault-Tolerant Gossip Graphs Based on Circulant Graphs

In this section, we assume that the number n of vertices is not a power of two.

Definition 1. *The ordered graph $(R(n), \rho_n)$ is defined as follows:*
$$\begin{cases} V(R(n)) = \{0, 1, \ldots, n-1\}, \\ E(R(n)) = \{\{u, v\} \mid v \equiv u + 2^i \pmod{n},\ 0 \le i \le \lfloor\log n\rfloor - 1\}. \end{cases}$$

$\rho_n(\{u, u + 2^i\}) = n(\lfloor\log_2 n\rfloor - i - 1) + u + 1$, *for* $0 \le u < n$, $0 \le i \le \lfloor\log n\rfloor - 1$.

Note that $R(n)$ is $2\lfloor\log n\rfloor$-regular. We denote a path in $R(n)$ by a sequence of vertices instead of a sequence of edges. The *span* $sp(c)$ of an edge $c = \{u, u + 2^i\pmod{n}\}$ in $R(n)$ is defined to be i. By definition, it holds that for any two edges e_1 and e_2, if $sp(e_1) > sp(e_2)$, then $\rho_n(e_1) < \rho_n(e_2)$. In what follows, the ordered graph $(R(n), \rho_n)$ is simply abbreviated to $R(n)$.

Let π_c be the cyclic permutation $(0\ 1\ \cdots\ n-1)$ on $V(R(n))$. Also, let π_m be the permutation $(0)(1\ n-1)(2\ n-2)\cdots(\tfrac{n-1}{2}\ \tfrac{n+1}{2})$ if n is odd, $(0)(1\ n-1)(2\ n-2)\cdots(\tfrac{n-2}{2}\ \tfrac{n+2}{2})(\tfrac{n}{2})$ if n is even. Then, it can be easily checked that π_c and π_m are automorphisms on $R(n)$. Thus, without loss of generality, it is sufficient to consider edge-disjoint paths from 0 to v for any $0 < v \le n/2$. Let $v \in V(R(n))$ such that $v = \sum_{i=1}^{\ell} s_i 2^{t_i}$, where $t_i \ge 0$ and $s_i \in \{-1, 1\}$ for $1 \le i \le \ell$. We denote by $P(0; s_1 t_1, s_2 t_2, \ldots, s_\ell t_\ell; v)$ the path $(0, s_1 2^{t_1}\pmod{n}, s_1 2^{t_1} + s_2 2^{t_2}\pmod{n}, \ldots, \sum_{i=1}^{\ell} s_i 2^{t_i})$ from 0 to v.

Now define the operations L_i, R_i, and X on paths in $R(n)$ as follows: for a given path $P = P(0; s_1 t_1, s_2 t_2, \ldots, s_\ell t_\ell; v)$,
$L_i(P) = P(0; -i, s_1 t_1, s_2 t_2, \ldots, s_\ell t_\ell, i; v)$, for $0 \le i \le \lfloor\log_2 n\rfloor - 1$,
$R_i(P) = P(0; i, s_1 t_1, s_2 t_2, \ldots, s_\ell t_\ell, -i; v)$, for $0 \le i \le \lfloor\log_2 n\rfloor - 1$,
$X(P) = P(0; s_\ell t_\ell, s_2 t_2, \ldots, s_{\ell-1} t_{\ell-1}, s_1 t_1; v)$.
Note that $X(P)$ is obtained from P by exchanging $s_1 t_1$ and $s_\ell t_\ell$.

Let $\mathcal{S}(P) = \{P\} \cup \{L_i(P) \mid i \ne t_1, t_\ell\} \cup \{R_i(P) \mid i \ne t_1, t_\ell\} \cup \{X(P)\}$. Then, we can check that the following lemma holds.

Lemma 2. *Let $P = P(0; s_1t_1, s_2t_2, \ldots, s_\ell t_\ell; v)$ be a path in $R(n)$. If all t_i's are distinct and every element in $\mathcal{S}(P)$ is a path, then all the elements in $\mathcal{S}(P)$ are pairwise edge-disjoint paths.*

In order to prove our upper bound result, we construct $2\lfloor \log_2 n \rfloor$ edge-disjoint paths from 0 to v in $R(n)$ for $0 < v \leq n/2$ so that the sum of their folded numbers is at most $4\lfloor \log_2 n \rfloor - 2$. Let $v = \sum_{i=1}^{\ell} 2^{t_i}$, where $\lfloor \log_2 n \rfloor - 1 \geq t_1 > t_2 > \cdots > t_\ell \geq 0$. Our constructions for edge-disjoint paths are divided into the following seven cases. We will basically explain each construction in this order.

- Case 1: $\ell \geq 2$ and $t_1 = \lfloor \log_2 n \rfloor - 1$,
- Case 2: $\ell \geq 2$, $t_1 \leq \lfloor \log_2 n \rfloor - 2$ and $v \neq 2^{t_\ell}(2^{t_1 - t_\ell + 1} - 1)$,
- Case 3: $\ell \geq 2$, $t_1 = \lfloor \log_2 n \rfloor - 2$ and $v = 2^{t_\ell}(2^{t_1 - t_\ell + 1} - 1)$,
- Case 4: $\ell \geq 2$, $t_1 \leq \lfloor \log_2 n \rfloor - 3$ and $v = 2^{t_\ell}(2^{t_1 - t_\ell + 1} - 1)$,
- Case 5: $\ell = 1$ and $t_1 = \lfloor \log_2 n \rfloor - 1$,
- Case 6: $\ell = 1$ and $t_1 = \lfloor \log_2 n \rfloor - 2$,
- Case 7: $\ell = 1$ and $t_1 \leq \lfloor \log_2 n \rfloor - 3$.

Define $P^* = P(0; t_1, t_2, \ldots, t_\ell; v)$ and call it the *standard path* from 0 to v. Note that P^* is an ascending path from 0 to v. Clearly, $X(P^*)$ is a path. Also, any $L_i(P^*)$ (respectively, $R_i(P^*)$) in $\mathcal{S}(P^*)$ is also a path, since $2^i \neq 2^{t_1} + 2^{t_2} + \cdots + 2^{t_k}$ $(1 < k \leq \ell)$ (respectively, $2^i \neq 2^{t_{k'}} + \cdots + 2^{t_\ell}$ $(1 \leq k' < \ell)$). Thus, every element in $\mathcal{S}(P^*)$ is a path from 0 to v. Then, we call $L_i(P^*)$, $R_i(P^*)$, and $X(P^*)$, the *i-left-shift path* of P^*, the *i-right-shift path* of P^*, and the *exchange path* of P^*, respectively. From Lemma 2, the following lemma holds.

Lemma 3. *All the paths in $\mathcal{S}(P^*)$ are pairwise edge-disjoint.*

By definition, P^* is an ascending path and the folded number of any other path in $\mathcal{S}(P^*)$ is at most two. Next, we define the *opposite path* OP^* as follows:

$$
OP^* = \begin{cases}
P(0; -(\lfloor \log_2 n \rfloor - 1), -y_1, -y_2, \ldots, -y_{\ell'}, -(\lfloor \log_2 n \rfloor - 1); v) \\
\quad \text{if } v + 2^{\lfloor \log n \rfloor} \leq n \text{ and } n - v - 2^{\lfloor \log n \rfloor} = \sum_{1 \leq i \leq \ell'} 2^{y_i} \\
\quad \text{where } \lfloor \log_2 n \rfloor - 1 \geq y_1 > y_2 > \cdots > y_{\ell'} \geq 0, \\
P(0; -(\lfloor \log_2 n \rfloor - 1), z_1, z_2, \ldots, z_{\ell''}, -(\lfloor \log_2 n \rfloor - 1); v) \\
\quad \text{if } v + 2^{\lfloor \log n \rfloor} > n \text{ and } v + 2^{\lfloor \log n \rfloor} - n = \sum_{1 \leq i \leq \ell''} 2^{z_i} \\
\quad \text{where } \lfloor \log_2 n \rfloor - 1 \geq z_1 > z_2 > \cdots > z_{\ell''} \geq 0.
\end{cases}
$$

Lemma 4. *Let $\ell \geq 2$. If $t_1 = \lfloor \log_2 n \rfloor - 1$, then OP^* is edge-disjoint to any path in $\mathcal{S}(P^*)$.*

When $t_1 \leq \lfloor \log_2 n \rfloor - 2$, we can define the *jumping path* JP^* as follows:
$JP^* = P(0; -t_1, t_1 + 1, t_1, t_2, \ldots, t_\ell, -t_1; v)$.

Lemma 5. *Let $\ell \geq 2$. If $t_1 < \lfloor \log_2 n \rfloor - 1$, then JP^* is edge-disjoint to any path in $\mathcal{S}(P^*)$.*

Note that the folded numbers of OP^* and JP^* are at most two. From Lemmas 3,4, and 5, it is sufficient to present one more path edge-disjoint to any path in $\mathcal{S}(P^*) \cup \{OP^*\}$ if $\ell \geq 2$ and $t_1 = \lfloor \log_2 n \rfloor - 1$, $\mathcal{S}(P^*) \cup \{JP^*\}$ if $\ell \geq 2$, $t_1 \leq \lfloor \log_2 n \rfloor - 2$, and $v \neq 2^{t_\ell}(2^{t_1 - t_\ell + 1} - 1)$. By defining the following path called the *merged path*, our constructions are completed for Cases 1 and 2.

$$MP^* = \begin{cases} P(0; -t_\ell, t_1, \ldots, t_k, t_\ell, t_{k+1} + 1, -t_\ell; v), \\ \quad \text{if there is a positive integer } j \text{ such that } t_j > t_{j+1} + 1 \\ \quad \text{and } k = \max\{j \mid t_j > t_{j+1} + 1, 1 \leq j < \ell\}, \\ P(0; -t_\ell, t_1, t_\ell, t_1, -t_\ell; v), \text{ otherwise.} \end{cases}$$

Note that MP^* is a 2-folded ascending path.

Proposition 1. *Suppose that $\ell \geq 2$. If $t_1 = \lfloor \log_2 n \rfloor - 1$, or $t_1 \leq \lfloor \log_2 n \rfloor - 2$ and $v \neq 2^{t_\ell}(2^{t_1 - t_\ell + 1} - 1)$, then there are $2\lfloor \log_2 n \rfloor$ edge-disjoint paths from 0 to v in $R(n)$ such that the folded number of one path is 0 and the folded numbers of remaining paths are at most two.*

Next, we consider Cases 3 and 4, i.e., $\ell \geq 2$, $t_1 \leq \lfloor \log_2 n \rfloor - 2$ and $v = 2^{t_\ell}(2^{t_1 - t_\ell + 1} - 1)$. In these cases, we cannot employ the set of paths for Case 2, since JP^* and MP^* are no longer edge-disjoint under the condition that $v = 2^{t_\ell}(2^{t_1 - t_\ell + 1} - 1)$. Then, we consider the path $P' = P(0; t_1 + 1, -t_\ell; v)$ from 0 to v. Let $\mathcal{S}(P') = \{P'\} \cup \{L_i(P') \mid i \neq t_1 + 1, t_\ell\} \cup \{R_i(P') \mid i \neq t_1 + 1, t_\ell\} \cup \{X(P')\}$. Clearly, P' is an ascending path and the folded number of any other path in $\mathcal{S}(P')$ is at most two. From Lemma 2, all the paths in $\mathcal{S}(P')$ are pairwise edge-disjoint.

Lemma 6. $2\lfloor \log_2 n \rfloor - 2$ paths in $\mathcal{S}(P')$ are pairwise edge-disjoint.

Similarly to Cases 1 and 2, we can add OP^* or JP' to $\mathcal{S}(P')$ while preserving edge-disjointness, where JP' is defined as follows:

$$JP' = P(0; -(t_1 + 1), t_1 + 2, t_1 + 1, -t_\ell, -(t_1 + 1); v).$$

The folded number of JP' is two.

Lemma 7. *Suppose that $\ell \geq 2$, $t_1 = \lfloor \log_2 n \rfloor - 2$ and $v = 2^{t_\ell}(2^{t_1 - t_\ell + 1} - 1)$. Then OP^* is edge-disjoint to any path in $\mathcal{S}(P')$.*

Lemma 8. *Suppose that $\ell \geq 2$, $t_1 \leq \lfloor \log_2 n \rfloor - 3$ and $v = 2^{t_\ell}(2^{t_1 - t_\ell + 1} - 1)$. Then JP' is edge-disjoint to any path in $\mathcal{S}(P')$.*

From Lemmas 6,7 and 8, it is sufficient to construct one more path edge-disjoint to any path in $\mathcal{S}(P') \cup \{OP^*\}$ if $\ell \geq 2$, $t_1 = \lfloor \log_2 n \rfloor - 2$ and $v = 2^{t_\ell}(2^{t_1 - t_\ell + 1} - 1)$, $\mathcal{S}(P') \cup \{JP'\}$ if $\ell \geq 2$, $t_1 \leq \lfloor \log_2 n \rfloor - 3$ and $v = 2^{t_\ell}(2^{t_1 - t_\ell + 1} - 1)$. Define the *right-cyclic-shift path* CP^* of the standard path P^* from 0 to v as $CP^* = P(0; t_\ell, t_1, \ldots, t_{\ell-2}, t_{\ell-1}; v)$. Besides, define the *divided path* DP^* of CP^* as $DP^* = P(0; t_\ell, t_1, \ldots, t_{\ell-2}, t_\ell, t_\ell; v)$. The folded number of DP^* is two. It can be checked that DP^* is edge-disjoint to any path in $\mathcal{S}(P') \cup \{OP^*\}$ ($\mathcal{S}(P') \cup \{JP'\}$) except for $L_{t_{\ell-1}}(P')$. Since DP^* and $L_{t_{\ell-1}}(P')$ have an edge with span t_ℓ in common, instead of $L_{t_{\ell-1}}(P')$, we employ the $t_{\ell-1}$-left-shift path of CP^*: $L_{t_{\ell-1}}(CP^*) = P(0; -t_{\ell-1}, t_\ell, t_1, \ldots, t_{\ell-2}, t_{\ell-1}, t_{\ell-1}; v)$. The folded number of $L_{t_{\ell-1}}(CP^*)$ is two.

Proposition 2. *Suppose that $\ell \geq 2$. If $t_1 \leq \lfloor \log_2 n \rfloor - 2$ and $v = 2^{t_\ell}(2^{t_1-t_\ell+1}-1)$, then there are $2\lfloor \log_2 n \rfloor$ edge-disjoint paths from 0 to v in $R(n)$ such that the folded number of one path is 0 and the folded numbers of remaining paths are at most two.*

Finally, we consider the case that $\ell = 1$ which is divided into Cases 5,6, and 7. In these cases, the standard path from 0 to v consists of one edge: $P^* = (0; t_1; v)$. From Lemma 2, $2\lfloor \log_2 n \rfloor - 1$ paths in $\mathcal{S}(P^*)$ are pairwise edge-disjoint. Thus, we need one more path edge-disjoint to any path in $\mathcal{S}(P^*)$. When $t_1 = \lfloor \log_2 n \rfloor - 1$, we can add OP^* to $\mathcal{S}(P^*)$ while preserving edge-disjointness. When $t_1 \leq \lfloor \log_2 n \rfloor - 3$, we can employ the *wide jumping path*:

$$WP^* = P(0; -t_1, t_1 + 2, t_1, -(t_1+1), -t_1).$$

The folded number of WP^* is two. For two vertices x and y, if $(x-y)(\bmod\ n) \leq n/2$ and $(x-y)(\bmod\ n) = \sum_{i=1}^{m} 2^{u_i}$, where $\lfloor \log_2 n \rfloor - 1 \geq u_1 > u_2 > \cdots > u_m \geq 0$, we can define the *opposite standard path* $\tilde{P}^*(x; y)$ from x to y as follows:

$$\tilde{P}^*(x; y) = P(x; -u_1, -u_2, \ldots, -u_m; y)$$

When $t_1 = \lfloor \log_2 n \rfloor - 2$, we use the *divided opposite path* OP^{**} defined as follows:

$$OP^{**} = \begin{cases} \tilde{P}^*(0; -2^{t_1+1}+1) \odot P(-2^{t_1+1}+1; -(t_1+1); 2^{t_1+2}) \odot \tilde{P}^*(2^{t_1+2}; v) \\ \qquad \text{if } n = 2^{t_1+3} - 1, \\ \tilde{P}^*(0; -2^{t_1+1}+1) \odot P(-2^{t_1+1}+1; -t_1, -(t_1+1), -t_1; v) \\ \qquad \text{if } n = 7 \cdot 2^{t_1} - 1, \\ \tilde{P}^*(0; -2^{t_1+1}+1) \odot \tilde{P}^*(-2^{t_1+1}+1; 2^{t_1+1}) \odot P(2^{t_1+1}; -t_1; v) \\ \qquad \text{if } n > 5 \cdot 2^{t_1} \text{ and } n \neq 7 \cdot 2^{t_1} - 1, 2^{t_1+3} - 1, \\ P(0; -t_1, -(t_1+1), -t_1; v) \qquad \text{if } n = 5 \cdot 2^{t_1}, \\ P(0; -t_1, -(t_1+1), 0, -t_1; v) \qquad \text{if } n = 5 \cdot 2^{t_1} - 1, \\ \tilde{P}^*(0; -2^{t_1+1}+1) \odot \tilde{P}^*(-2^{t_1+1}+1; 2^{t_1+1}) \odot P(2^{t_1+1}, -t_1; v) \\ \qquad \text{if } n \leq 5 \cdot 2^{t_1} - 2. \end{cases}$$

The folded number of OP^{**} is at most two. We can show that $\mathcal{S}(P^*) \cup \{OP^*\}$, $\mathcal{S}(P^*) \cup \{OP^{**}\}$, and $\mathcal{S}(P^*) \cup \{WP^*\}$ are sets of $2\lfloor \log_2 n \rfloor$ edge-disjoint paths for Cases 5,6, and 7, respectively.

Proposition 3. *Suppose that $\ell = 1$. Then, there are $2\lfloor \log_2 n \rfloor$ edge-disjoint paths from 0 to v in $R(n)$ such that the folded number of one path is 0 and the folded numbers of remaining paths are at most two.*

From Propositions 1, 2, and 3, it follows that in $R(n)$, there are $2\lfloor \log n \rfloor$ edge-disjoint paths from any vertex to any other vertex in which one path is an ascending path and the folded number of any other path is at most two. By letting F_i be the set of edges with span $\lfloor \log n \rfloor - i - 1$ for $0 \leq i < \lfloor \log n \rfloor$, where $|F_i| = n$, the last edges of two paths are in F_i, i.e., $r_i = 2$, for each i. Since the sum of the folded numbers of paths is at most $2(2\lfloor \log_2 n \rfloor - 1)$, from Theorem 1, we have the following theorem.

Theorem 4. $\tau(n, k) \leq 2n\lfloor \log_2 n \rfloor + n\lceil \frac{k-1}{2} \rceil$ *if n is not a power of two.*

When k is even, we can slightly improve the number of edges by the following observation. First we construct k edge-disjoint ascending paths for which we need $2n\lfloor \log_2 n \rfloor + \frac{n(k-2)}{2}$ edges. Since the spanning subgraph with edge set $F_{(k/2-1) \bmod \lfloor \log_2 n \rfloor}$ is 2-regular, it consists of t disjoint cycles with $\frac{n}{t}$ vertices. By adding edges in each cycle alternately, i.e., at most $t\lceil \frac{n}{2t} \rceil \le \frac{2n}{3}$ edges in $F_{(k/2-1) \bmod \lfloor \log_2 n \rfloor}$, we can obtain one more ascending path which is edge-disjoint to any path in the k edge-disjoint ascending paths.

5   A Lower Bound

In order to show our lower bound, we need two results on broadcasting. A *k-fault-tolerant broadcast graph* is an ordered graph in which there are $k+1$ edge-disjoint ascending paths from a vertex to any other vertex. Let $\mu(n, k)$ be the minimum number of edges in a k-fault-tolerant broadcast graph. Berman and Hawrylycz [2] determined $\mu(n, k)$ for any n and k.

Theorem 5. (Berman and Hawrylycz [2])

$$\mu(n, k) = \begin{cases} \left\lceil \left(\frac{k+2}{2}\right)(n-1)\right\rceil, & \text{if } k \le n-2, \\ \left\lceil \left(\frac{k+1}{2}\right)n\right\rceil, & \text{if } k > n-2. \end{cases}$$

For a vertex v in an ordered graph (G, ρ), the *v-broadcast number* of (G, ρ) is the number of vertices $w(\ne v)$ such that there is an ascending path from v to w. The *broadcast number* of (G, ρ) is the minimum v-broadcast number over all vertices v of (G, ρ). Berman and Paul presented an upper bound on the broadcast number of an ordered tree. (Besides, they showed that there exists an ordered tree with n vertices whose broadcasting number is equal to $\lfloor \log_2 n \rfloor$.)

Theorem 6. (Berman and Paul [3]) *Let (T, ρ) be an ordered tree with n vertices. Then, the broadcasting number of (T, ρ) is at most $\lfloor \log_2 n \rfloor$.*

For each ordered pair of vertices u and v in (G, ρ), define $\eta_{(G,\rho)}(u, v)$ to be the maximum number of edge-disjoint ascending paths from u to v in (G, ρ). Moreover, define $\psi_{(G,\rho)}(v; k) = \max\{0,\ k+1 - \min_{u \in V(G)-\{v\}} \eta_{(G,\rho)}(u, v)\}$ and we call $\psi_{(G,\rho)}(v; k)$ the *defect number* of v with respect to k.

Theorem 7. $\tau(n, k) \ge \left\lceil \frac{3n-5}{2} \right\rceil + \left\lceil \frac{1}{2}\left(nk + \lfloor \frac{n+1}{2} \rfloor - \lfloor \log n \rfloor\right)\right\rceil$ *for $n \ge 3$ and $k \ge 1$.*

Proof. Suppose that (G, ρ) is a k-fault-tolerant gossip with n vertices, where $n \ge 3$ and $k \ge 1$. Let $(G, \rho)_i$ be the ordered spanning subgraph of (G, ρ) having all the edges of order at most i. Now, let $(G, \rho^R)_{\lceil \frac{3}{2}(n-1)\rceil - 1}$ be the ordered graph obtained from $(G, \rho)_{\lceil \frac{3}{2}(n-1)\rceil - 1}$ by reversing the orders of edges, i.e., the edge-ordering ρ^R is defined as $\rho^R(e) = \lceil \frac{3}{2}(n-1) \rceil - \rho(e)$ for each edge e. From Theorem 5, $(G, \rho^R)_{\lceil \frac{3}{2}(n-1)\rceil - 1}$ is not a 1-fault-tolerant broadcast graph. Thus, for any vertex v in $(G, \rho^R)_{\lceil \frac{3}{2}(n-1)\rceil - 1}$, there exists a vertex u such that

$\eta_{(G,\rho^R)_{\lceil \frac{3}{2}(n-1)\rceil-1}}(v,u) \leq 1$. This means that for any vertex v in $(G,\rho)_{\lceil \frac{3}{2}(n-1)\rceil-1}$, there exists a vertex u such that $\eta_{(G,\rho)_{\lceil \frac{3}{2}(n-1)\rceil-1}}(u,v) \leq 1$. Thus, for any vertex v in $(G,\rho)_{\lceil \frac{3}{2}(n-1)\rceil-1}$, $\psi_{(G,\rho)_{\lceil \frac{3}{2}(n-1)\rceil-1}}(v;k) \geq k$.

There exists a connected component that is a tree in $(G,\rho)_{n-1}$. Thus, from Theorem 6, the broadcast number of $(G,\rho)_{n-1}$ is at most $\lfloor \log_2 n \rfloor$. Let x be a vertex in $(G,\rho)_{n-1}$ such that the x-broadcast number is equal to the broadcast number of $(G,\rho)_{n-1}$. Also, let $B_i(x)$ be the set of vertices to which there is an ascending path from x in $(G,\rho)_i$, where $i \geq n-1$. Then, $|B_{n-1}(x)| \leq \lfloor \log_2 n \rfloor$. Adding the edge with order $i+1$ into $(G,\rho)_i$, at most one vertex newly receives the message originated from x. Hence it holds that $|B_{i+1}(x)| \leq |B_i(x)| + 1$. Therefore, $|B_{(G,\rho)_{\lceil \frac{3}{2}(n-1)\rceil-1}}(x)| \leq \lfloor \log_2 n \rfloor + \lceil \frac{3}{2}(n-1) \rceil - n$. Let n_k be the number of vertices with defect number k in $(G,\rho)_{\lceil \frac{3}{2}(n-1)\rceil-1}$. For a vertex w in $(G,\rho)_{\lceil \frac{3}{2}(n-1)\rceil-1}$, if the defect number of w is k, then there must exist an ascending path from x to w unless $w = x$, i.e., $w \in B_{(G,\rho)_{\lceil \frac{3}{2}(n-1)\rceil-1}}(x) \cup \{x\}$. Hence, $n_k \leq \lfloor \log_2 n \rfloor + \lceil \frac{3}{2}(n-1) \rceil - n + 1$.

For a vertex w in $(G,\rho)_{\lceil \frac{3}{2}(n-1)\rceil-1}$, if the defect number of w is k (respectively, $k+1$), then there must exist at least k (respectively, $k+1$) edges with order at least $\lceil \frac{3}{2}(n-1) \rceil$ which are incident to w in (G,ρ). Therefore, $|E(G)| \geq \lceil \frac{3}{2}(n-1) \rceil - 1 + \lceil \frac{1}{2}(kn_k + (k+1)(n-n_k)) \rceil \geq \lceil \frac{3}{2}(n-1) \rceil - 1 + \lceil \frac{1}{2}((k+1)n - n_k) \rceil \geq \lceil \frac{3}{2}(n-1) \rceil - 1 + \lceil \frac{1}{2}((k+1)n - (\lfloor \log_2 n \rfloor + \lceil \frac{3}{2}(n-1) \rceil - n + 1)) \rceil = \lceil \frac{3n-5}{2} \rceil + \lceil \frac{1}{2}(kn - \lfloor \log_2 n \rfloor + \lfloor \frac{1}{2}(n+1) \rfloor) \rceil$. Hence, $\tau(n,k) \geq \lceil \frac{3n-5}{2} \rceil + \lceil \frac{1}{2}(kn - \lfloor \log_2 n \rfloor + \lfloor \frac{1}{2}(n+1) \rfloor) \rceil$.

$\square$

References

1. Baker, B., Shostak, R.: Gossips and telephones. Discrete Math. 2, 191–193 (1972)
2. Berman, K.A., Hawrylycz, M.: Telephone problems with failures. SIAM J. Alg. Disc. Meth. 7, 13–17 (1986)
3. Berman, K.A., Paul, J.L.: Verifiable broadcasting and gossiping in communication networks. Discrete Applied Math 118, 293–298 (2002)
4. Bumby, R.T.: A problem with telephones. SIAM J. Alg. Disc. Meth. 2, 13–18 (1981)
5. Haddad, R.W., Roy, S., Schäffer, A.A.: On gossiping with faulty telephone lines. SIAM J. Alg. Disc. Meth. 8, 439–445 (1987)
6. Hajnal, A., Milner, E.C., Szemerédi, E.: A cure for the telephone disease. Canad. Math. Bull. 15, 447–450 (1976)
7. Hedetniemi, S.M., Hedetniemi, S.T., Liestman, A.L.: A survey of gossiping and broadcasting in communication networks. Networks 18, 319–349 (1988)
8. Hou, Z., Shigeno, M.: New bounds on the minimum number of calls in failure-tolerant gossiping. Networks 53, 35–38 (2009)
9. Pelc, A.: Fault-tolerant broadcasting and gossiping in communication networks. Networks 28, 143–156 (1996)
10. Seress, Á.: Quick gossiping by conference calls. SIAM J. Disc. Math. 1, 109–120 (1988)
11. Tijdeman, R.: On a telephone problem. Nieuw Arch. Wisk. 3, 188–192 (1971)
12. West, D.B.: Gossiping without duplicate transmissions. SIAM J. Alg. Disc. Meth. 3, 418–419 (1982)

Parameterized Two-Player Nash Equilibrium

Danny Hermelin[1], Chien-Chung Huang[2],
Stefan Kratsch[3], and Magnus Wahlström[1]

[1] Max-Planck-Institute for Informatics, Saarbrücken, Germany
{hermelin,wahl}@mpi-inf.mpg.de
[2] Humboldt-Universität zu Berlin, Germany
villars@informatik.hu-berlin.de
[3] Utrecht University, Utrecht, The Netherlands
s.kratsch@uu.nl

Abstract. We study the problem of computing Nash equilibria in a
two-player normal form (bimatrix) game from the perspective of parameterized complexity. Recent results proved hardness for a number of
variants, when parameterized by the support size. We complement those
results, by identifying three cases in which the problem becomes fixed-
parameter tractable. Our results are based on a graph-theoretic representation of a bimatrix game, and on applying graph-theoretic tools on this
representation.

1 Introduction

Algorithmic game theory is a quite recent yet rapidly developing discipline that
lies at the intersection of computer science and game theory. The emergence of
the internet has given rise to numerous applications in this area such as online
auctions, online advertising, and search engine page ranking, where humans and
computers interact with each other as selfish agents negotiating to maximize
their own payoff utilities. The amount of research spent in attempting to devise
computational models and algorithms for studying these types of interactions has
been overwhelming in recent years; unsurprisingly perhaps, when one considers
the economical rewards available in this venture.

The central problem in algorithmic game theory is that of computing a *Nash
equilibrium*, a set of strategies for each player in a given game, where no player
can gain by changing his strategy when all other players strategies remain fixed.
This problem is so important because Nash equilibria provide a good way to pre-
dict the outcomes of many of the scenarios described above, and other scenarios
as well. Furthermore, Nash's Theorem states that for any finite game a mixed
Nash equilibrium always exists. However, for this concept to be meaningful for
predicting behaviors of rational agents which are in many cases computers, a
natural prerequisite is for it to be computable. This led researchers such as Papadimitriou to dub the problem of computing Nash equilibria as one of the most
important complexity problems of our time [31].

The initial breakthrough in determining the complexity of computing Nash
equilibria was made by Daskalakis, Goldberg, and Papadimitriou [12,24]. These

P. Kolman and J. Kratochvíl (Eds.): WG 2011, LNCS 6986, pp. 215–226, 2011.
© Springer-Verlag Berlin Heidelberg 2011

two papers introduced a reduction technique which was used by the authors for showing that computing a Nash equilibrium in a four player game is PPAD-complete. Shortly afterwards, this hardness result was simultaneously extended to three player games by Daskalakis and Papadimitriou [16], and by Chen and Deng [6]. The case of two player (bimatrix) games was finally cracked a year later by Chen and Deng [7], who proved it to be PPAD-complete. This implied that the existence of a polynomial-time algorithm for the core case of bimatrix games is unlikely.

Since the result of Chen and Deng [7], the focus on computing Nash equilibria in bimatrix games was directed either towards finding approximate Nash equilibria [4,8,9,10,13,14,27,28,29,32], or towards finding special cases where exact equilibria can be computed in polynomial time [2,9,11,26,28,29]. Nevertheless, for general bimatrix games the best known algorithm for computing either approximate or exact equilibria still requires exponential time in the worst case. Interestingly enough, if the support of both players is known, one can compute a Nash equilibrium by solving a linear program. This implies the following:

Theorem 1 ([30]). *A Nash equilibrium in a bimatrix game, where the support sizes are bounded by k, can be computed in $n^{O(k)}$ time.*

Due to the central role that the algorithm of Theorem 1 plays in computing exact and approximate Nash equilibria, it is natural to ask whether one can improve on its running-time substantially. In particular, can we remove the dependency on the support size from the exponent? The standard framework for answering such questions is that of parameterized complexity theory [17,21]. Estivill-Castro and Parsa initiated the study of computing Nash equilibria in this context [19,20]. They showed that when the support size is taken as a parameter, the problem is W[2]-hard even in certain restricted settings. The implication of their result is a negative answer to the above question. In particular, combining their reduction with the results of Chen *et al.* [5] gives a sharp contrast to Theorem 1 above.

Theorem 2 ([19]). *Unless* FPT=W[1], *there is no $n^{o(k)}$ time algorithm for computing a Nash equilibrium with support size at most k in a bimatrix game.*

The consequence of Theorem 2 above is devastating in the sense that for large enough games that have equilibriums with reasonably small supports, the task of computing equilibria already becomes infeasible. The main motivation of this paper is to find scenarios where one can circumvent this. Our goal is thus to identify natural parameters which govern the complexity of computing Nash equilibria, and which can help in devising feasible algorithms. We believe that this direction can prove to be fruitful in the quest for understanding the computational limitations of this fundamental problem. Indeed, prior to our work, Kalyanaraman and Umans [26] provided a fixed-parameter algorithm for finding equilibrium in bimatrix games whose matrices have small rank (and some additional constraints).

Our techniques are based on considering a natural graph-theoretic representation of bimatrix games. This is done by taking the union of the underlying boolean matrix of the two given payoff matrices, and considering this matrix as the biadjacency matrix of a bipartite graph. A similar approach was taken in [11], and in particular in [2] where games that have an underlying planar graph structure were considered. Our work complements both these results as will be explained further on, and further exemplifies the strength of a graph-theoretical approach when computing Nash equilibria in bimatrix games.

A natural class of games that has a convenient interpretation in the graph-theoretic context is the class of ℓ-sparse games [9,11,15]. Here each column and row in both payoff matrices of the game have at most ℓ non-zero entries. An initial tempting approach in these types of games would be to try to devise a parameterized algorithm with ℓ taken as a single parameter. However, Chen, Deng, and Teng [9] showed that unless PPAD = P, there is no algorithm for computing an ε-approximate equilibrium for a 10-sparse game in time polynomial both in ε and n (although algorithms polynomial only in n do in fact exist [9,15]). Thus, such an FPT algorithm cannot exist unless PPAD is in P. We complement this result by showing that if ℓ is taken as a parameter, and the size of the supports is taken as an additional parameter, then computing Nash equilibrium is fixed-parameter tractable.

Theorem 3. *A Nash equilibrium in an ℓ-sparse bimatrix game, where the support sizes is bounded by k, can be computed in $\ell^{O(k\ell)} \cdot n^{O(1)}$ time.*

Note that the above result also complements the polynomial time algorithms given in [9,11] for 2-sparse games. While in these algorithms there was no assumption made on the size of support of the equilibrium to be found, both algorithms could handle only *win-lose* games [1,10], games with boolean payoff matrices. Theorem 3, on the other hand, holds for arbitrary payoffs.

Our second result is concerned with k-*unbalanced games*, games where the row player has a small set of k strategies [26,29]. Lipton, Markakis, and Mehta [29] observed that in such games there is always an equilibrium where the row player plays a strategy with support size at most $k + 1$. Thus, by applying Theorem 1 one can find a Nash equilibrium in $n^{O(k)}$ time for these types of games. Can this result be improved to an algorithm running in $f(k) \cdot n^{O(1)}$ time? We give a partial answer to this question, by showing that if the number ℓ of different payoffs of the row player is taken as an additional parameter, the problem indeed becomes fixed-parameter tractable.

Theorem 4. *A Nash equilibrium in a k-unbalanced bimatrix game, where the row player has ℓ different payoff values, can be computed in $\ell^{O(k^2)} \cdot n^{O(1)}$ time.*

In our last result, examining the borderline of FPT cases, we consider a structural property that simultaneously extends both the previous cases (albeit only in the case of a bounded number of different payoffs). We show that for bimatrix games whose corresponding graph has locally bounded treewidth, and where the payoff matrices have at most ℓ different values, we can compute a Nash equilibrium of support size at most k in time $f(k, \ell) \cdot n^{O(1)}$. In addition to the above cases of sparse and unbalanced games, this also includes many other cases including games where the underlying graph structure is planar, as considered in [2]. However, as this class is quite general, the running-time dependency on both parameters is worse.

Theorem 5. *A Nash equilibrium in a locally bounded treewidth game, where the support sizes are bounded by k, and the payoff matrices have at most ℓ different values, can be computed in $f(k, \ell) \cdot n^{O(1)}$ time for some computable function $f()$.*

The paper is organized as follows: We begin with some preliminaries in Section 2. In Section 3 we consider ℓ-sparse games and prove Theorem 3. Section 4 addresses locally bounded treewidth games and proves Theorem 5. Finally, in Section 5 we prove Theorem 4 regarding k-unbalanced games. Due to space restrictions all proofs are deferred to the full version [25].

2   Preliminaries

Let $\mathcal{G} := (A, B)$ be a bimatrix game, where $A, B \in \mathbb{Q}^{n \times n}$ are the *payoff matrices* of the *row* and the *column* players respectively (note that the entries are allowed to be negative). The row (column) player has a strategy space consisting of the rows (columns) $[n] := \{1, \ldots, n\}$. (For ease of notation, except in unbalanced games, we assume that both players have the same number of strategies; different numbers of strategies do not affect any of our results.) The row (column) player chooses a strategy profile x (*resp.* y), which is a probability distribution over his strategy space. That is, $x_i, y_j \geq 0$ for all $i, j \in [n]$, and furthermore $\sum_{i=1}^{n} x_i = 1$ and $\sum_{j=1}^{n} y_j = 1$. The expected outcomes of the game for the row and the column players are $x^T A y$ and $x^T B y$ respectively.

The players are rational, always aiming for maximizing their expected payoffs. They have reached a *Nash equilibrium* if the current strategies x and y are such that neither player has a deviating strategy $\hat{x}$ or $\hat{y}$ such that $\hat{x}^T A y > x^T A y$ or $x^T B \hat{y} > x^T B y$. In other words, if neither of them can improve his payoff independently of the other. The following proposition gives an equivalent condition for a pair of strategies to be an equilibrium.

Lemma 1. ([30, Chapter 3]) *The pair of strategy vectors (x, y) is a Nash equilibrium for the bimatrix game (A, B) if and only if*

  (i) $x_s > 0 \Rightarrow (Ay)_s \geq (Ay)_j$ *for all $j \neq s$;*
  (ii) $y_s > 0 \Rightarrow (x^T B)_s \geq (x^T B)_j$ *for all $j \neq s$.*

The *support* of a strategy vector x is defined as the set $S(x) = \{i : x_i > 0\}$. Note that the above proposition implies that if (x, y) is a Nash equilibrium, then in the column vector Ay, the values in positions $S(x)$ are mutually equal, and no less than the values in positions not in $S(x)$; symmetrically, in the row vector $x^T B$, the values in positions $S(y)$ are equal and no less than the values in positions not in $S(y)$. It is known that, given possible supports $I, J \subseteq [n]$ it can be efficiently decided whether there is a matching Nash equilibrium, and the corresponding strategy vectors can be computed via linear programming.

The following graph associated with a bimatrix game is useful for presenting our algorithms in Sections 3 and 4.

Definition 1. *Let $\mathcal{G} = (A, B)$ be a bimatrix game with $A, B \in \mathbb{Q}^{n \times n}$. The undirected bipartite graph $G := G(\mathcal{G})$ associated with $\mathcal{G}$ is defined to be the bipartite graph with vertex classes $V_r := \{r_1, \ldots, r_n\}$ and $V_c := \{c_1, \ldots, c_n\}$, referred to as row resp. column vertices, where $r_i \in V_r$ and $c_j \in V_c$ are adjacent in G iff $A_{i,j} \neq 0$ or $B_{i,j} \neq 0$. For a component C of G, we let $V_r(C)$ (resp. $V_c(C)$) be the set of row vertices (resp. column vertices) occurring in C.*

As a last bit of notation: For $I, J \subseteq [n]$, and any $n \times n$ matrix A, we use $A_{I,J}$ to denote the submatrix composed of rows in I and columns in J. We also use $A_{I,*}$ as a shorthand for $A_{I,[n]}$. Thus, $A_{i,*}$ means the i'th row of A.

3   Sparse Games

In this section we present the proof for Theorem 3. Throughout the section we let $\mathcal{G} := (A, B)$ denote our given bimatrix game, where A and B are rational value matrices with at most ℓ non-zero entries per row or column. For ease of notation we let the vertices correspond directly to the respective row or column indices. We will present an algorithm for finding a Nash equilibrium where the support sizes of both players are at most k (and k is taken as a parameter). The high-level strategy is to show that it suffices to search for equilibria that induce one or two connected components in the associated graph $G = G(\mathcal{G})$. This permits us to find candidate support sets by enumerating subgraphs of G (on one or two components). Central to this strategy is the notion of minimal equilibria:

Definition 2. *A Nash equilibrium (x, y) is minimal if for any Nash equilibrium (x', y') with $S(x') \subseteq S(x)$ and $S(y') \subseteq S(y)$, we have $S(x') = S(x)$ and $S(y') = S(y)$.*

Our algorithm iterates through all possible support sizes $k_1, k_2 \leq k$ in increasing order to determine whether there exists an equilibrium (x, y) with $|S(x)| = k_1$ and $|S(y)| = k_2$. To avoid cumbersome notation, we will assume that $k_1 = k_2 = k$ (extending this to general case will be immediate). Thus at a given iteration, the algorithm can assume that no equilibrium exists with smaller supports, which means it can restrict its search to minimal equilibriums. This fact will prove crucial later on. Furthermore, observe that we can assume $n > \ell k$ since otherwise obtaining the running time required by Theorem 3 is trivial by enumerating all possible supports (which is dominated by $2^{O(n)}$). Therefore, since our game is ℓ-sparse, our algorithm only needs to search for equilibriums where both players receive non-negative payoffs.

Lemma 2. *If $\mathcal{G} = (A, B)$ is an ℓ-sparse game, where $A, B \in \mathbb{Q}^{n \times n}$ and $n > \ell k$, then in any Nash equilibrium where both players have support size at most k, both players receive non-negative payoffs.*

For an equilibrium (x, y), let the *extended support* of x be the rows $S(x) \cup N(S(y))$, and similarly for y, where the neighborhood $N(I)$ is taken over the graph $G := G(\mathcal{G})$ of the game. Note that any row not in the extended support of x would have payoff constantly zero given the current strategy of y, and thus is not important for the existence of an equilibrium. We will show that for a minimal equilibrium (x, y), the extended supports of x and y induce a subgraph of G which has at most two connected components. This will be done in two steps: The first is the special case where $A_{S(x),S(y)} = B_{S(x),S(y)} = \mathbf{0}$, while the second corresponds to the remaining cases.

Lemma 3. *If (x, y) is a minimal Nash equilibrium for a game (A, B) with $A_{S(x),S(y)} = B_{S(x),S(y)} = \mathbf{0}$, then the subgraphs induced by $N[S(x)]$ and $N[S(y)]$ in the graph associated with the game are both connected[1].*

Lemma 4. *If (x, y) is a minimal Nash equilibrium for a game (A, B) with either $A_{S(x),S(y)} \neq \mathbf{0}$ or $B_{S(x),S(y)} \neq \mathbf{0}$, and with a non-negative payoff for both players, then the subgraph induced by $N[S(x) \cup S(y)]$ is connected.*

As an immediate corollary of Lemmas 3 and 4, we get that the subgraph in $G(\mathcal{G})$ induced by the extended support of a minimal equilibrium has at most two connected components. In the following lemma we show that in graphs of small maximum degree, we can find all such subgraphs quite efficiently. This will allow us to find a small, minimal equilibrium by checking all sets of rows and columns that would be candidates for being the extended supports of one.

Lemma 5. *Let G be a graph on n vertices and with maximum degree $\Delta = \Delta(G)$. In time $(\Delta + 1)^{2t} \cdot n^{c+O(1)}$ one can enumerate all subgraphs on t vertices that consist of c connected components.*

We are now in position to describe our entire algorithm. It first iterates through all possible sizes of extended support in increasing order. In each iteration, it enumerates all subgraphs that might correspond to the extended support of a minimal equilibrium. It then checks all ways of selecting a support from the given subgraph, and for each such selection it uses the algorithm behind Theorem 1 to check whether there is an equilibrium on the support. If no equilibrium is found throughout the whole process, the algorithm reports that there exists no equilibrium with support size at most k in $\mathcal{G}$. The running time is bounded by $\ell^{O(k\ell)} n^{O(1)}$ from Lemma 5, times $\binom{k\ell+k}{k}^2 = 2^{O(k\ell)}$ ways of selecting the support, times $n^{O(1)}$ for checking for an equilibrium. In total, we get a running time of $\ell^{O(k\ell)} n^{O(1)}$.

Finally, completeness comes from the exhaustiveness of Lemma 5 and the structure given by Lemmas 3 and 4.

3.1   Non-negative Payoffs

In the case that the payoffs of our games are non-negative, i.e., $A, B \in \mathbb{Q}_{\geq 0}^{n \times n}$, we can reduce our running time to be polynomial in ℓ, for ℓ-sparse games. We get a strengthening of Lemmas 3 and 4.

---

[1] Note that $N[I] = N(I) \cup I$ denotes the closed neighborhood if I

Lemma 6. *Let $\mathcal{G} = (A, B)$ be a bimatrix game with $A, B \in \mathbb{Q}_{\geq 0}^{n \times n}$, and G be the graph associated with $\mathcal{G}$. If (x, y) is a minimal Nash equilibrium for $\mathcal{G}$, then either $|S(x)| = |S(y)| = 1$, or $G[S(x) \cup S(y)]$ is connected.*

Thus, to find an equilibrium in $\mathcal{G} = (A, B)$, it suffices to search for occurrences of the support, rather than the extended support. Invoking Lemma 5 directly with a bound of $2k$ vertices gives a running time of $\ell^{O(k)} n^{O(1)}$.

3.2   No Polynomial Kernels

We next show another interesting application of the connectivity lemmas described in the section above. In particular, we will use Lemma 6 to rule out the possibility of effectively compressing our problem instances.

Polynomial kernelization is a central concept in parameterized complexity, formalizing the notions of compression and data reduction. In our setting, a polynomial kernel is an algorithm that receives as input an ℓ-sparse game $\mathcal{G}$ and an integer k, and outputs in polynomial time an ℓ'-sparse game $\mathcal{G}'$ and an integer k', with $|\mathcal{G}'| + k' + \ell' \leq p(\ell + k)$ for some polynomial p, such that $\mathcal{G}$ has an equilibrium with support sizes bounded by k iff $\mathcal{G}'$ has an equilibrium with support sizes bounded by k'. Thus, a polynomial kernel outputs an equivalent game whose size is bounded by a polynomial in k and ℓ. This clearly implies an FPT algorithm for computing the equilibrium in $\mathcal{G}$, but it also gives something better since one can use other techniques and heuristics on the reduced instance. In the remainder of the section we prove the following theorem.

Theorem 6. *Unless* co-NP $\subseteq$ NP/poly, *the problem of determining whether a Nash equilibrium with support sizes at most k exists in an ℓ-sparse bimatrix game has no kernel which is polynomial in $k + \ell$.*

A framework for providing evidence that polynomial kernels do not exist for a given problem was given in [3]. There, the concept of composition algorithms plays a central role. A *composition algorithm* for a parameterized problem Π is an algorithm that receives as input a sequence of instances $(x_1, k), \ldots, (x_t, k)$ of Π, all sharing the same parameter, and outputs in time polynomial in $\sum_i |x_i|$ an instance (y, k') of Π such that $(y, k') \in \Pi \iff \exists i : (x_i, k) \in \Pi$ and $k' \leq p(k)$ for some polynomial p. A parameterized problem admitting a composition algorithm is called *compositional*. The following connection between compositional problems and polynomial kernels was proven in [3] and [22]:

Theorem 7 ([3,22]). *Unless* co-NP $\subseteq$ NP/poly, *no compositional parameterized problem whose unparameterized variant is NP-complete has a polynomial kernel.*

Observe that the unparameterized decision variant of our problem is the problem of determining whether there exists an equilibrium in a given bimatrix game with support sizes bounded by k (since ℓ is part of the input, the sparseness does not come into effect here). This problem was shown to be NP-complete in [23]. Thus, in order to show the non-existence of a polynomial kernel in our setting, we need to show that our problem is compositional. We will restrict ourselves to instances with non-negative payoffs. Given a sequence of ℓ-sparse games $(A_1, B_1), \ldots, (A_t, B_t)$ with non-negative payoffs, and a parameter k, our algorithm outputs the game $\mathcal{G} := (A, B)$ defined by

$$
A := \begin{pmatrix} A_1 & 0 & \cdots & 0 \\ 0 & A_2 & \cdots & 0 \\ \vdots & \vdots & \ddots & \vdots \\ 0 & \cdots & 0 & A_t \end{pmatrix} \quad \text{and } B := \begin{pmatrix} B_1 & 0 & \cdots & 0 \\ 0 & B_2 & \cdots & 0 \\ \vdots & \vdots & \ddots & \vdots \\ 0 & \cdots & 0 & B_t \end{pmatrix}.
$$

It is clear that our algorithm runs in polynomial time, and that $\mathcal{G}$ is ℓ-sparse. Furthermore, it is not difficult to see that due to Lemma 6, we know that $\mathcal{G}$ has an equilibrium with support sizes bounded by k iff (A_i, B_i) has an equilibrium with support sizes bounded by k, for some $i \in \{1, \ldots, t\}$. It follows that the above algorithm is a composition algorithm, and so according to Theorem 7, we can rule out the possibility of polynomial kernels for our problem.

4   Locally Bounded Treewidth Games

Let $\mathcal{G} = (A, B)$ be a given game with $A, B \in P^{n \times n}$, with $P \subset \mathbb{Q}$, $|P| \leq \ell$, and let $G = G(\mathcal{G})$ the graph associated with $\mathcal{G}$. In this section we will present an algorithm that finds an equilibrium with support sizes at most k when G comes from a graph class with locally bounded treewidth. Note that this is a partial extension of the results of the previous section, as graphs of bounded degree have locally bounded treewidth, while on the other hand we assume that there is a bounded set P of only ℓ different payoff values which can occur in the games. (The case $P = \{0, 1\}$ would correspond to win-lose games.)

Definition 3 ([18]). *A graph class has* locally bounded treewidth *if there is a function $f : \mathbb{N} \to \mathbb{N}$ such that for every graph $G := (V, E)$ of the class, any vertex $v \in V$, and any $d \in \mathbb{N}$, the subgraph of G induced by all vertices within distance at most d from v has treewidth at most $f(d)$.*

The crucial property of locally bounded treewidth graphs in our context is that first-order queries can be answered in FPT time on such graphs when the parameter is the size of first-order formula [21, Chapter 12.2]. Below we show how to use a bounded number of first-order queries in order to compute an equilibrium with support sizes at most k (assuming one exists) in $\mathcal{G}$.

For ease of presentation we show how to find an equilibrium where both players have support size k (the algorithm can be easily adapted to support sizes $k_1, k_2 \leq k$). Let I and J be two subsets of k elements in $[n]$. We say that two matrices $A^*, B^* \in \mathbb{Q}^{k \times k}$ *occur* in $\mathcal{G}$ at (I, J) if $A^* = A_{I,J}$ and $B^* = B_{I,J}$. The pair (A^*, B^*) forms an *equilibrium pattern* if there exists an equilibrium (x, y) where (A^*, B^*) occur in $\mathcal{G}$ at $(S(x), S(y))$. Our algorithm will try all possible ℓ^{2k^2}

pairs of matrices (A^*, B^*), and for each such pair it will determine whether it is an equilibrium pattern.

When does a pair of matrices (A^*, B^*) form an equilibrium pattern? The first obvious condition is that it occurs in $\mathcal{G}$ at some pair of position sets (I, J). Furthermore, by definition of a Nash equilibrium, there is a pair of strategies (x, y) with $S(x) = I$ and $S(y) = J$, such that neither player has a better alternative. The difficulty here lies in the fact that, even given the support $S(y)$ of the column player, there may be too many possible strategies for the row player that have supports different from I. To circumvent this, we define equivalence of rows with respect to supports $S(y)$, and of columns with respect to supports $S(x)$.

Definition 4. *Let $I, J \subseteq [n]$. Two rows $i_1, i_2 \in [n]$ are J-equivalent if $A_{i_1, J} = A_{i_2, J}$. Similarly, two columns $j_1, j_2 \in [n]$ are I-equivalent if $B_{I, j_1} = B_{I, j_2}$.*

Lemma 7. *Let J be the support of the column player. For any row strategy x there is a row strategy $\hat{x}$ such that:*

(i) the support $S(\hat{x})$ contains at most one row from each J-equivalence class
(ii) and for any column strategy y with support J we have $\hat{x}^T A y = x^T A y$.

The same is true for column strategies, given a support I of the row player.

For each possible equilibrium pattern (A^*, B^*) we do the following. For each choice of rows $A^\dagger \subseteq P^{1 \times k}$ that do not occur in A^* and each choice of columns $B^\dagger \subseteq P^{k \times 1}$ that do not occur in B^*, we create two matrices

$$C = \begin{pmatrix} A^* & 0 \\ A^\dagger & 0 \end{pmatrix} \text{ and } D = \begin{pmatrix} B^* & B^\dagger \\ 0 & 0 \end{pmatrix}.$$

We then check if there is an equilibrium (x, y) in the game (C, D) with $S(x) = S(y) = [k]$ using linear programming. If there is such an equilibrium, then we proceed as follows to find an occurrence of (A^*, B^*) that avoids the rows and columns which were not chosen. For this let F_1 be the rows which occur neither in A^* nor in $A^\dagger$ and let F_2 be the columns which occur neither in B^* nor in $B^\dagger$. We say that F_1 and F_2 are *forbidden* for (A^*, B^*). We note that given (A^*, B^*), a set of rows $F_1 \subseteq P^{1 \times k}$, and a set of columns $F_2 \subseteq P^{k \times 1}$, one can write a first-order formula of size bounded by some function in k and $|P| = \ell$ to determine whether (A^*, B^*) has an occurrence which avoids F_1 and F_2. Since the number of possible choices for F_1 and F_2 is bounded by some function in k and ℓ, and for each such choice we can determine whether F_1 and F_2 is a forbidden pair for (A^*, B^*) in polynomial time, the total time for determining whether (A^*, B^*) is an equilibrium pattern is FPT in k and ℓ. Since the number of pairs (A^*, B^*) is also bounded by a function in k, the total running time of our entire algorithm is also FPT in k and ℓ.

To complete the proof of Theorem 5, let us briefly argue completeness. Assume that there is any equilibrium with support sizes equal to k, let I and J be the supports, and let A^* and B^* be corresponding sub-matrices. Observe that we may set all entries in columns outside J of A to zero without harm, ditto for rows outside I in B. According to Lemma 7 it suffices to keep one copy of each row outside A^* in A (also discard the corresponding zero-row in B to keep the size the same). The same is of course true for columns outside B^* in B. Except for a permutation this is equal to one of the games (C, D) that we considered. Therefore our algorithm will find such an equilibrium if one exists.

5   Unbalanced Games

In this section we briefly consider k-unbalanced bimatrix games. A bimatrix game (A, B) is *k-unbalanced* if $A, B \in \mathbb{Q}_{\geq 0}^{k \times n}$ for some $k << n$ [26,29] (i.e., the row player has a significantly smaller number of strategies than the column player). We will show that a Nash equilibrium can be computed in FPT-time with respect to k and ℓ, where ℓ denotes the number of different payoffs that the row player has, i.e., $\ell := |\{A_{i,j} : 1 \leq i \leq k, 1 \leq j \leq n\}|$.

Similar to Definition 4 we define two column strategies $i, j \in [n]$ to be equivalent if $A_{*,i} = A_{*,j}$. (However, notice that unlike Definition 4, here equivalence of column strategies is defined with respect to the row player payoff.)

Lemma 8. *For each equilibrium there is an equilibrium where the column player plays at most one column from each equivalence class.*

Using Lemma 8 we can easily devise an FPT algorithm for computing a Nash equilibrium in our setting. The algorithm simply guesses the support of the row player and column player, and then uses the method of Theorem 1 to determine whether there exists a Nash equilibrium corresponding to these sets of supports. Observe that there are at most ℓ^k column-strategy equivalence classes. Furthermore, according to Lipton, Markakis, and Mehta [29], in a k-unbalanced game there always exists an equilibrium where the column player has support size at most $k + 1$. Thus the number of guesses the algorithm makes is bounded by $2^k \cdot \binom{\ell^k}{k+1} = \ell^{O(k^2)}$, and for each such guess, the amount of time required is polynomial. This completes the proof of Theorem 4.

6   Conclusions

This paper is among the first attempts at applying parameterized complexity techniques in algorithmic game theory. This seems surprising when considering the potential benefit both fields can enjoy from each other. Our paper focused on the fundamental game-theoretical problem of computing a Nash equilibrium in bimatrix game. Three parameterized algorithms were presented, each corresponding to a different parameterization of the problem. In two cases, our algorithms utilized the graph-theoretical structure inherited in bimatrix games, and we believe this perspective will be useful in other settings as well.

Our work is only the first step towards completely understanding the multivariate complexity of computing Nash equilibria in bimatrix games. There are still several parameters of the problems which were left unexplored, and we consider the problem of identifying new parameterizations to be the central open problem of this paper. Other questions which would be interesting to explore include the existence of a polynomial-time algorithm for computing Nash equilibria in games whose bipartite graph representation has bounded treewidth, and whether the dependency on the number of different values can be removed from the parameter in Theorems 4 and 5. For the latter question, it is also interesting to ask what strong negative evidence would look like (note that the concept of PPAD-completeness does not immediately apply).

References

1. Abbott, T., Kane, D., Valiant, P.: On the complexity of two-player win-lose games. In: Proc. of the 46th Annual IEEE symposium on Foundations of Computer Science (FOCS), pp. 113–122 (2005)
2. Addario-Berry, L., Olver, N., Vetta, A.: A polynomial time algorithm for finding Nash equilibria in planar win-lose games. Journal of Graph Algorithms and Applications 11(1), 309–319 (2007)
3. Bodlaender, H., Downey, R., Fellows, M., Hermelin, D.: On problems without polynomial kernels. Journal of Computer and System Sciences 75(8), 423–434 (2009)
4. Bosse, H., Byrka, J., Markakis, E.: New algorithms for approximate Nash equilibria in bimatrix games. Theoretical Computer Science 411(1), 164–173 (2010)
5. Chen, J., Chor, B., Fellows, M., Huang, X., Juedes, D., Kanj, I., Xia, G.: Tight lower bounds for certain parameterized NP-hard problems. Information and Computation 201(2), 216–231 (2005)
6. Chen, X., Deng, X.: 3-NASH is PPAD-complete. Electronic Colloquium on Computational Complexity (134) (2005)
7. Chen, X., Deng, X.: Settling the complexity of two-player Nash equilibrium. In: Proc. of the 47th Annual IEEE Symposium on Foundations of Computer Science (FOCS), pp. 261–272 (2006)
8. Chen, X., Deng, X., Teng, S.-H.: Computing Nash equilibria: Approximation and smoothed complexity. In: Proc. of the 47th Annual IEEE Symposium on Foundations of Computer Science (FOCS), pp. 603–612 (2006)
9. Chen, X., Deng, X., Teng, S.-H.: Sparse Games are Hard. In: Spirakis, P.G., Mavronicolas, M., Kontogiannis, S.C. (eds.) WINE 2006. LNCS, vol. 4286, pp. 262–273. Springer, Heidelberg (2006)
10. Chen, X., Teng, S.-H., Valiant, P.: The approximation complexity of win-lose games. In: Proc. of the 18th Annual ACM-SIAM Symposium on Discrete Algorithms (SODA), pp. 159–168 (2007)
11. Codenotti, B., Leoncini, M., Resta, G.: Efficient Computation of Nash Equilibria for Very Sparse Win-Lose Bimatrix Games. In: Azar, Y., Erlebach, T. (eds.) ESA 2006. LNCS, vol. 4168, pp. 232–243. Springer, Heidelberg (2006)
12. Daskalakis, C., Goldberg, P.W., Papadimitriou, C.H.: The complexity of computing a Nash equilibrium. Commun. ACM 52(2), 89–97 (2009)
13. Daskalakis, C., Mehta, A., Papadimitriou, C.: Progress in approximate Nash equilibria. In: Proc. of the 8th ACM Conference on Electronic Commerce (EC), pp. 355–358 (2007)
14. Daskalakis, C., Mehta, A., Papadimitriou, C.: A note on approximate Nash equilibria. Theoretical Computer Science 410(17), 1581–1588 (2009)
15. Daskalakis, C., Papadimitriou, C.: On oblivious PTAS's for Nash equilibrium. In: Proc. of the 41st Annual ACM Symposium on Theory of Computing (STOC), pp. 75–84 (2009)

16. Daskalakis, K., Papadimitriou, C.: Three-player games are hard. Electronic Colloquium on Computational Complexity (139) (2005)
17. Downey, R., Fellows, M.: Parameterized Complexity. Springer, Heidelberg (1999)
18. Eppstein, D.: Subgraph isomorphism in planar graphs and related problems. Journal of Graph Algorithms and Applications 3(3) (1999)
19. Estivill-Castro, V., Parsa, M.: Computing Nash equilibria gets harder: New results show hardness even for parameterized complexity. In: Proc. of the 15th Computing: the Australasian Theory Symposium (CATS), vol. 94, pp. 81–87 (2009)
20. Estivill-Castro, V., Parsa, M.: Single Parameter FPT-Algorithms for Non-Trivial Games. In: Iliopoulos, C.S., Smyth, W.F. (eds.) IWOCA 2010. LNCS, vol. 6460, pp. 121–124. Springer, Heidelberg (2011)
21. Flum, J., Grohe, M.: Parameterized Complexity Theory. Springer, Heidelberg (2006)
22. Fortnow, L., Santhanam, R.: Infeasibility of instance compression and succinct PCPs for NP. J. Comput. Syst. Sci. 77(1), 91–106 (2011)
23. Gilboa, I., Zemel, E.: Nash and correlated equillbrla: Some complexity considerations. Games and Economic Behavior (1989)
24. Goldberg, P., Papadimitriou, C.: Reducibility among equilibrium problems. In: Proc. of the 38th Annual ACM Symposium on Theory of Computing (STOC), pp. 61–70 (2006)
25. Hermelin, D., Huang, C.-C., Kratsch, S., Wahlström, M.: Parameterized two-player Nash equilibrium. CoRR (2010), http://arxiv.org/abs/1006.2063
26. Kalyanaraman, S., Umans, C.: Algorithms for playing games with limited randomness, pp. 323–334 (2007)
27. Kannan, R., Theobald, T.: Games of fixed rank: A hierarchy of bimatrix games. In: Proc. of the 18th Annual ACM-SIAM Symposium on Discrete Algorithms (SODA), pp. 1124–1132 (2007)
28. Kontogiannis, S., Spirakis, P.: Exploiting concavity in bimatrix games: New polynomially tractable subclasses. In: Serna, M., Shaltiel, R., Jansen, K., Rolim, J. (eds.) APPROX 2010, LNCS, vol. 6302, pp. 312–325. Springer, Heidelberg (2010)
29. Lipton, R., Markakis, E., Mehta, A.: Playing large games using simple strategies. In: Proc. of the 4th ACM Conference on Electronic Commerce (EC), pp. 36–41 (2003)
30. Nisan, N., Roughgarden, T., Tardos, E., Vazirani, V.: Algorithmic Game Theory. Cambridge University Press (2007)
31. Papadimitriou, C.: Algorithms, games, and the internet. In: Proc. of the 33rd Annual ACM Symposium on Theory of Computing (STOC), pp. 749–753 (2001)
32. Tsaknakis, H., Spirakis, P.G.: An Optimization Approach for Approximate Nash Equilibria. In: Deng, X., Graham, F.C. (eds.) WINE 2007. LNCS, vol. 4858, pp. 42–56. Springer, Heidelberg (2007)

Counting Independent Sets in Claw-Free Graphs

Konstanty Junosza-Szaniawski, Zbigniew Lonc, and Michał Tuczyński

Warsaw University of Technology
Faculty of Mathematics and Information Science
Pl. Politechniki 1 / 207, 00-661 Warsaw, Poland
{k.szaniawski,zblonc,m.tuczynski}@mini.pw.edu.pl

Abstract. In this paper we give an algorithm for counting the number of all independent sets in a claw-free graph which works in time $O^*(1.08352^n)$ for graphs with no vertices of degree larger than 3 and $O^*(1.23544^n)$ for arbitrary claw-free graphs, where n is the number of vertices in the instance graph.

1 Introduction

Recently much attention has been paid to algorithmic aspects of some counting problems. Although many of the problems (e.g. counting independent sets or matchings in a graph) are known to be #P-Complete (see Vadhan [9]), a remarkable progress has been done in designing exponential time algorithms solving them. Dahllöf, Jonsson, Wahlström [2] constructed algorithms that count maximum weight models of 2-SAT and 3-SAT formulas in time $O^*(1.2561^n)$ and $O^*(1.6737^n)$, respectively. The former bound was later improved to $O^*(1.2461^n)$ by Fürer and Kasiviswanathan [4] and subsequently to $O^*(1.2377^n)$ by Wahlström [10]. The latter bound was improved by Kutzkov [6] to $O^*(1.6423^n)$.

Independent sets in a graph naturally correspond to models of 2-SAT formulas with all variables negated. In particular the algorithm of Wahlström [10] can be applied to count all independent sets and all independent sets of maximum size in a graph. In fact this algorithm was used by Björklund, Husfeldt and Koivisto [1] as a subroutine in their (based on the inclusion-exclusion principle) algorithm for graph coloring. This algorithm is currently the fastest exact algorithm for graph coloring.

In this paper we present an algorithm counting independent sets in claw-free graphs whose time complexity is $O^*(1.23544^n)$. Notice that matchings in a graph correspond to independent sets in its line graph. The problem of counting matchings is also #P-Complete (see Vadhan [9]). Hence, as line graphs are claw-free, the problem of counting independent sets in a graph stays #P-Complete in the class of claw-free graphs. Using the algorithm presented in this paper, we can count matchings in a graph with m edges in time $O^*(1.23544^m)$. On the other hand, matchings can be counted in time $O^*(2^n)$, where n is the number of vertices of a graph, thanks to Ryser's formula [8]. So our algorithm applied for counting matchings is faster for graphs where $m \leq \frac{1}{\log_2 1.23544} n \approx 3.278n$. If we use our algorithm as a subroutine in the algorithm of Björklund, Husfeldt

P. Kolman and J. Kratochvíl (Eds.): WG 2011, LNCS 6986, pp. 227–237, 2011.
© Springer-Verlag Berlin Heidelberg 2011

and Koivisto [1] for graph coloring, we obtain a faster algorithm for coloring claw-free graphs, which running time is $O^*(2.23544^n)$.

The algorithm presented in this paper is a modification of the algorithm of Dahllöf *et al.* [2]. In our algorithm we introduce an extra simplification (we call it FOLDING) for graphs with maximum degree of a vertex at most 3 that removes vertices with the same closed neighborhoods. This simplification guarantees a better choice of the next vertex for branching than in the algorithm in [2]. The rest of the algorithm is just an adaptation of the algorithm of Dahllöf *et al.* [2] so that it counts independent sets in graphs instead of models of 2-SAT formulas. Consequently, the correctness of our algorithm (except for the procedure FOLDING) follows directly from the correctness of the algorithm of Dahllof *et al.* [2].

In analyzing the running time of recursive algorithms a method sometimes called the "measure and conquer" method (introduced by Kullman [5] and developed by Fomin *et al.* [3]) is very useful. In particular this method was used in analyzing the running time of the algorithm in Dahllof *et al.* [2]. They defined the measure of a graph with maximum degree of a vertex at most 3 to be equal to the number of vertices of degree 3 in this graph. It allowed them to obtain the time complexity bound of $O^*(1.1892^n)$ for such graphs. An extension of this measure to some piecewise linear function depending on the number of vertices and the number of edges is the measure of graphs with larger degree. It allowed to prove the bound of $O^*(1.2561^n)$ for graphs with arbitrary maximum degree. Fürer and Kasiviswanathan [4] observed that piecewise linear measure can be applied already to graphs with degree maximum of a vertex at most 3. This way they obtained the complexity bound of $O^*(1.1504^n)$ for such graphs and $O^*(1.2461^n)$ for general graphs. Wahlström [10] considered still more complicated measure for graphs with the maximum degree of a vertex larger than 3 (depending on the degrees of vertices) and obtained the complexity bound of $O^*(1.2377^n)$. In all these three cases, any improvement of the complexity bound for graphs with the maximum degree of a vertex at most 3 implies a better bound for all graphs.

The algorithm of Wahlström works for claw-free graphs with maximum degree of a vertex at most 3 in time $O^*(1.1406^n)$ while our algorithm for such graphs works in time $O^*(1.08352^n)$. There are two reasons of this. One of them is introduction of the procedure FOLDING. The other one is a new measure that we use for claw-free graphs in the discussion of the running time of our algorithm. The measure is defined as the number of vertices of degree equal to 3 contained in triangles whose all vertices have degrees equal to 3. An appropriate extension of this new measure to some piecewise linear function defined for all graphs gives the complexity bound of our algorithm $O^*(1.23544^n)$, for all claw-free graphs. This bound can be slightly improved with the method used by Wahlström in [10]. It would require, however, consideration of a much larger number of cases.

Our algorithm and its analysis can be easily extended, in the way described in [2], to a version allowing to count weighted independent sets in a claw-free graph. Nevertheless, we do not do it for clarity.

We observe that a set of vertices is independent in a graph if and only if its complement is a transversal (also called a vertex cover) in this graph. Therefore, our algorithm applies not only for counting independent sets in claw-free graphs but for counting transversals in such graphs as well.

2   Preliminaries

We denote by $V(G)$ the vertex set of a graph G and by $E(G)$ its edge set. Let $n(G)$ and $m(G)$ be the number of vertices and the number of edges of G, respectively. We write n instead of $n(G)$ and m instead of $m(G)$ whenever it does not lead to a confusion.

An *open neighborhood* of a vertex v is the set of vertices $N(v) = \{u \in V(G) : uv \in E(G)\}$ and a *closed neighborhood* of v is $N[v] = N(v) \cup \{v\}$. Let $d(v) = |N(v)|$ be the *degree* of a vertex v. By $n_i(G)$ and $n_{\geq i}(G)$ we denote the number of vertices of degree i and at least i in G, respectively. A vertex of degree 0 is called *isolated* and a vertex of degree 1 is called a *leaf*. By $\Delta(G)$ we mean the maximum degree of a vertex in G. Let $S(v) = \sum_{u \in N[v]} d(u)$.

For a vertex set $U \subset V(G)$, $G[U]$ is the subgraph induced by U and $G - U = G[V(G) - U]$. If $U = \{u\}$, then we write $G - u$ instead of $G - \{u\}$. For a subgraph H of G instead of $G - V(H)$, we write $G - H$. A graph is *claw-free* if it does not contain the complete bipartite graph $K_{1,3}$ as an induced subgraph. For the empty graph $(\emptyset, \emptyset)$ we simple write $\emptyset$.

A set U of vertices of G is a *cut set* if $G - U$ has more components than G. A vertex u is a *cut vertex*, if $U = \{u\}$ is a cut set. A graph G is called *k-connected* if $n(G) > k$ and $G - U$ is connected for every set $U \subset V(G)$ such that $|U| < k$.

A set $S \subset V(G)$ is an *independent set* (or an *IS* for short) in G, if no edge in G has both ends in S.

For some technical reasons in our algorithm we use a function $\mathbf{c} : \{0,1\} \times V(G) \to \{0,1,\ldots\}$ called the *cardinality function*. For convenience we write $\mathbf{c_0}(v)$ and $\mathbf{c_1}(v)$ instead of $\mathbf{c}(0,v)$ and $\mathbf{c}(1,v)$, respectively. Given a cardinality function $\mathbf{c}$ and an independent set S, we define

$$C(S, \mathbf{c}) = \prod_{v \notin S} \mathbf{c_0}(v) \prod_{v \in S} \mathbf{c_1}(v).$$

Let $IS(G)$ denote the set of independent sets of G. We define

$$\#IS(G, \mathbf{c}) = \sum_{S \in IS(G)} C(S, \mathbf{c}).$$

Notice that if $\mathbf{c_0}(v) = 1$ and $\mathbf{c_1}(v) = 1$, for every vertex $v \in V(G)$, then $C(S, \mathbf{c}) = 1$ for any IS in G and, thus, $\#IS(G, \mathbf{c})$ is equal to the number of independent sets in G.

Our algorithm solves the problem of computing the number $\#IS(G, \mathbf{c})$ for a given graph G and a cardinality function $\mathbf{c}$.

For functions f and g we write $f(n) = O^*(g(n))$ if $f(n) = O(g(n)p(n))$, where p is a polynomial.

For positive real numbers $t_0, \ldots, t_d$ we denote by $\tau(t_0, \ldots, t_d)$ the unique solution $\tau > 1$ of the equation

$$\sum_{i=0}^{d} \frac{1}{\tau^{t_i}} = 1.$$

One can readily verify that

$$\text{if } t_i \leq t_i' \text{ for } i \in \{0, \ldots, d\}, \text{ then } \tau(t_0', \ldots, t_d') \leq \tau(t_0, \ldots, t_d). \tag{1}$$

In the complexity analysis we use a measure μ, which is a function assigning a nonnegative real number to every graph. Consider an arbitrary graph G' which labels some internal vertex of this tree. Let the children of G' be vertices labeled with the subgraphs $G_0, \ldots, G_d$ of G' for which our algorithm is next called. Assume that $\Delta_i \mu(G') = \mu(G') - \mu(G_i) > 0$ for $i = 0, \ldots, d$. Then the number $\tau(\Delta_0 \mu(G'), \ldots, \Delta_d \mu(G'))$ is well defined and we call this number the *branching number* for G' (with respect to the measure μ). Kullmann [5] proved that if this assumption is satisfied for all internal vertices of the tree of recursive calls of the algorithm for a graph G, then the number of leaves of this tree is bounded by $O(\tau_0^{\mu(G)})$, where τ_0 is the largest branching number for the internal vertices of the tree. In our analysis it is convenient to consider a subtree of the tree of recursive calls of our algorithm whose internal vertices are restricted to vertices labeled with graphs whose measure is larger than some constant, say c. The leaves of this tree are labeled with graphs whose measure is at most c. We show that our algorithm applied to each graph which labels a leaf works in polynomial time. Moreover, in our algorithm, the height of the tree of recursive calls is bounded by the number of vertices of the instance graph. Therefore the number of internal vertices of the tree is bounded by a linear function of the number of leaves. We also show that the amount of time between one recursive call of our algorithm and the next is polynomial with respect to the order of the graph. Hence the running time of the considered algorithm applied to a graph G is bounded by $O^*(\tau_0^{\mu(G)})$.

3   Procedures

Our main algorithm TCOUNT returns $\#IS(G, \mathbf{c})$ for a given pair $(G, \mathbf{c})$. It calls two subalgorithms TCOUNT_3, TCOUNT_6 that are applied to graphs with maximum degree of a vertex at most 3 and between 4 and 6, respectively. They use auxiliary procedures introduced below. In the procedures REDUCTION and BRANCHING we assume, abusing the notation a little bit, that calls of TCOUNT refer to the one of the algorithms TCOUNT_3, TCOUNT_6, or TCOUNT that called this procedure.

3.1 FOLDING

The procedure FOLDING simplifies an input graph by removing vertices of degree 0 and 1 and identifying vertices with common closed neighborhoods.

Algorithm 1. FOLDING$(G, \mathbf{c})$

---

1 **while** *any of the conditions 2, 5, or 8 is applicable* **do**
2 | **if** *$n(G) > 1$ and there exists an isolated vertex v* **then**
3 | | Let $u \neq v$ be any vertex of G.   **(F1)**
4 | | $\mathbf{c_0}(u) \leftarrow \mathbf{c_0}(u)(\mathbf{c_0}(v) + \mathbf{c_1}(v))$, $\mathbf{c_1}(u) \leftarrow \mathbf{c_1}(u)(\mathbf{c_0}(v) + \mathbf{c_1}(v))$
5 | **else if** *there exists a leaf v* **then**
6 | | Let u be the neighbor of v.   **(F2)**
7 | | $\mathbf{c_0}(u) \leftarrow \mathbf{c_0}(u)(\mathbf{c_0}(v) + \mathbf{c_1}(v))$, $\mathbf{c_1}(u) \leftarrow \mathbf{c_1}(u)\mathbf{c_0}(v)$
8 | **else if** *there are distinct vertices u and v such that $N[u] = N[v]$* **then**
9 | | $\mathbf{c_0}(u) \leftarrow \mathbf{c_0}(u)\mathbf{c_0}(v)$, $\mathbf{c_1}(u) \leftarrow \mathbf{c_0}(u)\mathbf{c_1}(v) + \mathbf{c_1}(u)\mathbf{c_0}(v)$ **(F3)**
10 | $G \leftarrow G - v$
11 **return** $(G, \mathbf{c})$

Lemma 1. *Let $(\tilde{G}, \tilde{\mathbf{c}}) = $ FOLDING$(G, \mathbf{c})$. Then $\#IS(G, \mathbf{c}) = \#IS(\tilde{G}, \tilde{\mathbf{c}})$.*

Sketch of the proof. (For detailed proof see Appendix B.)

We proceed by induction on $n(G)$. Clearly, the lemma holds for graphs with at most 1 vertex. We sketch the proof of the induction step in the case when the condition in line 8 is satisfied but the conditions in lines 2 and 5 are not (the step **(F3)** is executed).

Notice that every IS in G contains at most one of the vertices v and u, because $N[v] = N[u]$, so uv is an edge in G. Moreover $\{S \in IS(G) : u, v \notin S\} = \{S \in IS(G - v) : u \notin S\}$ and $\{S \in IS(G) : u \in S$ or $v \in S\} = \{S, S \cup \{v\} - \{u\} : S \in IS(G - v), u \in S\}$. Hence the value of the function $\#IS$ stays the same if we change $\mathbf{c_0}(u)$ to $\mathbf{c_0}(u)\mathbf{c_0}(v)$, $\mathbf{c_1}(u)$ to $\mathbf{c_1}(u)\mathbf{c_0}(v) + \mathbf{c_0}(u)\mathbf{c_1}(v)$, and remove v.

We proceed similarly in the remaining cases, i.e. when line 4 or 7 of the algorithm is executed.   □

Lemma 2. *Procedure FOLDING runs in polynomial time.*   □

We say that a graph G is *folded* if none of the conditions in lines 2, 5, or 8 in the FOLDING procedure is satisfied, i.e. if FOLDING$(G, \mathbf{c}) = (G, \mathbf{c})$.

3.2 REDUCTION

The procedure REDUCTION is applied when there is a cut vertex, say v, in the input graph. Let H be a component of $G - v$. Instead of working with the whole G we can compute $\#IS(G, \mathbf{c})$ as follows: calculate separately the numbers $\sum_{S \in IS(G[H \cup \{v\}]):v \notin S} C(S, \mathbf{c})$ and $\sum_{S \in IS(G[H \cup \{v\}]):v \in S} C(S, \mathbf{c})$, save them in $\mathbf{c_0}(v)$ and $\mathbf{c_1}(v)$, respectively and then call TCOUNT for $G - H$. The correctness of the REDUCTION procedure follows from Lemma 2 in [2]

Algorithm 2. REDUCTION$(G, \mathbf{c}, v, H)$

---

1   $H_1 \leftarrow G[V(H) \cup \{v\}]$, $H_2 \leftarrow G - H$

2   $c_0 \leftarrow$ TCOUNT(FOLDING($H_1 - v, \mathbf{c}$))

3   $c_1 \leftarrow$ TCOUNT(FOLDING($H_1 - N[v], \mathbf{c}$))

4   $\mathbf{c_0}(v) \leftarrow c_0 \mathbf{c_0}(v)$, $\mathbf{c_1}(v) \leftarrow c_1 \mathbf{c_1}(v) \prod_{u \in N_{H_1}(v)} \mathbf{c_0}(u)$

5   **return** TCOUNT(FOLDING($H_2, \mathbf{c}$))

3.3   BRANCHING

The BRANCHING procedure is the main step of the algorithm. It recursively counts the number of independent sets containing and not containing a chosen vertex v. Its correctness follows from Lemma 3 in [2].

Algorithm 3. BRANCHING$(G, \mathbf{c}, v)$

---

1   **return** $\mathbf{c_0}(v) \cdot$ TCOUNT(FOLDING($G - v, \mathbf{c}$))+

2                      $+\mathbf{c_1}(v) \prod_{u \in N(v)} \mathbf{c_0}(u) \cdot$ TCOUNT(FOLDING($G - N[v], \mathbf{c}$))

---

4   Algorithm TCOUNT$_3$

Let G be a claw-free graph with $\Delta(G) \leq 3$. We say that a triangle in G is a *3-triangle* (respectively *2-triangle*, *1-triangle*), if it contains 3 (respectively 2, 1) vertices of degree 3. We observe that if G is folded, then all triangles in G are disjoint. Indeed, suppose some two triangles have a common edge uv in G. Then $N[u] = N[v]$, because $\Delta(G) \leq 3$, and we obtain a contradiction with the fact that G is folded. Moreover, since $\Delta(G) \leq 3$, there is no pair of triangles in G with one common vertex.

Let $t_3(G)$ be the number of 3-triangles in G. In the complexity analysis of our algorithm we use a measure of a graph G defined by $\mu(G) = 3t_3(H)$, where H is the graph obtained by applying the procedure FOLDING to the graph G. Notice that $\mu(G) \leq n_3(G) \leq n(G)$.

Algorithm 4. TCOUNT$_3(G, \mathbf{c})$

1 $(G, \mathbf{c}) \leftarrow$ FOLDING$(G, \mathbf{c})$
2 if G *is empty* **then return** 1
3 if G *consists of only one vertex* v **then return** $\mathbf{c}_1(v) + \mathbf{c}_0(v)$
4 if G *is disconnected* **then**

5 $\quad$ **return** $\prod_{i=1}^{s} c_i$, where $c_i \leftarrow$ TCOUNT$_3(C_i, \mathbf{c})$, for the components $\qquad$ **(R0)**

6 $\quad$ $C_1, \ldots, C_s$ of G

7 if *there exists a cut vertex v in G* **then**

8 $\quad$ **return** REDUCTION$(G, \mathbf{c}, v, H)$, where H is the component $\qquad$ **(R1)**

9 $\quad$ of $G - v$ with minimum $n(H)$

10 if $t_3(G) = 0$ **then** let v be a vertex of the maximum degree in G
11 else let v be a vertex contained in a 3-triangle
12 return BRANCHING$(G, \mathbf{c}, v)$ $\qquad$ **(B)**

Theorem 3. *The algorithm* TCOUNT$_3$ *counting the number of independent sets in claw-free graphs with maximum degree of a vertex at most 3 runs in time* $O^*(\tau(6, 12)^n) = O^*(1.08352^n)$.

Proof. It can be shown that the asymptotic behavior of the running time of the algorithm is determined by the step **(B)**. Therefore we do not consider the steps **(R0)**, **(R1)** in the complexity analysis. We only have to analyze the changes of measure when applying the branching procedure **(B)**. This procedure is applied for graphs G which are folded and 2-connected so we assume that G is such a graph. In particular G contains no 1-triangles.

Let u be any vertex of a 3-triangle in G and let P be a maximal path starting with u which contains no edge of a 3-triangle. Since G is claw-free and folded, every internal vertex of P is either a vertex of degree 2 or a vertex of degree 3 contained in a 2-triangle and the last vertex of P is a vertex of a 3-triangle. We call such paths *f-paths*. A 2-connected claw-free graph with no 3-triangles is called an *f-cycle*. Clearly, f-cycle is a cycle or it consists of 2-triangles joined by paths with all internal vertices of degree 2. One can easily see that G consists of pairwise disjoint 3-triangles joined by f-paths or it is an f-cycle.

Let us observe, that an f-path cannot have both ends in the same 3-triangle. Otherwise the third vertex of this 3-triangle is a cut vertex in G which contradicts 2-connectivity of G.

Let us consider the subtree of the tree of recursive calls of our algorithm whose internal vertices are graphs whose measure is larger than 6. We show first that if the measure of the input graph is at most 6 then the algorithm works in a polynomial time.

Let $\mu(G) = 3t_3(G) \le 6$. By the handshaking lemma $t_3(G)$ is even, so $t_3(G) = 0$ or $t_3(G) = 2$. In the former case, as G is folded and 2-connected, G is an f-cycle. The procedure FOLDING (which is called in BRANCHING$(G, \mathbf{c}, v)$ for some vertex v) applied to graphs $G - N[v]$ and $G - v$ returns graphs with only one vertex. Hence, by Lemma 2, for graphs whose measure is 0, our algorithm works in polynomial time.

Let us assume now that $t_3(G) = 2$. As G is 2-connected and folded, the three vertices of one of the 3-triangles in G are joined with the three vertices of the other 3-triangle with three disjoint f-paths and there are no more vertices in G. The procedure FOLDING (which is called in BRANCHING$(G, \mathbf{c}, v)$ for some vertex v of a 3-triangle) applied to the graph $G - N[v]$ returns a graph with only one vertex and applied to the graph $G - v$ returns an f-cycle. Polynomiality of our algorithm in this case follows from Lemma 2 and the statement proved in the preceding paragraph.

We assume now that $\mu(G) = 3t_3(G) > 6$. Let v be the vertex chosen in line 11 of the algorithm TCOUNT$_3$ and let x and y be the remaining two vertices of the 3-triangle T containing v. We denote by P_v, P_x, and P_y the f-paths starting at v, x, and y, respectively and ending at v', x', and y', respectively. Let T_v, T_x, and T_y be the 3-triangles containing v', x', and y', respectively. Let G_0 and G_1 be the graphs obtained by applying the procedure FOLDING (called in BRANCHING$(G, \mathbf{c}, v)$) for graphs $G - v$ and $G - N[v]$, respectively. We shall compute $\Delta_j \mu(G) = 3t_3(G) - 3t_3(G_j)$, for $j = 0, 1$.

First, we consider the graph $G - v$. Clearly $G - v$ does not contain the 3-triangle T. Moreover, the procedure FOLDING removes all the remaining vertices of P_v except v', so in G_0 the triangle T_v is no longer a 3-triangle. Hence, $t_3(G_0) = t_3(G) - 2$, so $\Delta_0 \mu(G) = 6$.

Let us consider the graph $G - N[v]$ now. Clearly $G - N[v]$ does not contain the 3-triangle T. Moreover, the procedure FOLDING removes all the remaining vertices of P_v, P_x, and P_y except v', x', and y', respectively. Thus, the triangles T_v, T_x, and T_y are no longer a 3-triangles in G_1.

If $T_v = T_x = T_y$, then, by connectivity of G, the triangles T and T_v are the only two 3-triangles in G, a contradiction with the assumption $\mu(G) > 6$.

Suppose now that exactly two of the triangles T_v, T_x, and T_y, are the same, say $T_v = T_x$. Then, after removing the paths $P_v - v'$ and $P_x - x'$ by the procedure FOLDING, the triangle T_v becomes a 1-triangle. Let w be the vertex of T_v different from v' and x'. Since $N[v'] = N[x']$ in this graph, FOLDING removes v', x', and all vertices of the f-path P_w starting at w except its last vertex, say w'. Let T_w be the 3-triangle containing w'. Notice that $T_w \neq T_y$, because otherwise the vertex of $T_w = T_y$ different from w' and y' is a cut-vertex in G, a contradiction. Thus, none of the 3-triangles T, T_v, T_y, and T_w in G is a 3-triangle in G_1. Consequently, $\Delta_1 \mu(G) = 12$ in this case.

Finally, if T_v, T_x, and T_y are pairwise different 3-triangles in G, then none of the 3-triangles T, T_v, T_x, and T_y is a 3-triangle in G_1, so $\Delta_1 \mu(G) = 12$.

We have shown that the branching number of every graph G, for which the procedure BRANCHING is called, is equal to $\tau(6, 12)$, if $\mu(G) > 6$. Moreover, the algorithm TCOUNT$_3$ works in polynomial time for all graphs G such that $\mu(G) \leq 6$. By the remarks in the last paragraph of Section 2 and the fact that $\mu(G) \leq n(G)$ it follows that the running time of the algorithm TCOUNT$_3$ is $O^*(\tau(6, 12)^{\mu(G)}) = O^*(1.08352^n)$.   □

5   Algorithm TCOUNT$_6$

In this section we consider graphs G in which all vertices have degrees at most 6. Following [2], we define the *average degree* $d_{av}(v) = \frac{d(v)+|\{w\in N(v):d(w)<\Delta(G)\}|}{1+\sum\limits_{\{w\in N(v):d(w)<\Delta(G)\}} \frac{1}{d(w)}}$, for every vertex v in G of degree $\Delta(G)$.

Algorithm 5. TCOUNT$_6(G, \mathbf{c})$

---

1   **if** $\Delta(G) \leq 3$ **then return** TCOUNT$_3(G, \mathbf{c})$
2   **if** G *is disconnected* **then**
3    $\qquad$ **return** $\prod\limits_{i=1}^{s} c_i$, where $\qquad\qquad\qquad\qquad\qquad\qquad\qquad\qquad$ (R0)
4    $\qquad$ $c_i \leftarrow$ TCOUNT$_6(C_i, \mathbf{c})$ for the components $C_1, \ldots, C_s$ of G
5   **if** *there exists a cut vertex v in G* **then**
6    $\qquad$ **return** REDUCTION$(G, \mathbf{c}, v, H)$, where $\qquad\qquad\qquad\qquad$ (R1)
7    $\qquad$ H is the component of $G - v$ with the least $n(H)$
8   Let v be a vertex of degree $\Delta(G)$ with the largest average degree.
9   **if** *there is a 2-element cut set $\{a, b\}$ in G, such that $n_{\geq 3}(L) \leq n_{\geq 3}(G - L - a)$ and $d_{G-L}(a) \geq d_{G-L}(b) \geq 2$, where L is the component of $G - \{a, b\}$ containing v* **then return** BRANCHING$(G, \mathbf{c}, a)$ $\qquad\qquad\qquad\qquad$ (B+R)
10   **else return** BRANCHING$(G, \mathbf{c}, v)$ $\qquad\qquad\qquad\qquad\qquad\qquad\qquad$ (B)

It can be shown that the asymptotic behavior of the running time of the algorithm TCOUNT$_6$ is determined by the calls of the procedure BRANCHING. Recall that for graphs G with maximum degree of a vertex at most 3, the measure $\mu(G) = 3t_3(G)$. Let us consider now graphs which have a vertex of degree larger than 3. In this case, as in [2], we define a measure of a connected graph G which depends on $n(G)$ and $m(G)$ only, i.e $\mu(G) = \mu'(n(G), m(G))$. For a disconnected graph G let $\mu(G) = \sum\limits_{C:C \text{ is a component of } G} \mu(C)$.

The function $\mu'(n, m)$ is a piecewise linear function defined as follows. We partition the interval $(0, 6]$ into subintervals $(k_i, k_{i+1}]$ for $i = 0, \ldots, 15$ (see Table 1 for the values of the k_is). We define $\mu'(n, m) = \mu_i(n, m) = a_i n + b_i m$, if $\frac{2m}{n} \in (k_i; k_{i+1}]$. We observe that $\frac{2m(G)}{n(G)} \leq 6$, for graphs G whose vertices have degrees at most 6, so the measure μ has been defined for all such graphs. The coefficients a_i, b_i (whose approximate values are given in Table 1) are chosen so that the function μ' is *continuous* (i.e. $\mu_{i-1}(n, m) = \mu_i(n, m)$ when $\frac{2m}{n} = k_i$, for $i = 1, \ldots, 15$), the largest branching numbers of a graph G such that $\frac{n(G)}{M(G)} \in (k_i; k_{i+1}]$ are equal and for every $i = 1 \ldots 15$ the inequality $3t_3(G) \leq \mu(G)$ holds for any claw-free graph G with maximum degree at most 3. For convenience we introduce some auxiliary numbers χ_i, for $i = 0, \ldots, 15$. The approximate values of a_i, b_i, k_i, χ_i are given in the Table 1.

Table 1.

i	k_i	k_{i+1}	a_i	b_i	χ_i	$O^*(\tau_0^{\chi_i n})$
0	0	2	0	0	0	$O^*(1)$
1	2	3.2	-2	2	1.2	$O^*(1.127128465^n)$
2	3.2	3.5	-1.391248	1.61953	1.4429295	$O^*(1.154768641^n)$
3	3.5	3.75	-0.9141105	1.34688	1.6112895	$O^*(1.174321067^n)$
4	3.75	4	-0.53470425	1.14453	1.75435575	$O^*(1.191195984^n)$
5	4	$4\frac{4}{29}$	0.29185575	0.73125	1.804786784	$O^*(1.19720204^n)$
6	$4\frac{4}{29}$	$4\frac{4}{9}$	0.4833964393	0.638672	1.90266755	$O^*(1.208945658^n)$
7	$4\frac{4}{9}$	$4\frac{4}{7}$	0.657876439	0.560156	1.93823301	$O^*(1.213241233^n)$
8	$4\frac{4}{7}$	4.8	0.7989484384	0.498437	1.995197239	$O^*(1.220153162^n)$
9	4.8	5	0.924571639	0.446094	2.039806639	$O^*(1.225593462^n)$
10	5	$5\frac{5}{47}$	1.546641639	0.197266	2.050299511	$O^*(1.226876631^n)$
11	$5\frac{5}{47}$	$5\frac{1}{3}$	1.587533554	0.18125	2.070866887	$O^*(1.229395705^n)$
12	$5\frac{1}{3}$	5.5	1.627117553	0.166406	2.084734054	$O^*(1.231097063^n)$
13	5.5	5.625	1.662565054	0.153516	2.094328804	$O^*(1.232275617^n)$
14	5.625	$5\frac{5}{6}$	1.695524742	0.141797	2.109099325	$O^*(1.234092133^n)$
15	$5\frac{5}{6}$	6	1.726286825	0.13125	2.120036825	$O^*(1.23543898^n)$

We assume that the numbers a_i, b_i, χ_i satisfy the following conditions:

$$a_0 = b_0 = \chi_0 = 0 \tag{2}$$

$$\chi_i = \chi_{i-1} + \frac{b_i}{2}(k_{i+1} - k_i), \text{ for } i = 1 \ldots, 15, \tag{3}$$

$$a_i = \chi_{i-1} - \frac{k_i b_i}{2}, \text{ for } i = 1 \ldots, 15 \tag{4}$$

$$\mu_i(n, m) = a_i n + b_i m = \chi_{i-1} n + (m - \frac{k_i n}{2})b_i, \text{ for } i = 1 \ldots, 15. \tag{5}$$

Using (2)-(5) and the fact that $b_1 \geq b_2 \geq \ldots \geq b_{15}$ one can easily show that the function $\mu'(n, m)$ has the following properties:

(**P1**) $\mu'(n, m)$ is continuous.
(**P2**) $\mu'(n, m)$ is concave, i.e. $\mu'(n, m) \leq \mu_i(n, m)$, for $m \geq n$ and $i = 1 \ldots, 15$.
(**P3**) If $\frac{2m}{n} \in (k_i; k_{i+1}]$, then $0 \leq \mu_i(n, m) \leq \chi_i n$, for $i = 0 \ldots, 15$.

By an exhaustive case analysis we prove that the largest branching number defined for the algorithm TCOUNT$_6$ and the measure μ defined in this section is $\tau(6, 8) \approx 1.10488$ (see Appendix A for details).

Let C be a component of a graph G and let $j = 0, \ldots, 15$ be such that $\frac{2m(C)}{n(C)} \in (k_j, k_{j+1}]$. Then, by properties (**P2**) and (**P3**), $\mu(C) = \mu'(n(C), m(C)) \leq \mu_j(n(C), m(C)) \leq \chi_j n(C) \leq \chi_{15} n(C)$ so, consequently, $\mu(G) \leq \chi_{15} n(G)$. This way we obtain the following result.

Theorem 4. *The algorithm* TCOUNT_6 *counting the number of independent sets in claw-free graphs with maximum degree of a vertex at most 6 runs in time* $O^*(\tau(6,8)^{\chi_{15}n}) = O^*(1.23544^n)$. $\qquad\qquad\square$

6   Algorithm TCOUNT

We extend the measure μ defined in Section 5 to $\mu(G) = \chi_{15}n(G)$, for claw-free graphs G that have a vertex of degree larger than 6. By **(P3)**, $\mu(G') \leq \chi_{15}n(G')$, for every claw-free graph G'. Thus, for graphs G that have a vertex of degree larger than 6, $\Delta_0\mu(G) = \mu(G) - \mu(G-v) \geq \chi_{15}n(G) - \chi_{15}n(G-v) = \chi_{15}$ and $\Delta_1\mu(G) = \mu(G) - \mu(G-N[v]) \geq \chi_{15}n(G) - \chi_{15}n(G-N[v]) \geq 8\chi_{15}$. Hence, the branching number of the graph G is not larger than $\tau(\chi_{15}, 8\chi_{15}) = \tau(1,8)^{\frac{1}{\chi_{15}}} \leq 1.10344 < \tau(6,8)$.

Algorithm 6. $\text{TCOUNT}(G, \mathbf{c})$

1 **if** $\Delta(G) \leq 6$ **then return** $\text{TCOUNT}_6(G, \mathbf{c})$
2 Let v be a vertex with maximum degree. **return** $\text{BRANCHING}(G, \mathbf{c}, v)$.   **(B)**

Since $\mu(G) \leq \chi_{15}n(G)$, we get the final result of this paper.

Theorem 5. *The algorithm* TCOUNT *counting the number of independent sets in an arbitrary claw-free graph runs in time* $O^*(\tau(6,8)^{\chi_{15}n}) = O^*(1.23544^n)$. $\square$

References

1. Björklund, A., Husfeldt, T., Koivisto, M.: Set partitioning via inclusion-exclusion. SIAM J. Comput. 39(2), 546–563 (2009)
2. Dahllöf, V., Jonsson, P., Wahlström, M.: Counting models for 2SAT and 3SAT formulae. Theor. Comput. Sci. 332, 265–291 (2005)
3. Fomin, F.V., Grandoni, F., Kratsch, D.: Measure and Conquer: Domination – A Case Study. In: Caires, L., Italiano, G.F., Monteiro, L., Palamidessi, C., Yung, M. (eds.) ICALP 2005. LNCS, vol. 3580, pp. 191–203. Springer, Heidelberg (2005)
4. Fürer, M., Kasiviswanathan, S.P.: Algorithms for counting 2-SAT solutions and colorings with applications, Tech. Rep. 05-033, ECCC (2005)
5. Kullmann, O.: New methods for 3-SAT decision and worst-case analysis. Theor. Comput. Sci. 223(1-2), 1–72 (1999)
6. Kutzkov, K.: New upper bound for the #3-SAT problem. Inform. Process. Lett. 105, 1–5 (2007)
7. Lonc, Z., Truszczynski, M.: Computing minimal models, stable models and answer sets. Theory and Practice of Logic Prog. 6(4), 395–449 (2006)
8. Ryser, H.J.: Combinatorial Mathematics. The Mathematical Association of America, Washington (1963)
9. Vadhan, S.P.: The Complexity of Counting in Sparse, Regular, and Planar Graphs. SIAM J. on Comput. 31, 398–427 (1997)
10. Wahlström, M.: A Tighter Bound for Counting Max-Weight Solutions to 2SAT Instances. In: Grohe, M., Niedermeier, R. (eds.) IWPEC 2008. LNCS, vol. 5018, pp. 202–213. Springer, Heidelberg (2008)

On the Independence Number of Graphs
with Maximum Degree 3

Iyad A. Kanj[1] and Fenghui Zhang[2]

[1] School of Computing, DePaul University, 243 S. Wabash Avenue, Chicago, IL 60604
ikanj@cs.depaul.edu
[2] Google Kirkland, 747 6th Street South, Kirkland, WA 98033
fhzhang@gmail.com

Abstract. Let G be an undirected graph with maximum degree at most
3 such that G does not contain any of the three graphs shown in Figure 1
as a subgraph. We prove that the independence number of G is at least
$n(G)/3 + nt(G)/42$, where $n(G)$ is the number of vertices in G and $nt(G)$
is the number of nontriangle vertices in G. This bound is tight as implied
by the well-known tight lower bound of $5n(G)/14$ on the independence
number of triangle-free graphs of maximum degree at most 3. We then
proceed to show some algorithmic applications of the aforementioned
combinatorial result to the area of parameterized complexity. We present
a linear-time kernelization algorithm for the independent set problem on
graphs with maximum degree at most 3 that computes a kernel of size at
most $140k/47 < 3k$, where k is the given parameter. This improves the
known $3k$ upper bound on the kernel size for the problem, and implies a
lower bound of $140k/93$ on the kernel size for the vertex cover problem
on graphs with maximum degree at most 3.

1 Introduction

We study the independence number of graphs with maximum degree at most 3.
This study is motivated by the importance of the independent set problem on
graphs with maximum degree at most 3, abbreviated IS-3: Given an undirected
graph G with maximum degree at most 3 and a nonnegative integer k, decide if G
has an independent set of cardinality at least k. This problem is known to be NP-
complete [7], and the optimization version of the problem has received significant
interest from both areas of approximation and exact algorithms. After a long
sequence of results in each area, up to the authors' knowledge, the currently-best
approximation algorithm for the problem achieves a ratio that is arbitrarily close
to 6/5 [1], and the currently-best exact algorithm runs in time $O(1.0885^{n(G)})$,
where $n(G)$ is the number of vertices in G [13].

We take a combinatorial approach, establishing lower bounds on the independence number of a graph with maximum degree at most 3 that excludes the
three obstacle-graphs depicted in Figure 1 as subgraphs. Combinatorial results
of a similar nature are very common in the literature. Brook's theorem [2], published as early as 1941, implies that the independence number of a K_4-free graph

P. Kolman and J. Kratochvíl (Eds.): WG 2011, LNCS 6986, pp. 238–249, 2011.
© Springer-Verlag Berlin Heidelberg 2011

G with maximum degree at most 3 is at least $n(G)/3$. Staton showed in 1979 [11] that the independence number of a triangle-free graph G with maximum degree at most 3 is at least $5n(G)/14$. Staton's lower bound for triangle-free graphs is tight, as shown by the example given in [5]. A simpler proof of Staton's result was given by Jones in 1990 [10], and an even simpler proof was given by Heckman and Thomas in 2001 [9]. In their result [9], Heckman and Thomas define the notion of a *difficult component* in a graph, based on some "obstacle" subgraphs. They then prove that every triangle-free graph with maximum degree at most 3 has an independence number of at least $(4n(G) - e(G) - \lambda(G))/7$, where $e(G)$ and $\lambda(G)$ are the number of edges and the number of difficult components in G, respectively. They showed how their result implies Staton's result [11]. We mention that for a *connected* triangle-free graph with maximum degree at most 3, Fraughnaugh and Locke proved a lower bound of $(11n(G) - 4)/30$ on its independence number, which is strictly larger than $5n(G)/14$ for $n(G) \geq 15$ [6].

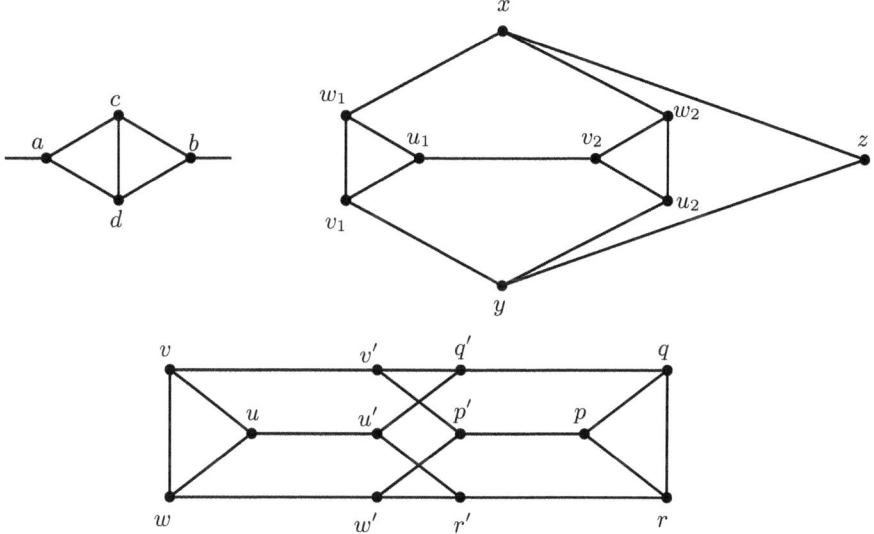

Fig. 1. The obstacle graphs. The graph on the top-left is referred to as the *small obstacle*; the graph on the top-right is referred to as the *medium obstacle*; and the graph at the bottom is referred to as the *large obstacle*. The degree of vertex z in the medium obstacle could be either 2 or 3, and the degree of vertices a and b in the small obstacle is 3, and a and b could be adjacent.

Very recently, Harant et al. [8] generalized Heckman and Thomas' result to graphs of maximum degree at most 3 that may contain triangles. They define the notion of a *difficult block* and use it to define the *bad components* of a graph: the components in which every block is either a difficult block or an edge between two difficult blocks. They then prove that the independence number of a K_4-free graph with maximum degree at most 3 is at least $(4n(G)-e(G)-\lambda(G)-tr(G))/7$, where $\lambda(G)$ and $tr(G)$ are the number of bad components and the number of vertex-disjoint triangles in G, respectively.

Since at most one vertex from a triangle can be in any maximum independent set of G, the presence of a lot of triangles poses an obstacle for obtaining a lower bound on the independence number that is larger than $n(G)/3$. Intuitively, one would think that we should be able to "gain" a certain fraction of the number of nontriangle vertices in G (i.e., vertices that do not appear in any triangle), above the "guaranteed" $n(G)/3$ lower bound on the independence number. This impression, however, is incorrect, as can be seen from the top-right and bottom graphs in Figure 1: the independence number of each of these two graphs is precisely $n(G)/3$, despite the presence of nontriangle vertices in it. A natural question to ask then is whether there are certain "obstacle" subgraphs that can be excluded from G, so that a lower bound of the form $n(G)/3 + nt(G)/c$ on the independence number can be derived, where $nt(G)$ is the number of nontriangle vertices in G, and c is some fixed (proper) constant; such a result can be interpreted as we are gaining a fraction $n(G)/c$ of the nontriangle vertices in G above the guaranteed value of $n(G)/3$.

In the current paper we prove the following combinatorial result: if G is a graph with maximum degree at most 3 that does not contain any of the three obstacle graphs depicted in Figure 1 as a subgraph, then the independence number of G is at least $n(G)/3 + nt(G)/42$. This lower bound on the independence number in terms of the number of vertices and the number of nontriangle vertices in the graph is tight, as implied by the well-known tight lower bound of $5n(G)/14 = n(G)/3 + nt(G)/42$ on the independence number of triangle-free graphs of maximum degree at most 3. To prove the aforementioned result, we use a discharging argument coupled with an amortized analysis. Up to the authors' knowledge, the technique of using amortized analysis for establishing lower bounds on the independence number is new, and could have wider applications.

The fact that the three obstacle graphs can be pre-processed (removed) in polynomial time allows this result to have algorithmic applications. As shown in Section 5, by introducing some reduction rules that enable us to lower bound the value of $nt(G)$ by $n(G)/10$ in the resulting graph, and then using the combinatorial result in the current paper, we obtain a kernelization algorithm for the IS-3 problem that produces a kernel of size at most $140k/47$ in $O(k)$ time. We note that since a K_4 subgraph must appear as a separate component in a graph of maximum degree 3, Brook's theorem [2] implies a kernel of size at most $3k$ for IS-3. The current result is a linear improvement over the $3k$ upper bound.

Using the notion of *duality* introduced in [3], the $140k/47$ upper bound on the kernel size for IS-3 implies a lower bound of $140k/93$ on the kernel size for the vertex cover problem on graphs with maximum degree at most 3, abbreviated VC-3.

Most of the proofs are omitted for lack of space.

2   Preliminaries

We assume familiarity with the basic notations and terminologies about graphs. For more information, we refer the reader to [12]. We consider only simple undirected graphs.

For a graph G we denote by $V(G)$ and $E(G)$ the set of vertices and edges of G, respectively; $n(G) = |V(G)|$ and $e(G) = |E(G)|$ are the number of vertices and edges in G. A set of vertices in $V(G)$ is said to be an *independent set* if no edge in $E(G)$ exists between any two vertices in this set. By $\alpha(G)$ we denote the *independence number* of G, that is, the cardinality/size of a maximum independent set in G.

For a set of vertices $S \subseteq V(G)$, we denote by $G[S]$ the subgraph of G induced by the set of vertices in S. For a vertex $v \in G$, $G - v$ denotes $G[V(G) \setminus \{v\}]$, and for a subset of vertices $S \subseteq V(G)$, $G - S$ denotes $G[V(G) \setminus S]$. By *removing* a subgraph H of G we mean removing $V(H)$ from G to obtain $G - V(H)$. For two vertices $u, v \in V(G)$, we denote by $G - (u, v)$ the graph $(V(G), E(G) \setminus \{(u, v)\})$, and by $G + (u, v)$ the *simple* graph $(V(G), E(G) \cup \{(u, v)\})$.

The *degree* of a vertex v in G, denoted $d(v)$, is the number of edges in G that are incident to v. The *degree* of G, denoted $\Delta(G)$, is $\Delta(G) = \max\{d(v) : v \in G\}$.

Call a vertex $v \in G$ a *triangle vertex* if v is a vertex of some triangle in G; otherwise call v a *nontriangle vertex*. We denote the number of vertex-disjoint triangles in G by $tr(G)$, and the number of nontriangle vertices in G by $nt(G)$.

Two triangles in a graph G are said to *share an edge* if the two triangles have exactly two vertices in common. Two triangles are said to be *adjacent* if the two triangles do not have any common vertex and a vertex in one of the triangles is adjacent to a vertex in the other triangle. Note that if a graph has maximum degree at most 3, then no two triangles in the graph can have exactly one vertex in common.

The *blocks* of a graph G are its maximal 2-connected subgraphs, its cut-edges, and its isolated vertices. Two blocks may only intersect at a cut-vertex of G. The *block-cutpoint tree* of a connected graph G is the tree whose vertices are the blocks and cut-vertices of G, with an edge from cut-vertex to each block that contains it. A connected graph that is not 2-connected has a nontrivial block-cutpoint tree; its *leaf blocks* are its blocks that are leaves in its block-cutpoint tree.

A *parameterized problem* is a set of instances of the form (x, k), where $x \in \Sigma^*$ for a finite alphabet set Σ, and k is a non-negative integer called the *parameter* [4]. A parameterized problem Q is *kernelizable* [4] if there exists a polynomial-time computable reduction that maps an instance (x, k) of Q to another instance (x', k') of Q such that: (1) $|x'| \leq g(k)$ for some recursive function g, (2) $k' \leq k$, and (3) (x, k) is a yes-instance of Q if and only if (x', k') is a yes-instance of Q. The instance x' is called the *kernel* of x. For more information on parameterized complexity and kernelization we refer the reader to [4].

The INDEPENDENT SET problem on graphs of maximum degree at most 3, abbreviated IS-3, is defined as follows: Given an undirected graph G with

$\Delta(G) \leq 3$, and a nonnegative integer k, determine if G has an independent set of size at least k.

3   Structural Results

We present in this section some structural results that will be used in the remaining sections of the paper. Let G be a graph such that $\Delta(G) \leq 3$.

Fact 1 (A triangle with a degree-2 vertex). *Let (u, v, w) be a triangle in G such that $d(u) = 2$. Then there exists a maximum independent set of G that contains u.*

Fact 2 (Two triangles sharing an edge). *Let (u, v, w) and (p, v, w) be two triangles in G that share an edge (v, w). Then there exists a maximum independent set of G that excludes v (or w).*

We assume in the remaining discussion in this section that no triangle in G contains a vertex of degree 2, and that no two triangles in G share an edge. Therefore, any two triangles in G are vertex-disjoint.

A sequence of distinct triangles $T_1, \ldots, T_\ell$, $\ell \geq 1$, in G is said to form a *path of triangles* if either $\ell = 1$, or if $\ell > 1$ and triangle T_i is adjacent to T_{i+1}, for $i = 1, \ldots, \ell - 1$. A path of triangles $T_1, \ldots, T_\ell$ is said to be a *cycle of triangles* if either $\ell > 2$ and T_1 and T_ℓ are adjacent, or $\ell = 2$ and (some) two vertices of T_1 are neighbors of two vertices of T_2 (i.e., there are at least two edges between the vertices of T_1 and the vertices of T_2). The *length* of a path/cycle of triangles is the number of triangles in it. A path of triangles is *maximal* if it is maximal under containment. We have the following lemmas:

Lemma 1 (Cycle of triangles). *Let $T_1, \ldots, T_\ell$ be a cycle of triangles in G, where $T_i = (u_i, v_i, w_i)$ for $i = 1, \ldots, \ell$, u_i is adjacent to v_{i+1} for $i = 1, \ldots, \ell - 1$, and u_ℓ is adjacent to v_1. Then there exists a maximum independent set of G that contains $\{v_1, \ldots, v_\ell\}$.*

Lemma 2 (Medium obstacle). *Let $T_1 = (u_1, v_1, w_1)$ and $T_2 = (u_2, v_2, w_2)$ be two adjacent triangles in G where u_1 is adjacent to v_2. Suppose that w_1 and w_2 share a common neighbor x, v_1 and u_2 share a common neighbor y, and x and y share a common neighbor z; that is, the subgraph of G induced by $V(T_1) \cup V(T_2) \cup \{x, y, z\}$ is a medium obstacle graph (see Figure 1). Then there exists a maximum independent set of G containing the set of vertices $\{x, y, v_2\}$.*

4   A Combinatorial Result

We will say that a graph G is *obstacle-free* if G does not contain any of the three obstacle graphs depicted in Figure 1 as a subgraph. Let G be an obstacle-free graph such that $\Delta(G) \leq 3$. This section is devoted to proving that $\alpha(G) \geq n(G)/3 + nt(G)/12$. The proof proceeds in three phases. In the first phase, we apply a set of graph operations to G to obtain a "simplified" graph. The operations performed in the first phase reduce the number of triangles in G without affecting the nontriangle vertices; these operations also ensure that given any independent set I' of the resulting graph, an independent set I of G containing I' can be obtained by adding a vertex from each of the triangles removed by one of these operations. Let G_1 be the graph resulting from G at the end of the first phase. In the second phase we apply more operations to G_1 to simplify its structure further. In contrast to the operations performed in the first phase, the operations performed in the second phase may remove nontriangle vertices. Each of these operations removes a subgraph H from G_1 to obtain the subgraph $G_1 - V(H)$ such that there exists a subset of vertices $S_H \subseteq V(H)$ that is an independent set satisfying: (1) $\alpha(G_1) \geq |S_H| + \alpha(G_1 - V(H))$, (2) $|S_H| \geq n(H)/3 + nt(H)/42$, and (3) $nt(G_1) = nt(H) + nt(G_1 - V(H))$. Let G_2 be the graph resulting from G_1 at the end of the second phase. In the third phase we prove that $\alpha(G_2) \geq n(G_2)/3 + nt(G_2)/42$. To do so, we prove using an amortized analysis technique that $\alpha(G_2) \geq (23n(G_2) - 6e(G_2) + nt(G_2))/42$. Since $e(G_2)$ is at most $3n(G_2)/2$, the desired statement follows. The following simple observation will be useful:

Observation 3. *Let G be a graph with $\Delta(G) \leq 3$ such that G is obstacle-free. Then for any subset of vertices $S \subseteq V(G)$, the subgraph $G - S$ of G has maximum degree at most 3 and is obstacle-free.*

4.1   The First Phase

In what follows we introduce a set of graph operations to be applied to the graph G to obtain a simpler graph in which every triangle is contained in one of two specific structures. We will need to keep track of how each operation affects the number of vertices, the number of nontriangle vertices, and the independence number of the graph G. For convenience, if an operation, or a set of operations, is applied to G to obtain a graph G', we will denote by $\delta_{n(G)}$, $\delta_{nt(G)}$, and $\delta_{\alpha(G)}$ the entities $n(G) - n(G')$, $nt(G) - nt(G')$, and $\alpha(G) - \alpha(G')$, respectively. The operations are considered/applied in the listed order.

Operation 4.1. *Let (u, v, w) be a triangle in G such that one of its vertices is of degree 2. Then set $G := G - \{u, v, w\}$.*

Operation 4.2. *Let (u, v, w) be a triangle in G such that $d(u) = d(v) = d(w) = 3$. Let u', v', and w' be the neighbors of u, v, w, respectively that are nontriangle vertices. If two vertices in $\{u', v', w'\}$ are adjacent then set $G := G - \{u, v, w\}$.*

Operation 4.3. *Let $T_1, \ldots, T_\ell$ be a cycle of triangles, where $T_i = (u_i, v_i, w_i)$, $i = 1, \ldots, \ell$, u_i is adjacent to v_{i+1} for $i = 1, \ldots, \ell - 1$, and u_ℓ is adjacent to v_1. Then set $G := G - \bigcup_{i=1}^{\ell} V(T_i)$.*

Operation 4.4. *Let $T_1, \ldots, T_\ell$, $\ell > 2$ be a maximal path of triangles, where $T_i = (u_i, v_i, w_i)$ for $i = 1, \ldots, \ell$, and u_i is adjacent to v_{i+1}, for $i = 1, \ldots, \ell - 1$.*

If w_1 and w_ℓ share a common neighbor x, v_1 and u_ℓ share a common neighbor y, and x and y share a common neighbor z, then set $G := (G - (V(T_1) \cup \bigcup_{i=3}^{\ell} V(T_i))) + (x, v_2) + (y, u_2)$.

Operation 4.5. *Let $T_1, \ldots, T_\ell$, $\ell > 1$, be a maximal path of triangles, where $T_i = (u_i, v_i, w_i)$ for $i = 1, \ldots, \ell$, and u_i is adjacent to v_{i+1}, for $i = 1, \ldots, \ell - 1$. Suppose that w_1 and w_ℓ share a common neighbor x and v_1, u_ℓ share a common neighbor y, and x and y do not share a neighbor. Then set $G := (G - \bigcup_{i=1}^{\ell} V(T_i)) + (x, y)$.*

Operation 4.6. *Let $T_1, \ldots, T_\ell$, $\ell > 1$, be a maximal path of triangles, where $T_i = (u_i, v_i, w_i)$ for $i = 1, \ldots, \ell$, and u_i is adjacent to v_{i+1}, for $i = 1, \ldots, \ell - 1$. Suppose that a vertex in T_ℓ, say w_ℓ, does not share a common neighbor with v_1 and does not share a common neighbor with w_1. Let w_ℓ' be the nontriangle vertex that is a neighbor of w_ℓ. Then set $G := (G - \bigcup_{i=2}^{\ell} V(T_i)) + (w_\ell', u_1)$.*

Operation 4.7. *Suppose that no two triangles in G are adjacent, and let (u, v, w) be a triangle in G such that $d(u) = d(v) = d(w) = 3$. Let u', v', and w' be the neighbors of u, v, w, respectively that are nontriangle vertices. If there are two vertices in $\{u', v', w'\}$, say u' and v', that do not share a common neighbor in G, then set $G' := (G - \{u, v, w\}) + (u', v')$.*

Proposition 1. *Let G be a graph with $\Delta(G) \leq 3$ such that G is obstacle-free. Let G_1 be the graph resulting from the application of Operations 4.1– 4.7 to G until none of these operations is applicable. Then the following are true:*

(i) $\Delta(G_1) \leq 3$ and G_1 is obstacle-free.
(ii) Every triangle vertex in G_1 has degree 3 (in G_1).
(iii) $\delta_{nt(G)} \leq 0$, and hence $nt(G_1) \geq nt(G)$.
(iv) $\delta_{\alpha(G)} \geq \delta_{n(G)}/3$.
(v) No two triangles in G_1 share an edge or are adjacent.
(vi) If (u, v, w) is a triangle in G_1 then each of u, v, w has exactly one neighbor u', v', w', respectively, that is a nontriangle vertex. Moreover, vertices u', v', w' are distinct, no two of them are adjacent, and every two of them share a neighbor.

4.2   The Second Phase

Let G_1 be the graph resulting from G after the first phase. In the second phase we apply more operations to simplify G_1 further. Each of the operations introduced in the second phase removes a subgraph H from G_1 satisfying a *local ratio* property. More formally, there exists a subset of vertices $S_H \subseteq V(H)$ that is an independent set, and such that the following are true: (1) $\alpha(G_1) \geq |S_H| + \alpha(G_1 - V(H))$, (2) $|S_H| \geq n(H)/3 + nt(H)/42$, and (3) $nt(G_1) = nt(H) + nt(G_1 - V(H))$. Since each of the operations introduced in this phase removes a set of vertices from G_1, by Observation 3, the graph resulting after each operation is obstacle-free and has maximum degree at most 3.

By part (v) of Proposition 1, no two triangles in G_1 share an edge or are adjacent; therefore, any two triangles in G_1 are disjoint. Moreover, by part (vi) of Proposition 1, every triangle vertex in G_1 is of degree 3 and has exactly one neighbor that is a nontriangle vertex. For a triangle vertex u, we denote its nontriangle neighbor by u'. Note that for two distinct triangle vertices u, v that are not vertices of the same triangle, u' can be identical to v'. For any triangle (u, v, w) in G_1, by part (vi) of Proposition 1, the vertices u', v', w' are distinct, no two of them are adjacent, and every two of them share a common neighbor that is a nontriangle vertex. Note that the three vertices u', v', w' could all share the same common neighbor (or two common neighbors).

Let u' be a vertex that is adjacent to some triangle vertex. It is not difficult to verify that u' has exactly one neighbor that is a triangle vertex, unless the graph G_1 has a component H of exactly 10 vertices, 4 of which are nontriangle vertices, and an independent set of size 4; in this case we have $\alpha(H) \geq n(H)/3 + nt(H)/42$. Therefore, we can assume that each such vertex u' is adjacent to exactly one triangle vertex.

From the above discussion, every triangle in G_1 must be contained in one of two subgraphs that we call *type-I steeple* and *type-II steeple*. Each of the two subgraphs consists of *triangle vertices* u, v, w, and their (distinct) neighbors u', v', w', respectively, that we call the *middle vertices* of the steeple. In a type-I steeple the three vertices u', v', w' all share at least one neighbor x, and in a type-II steeple each pair of vertices in $\{u', v', w'\}$ shares a distinct neighbor; we call the common neighbors of the vertices u', v', w' (or any two of them) the *top vertices* of the steeple. Since no two triangles in G_1 are adjacent, no edge exists between two top vertices of a type-II steeple. Therefore, G_1 contains a steeple as a subgraph if and only if it contains it as an induced subgraph. The vertices u', v', w' in a type-I steeple can be of degree 2 or 3 in the graph, and so can the vertices x, y, z in a type-II steeple. Moreover, the vertices u', v', w' in a type-I steeple can have a common neighbor(s) other than x.

Since G_1 is obstacle-free, it is not difficult to show using the properties of G_1 described in Proposition 1 that any two steeples in G_1 must be vertex-disjoint, unless the three middle vertices of one steeple all share another common neighbor beside the top vertex of the steeple, thus forming a second steeple with a different top vertex. We now apply more operations to simplify G_1. Each of these operations removes a subgraph H from G_1 to obtain a subgraph $G_1 - V(H)$ of G_1 such that there exists a subset of vertices $S_H \subseteq V(H)$ that is an independent set (the set of black vertices), and such that the following holds true: (1) $\alpha(G_1) \geq |S_H| + \alpha(G_1 - V(H))$, (2) $|S_H| \geq n(H)/3 + nt(H)/42$, and (3) $nt(G_1) = nt(H) + nt(G_1 - V(H))$. The purpose behind those operations is to make the steeples further apart.

Definition 1. Let S and S' be two steeples. The *distance* between S and S' is defined as follows. If S and S' are not vertex disjoint, then the distance between S and S' is zero. Otherwise, the distance between S and S' is the length of a shortest path between a middle vertex of S and a middle vertex of S' if both S and S' are type-I steeples, the length of a shortest path between a middle vertex of S and a top vertex of S' if S is a type-I steeple and S' is a type-II steeple, and the length of a shortest path between a top vertex of S and a top vertex of S' if both S and S' are type-II steeples.

Proposition 2. *There is a sequence of operations that can be applied to G_1 to obtain a graph G_2 such that G_2 satisfies the following properties:*

(i) *Every triangle in G_2 appears either in a type-I or a type-II steeple.*
(ii) *The distance between any type-I steeple and any other steeple in G_2 is at least 4, and the distance between any two type-II steeples in G_2 is at least 3.*
(iii) *For any type-I steeple in G_2, the neighbors of its middle vertices (if exist) are all distinct degree-3 vertices, and no two of them are adjacent.*
(iv) *For any type-II steeple in G_2, the neighbors of its top vertices (if exist) are all distinct degree-3 vertices.*
(v) *For any type-II steeple in G_2, none of its top vertices is adjacent to a vertex of a K_4^*, and no two of its top vertices are adjacent to two nonadjacent vertices of a C_5.*
(vi) *No C_5 in G_2 has two degree-2 nonadjacent vertices, and no K_4^* in G_2 has a degree-2 vertex.*
(vii) *If $\alpha(G_2) \geq n(G_2)/3 + nt(G_2)/42$ then $\alpha(G) \geq n(G)/3 + nt(G)/42$.*

4.3   The Third Phase

Let G_2 be the resulting graph after the second phase. Then G_2 satisfies all the properties described in Proposition 2. By part (vii) of Proposition 2, it suffices to show that $\alpha(G_2) \geq n(G_2)/3 + nt(G_2)/42$ to conclude that $\alpha(G) \geq n(G)/3 + nt(G)/42$.

As in the previous two phases, we will apply some operations to G_2. The purpose of the operations applied in this phase is the removal of all triangles from G_2. Each of these operations removes a subgraph H of G_2; however, in contrast to the operations performed in phase 2, the removed subgraph does not satisfy the local ratio property, namely that we can always add to any independent set of $G - V(H)$ an independent set of H of cardinality at least $n(H)/3 + nt(H)/42$. To show that $\alpha(G_2) \geq n(G_2)/3 + nt(G_2)/42$, we use a charging argument to measure the impact of each of these operations on the resulting graph, in addition to amortized analysis.

A block of a graph is called *difficult* [8] if it is isomorphic to one of the following four graphs: K_3, C_5, K_4^* (K_4 with two of its edges each subdivided twice), and a graph arising from C_5 by adding a new vertex and connecting it to three consecutive vertices of C_5. A connected graph is called *bad* [8] if every block of the graph is either a difficult block or an edge between two difficult blocks. Harant et al. [8] showed that if a graph G' is a K_4-free graph with $\Delta(G') \leq 3$ then $\alpha(G') \geq (4n(G') - e(G') - \lambda(G') - tr(G'))/7$, where $\lambda(G')$ is the number of components of G' that are bad. It follows that:

Lemma 3 ([8]). *Let G' be a triangle-free graph such that $\Delta(G') \leq 3$. If G' does not contain bad components then $\alpha(G') \geq (4n(G') - e(G'))/7$.*

Definition 2. Let H be a subgraph of G_2. Call an edge e with exactly one endpoint in H a *fringe edge* to H. Call a vertex $v \in V(H)$ a *boundary vertex* if v is an endpoint of a fringe edge to H; otherwise, call v an *internal vertex*. Let $e^+(H)$ denote the number of edges in H plus the number of fringe edges to H.

Suppose that we can apply some operations to G_2 to remove all triangles such that the following conditions are satisfied: (1) each operation removes a subgraph H such that there exists an independent set consisting of internal vertices in H of size at least $(23n(H) - 6e^+(H) + nt(H))/42$; and (2) the subgraph resulting from G_2 at the end of these operations is triangle-free and contains no bad components. Suppose that all these operations remove a subgraph G_2^- from G_2. Then by Lemma 3 we have $\alpha(G_2 - V(G_2^-)) \geq (4n(G_2 - V(G_2^-)) - e(G_2 - V(G_2^-)))/7 = (23n(G_2 - V(G_2^-)) - 6e(G_2 - V(G_2^-)) + nt(G_2 - V(G_2^-)))/42$; the last equality is true because $G_2 - V(G_2^-)$ is triangle-free, and hence $nt(G_2 - V(G_2^-)) = n(G_2 - V(G_2^-))$. Since the operations performed satisfy condition (1) above, we can add to any independent set of $G_2 - V(G_2^-)$ an independent set of G_2^- of size at least $(23n(G_2^-) - 6e^+(G_2^-) + nt(G_2^-))/42$. Therefore, the independence number of G_2 satisfies: $\alpha(G_2) \geq (23n(G_2) - 6e(G_2) + nt(G_2))/42$. Since $\Delta(G_2) \leq 3$, $e(G_2) \leq 3n(G_2)/2$, and $\alpha(G_2) \geq (14n(G_2) + nt(G_2))/42 = n(G_2)/3 + nt(G_2)/42$.

It follows that it is sufficient to show that each of the operations that we apply satisfies conditions (1) and (2) above. For a subgraph H of G_2, let $\phi(H) = |S_H| - (23n(H) - 6e^+(H) + nt(H))/42$, where S_H is a maximum independent set consisting of internal vertices in H. Then an operation that removes a subgraph H such that $\phi(H) \geq 0$ satisfies condition (1) above. We would like to show that each introduced operation that removes a subgraph H satisfies $\phi(H) \geq 0$. This will be the case for most of the operations that we apply except few. To circumvent this issue, we use amortized analysis. To implement this concept, for each operation that removes a subgraph H, we introduce a parameter $c(H)$, where $c(H)$ is the *cost* (or debit) of the operation, meant to possibly pay off the deficit of some later operations. We have the following definition:

Definition 3. Let H be a subgraph of G_2, and let S_H be a maximum independent set in H consisting of internal vertices to H. Let $E_1(H)$ be the set of fringe edges to H whose endpoint in $G_2 - V(H)$ is a neighbor of a top vertex in some type-II steeple, and let $E_2(H)$ be the set of remaining fringe edges to H. Let $s = 1/14$. Define the functions $\phi(H) = |S_H| - (23n(H) - 6e^+(H) + nt(H))/42$ and $\Phi(H) = \phi(H) - c(H)$, where $c(H) = (s/2)|E_1(H)| + (s/4)|E_2(H)|$.

Our task now becomes to remove all triangles in G_2 by applying operations, each of which removes a subgraph H from the graph, such that the sum of $\Phi(H)$ over all operations is nonnegative. At each point we consider a triangle in the resulting graph. This triangle will always be contained in a type-I or a type-II steeple. Since every triangle in G_2 is contained in a steeple, and since each steeple is 2-connected, no bad component of G_2 contains a triangle, and every bad component C in G_2 is triangle-free. Now if a component C of G_2 is triangle-free, then by [9], we have $\alpha(C) \geq 5n(C)/14 = n(C)/3 + n(C)/42$. Therefore, since our goal is to prove that $\alpha(G_2) \geq n(G_2)/3 + nt(G_2)/42$, by additivity, we can assume that at the beginning of this phase every component of G_2 contains some triangle, and that G_2 does not contain bad components. We will ensure that none of the operations applied to G_2 in this phase introduces bad components. We can prove that:

Proposition 3. *There exists a sequence of operations, each removing a subgraph H of G_2, that can be applied to G_2 to obtain a graph G_3 such that G_3 is triangle-free and contains no bad components, and such that the sum of $\Phi(H)$ over all removed subgraphs is nonnegative.*

Theorem 4. *Let G be an obstacle-free graph with $\Delta(G) \leq 3$. Then $\alpha(G) \geq n(G)/3 + nt(G)/42$.*

5   The Kernel

Let (G, k) be an instance of IS-3. The validity of Reduction Rules 1 – 4 follows from Fact 1, Fact 2, Lemma 1, and Lemma 2, respectively. The validity of Reduction Rule 5 can be easily verified.

Reduction Rule 1. *If (u, v, w) is a triangle in G such that $d(u) = 2$ then set $G := G - \{u, v, w\}$ and $k := k - 1$.*

Reduction Rule 2. *If (u, v, w) and (p, v, w) are two triangles in G that share an edge (v, w) then set $G := G - v$.*

Reduction Rule 3. *If $T_1, \ldots, T_\ell$ is a cycle of triangles in G then set $G := G - \bigcup_{i=1}^{\ell} V(T_i)$ and $k := k - \ell$.*

Reduction Rule 4. *If H is a subgraph of G that is a medium obstacle then set $G := G - V(H)$ and $k := k - 3$.*

Reduction Rule 5. *If C is a component in G that is a large obstacle then set $G := G - V(C)$ and $k := k - 4$.*

Definition 4. Call a graph *reduced* if none of Reduction Rules 1 – 5 applies to it.

Lemma 4. *Let G be a reduced graph. Then the number of nontriangle vertices $nt(G)$ satisfies $nt(G) \geq n(G)/10$.*

Combining Lemma 4 with Theorem 4 we obtain:

Theorem 5. *Given an instance (G, k) of IS-3, the algorithm **Kernelize** given in Figure 2 either accepts the instance (G, k) correctly, or returns an equivalent instance (G', k') of IS-3 such that $n(G') \leq 140k'/47$. The running time of the algorithm **Kernelize** is $O(k)$.*

Using the notion of *duality* introduced in [3], Theorem 5 yields a lower bound on the kernel size of the vertex cover problem on graphs with maximum degree at most 3 (VC-3):

Theorem 6. *Unless P=NP, the VC-3 problem does not have a kernel of size at most $140k/93$.*

Algorithm **Kernelize**

INPUT: *An instance (G, k) of* IS-3
OUTPUT: *An instance (G', k') of* IS-3

1. **if** $k \leq n(G)/4$ **then accept** the instance (G, k);
2. Apply Reduction Rules $1 - 5$ to (G, k) until none of them applies;
3. let (G', k') be the resulting instance;
4. **if** $k' \leq 47n(G')/140$ **then accept** the instance (G, k);
   else return the instance (G', k').

Fig. 2. The algorithm **Kernelize**

References

1. Berman, P., Fujito, T.: On approximation properties of the independent set problem for low degree graphs. Theory Comput. Syst. 32(2), 115–132 (1999)
2. Brooks, R.: On colouring the nodes of a network. Math. Phys. Sci. 37(4), 194–197 (1941)
3. Chen, J., Fernau, H., Kanj, I., Xia, G.: Parametric duality and kernelization: Lower bounds and upper bounds on kernel size. SIAM Journal on Computing 37(4), 1077–1106 (2007)
4. Downey, R., Fellows, M.: Parameterized Complexity. Springer, New York (1999)
5. Fajtlowicz, S.: On the size of independent sets in graphs. Congr. Numer. 21, 269–274 (1978)
6. Fraughnaugh, K., Locke, S.: Finding large independent sets in connected triangle-free 3-regular graphs. Journal of Combinatorial Theory B 65, 51–72 (1995)
7. Garey, M., Johnson, D., Stockmeyer, L.: Some simplified NP-complete problems. In: STOC, pp. 47–63. ACM (1974)
8. Harant, J., Henning, M., Rautenbach, D., Schiermeyer, I.: The independence number in graphs of maximum degree three. Discrete Mathematics 308(23), 5829–5833 (2008)
9. Heckman, C., Thomas, R.: A new proof of the independence ratio of triangle-free cubic graphs. Discrete Mathematics 233(1-3), 233–237 (2001)
10. Jones, K.: Size and independence in triangle-free graphs with maximum degree three. Journal of Graph Theory 14(5), 525–535 (1990)
11. Staton, W.: Some Ramsey-type numbers and the independence ratio. Transactions of the American Mathematical Society 256, 353–370 (1979)
12. West, D.: Introduction to graph theory. Prentice-Hall, NJ (1996)
13. Xiao, M.: A simple and fast algorithm for maximum independent set in 3-degree graphs. In: Rahman, M. S., Fujita, S. (eds.) WALCOM 2010. LNCS, vol. 5942, pp. 281–292. Springer, Heidelberg (2010)

On Computing an Optimal Semi-matching

František Galčík, Ján Katrenič, and Gabriel Semanišin[*]

Institute of Computer Science,
P.J. Šafárik University, Faculty of Science,
Jesenná 5, 041 54 Košice, Slovak Republic
{frantisek.galcik,jan.katrenic,gabriel.semanisin}@upjs.sk

Abstract. The problem of finding an optimal semi-matching is a generalization of the problem of finding classical matching in bipartite graphs. A *semi-matching* in a bipartite graph $G = (U, V, E)$ with n vertices and m edges is a set of edges $M \subseteq E$, such that each vertex in U is incident to at most one edge in M. An *optimal semi-matching* is a semi-matching with $deg_M(u) = 1$ for all $u \in U$ and the minimal value of $\sum_{v \in V} \frac{deg_M(v) \cdot (deg_M(v)+1)}{2}$. We propose a schema that allows a reduction of the studied problem to a variant of the maximum bounded-degree semi-matching problem. The proposed schema yields to two algorithms for computing an optimal semi-matching. The first one runs in time $O(\sqrt{n} \cdot m \cdot \log n)$ that is the same as the time complexity of the currently best known algorithm. However, our algorithm uses a different approach that enables some improvements in practice (e.g. parallelization, faster algorithms for special graph classes). The second one is randomized and it computes an optimal semi-matching with high probability in $O(n^\omega \cdot \log^{1+o(1)} n)$, where ω is the exponent of the best known matrix multiplication algorithm. Since $\omega \le 2.38$, this algorithms breaks through $O(n^{2.5})$ barrier for dense graphs.

1 Introduction

We consider unweighted bipartite graphs without loops, multiple edges, and isolated vertices. In general we use standard graph theory terminology and notations. If it is not stated otherwise, the symbols n and m stand for the number of vertices and the number of edges of a graph respectively. By $N(X)$ we understand the open neighborhood of a set X, subset of the set of vertices. If v is a vertex belonging to a set W then we refer it as a W-vertex. A *semi-matching M* of a bipartite graph $G = (U, V, E)$ is a set of edges such that each vertex of U is incident to at most one edge of M. A semi-matching M is a *maximum semi-matching* if each vertex of U is incident to exactly one edge of M. We denote by $SM(G)$ the set of all maximum semi-matchings of G. It is easy to see that any

[*] The research of the authors was supported in part by the Slovak Research and Development Agency under contracts APVV-0035-10 "Algorithms, Automata, and Discrete Data Structures" and SK-SI-0014-10 "Contemporary graph invariants and applications" and in part by the VVGS grant No. I-10-003-10.

P. Kolman and J. Kratochvíl (Eds.): WG 2011, LNCS 6986, pp. 250–261, 2011.
© Springer-Verlag Berlin Heidelberg 2011

bipartite graph without isolated vertices contains a maximum semi-matching and therefore $SM(G)$ is non-empty.

The studied problem is motivated by the following off-line load balancing scenario: Given a set of tasks and a set of machines, each of which can process a subset of tasks. Each task requires one unit of processing time and must be assigned to some machine that can process it. The tasks has to be assigned in such a manner that minimizes some optimization objective. One natural goal is to process all tasks with the *minimum total completion time*. Another goal is to minimize the *average completion time*, or *total flow time*, which is the sum of time units necessary for completion of all jobs (including the units while a job is waiting in the queue).

Let M be a semi-matching. Let us denote by $deg_M(v)$ the degree of a vertex $v \in U \cup V$ in the subgraph of a graph G induced by M. (The definition of a maximum semi-matching immediately implies that if $M \in SM$, then $|M| = |U| = \sum_{v \in V} deg_M(v)$.) For any maximum semi-matching M we define the *cost* of M, denoted by $cost(M)$, as follows: For any machine $v \in V$, its cost with respect to the semi-matching M is

$$cost_M(v) = \sum_{i=1}^{deg_M(v)} i = \frac{deg_M(v) \cdot (deg_M(v) + 1)}{2}.$$

Intuitively, $cost_M(v)$ is the total completion time of jobs assigned to the machine v. The cost of the maximum semi-matching M is the sum of costs taken over all machines:

$$cost(M) = \sum_{v \in V} cost_M(v).$$

An *optimal semi-matching* is a maximum semi-matching with the minimum possible cost among all maximum semi-matchings.

The problem of finding an *optimal semi-matching* is studied from early 70s when an $O(n^3)$-algorithm was developed independently by Horn [5] and Bruno et al. [2]. In the next period, no significant progress has been reported besides the results related to some special cases of the problem and its variations. The problem received considerable attention in the past few years. Harvey et al. [4] shown that an optimal semi-matching minimizes simultaneously the maximum number of tasks assigned to a machine, the flow time and the variance of loads on the machines. They gave also a characterization of an optimal assignment based on cost-reducing paths and provided two algorithms for finding an optimal semi-matching in time $O(n \cdot m)$. Their first algorithm builds the semi-matching step by step starting with an empty semi-matching. Subsequently, in each step, it finds an augmenting path from a free U-vertex to a vertex in V with the smallest possible degree. The second proposed algorithm starts with an arbitrary semi-matching M, where in each step, the algorithm finds a cost-reducing path. The authors provided the $O(\min\{n^{3/2}, m \cdot n\} \cdot m)$ upper bound for running time of this algorithm [4].

The unweighted semi-matching problem was recently generalized to the quasi-matching problem by Bokal et al. [1]. In this problem, a function g is provided, and each vertex $u \in U$ is required to be connected to at least $g(u)$ vertices in V. Observe that the maximum semi-matching problem corresponds to the case when $g(u) = 1$ for each $u \in U$. Bokal et al. [1] used another definition of the optimality and shown that a maximum semi-matching is an optimal semi-matching if and only if its degree distribution is lexicographically minimal. More precisely, let $G = (U, V, E)$ be a bipartite graph, $X \subseteq V$, and $M \subseteq E$. Denote by $d_M(X)$ the sequence $(d_1, d_2, \ldots, d_{|X|})$, $d_1 \geq d_2 \geq \ldots \geq d_{|X|}$, of degrees of vertices from X in the subgraph induced by M. We say that a maximum semi-matching M_1 is lexicographically smaller than a maximum semi-matching M_2 (denoted by $M_1 <_{lex} M_2$), if the sequence $d_{M_1}(V)$ is lexicographically smaller than the sequence $d_{M_2}(V)$. Analogously we define the relation $M_1 \leq_{lex} M_2$. A maximum semi-matching $M \in SM(G)$ is a *lexicographically minimal semi-matching*, if $M \leq_{lex} M'$ for all $M' \in SM(G)$. A set of all lexicographically minimal semi-matchings of the graph G is denoted as $LSM(G)$.

Very recently, Fakcharoenphol, Laekhanukit and Nanongkai [3] presented $O(\sqrt{n} \cdot m \cdot \log n)$ algorithm for the optimal semi-matching problem, which improves the previous $O(n \cdot m)$ algorithm by Harvey et al. [4]. Their algorithm uses the same reduction to the min-cost flow problem as in [4]. However, instead of canceling one negative cycle in each iteration, the algorithm exploits the structure of the graphs and the cost functions to cancel many negative cycles in a single iteration.

In this paper, we present a new algorithm for finding an optimal semi-matching with respect to optimality criterion due to [1]. Our algorithm is based on the divide-and-conquer strategy that yields to some practically useful properties of the algorithm. In the algorithm, we reduce the problem of computing a lexicographically minimal semi-matching to several computations of a specific variant of the maximum bounded-degree semi-matching problem. This problem is referred as a $BDSM$ problem. The proposed algorithm runs in time $O(T_{BDSM}(n, m) \cdot \log n)$, where $T_{BDSM}(n, m)$ is the time complexity in which the $BDSM$ problem can be solved. Hence our algorithm can viewed also as a schema that utilizes an algorithm for the $BDSM$ problem as a black-box subroutine, i.e. each algorithm for $BDSM$ problem results in an algorithm for computing an optimal semi-matching.

As a consequence, applying known algorithms we get two algorithms for computing an optimal semi-matching. The first one is deterministic and runs in time $O(\sqrt{n} \cdot m \cdot \log n)$ that is the same as the running time of the algorithm from [3]. The second one is randomized and computes an optimal semi-matching with high probability in time $O(n^{\omega} \cdot \log^{1+o(1)} n)$, where ω is the exponent of the best known matrix multiplication algorithm. Since $\omega < 2.38$, for dense graphs this algorithm breaks the $O(n^{2.5})$ upper-bound known before.

2   Preliminaries

The algorithm proposed in this paper is in fact a schema that reduces the problem of computing a lexicographically minimal semi-matching to several instances of a variant of the maximum bounded-degree semi-matching problem (shortly BDSM) that can be stated as follows:

PROBLEM [BDSM]
INSTANCE: A bipartite graph $G = (U, V, E)$ with $n = |U| + |V|$ vertices and $m = |E|$ edges; a capacity mapping $c : V \to \mathbb{N}$ satisfying $\sum_{v \in V} c(v) \leq 2 \cdot n$.
QUESTION: Find a semi-matching M in G with maximum number of edges such that $deg_M(v) \leq c(v)$ for all $v \in V$.

By $ComputeBDSM(G, c)$ we shall denote an algorithm for solving the instance of $BDSM$ problem.

2.1 Balancing Subroutine

In this subsection, we describe a balancing subroutine. The balancing subroutine provides a method for a transformation of a semi-matching M to a new semi-matching M^* in such a way that for a given semi-matching M_f and a subset W of V we guarantee that all its vertices are loaded at least as in the semi-matching M_f. Moreover, the proposed method is time-efficient and in some sense guarantees that the transformation affects only a small number of V-vertices.

Function Balance(G, M, M_f, W)

Input: a bipartite graph $G = (U, V, E)$, semi-matchings M and M_f in G,
and a subset $W \subseteq V$
Output: a semi-matching M_b satisfying properties stated in Lemma 1

foreach $w \in W$ **do** $R_w \leftarrow \{u \in U | (u, w) \in M_f \setminus M\}$;
foreach $u \in U$ **do** $p(u) \leftarrow nil$;
foreach $(u, v) \in M$ **do** $p(u) \leftarrow v$;
$Q \leftarrow \{w \in W | deg_M(w) < deg_{M_f}(w)\}$;
$X \leftarrow M$;
while $Q \neq \emptyset$ **do**
  $w \leftarrow$ arbitrary vertex from Q;
  $Q \leftarrow Q \setminus \{w\}$;
  $u \leftarrow$ arbitrary vertex from R_w;
  $R_w \leftarrow R_w \setminus \{u\}$;
  if $p(u) \neq nil$ **then**
    $X \leftarrow X \setminus \{(u, p(u))\}$;
    if $p(u) \in W \wedge deg_X(p(u)) < deg_{M_f}(p(u))$ **then** $Q \leftarrow Q \cup \{p(u)\}$;
  end
  $X \leftarrow X \cup \{(u, w)\}$;
  $p(u) \leftarrow w$;
  if $deg_X(w) < deg_{M_f}(w)$ **then** $Q \leftarrow Q \cup \{w\}$;
end
return $X(= M_b)$

Lemma 1. *Let M and M_f be semi-matchings of a bipartite graph $G = (U, V, E)$ and let $V_{M_f} = \{v \in V | \exists u \in U, \exists v' \in V, (u, v) \in E \wedge (u, v') \in M_f\}$. Then for any subset $W \subseteq V$, a semi-matching M_b of G satisfying*

(1) *$deg_{M_b}(v) = deg_{M_f}(v)$, for each $v \in W$ with $deg_M(v) \leq deg_{M_f}(v)$,*
(2) *$deg_{M_f}(v) \leq deg_{M_b}(v) \leq deg_M(v)$, for each $v \in W$ with $deg_M(v) > deg_{M_f}(v)$,*
(3) *$deg_{M_b}(v) \leq deg_M(v)$, for each $v \in V \setminus W$ with $v \in V_{M_f}$,*
(4) *$deg_{M_b}(v) = deg_M(v)$, for each $v \in V \setminus W$ with $v \notin V_{M_f}$, and*
(5) *$|M_b| \geq |M|$*

can be obtained as a results of the function $Balance(G, M, M_f, W)$ in linear time $O(|V(G)| + |E(G)|)$.

2.2   Dividing Subroutine

In the proposed algorithm, we apply the divide-and-conquer strategy. The key element of the algorithm is a dividing construction that allows to reduce the problem of finding a lexicographically minimal semi-matching of a bipartite graph G into two problems of finding a lexicographically minimal semi-matchings of induced subgraphs of G.

The dividing strategy is based on the following observation: Let us assume that we have already computed a maximum semi-matching M of a bipartite graph $G = (U, V, E)$ in the situation when the maximum load (maximal number of incident matching edges) of each V-vertex is bounded from above by a constant cut (the value of cut will be determined in the main algorithm, where cut is used as a variable). Obviously, the computed semi-matching M is not necessary maximum. However, one can expect that if a V-vertex v is not fully loaded in M (i.e., $deg_M(v) < cut$), an enlargement of the upper bound for the load of the vertex v would not help to find a larger semi-matching. On the other hand, if $deg_M(v) = cut$, it could be possible to enlarge M after increasing the allowed capacity (maximum allowed load) of v. As we shall show later, according to a computed semi-matching and an actual load of V-vertices, we can even divide the graph into two disjoint subgraphs in such a way, that the loads of vertices in all lexicographically minimal semi-matchings in one of them are upper-bounded by a given constant cut, and the loads of vertices in all lexicographically minimal semi-matchings in the other subgraph are lower-bounded by cut.

The following theorem states a few properties of our main algorithm that are crucial for estimation of the total time complexity of the algorithm.

Theorem 1. *Let $down, cut, up$ be non-negative integers such that $down < cut < up$ and $cut \leq 2 \cdot down$. Let $G = (U, V, E)$ be a bipartite graph such that for any lexicographically minimal semi-matching $M \in LSM(G)$ of G it holds $down \leq deg_M(v) \leq up$ for all $v \in V$. Let M_f be a semi-matching in G that satisfies $deg_{M_f}(v) \geq down$ for each vertex $v \in V$. Then there exist bipartite graphs $G^- = (U^-, V^-, E^-)$, $G^+ = (U^+, V^+, E^+)$ and semi-matchings $M_f^- \subseteq E^-$ in G^-, $M_f^+ \subseteq E^+$ in G^+ such that the following properties hold*

(1) G^-, G^+ are disjoint and induced subgraphs of G, such that $U^- \cup U^+ = U$, $V^- \cup V^+ = V$, $U^- \cap U^+ = \emptyset$, $V^- \cap V^+ = \emptyset$,

(2) $\forall v \in V^-$: $deg_{M_f^-}(v) \geq down$,

(3) $\forall v \in V^+$: $deg_{M_f^+}(v) = cut$,

(4) $\forall M^- \in LSM(G^-)$, $\forall v \in V$: $down \leq deg_{M^-}(v) \leq cut$,

(5) $\forall M^+ \in LSM(G^+)$, $\forall v \in V$: $cut \leq deg_{M^+}(v) \leq up$, and

(6) $\forall M^- \in LSM(G^-)$, $\forall M^+ \in LSM(G^+)$: $M^- \cup M^+ \in LSM(G)$.

Moreover, graphs G^+ and G^- can be constructed in time $T_{BDSM}(n, m) + O(n + m)$, where $n = |U| + |V|$, $m = |E|$ and $T_{BDSM}(n, m)$ is time complexity for solving an instance of the Problem BDSM for a graph with n vertices and m edges.

By $Divide(G, down, cut, up, M_f)$ we shall denote an algorithm that constructs graphs G^-, G^+ and semi-matchings M^-, M^+ as claimed in Theorem 1.

3   The Main Algorithm

Our algorithm for computing a lexicographically minimal semi-matching is an implementation of the divide-and-conquer strategy whose division stage is based on Theorem 1. The key computational element of the proposed algorithm is a recursive subroutine $ComputeBLSM(G, down, up, M_f, l)$ that expects five parameters. First three parameters are a bipartite graph $G = (U, V, E)$ and two non-negative integers $down$ and up, such that for an arbitrary lexicographically minimal semi-matching M of G it holds that $down \leq deg_M(v) \leq up$ for all $v \in V$. We use here a special value $\diamond_\infty$ to denote that no upper-bound on the maximum degree of vertices in a lexicographically minimal semi-matching is given. In all inequalities, we assume that $x \leq \diamond_\infty$ is valid for all $x \in \mathbb{N}$. Observe that the whole computation starts with the parameter $down$ set to 0 and the parameter up set to $\diamond_\infty$. Note also that it follows from the algorithm that $\diamond_\infty$ can be practically set to the value $2 \cdot n$ without affecting the computation. The fourth parameter is a semi-matching M_f in G satisfying $deg_{M_f}(v) \geq down$ for all $v \in V$. This semi-matching is important for the $Divide$ subroutine. Moreover, it is utilized to transfer some results of already realized computation to subproblems. The last parameter l is not related to the computation. It is added as a supplementary element used in the time complexity analysis and denotes a level (depth of recursive call) of the current subproblem.

Algorithm 1. ComputeLSM(G)

Input: a bipartite graph $G = (U, V, E)$;
Output: a lexicographically minimal semi-matching M;
return $ComputeBLSM(G, 0, \diamond_\infty, \emptyset, 0)$;

Function ComputeBLSM(G, *down*, *up*, M_f, *l*)

> **Input**: a bipartite graph $G = (U, V, E)$, *down* $\in \mathbb{N}$, *up* $\in \mathbb{N} \cup \{\diamond_\infty\}$, s.t.,
> $\quad \forall M \in LSM(G), \forall v \in V : down \leq deg_M(v) \leq up$, and a
> $\quad$ semi-matching M_f in G, s.t., $\forall v \in V : deg_{M_f}(v) \geq down$;
> **Output**: a lexicographically minimal semi-matching M of G
>
> //the trivial case;
> **if** $G = \emptyset$ **then return** $\emptyset$;
> **if** $up \neq \diamond_\infty \wedge up - down \leq 1$ **then**
> $\quad\mid\quad M \leftarrow ComputeBDSM(G, c : \{c(v) \rightarrow up | v \in V\})$;
> $\quad\mid\quad M \leftarrow Balance(G, M, M_f, V)$;
> $\quad\mid\quad$ **return** M ;
> **end**
>
> //recursive case;
> **if** $up \neq \diamond_\infty$ **then**
> $\quad\mid\quad cut \leftarrow \lfloor (down + up)/2 \rfloor$;
> **else**
> $\quad\mid\quad cut \leftarrow max\{2 \cdot down, 1\}$;
> **end**
> $(G^-, M_f^-, G^+, M_f^+) \leftarrow Divide(G, down, cut, up, M_f)$;
> $M^- \leftarrow ComputeBLSM(G^-, down, cut, M_f^-, l+1)$;
> $M^+ \leftarrow ComputeBLSM(G^+, cut, up, M_f^+, l+1)$;
> **return** $M^- \cup M^+$;

Let us analyze the correctness of the algorithm. In what follows, we shall use the symbol ? to denote a parameter whose value is not important in an actual context.

Lemma 2. *Let $G = (U, V, E)$ be a bipartite graph and M be a maximum semi-matching of G such that $\forall v, w \in V: deg_M(v) - deg_M(w) \leq 1$. Then $M \in LSM(G)$.*

Proof. Follows from Theorem 3.1 in [4] which shows that a semi-matching is optimal if and only if no cost reducing path exists. $\qquad\square$

Lemma 3. *Let l_{max} be the maximal value of the parameter l that occurs in a call of the subroutine ComputeBLSM during the computation of ComputeLSM(G). Then, $l_{max} = O(\log n)$ where $n = |U| + |V|$.*

Proof. First we show, using mathematical induction with respect to l, that if a subproblem $ComputeBLSM(?, down, \diamond_\infty, ?, l)$, $l \geq 1$, occurs during the computation of $ComputeLSM(G)$, then it holds $down = 2^{l-1}$. The call of $ComputeBLSM(G, 0, \diamond_\infty, \emptyset, 0)$ can yield only to the computation of a subproblem $ComputeBLSM(?, 1, \diamond_\infty, ?, 1)$ and in this case the induction hypothesis is valid. For $l > 1$, a subproblem $ComputeBLSM(?, down, \diamond_\infty, \emptyset, l-1)$ can yield only to the computation of a subproblem $ComputeBLSM(?, 2 \cdot down, \diamond_\infty, \emptyset, l)$.

From the induction hypothesis, it follows $2 \cdot down = 2 \cdot 2^{l-2} = 2^{l-1}$. Similarly, we can show by induction on l and utilizing Theorem 1 that if a problem $ComputeBLSM(G', down, \diamond_\infty, ?, l)$ occurs in computation then G' is a subgraph of G. For $down \geq n = |U| + |V|$, the precondition of $ComputeBLSM$ implies that an input graph $G' = (U', V', E')$ is empty. Indeed, if for any semi-matching $M \in LSM(G')$ it holds $deg(v) \geq deg_M(v) \geq down \geq n$ for all $v \in V' \neq \emptyset$, we get a contradiction to $deg(v) < n$ (as G' is a subgraph of G). Hence, if a subproblem $ComputeBLSM(G', ?, \diamond_\infty, ?, l)$ with $G' \neq \emptyset$ is a part of the computation, then $2^{l-1} = down < n$ which implies $l \leq \log n + 1$.

In order to prove formally an upper-bound for l_{max}, we shall show, by induction on l, that if a subproblem $ComputeBLSM(G', down, up, ?, l)$ with $up \neq \diamond_\infty$, $l \geq 1$, and $G' \neq \emptyset$, is a part of the computation then $up - down \leq \frac{n}{2^{l-2-\log n}} + 2 = \frac{n^2}{2^{l-2}} + 2$. Since $ComputeBLSM(?, 0, 1, ?, l)$ is the only subproblem with $l = 1$ and $up \neq \diamond_\infty$ that can occur in the computation, the claim is obvious for $l = 1$. So we can assume that the induction hypothesis holds for $l = 1$. Let us assume now that $up - down \leq \frac{n^2}{2^{l-3}} + 2$ for all subproblems $ComputeBLSM(?, down, up, ?, l - 1)$ at level $l - 1$. Each subproblem $ComputeBLSM(?, down, up, ?, l - 1)$ can be divided into at most two subproblems $ComputeBLSM(?, down, cut, ?, l)$ and $ComputeBLSM(?, cut, up, ?, l)$ where $cut = \lfloor (down + up)/2 \rfloor$. By an application of the induction hypothesis, for a subproblem $ComputeBLSM(?, down, cut, ?, l)$, we get $cut - down = \lfloor (down + up)/2 \rfloor - down \leq (up - down)/2 \leq \frac{n^2}{2^{l-2}} + 1$. Similarly, for a subproblem $ComputeBLSM(?, cut, up, ?, l)$, we get $up - cut \leq (up - down)/2 + 1 \leq \frac{n^2}{2^{l-2}} + 2$. Hence, the claim holds for the both subproblems. It remains to analyze the case when a subproblem $ComputeBLSM(?, down, up, ?, l)$ is forced by $ComputeBLSM(?, down, \diamond_\infty, ?, l - 1)$. We already know that the call $ComputeBLSM(?, down, \diamond_\infty, ?, l - 1)$ implies $down - 2^{l-2}$ and $l - 1 \leq \log n + 1$. The problem $ComputeBLSM(?, down, \diamond_\infty, ?, l - 1)$ can force only a subproblem $ComputeBLSM(?, down, 2 \cdot down, ?, l)$. For this subproblem, we get $2 \cdot down - down = down = 2^{l-2} \leq n$. Since $l - 2 \leq \log n$, it implies $n \leq \frac{n^2}{2^{l-2}}$. It means that in this case the claim holds too.

Hence, if a subproblem $ComputeBLSM(?, down, up, ?, l)$ with $up \neq \diamond_\infty$ occurs during a computation, it holds $up - down \leq \frac{n}{2^{l-2-\log n}} + 2$. For $l > 2 + 2 \cdot \log n$, it follows $\frac{n}{2^{l-2-\log n}} + 2 < 3$ and $up - down \leq 2$. In this case the construction of subroutine $ComputeBLSM$ guarantees that if a new subproblem is forced then it is the trivial one. Since there are no subproblems with $up \neq \diamond_\infty$ for $l > 3 + 2 \cdot \log n$ and no subproblem with $up = \diamond_\infty$ for $l > \log n + 3$, the assertion follows.   □

Lemma 4. *The algorithm ComputeLSM computes a lexicographically minimal semi-matching of G.*

Proof. We have to show that the computation is finite and a semi-matching returned by $ComputeBLSM(G, 0, \diamond_\infty, \emptyset, 0)$ is lexicographically minimal. Since each problem $ComputeBLSM(?, ?, ?, ?, l)$ can yield to the computation of at most two subproblems $ComputeBLSM(?, ?, ?, ?, l + 1)$ and from Lemma 3 we have $l \in O(\log n)$, the computation is finite.

Now, we show the correctness of the algorithm. In particular, we show that the subroutine $ComputeBLSM(G, down, up, M_f, l)$ returns a semi-matching M such that $M \in LSM(G)$. We show it by a backward induction on l. In what follows, G refers to the first parameter of a given subproblem $ComputeBLSM$ and not to a parameter of $ComputeLSM$ which initiates the whole computation.

Let us analyze the induction base, i.e., when $l = l_{max}$ and l_{max} is the maximal value of the parameter l that occurs in the computation. Since no subproblem at level $l_{max} + 1$ is forced, either the input graph is empty or $up - down \leq 1$. In the former case is obvious. In the latter case when $up - down \leq 1$, Lemma 1 implies that $down \leq deg_M(v) \leq up$ for all $v \in V$. Let M' be an arbitrary semi-matching such that $M' \in LSM(G)$. Precondition of $ComputeBLSM$ implies $deg_{M'}(v) \leq up$ for all $v \in V$. If $|M'| > |M|$, we get a contradiction with maximality of a semi-matching computed by $ComputeBDSM$. Indeed, M' is a feasible solution for the $BDSM$ problem. Hence, due to maximality of M' we have $|M'| = |M|$. It follows that the semi-matching M is a maximum semi-matching for G. And therefore Lemma 2 implies that $M \in LSM(G)$.

Let us analyze the inductive step. If no subproblem is forced, the proof is the same as in the induction base. In the complementary case, which corresponds to a recursive step of the algorithm, two new subproblems are forced. Let G^-, G^+, M_f^-, and M^+ be the results of the subroutine $Divide$. Note that all assumptions of Theorem 1 are satisfied due to preconditions of $ComputeBDSM$ and the fact, which can be easily shown by induction on l, that $up \leq 2 \cdot down + 1$ in the case when $up \neq \diamond_\infty$. Therefore, properties (2) and (4) of Theorem 1 imply that preconditions of $ComputeBLSM(G^-, down, cut, M_f^-, l+1)$ are satisfied as well. Similarly, preconditions of $ComputeBLSM(G^+, cut, up, M_f^+, l+1)$ are satisfied due to properties (3) and (5) of Theorem 1. The new subproblems are at level $l+1$. Hence we can apply the induction hypothesis and we get $M^- \in LSM(G^-)$ and $M^+ \in LSM(G^+)$. Finally, due to the property (6) of Theorem 1 the returned semi-matching $M^- \cup M^+$ belongs to $LSM(G)$.   □

Lemma 5. *Let $\mathcal{G}_l$ be a collection of subgraphs of G such that $G' \in \mathcal{G}_l$ if and only if a subproblem $ComputeBLSM(G', ?, ?, ?, l)$ occurs in the computation of $ComputeLSM(G)$. Then, all graphs in $\mathcal{G}_l$ are disjoint and $\bigcup_{G' \in \mathcal{G}_l} G'$ is a subgraph of G.*

Proof. We show the assertion of the lemma by induction on l. For $l = 0$, the claim is trivial. Let us assume that the claim holds for $l - 1$. Clearly, if $G' \in \mathcal{G}_l$, then there is a subproblem $ComputeBLSM(G^*, ?, ?, ?, l-1)$ that forces a computation $ComputeBLSM(G', ?, ?, ?, l)$. As we can see in the subroutine $ComputeBLSM$, new subproblems are generated as the result of the $Divide$ subroutine. In particular, a subproblem $ComputeBLSM(G^*, ?, ?, ?, l-1)$ can force only subproblems $ComputeBLSM(G^-, ?, ?, ?, l)$ and $ComputeBLSM(G^+, ?, ?, ?, l)$. Then, due to the property (1) of Theorem 1, both graphs G^- and G^+ are disjoint subgraphs of G^* and the claim follows.   □

The proposed algorithm is a recursive algorithm that is normally executed in the DFS like manner. However, to complete the analysis of time complexity, let us consider a computation in which the computational tree is traversed in BFS-like manner. In particular, the computation works in stages. In a stage l, we process all subproblems $ComputeBLSM(?,?,?,?,l)$ at level l. Each subproblem is preprocessed and subproblems at level $l + 1$ are constructed. Consequently, execution in the stage l is interrupted and we start to solve subproblems at level $l + 1$ and the stage $l + 1$ starts. After the stage $l + 1$ is completed, we continue with the stage l. From results obtained at the stage $l + 1$ we construct results for all subproblems at level l and the stage l is completed.

The motivation for considering BFS-like computation is the following. Observe that in order to compute $ComputeBLSM(G',?,?,?,l)$, we execute the subroutine $ComputeBDSM(G',?)$ that computes a solution for the $BDSM$ problem. Let us denote the graph $\bigcup_{G' \in \mathcal{G}_l} G'$ by $G^l = (U^l, V^l, E^l)$. By Lemma 5, the graph G^l is a subgraph of G that consists of components corresponding to graphs in $\mathcal{G}_l$. Therefore, instead of executing the subroutine $ComputeBDSM$ for each graph $G' = (U', V', E') \in \mathcal{G}_l$ separately, we can execute the subroutine $ComputeBDSM$ with the graph G^l as an input. Since vertex sets of all graphs in $\mathcal{G}_l$ are disjoint, we can set capacities for all V-vertices separately. Observe that $\sum_{v \in V(G^l)} c(v) = \sum_{G' \in \mathcal{G}_l} \sum_{v \in V'} c(v) \leq \sum_{G' \in \mathcal{G}_l} 2 \cdot (|U'| + |V'|) = 2 \cdot (|U^l| + |V^l|)$. Hence the precondition of $ComputeBDSM$ is satisfied for input graph G^l. It means that using a BFS-like computational approach we can merge computations of bounded-degree semi-matchings for individual graphs occurring in subproblems at a given level to one computation of a maximum bounded-degree semi-matching at the given level.

Theorem 2. *Let $G = (U, V, E)$ be a bipartite graph. A lexicographically minimal semi-matching of G can be computed in time $O((n + m + T_{BDSM}(n,m)) \cdot \log n)$ where $n = |U| + |V|$, $m = |E|$, and $T_{BDSM}(n,m)$ is the time complexity of an algorithm for the $BDSM$ problem. Moreover, the algorithm for the $BDSM$ problem is applied $O(\log n)$ times during the computation.*

Proof. Let us analyze the modified algorithm that realizes a BFS-like computation of $ComputeLSM$. The correctness of the algorithm follows from Lemma 4. Note that, in order to allow sharing some computational parts, we changed only the order in which independent subproblems are computed. Lemma 3 implies that there are $O(\log n)$ stages in the computation. If we do not take into account time spent by the run of $ComputeBDSM$, due to Theorem 1 the time complexity of each subproblem is linear in size of the input subgraph. From Lemma 5, we have that $\bigcup_{G' \in \mathcal{G}_l} G'$ is a subgraph of G and graphs in $\mathcal{G}_l$ are disjoint. Hence $\sum_{G' \in \mathcal{G}_l} O(|V(G')| + |E(G')|) = O(|V(G)| + |E(G)|) = O(n + m)$. In each stage, the algorithm $ComputeBDSM$ is executed with a subgraph of G as an input. Therefore, the computation of $ComputeBDSM$ is in time $T_{BDSM}(n,m)$. It follows that the total time complexity of a stage is $O(n + m) + T_{BDSM}(n,m)$. Taking into account the number of stages, the claim follows.   □

Note that the previous theorem holds even in the case of the standard DFS-like computation under assumption that the time complexity $T_{BDSM}(n,m)$ satisfies that $\sum_i T_{BDSM}(n_i, m_i) = O(T_{BDSM}(n, m))$ for all n_i, m_i such that $\sum_i n_i = n$ and $\sum_i m_i = m$.

In what follows, we show how to construct efficient algorithms for finding optimal semi-matching by an application of known algorithms for other problems.

Theorem 3. *Let $G = (U, V, E)$ be a bipartite graph with n vertices and m edges. A lexicographically minimal semi-matching of G can be computed in time $O(\sqrt{n} \cdot m \cdot \log n)$.*

Proof. In [6], Katrenič and Seminišin investigated a problem of finding a maximum (f, g)-semi-matching of a given graph G. The special case of this problem is a computation of a maximum $(1, c)$-semi-matching. A semi-matching M is a $(1, c)$-semi-matching of a bipartite graph $G = (U, V, E)$ if and only if $deg_M(u) \leq 1$ for all $u \in U$ and $deg_M(v) \leq c(v)$ for all $v \in V$. It is easy to see that any instance (G, c) of the $BDSM$ problem can be solved by finding a maximum $(1, c)$-semi-matching of the graph G. The authors of [6] showed that a maximum $(1, c)$-semi-matching of a graph G can be constructed in time $O(\sqrt{n} \cdot m)$, hence $T_{BDSM}(n, m) = O(\sqrt{n} \cdot m)$. Finally, Theorem 2 implies that a lexicographically minimal semi-matching can be computed in time $O(\sqrt{n} \cdot m \cdot \log n)$.   □

Proposition 1. *For any bipartite graph $G = (U, V, E)$ with n vertices and a capacity mapping c, the $BDSM$ problem can be solved with high probability in time $O(n^\omega)$, where ω is the exponent of the best known matrix multiplication algorithm.*

Proof. We use a simple reduction to the problem of maximum matching in bipartite graphs. First, we create a new graph G' from the graph $G = (U, V, E)$ by splitting each vertex $v \in V$ exactly $c(v)$ times. The new graph has $|U| + \sum_{v \in V} c(v)$ vertices, which is at most $3 \cdot n$ due to precondition about the mapping c. Next, we compute the maximum matching M' in G'. In [7], the authors shown that a maximum matching can be computed in time $O(n^\omega)$ with high probability. Finally, from M' we construct a semi-matching M for G by merging the corresponding vertices.   □

Theorem 4. *Let $G = (U, V, E)$ be a bipartite graph with n vertices and m edges. A lexicographically minimal semi-matching of G can be computed with high probability in time $O(n^\omega \cdot \log n \cdot \log \log n)$, where ω is the exponent of the best known matrix multiplication algorithm.*

Proof. From Proposition 1, $T_{BDSM}(n, m) = O(n^\omega)$. Applying Theorem 2, there are $O(\log n)$ instances of the $BDSM$ problem solved during the computation. Executing the randomized algorithm for the $BDSM$ problem $\log \log n$ times on each problem instance, the claim follows from Theorem 2.   □

4   Conclusion

We presented a schema reducing the problem of computing an optimal semi-matching to a variant of the maximum bounded-degree semi-matching problem.

This problem can be efficiently solved utilizing known algorithms for maximum (f, g)-semi-matching or maximum matching using Gaussian elimination. The schema is based on a divide-and-conquer strategy. The problem is recursively divided into smaller independent subproblems that can be solved separately. This property can lead to a construction of simple and efficient parallel algorithms for computing an optimal semi-matching. Another its useful property follows from reduction to a variant of the maximum bounded-degree semi-matching problem. It shows that the time complexity of finding an optimal semi-matching is at most $O(\log n)$ times worse than the time complexity of solving the $BDSM$ problem, which matches best known complexity upper bounds for maximum matchings in bipartite graphs.

References

1. Bokal, D., Brešar, B., Jerebic, J.: A generalization of Hungarian method and Hall's theorem with applications in wireless sensor networks. IFMF, University of Ljubljana, Slovenia, Preprint Series 47(1102), 15 (2009)
2. Bruno, J., Coffman Jr., E.G., Sethi, R.: Scheduling independent tasks to reduce mean finishing time. Commun. ACM 17, 382–387 (1974)
3. Fakcharoenphol, J., Laekhanukit, B., Nanongkai, D.: Faster Algorithms for Semi-Matching Problems (Extended Abstract). In: Abramsky, S., Gavoille, C., Kirchner, C., Meyer auf der Heide, F., Spirakis, P.G. (eds.) ICALP 2010. LNCS, vol. 6198, pp. 176–187. Springer, Heidelberg (2010)
4. Harvey, N.J.A., Ladner, R.E., Lovász, L., Tamir, T.: Semi-matchings for bipartite graphs and load balancing. Journal of Algorithms 59(1), 53–78 (2006)
5. Horn, W.A.: Minimizing average flow time with parallel machines. Operations Research 21(3), 846–847 (1973)
6. Katrenič, J., Semanišin, G.: A generalization of Hopcroft-Karp algorithm for semi-matchings and covers in bipartite graphs, Technical report (2010)
7. Mucha, M., Sankowski, P.: Maximum matchings via gaussian elimination. In: Proceedings of the 45th Annual IEEE Symposium on Foundations of Computer Science, pp. 248–255. IEEE Computer Society Press, Washington, DC, USA (2004)

Planar k-Path in Subexponential Time and Polynomial Space

Daniel Lokshtanov[1], Matthias Mnich[2], and Saket Saurabh[3]

[1] University of California, San Diego, USA
daniello@ii.uib.no
[2] International Computer Science Institute, Berkeley, USA
mmnich@icsi.berkeley.edu
[3] The Institute of Mathematical Sciences, Chennai, India
saket@imsc.res.in

Abstract. In the k-PATH problem we are given an n-vertex graph G together with an integer k and asked whether G contains a path of length k as a subgraph. We give the first subexponential time, polynomial space parameterized algorithm for k-PATH on planar graphs, and more generally, on H-minor-free graphs. The running time of our algorithm is $O(2^{O(\sqrt{k}\log^2 k)}n^{O(1)})$.

1 Introduction

In the k-PATH problem we are given a n-vertex graph G and integer k and asked whether G contains a path of length k as a subgraph. The problem is a generalization of the classical HAMILTONIAN PATH problem, which is known to be NP-complete [16] even when restricted to planar graphs [15]. On the other hand k-PATH is known to admit a subexponential time parameterized algorithm when the input is restricted to planar, or more generally H-minor free graphs [4]. For the case of k-PATH a subexponential time parameterized algorithm means an algorithm with running time $2^{o(k)}n^{O(1)}$. More generally, in parameterized complexity problem instances come equipped with a *parameter k* and a problem is said to be *fixed parameter tractable* (FPT) if there is an algorithm for the problem that runs in time $f(k)n^{O(1)}$. The algorithm is said to be a subexponential time parameterized algorithm if $f(k) \leq 2^{o(k)}$. For an introduction to parameterized algorithms and complexity see the textbooks [10,11,22].

In this paper we pose the following question. Does k-PATH on planar graphs admit a subexponential time parameterized algorithm which only uses space polynomial in n? We give a positive answer to this question by presenting a polynomial space, $2^{O(\sqrt{k}\log^2 k)}n^{O(1)}$ time algorithm for k-PATH restricted to planar graphs. Our algorithm easily generalizes to any family of graphs which exclude a fixed graph H as a minor.

The fastest parameterized algorithm for k-PATH on planar graphs runs in $2^{O(\sqrt{k})}n^{O(1)}$ time and uses $2^{O(\sqrt{k})}\log n + n^{O(1)}$ space [9]. The algorithm we present uses polynomial space, but is slower by a factor of $O(\log^2 k)$ in the

P. Kolman and J. Kratochvíl (Eds.): WG 2011, LNCS 6986, pp. 262–270, 2011.

exponent of 2. Is the trade-off worth it? In general, does it make sense to settle for slightly slower algorithms if used space is reduced drastically? We believe that such a trade-off is reasonable because algorithms that use exponential time and space tend to run out of space long before they run out of time. In the survey paper on exponential time algorithms, Woeginger [24] states that "algorithms with exponential space complexities are absolutely useless for real life applications". This line of reasoning has opened up an interesting research direction; for which problems can we obtain space-efficient algorithms that are (almost) as time-efficient as the fastest ones? Some progress has been made – Fomin, Grandoni and Kratsch [12] gave a $6^k n^{O(\log k)}$ time, polynomial space algorithm for the STEINER TREE problem and showed how to use it to obtain a $O(1.60^n)$ time polynomial space algorithm. In a breakthrough paper, Nederlof [21] gave a $2^k n^{O(1)}$ time polynomial space algorithm for STEINER TREE. Subsequently, Lokshtanov and Nederlof [20] devised general sufficient conditions for turning exponential space dynamic programming algorithms into polynomial space algorithms based on algebraic transforms.

It is natural to ask which problems admit polynomial space FPT algorithms. While this might look like a whole new research program, the question has a surprising answer; *any* problem for which a polynomial space FPT algorithm is remotely feasible has one. In particular, a necessary condition for a problem to have a polynomial space FPT algorithm is to have an FPT algorithm. Another necessary condition is that the problem is solvable in polynomial space and time $n^{g(k)}$ for some function g (that is, a polynomial space XP algorithm for the problem). A well-known trick from parameterized complexity shows that these two conditions are not only necessary, but also sufficient.

Theorem 1. *If a parameterized problem Π has an algorithm $\mathcal{A}$ which uses $f(k)n^c$ time and space, and an algorithm $\mathcal{B}$ which uses $n^{g(k)}$ time and polynomial space, then Π can be solved in $n^{c+1} + f(k)^{g(k)}$ time and polynomial space.*

Proof. If $n \geq f(k)$ then run algorithm $\mathcal{A}$, this takes polynomial time and space. If $n \leq f(k)$ then run algorithm $\mathcal{B}$, this takes polynomial space and $f(k)^{g(k)}$ time. $\qquad\square$

Because of Theorem 1 the right question to ask is not which problems admit polynomial space FPT algorithms, but rather which problems admit *fast* polynomial space FPT algorithms. Obvious candidates for scrutiny are problems for which the fastest parameterized algorithms require exponential space. Graph problems restricted to planar and H-minor free graphs have this property – the Bidimensionality theory of Demaine et al. [4] gives $2^{O(\sqrt{k})}n^{O(1)}$ or $2^{O(\sqrt{k}\log k)}n^{O(1)}$ time algorithms for a multitude of graph problems on H-minor free graphs. However, these algorithms crucially depend on exponential space dynamic programming algorithms on graphs of bounded treewidth. In particular, the crux of all bidimensionality based algorithms is to first bound the treewidth of the input graph by $t = O(\sqrt{k})$, and then solve the problems in $2^{O(t)}n$ (or $t^{O(t)}n$) time and

$2^{O(t)} \log n$ (or $t^{O(t)} \log n$) space. We refer to following surveys for further details on the the Bidimensionality theory and its several applications [5,8].

Interestingly, using another simple trick we can make most of the algorithms for bidimensional problems on H-minor free graphs run in polynomial space, at the cost of a $O(\log k)$ factor in the exponent of the running time. The trick has two components, the first is that (almost) all of the $2^{O(t)}n$ time and $2^{O(t \log t)}n$ time dynamic programming algorithms on graphs of treewidth t can be turned into polynomial space divide and conquer algorithms with running time $n^{O(t)}$ and $n^{O(t \log t)}$. The second component is that most problems for which the Bidimensionality theory of Demaine et al. [4] gives fast algorithms admit *linear kernels* on planar and H-minor-free graphs [3,14]. A linear kernel for a parameterized graph problem is a polynomial time pre-processing algorithm that takes instances (G, k) and transforms them into equivalent instances (G', k') of the same problem such that $k' = O(k)$ and $V(G') = O(k)$. Now the subexponential time algorithms of Demaine et al. [4] can be made to run in polynomial space as follows. Run the pre-processing algorithm first, to ensure that the number of vertices in the input graph is at most $O(k)$. Bound the treewidth by $t = O(\sqrt{k})$ as before, but replace the $2^{O(t)}n$ (or $t^{O(t)}n$) time algorithms by $n^{O(t)}$ (or $n^{O(t \log t)}$) time polynomial space algorithms. Since $n = O(k)$ this second step uses only $2^{O(\sqrt{k} \log k)}$ (or $2^{O(\sqrt{k} \log^2 k)}$) time respectively. We remark that for many of these problems one can even remove the $\log k$ overhead in the exponent by using a Lipton-Tarjan separator approach [18,19] instead of the polynomial space algorithms for graphs of bounded treewidth after obtaining the linear kernel for the problem, as done in [1].

In the above argument it was crucial that the problem considered admits a linear kernel. What about the problems that do not? Notably, the k-PATH problem does have a subexponential time parameterized algorithm on H-minor-free graphs. At the same time k-PATH does not admit a linear (or polynomial size) kernel unless the polynomial hierarchy collapses to the third level [2], even when the input is restricted to planar graphs. Thus, with respect to parameterized polynomial space, subexponential time algorithms for planar graph problems, k-PATH stands out as a blank spot in an almost chartered map. Our polynomial space, $2^{O(\sqrt{k} \log^2 k)}n^{O(1)}$ time algorithm for k-PATH on H-minor free graphs fills this gap. Our approach for obtaining polynomial space subexponential time algorithm for k-PATH seems applicable to other problems also that do not admit polynomial kernels or not known to admit a polynomial kernel.

2   Definitions and Notations

In this section we give various definitions which we make use of in the paper. Let G be a graph then we use $V(G)$ and $E(G)$ to denote its vertex set and the edge set respectively. A graph G' is a *subgraph* of G if $V(G') \subseteq V(G)$ and $E(G') \subseteq E(G)$. The subgraph G' is called an *induced subgraph* of G if $E(G') = \{uv \in E(G) \mid u, v \in V(G')\}$, in this case, G' is also called the subgraph *induced by* V' and denoted with $G[V']$. By $N(u)$ we denote the (open) neighborhood of u, that is, the set of all vertices adjacent to u. The closed neighbourhood of u is $N[u] = N(u) \cup \{u\}$. Similarly, for a subset $D \subseteq V$, we define $N[D] = \bigcup_{v \in D} N[v]$ and $N(D) = N[D] \setminus D$.

Parameterized algorithms and Treewidth. A parameterized problem Π is a subset of $\Gamma^* \times \mathbb{N}$ for some finite alphabet Γ. An instance of a parameterized problem consists of (x, k), where k is called the parameter. A central notion in parameterized complexity is *fixed parameter tractability (FPT)* which means, for a given instance (x, k), solvability in time $f(k) \cdot p(|x|)$, where f is an arbitrary function of k and p is a polynomial in the input size.

A *tree decomposition* of a graph G is a pair $(T, \mathcal{B})$ where T is a tree and $\mathcal{B} = \{X_i \mid i \in V(T)\}$ is a collection of subsets of $V(G)$ such that

1. $\bigcup_{i \in V(T)} X_i = V(G)$,
2. for each edge $xy \in E$, $\{x, y\} \subseteq X_i$ for some $i \in V(T)$;
3. for each $x \in V(G)$ the set $\{i \mid x \in X_i\}$ induces a connected subtree of T.

The *width* of the tree decomposition is $\max_{i \in V_T} |X_i| - 1$. The *treewidth* of a graph G is the minimum width over all tree decompositions of G. A tree decomposition $(T, \mathcal{B})$ can be converted in linear time [17] into a *nice* tree decomposition of the same width: here, the tree T is rooted and binary, and its nodes are of four types:

- *Leaf nodes* h are leaves of T and have $|X_h| = 1$.
- *Introduce nodes* h have one child i with $X_h = X_i \cup \{v\}$ for some vertex $v \in V$.
- *Forget nodes* h have one child i with $X_h = X_i \setminus \{v\}$ for some vertex $v \in V$.
- *Join nodes* h have two children i, j with $X_h = X_i = X_j$.

We denote by $\mathbf{tw}(G)$ the treewidth of the graph G.

3   Polynomial Space Algorithm for the k-PATH Problem

In this section we prove our main result, which is a polynomial space subexponential time algorithm for the k-PATH problem. For a set $W \subseteq V(G)$ a set $S \subseteq V(G)$ is a *balanced separator* for W if $V(G)$ can be partitioned into L, S and R such that there is no edge from L to R and $|W \cap L| \leq \frac{2|W \setminus S|}{3}$ and $|W \cap R| \leq \frac{2|W \setminus S|}{3}$. In other words, W is evenly distributed between L and R. It is well-known that in any *tree* T, for every set $W \subseteq V(G)$ there is a balanced separator S for W with $|S| = 1$. This result has been generalized to graphs of bounded treewidth [23][11, Lemma 11.16] - in particular in a graph G of treewidth at most t, for any set W there is a balanced separator S of size at most $t + 1$. Lemma 1 is a subtle strengthening of this fact and states that given any tree-decomposition of G of width at most t the separator S can be chosen as one of the bags of the decomposition. In fact, the proofs given in [23][11, Lemma 11.16] already imply Lemma 1, we include a proof here for completeness.

For a graph G and a nice tree-decomposition $(T, \mathcal{B})$ of G and node $v \in V(T)$ let $X_v \in \mathcal{B}$ be the corresponding bag. Let T_v be the subtree of T rooted at v and let $A(v) = (\bigcup_{u \in V(T_v)} X_u) \setminus X_v$.

Lemma 1. *Let G be a graph, let $(T, \mathcal{B})$ be a nice tree-decomposition of G of width t and let $W \subseteq V$ be a vertex set of size at least 3. Then there exists a vertex v such that X_v is a balanced separator for W and in the corresponding partition $V(G) = L \cup X_v \cup R$, $L = A(v)$.*

Proof. Recall that T is a rooted tree with root r, and that each node of T has at most two children. Observe that $|A(r) \cap W| = |W \setminus X_r| \geq \frac{|W \setminus X_r|}{3}$. Choose a lowermost node v such that $|A(v) \cap W| \geq \frac{|W \setminus X_v|}{3}$. We prove that $|A(v) \cap W| \leq \frac{2|W \setminus X_v|}{3}$. If v is an introduce node or a forget node with child v' then $|A(v') \cap W| \leq \frac{|W \setminus X_{v'}|}{3}$ and hence $|A(v) \cap W| \leq \frac{2|W \setminus X_v|}{3}$. Here we used that $|W| \geq 3$ and that $|X_v| = |X_{v'}| + 1$ if v is an introduce node, and that $|A(v)| = |A(v')| + 1$ if v is a forget node. If v is a join node with children u and w then $X_v = X_u = X_w$ and $A(v) = A(u) \cup A(w)$. Finally $|A(u) \cap W| \leq \frac{|W \setminus X_u|}{3}$ and $|A(w) \cap W| \leq \frac{|W \setminus X_w|}{3}$ and hence $|A(v) \cap W| \leq \frac{2|W \setminus X_v|}{3}$. This concludes the proof. □

A key component of our algorithm is a new divide and conquer algorithm for the k-PATH problem on graphs of bounded treewidth.

Lemma 2. *There is an $(ntt)^{O(\log k)}$ time and polynomial space algorithm for k-PATH if a nice tree-decomposition $(T, \mathcal{B})$ of G of width t is given as input.*

Proof. We describe a divide and conquer algorithm for the problem. In order to handle the instances that are generated in the recursive steps of the algorithm we will solve a slightly more general problem. In the *generalized k-PATH* problem (k-GP) we are given a graph G, integer k, vertex set V_p and an edge set E_p such that every edge in E_p has both endpoints in V_p. The task is to determine whether there is a path P in G such that $V_p \subseteq V(P)$, $E_p \subseteq E(P)$ and $|V(P) \setminus V_p| = k$. One can think of the sets V_p and E_p as vertices and edges which are pre-determined to be in the path P. We are now ready to give an algorithm for k-GP in graphs of bounded tree-width with running time $O((nk^2(t+1)!3^{t+2})^{\log k + O(1)} \cdot (V_p + t \log k)!)$. Suppose that there is a path P such that $V_p \subseteq V(P)$, $E_p \subseteq E(P)$ and $|V(P) \setminus V_p| = k$. Let $W = V(P) \setminus V_p$, with $|W| = k$. If $k = 0$ the algorithm tries all the $|V_p|!$ possible orderings of V_p and checks whether any of the orderings is a path that contains all edges of E_p. If $k < 3$ the algorithm tries all possible ways to extend V_p by k vertices, and then proceeds to the case when $k = 0$. We now handle the case that $k \geq 3$.

By Lemma 1 there exists a node $v \in V(T)$ such that $X_v \in \mathcal{B}$ is a balanced separator for W. The algorithm guesses the correct vertex v by looping over all the n possible bags in the decomposition. Furthermore, by Lemma 1 there is a partition of $V(G)$ into $L \cup X_v \cup R$ such that $L = A(v)$ and $R = V(G) \setminus (X_v \cup R)$, there is no edge from L to R, $|W \cap L| \leq \frac{2|W \setminus X_v|}{3}$ and $|W \cap R| \leq \frac{2|W \setminus X_v|}{3}$. The algorithm computes L and R from v. Now the algorithm guesses $V(P) \cap X_v = X$ by trying all the 2^{t+1} possible subsets of X_v.

The path P visits the vertices of X in some order, say $x_1, x_2, \ldots x_q$. The algorithm guesses this order by trying all the (at most) $(t+1)!$ possible permutations of X. Now, for each $i < q$ the subpath of P from x_i to x_{i+1} either uses the edge $x_i x_{i+1}$, or has all its inner vertices in L, or has all its inner vertices in R. By trying each of the three possibilities for each $i < q$ the algorithm correctly guesses which of the possibilities it is. The algorithm also guesses whether the subpath of P attached to x_1 lies in its entirety in $L \cap X$ or in $R \cap X$. The same guess is performed for the subpath of P attached to x_q. Finally the algorithm guesses $k_L = |(V(P) \setminus V_p) \cap L|$ and $k_R = |(V(P) \setminus V_p) \cap R|$.

Using all the guesses the algorithm constructs two instances G_L, V_p^L, E_p^L, k_L and $G_R, V_p^R, E_p^R, L, k_R$ as follows. We set $V_p^L = (V_p \cap L) \cup X$ and $E_p^L = E_p \cap E(G[L \cup X])$. To construct G_L we start with $G[L \cup X]$ and remove all the edges with both endpoints in X. For every pair x_i, x_{i+1} such that we have guessed that the subpath of P from x_i to x_{i+1} uses the edge $x_i x_{i+1}$ or has all its internal vertices in R, we add the edge $x_i x_{i+1}$ to G_L and to E_p^L. Finally, if we have guessed that the subpath of P from the start point until x_1 lies entirely in $X \cup R$ we add a vertex p_{start} to G_L and to V_p^L, make p_{start} adjacent to x_1 and add $p_1 p_{start}$ to E_p^L. Similarly if we have guessed that the subpath of P from x_q to the end point of P lies entirely in $X \cup R$ we add a vertex p_{end} to G_L and to V_p^L, and add the edge $x_q p_{end}$ to G_L and to E_p^L. The graph G_R and set V_p^R is constructed symmetrically.

If the two instances G_L, V_p^L, E_p^L, k_L and G_R, V_p^R, E_p^R, k_R are both "yes" instances one can glue their solution paths together to form a solution path for G, V_p, E_p, k. In particular every edge $x_i x_{i+1}$ in G_L will correspond either to an edge $x_i x_{i+1}$ in the solution paths of both G_L, V_p^L, E_p^L, k_L and G_R, V_p^R, E_p^R, k_R, or it can be replaced by a subpath of the solution path of G_R, V_p^R, E_p^R, k_R. Edges $x_i x_{i+1}$ in G_R are handled symmetrically. In the reverse direction for the correct set of guesses the path P breaks up into solution paths to G_L, V_p^L, E_p^L, k_L and G_R, V_p^R, E_p^R, k_R respectively. On the side that contains p_{start} we add the edge $p_{start} x_1$ to the solution path, and on the side that contains p_{end} we add the edge $x_q p_{end}$.

Now we bound the running time of the algorithm. Observe that since we only add edges between vertices in $X \subseteq X_v$ and pendant vertices p_{start} and p_{end} of degree 1 attached to X, the treewidth of G_L and G_R is at most t. Let $r = |V_p|$, and let $T(k, r, n, t)$ be a function that upper bounds the running time of the algorithm. The function T is bounded by the following recurrence.

$$T(k, r, n, t) \le n \cdot (t+1)! \cdot 3^t \cdot 2^{t+3} \cdot k^2 \cdot 2T(k/2, r+t, n, t) \text{ when } k \ge 3$$

Here the factors in the recurrence reflect the number of possibilities for each guess, except for the factor 2 in $2T(k/2, n, t)$ which reflects that G_L and G_R are handled independently. Observe that n and t never increase throughout the recurrence. When $k < 3$ we have that $T(k, r, n, t) \le O(n^3 r!)$. Hence $T(k, r, n, t)$ can be bounded from above by

$$T(k, r, n, t) \le (nk^2(t+1)!6^{t+3})^{\log k} \cdot n^3 (t \log k)! \le (nt^t)^{O(\log k)}.$$

The space requirement of the algorithm is clearly polynomial. This concludes the proof. □

We are now in position to give the main result of this section.

Theorem 2. *For every fixed graph H, there is an algorithm for k-PATH on H-minor-free graphs running in time $2^{O(\sqrt{k}\log^2 k)}n^{O(1)}$ and using polynomial space.*

Proof. We use the fact that for any H there exists a constant h such that in any H-minor-free graph of treewidth at least hk there is a k-path of length at least k^2 [4]. Set $t = h\sqrt{k}$, then if $\mathbf{tw}(G) \geq t$ then G contains a k-path. We use the approximation algorithm of Diestel et al [6] to either compute a tree-decomposition of G of width at most $3t/2$, or to conclude that the treewidth of G is at least t, which implies that G has a k-path. Diestel et al's algorithm uses polynomial space and $2^{O(t)}n^{O(1)}$ time. If we obtain a tree-decomposition we proceed as follows. If $n \leq 2^{\sqrt{k}}$ apply the algorithm from Lemma 2 to solve the problem in time $2^{O(\sqrt{k}\log^2 k)}n^{O(1)}$ and using polynomial space. If, on the other hand $n \geq 2^{\sqrt{k}}$ the standard dynamic programming algorithm on graphs of bounded treewidth that uses at most $2^{O(\sqrt{k})}n^{O(1)}$ time and space [7], runs in polynomial time and space. This concludes the proof.   □

We remark that the algorithm presented in Theorem 2 for k-PATH can be made to run in time $2^{O(\sqrt{k}\log k)}n^{O(1)}$ and space polynomial in n on planar graphs using sphere cut decompositions introduced by Dorn et al. [9].

4   Conclusion and Discussion

We gave a subexponential time, polynomial space parameterized algorithm for the k-PATH problem on H-minor-free graphs. A key component of our algorithm is a new $(n \cdot t^t)^{O(\log k)}$ time and polynomial space algorithm for k-PATH in graphs of treewidth at most t. In general, it is possible to design similar $(n \cdot 2^t)^{O(\log k)}$ or $(n \cdot t^t)^{O(\log k)}$ time and polynomial space algorithms for many problems where one is looking for a specific vertex set of size k in a graph of treewidth t. A concrete example where this is useful is the k-PARTIAL VERTEX COVER problem. Here we are given a graph G, positive integers k and t and we look for a subset $S \subseteq V(G)$ such that $|S| \leq k$ and the number of edges incident to S is at least t. This problem is not known to admit a polynomial kernel, even on planar graphs. However using the approach described in this paper for k-PATH and combining it with an algorithm of Fomin et al. [13] that in polynomial time either finds a solution for an instance (G, k, t) or obtains an equivalent instance (G', k, t) such that $\mathbf{tw}(G') \leq O(\sqrt{k})$, one can give a subexponential time, polynomial space parameterized algorithm for k-PARTIAL VERTEX COVER on apex-minor-free graphs.

We conclude with two open problems. First, is there a polynomial space parameterized algorithm for the k-PATH problem on planar graphs with running time $2^{O(\sqrt{k})}n^{O(1)}$? Second, by combining the well known $2^t n^{O(1)}$ time and space algorithm for INDEPENDENT SET in graphs of treewidth t, and the folklore $n^{O(t)}$ time, polynomial space algorithm, Theorem 1 yields a $2^{O(t^2)} + n^2$ time and polynomial space algorithm for INDEPENDENT SET. Is there a polynomial space algorithm for INDEPENDENT SET on graphs of treewidth t with running time $2^{t^{2-\epsilon}} n^{O(1)}$ for some $\epsilon > 0$?.

References

1. Alber, J., Fernau, H., Niedermeier, R.: Graph separators: a parameterized view. J. Comput. System Sci. 67(4), 808–832 (2003); Special issue on parameterized computation and complexity
2. Bodlaender, H.L., Downey, R.G., Fellows, M.R., Hermelin, D.: On problems without polynomial kernels. J. Comput. System Sci. 75(8), 423–434 (2009)
3. Bodlaender, H.L., Fomin, F.V., Lokshtanov, D., Penninkx, E., Saurabh, S., Thilikos, D.M.: (Meta) Kernelization. In: Proc. 50th Annual IEEE Symposium on Foundations of Computer Science, pp. 629–638. IEEE Computer Society (2009)
4. Demaine, E.D., Fomin, F.V., Hajiaghayi, M., Thilikos, D.M.: Subexponential parameterized algorithms on bounded-genus graphs and H-minor-free graphs. J. ACM 52(6), 866–893 (electronic), (2005)
5. Demaine, E.D., Hajiaghayi, M.: The bidimensionality theory and its algorithmic applications. Comput. J. 51(3), 292–302 (2008)
6. Diestel, R., Jensen, T.R., Gorbunov, K.Y., Thomassen, C.: Highly connected sets and the excluded grid theorem. J. Comb. Theory, Ser. B 75(1), 61–73 (1999)
7. Dorn, F., Fomin, F.V., Thilikos, D.M.: Catalan structures and dynamic programming in H-minor-free graphs. In: Proc. 19th Annual ACM-SIAM Symposium on Discrete Algorithms, pp. 631–640. ACM
8. Dorn, F., Fomin, F.V., Thilikos, D.M.: Subexponential parameterized algorithms. Computer Science Review 2(1), 29–39 (2008)
9. Dorn, F., Penninkx, E., Bodlaender, H.L., Fomin, F.V.: Efficient Exact Algorithms on Planar Graphs: Exploiting Sphere Cut Branch Decompositions. In: Brodal, G.S., Leonardi, S. (eds.) ESA 2005. LNCS, vol. 3669, pp. 95–106. Springer, Heidelberg (2005)
10. Downey, R.G., Fellows, M.R.: Parameterized Complexity. Monographs in Computer Science. Springer, Heidelberg
11. Flum, J., Grohe, M.: Parameterized Complexity Theory. Texts in Theoretical Computer Science. An EATCS Series. Springer, Heidelberg
12. Fomin, F.V., Grandoni, F., Kratsch, D.: Faster Steiner Tree Computation in Polynomial-Space. In: Halperin, D., Mehlhorn, K. (eds.) ESA 2008. LNCS, vol. 5193, pp. 430–441. Springer, Heidelberg (2008)
13. Fomin, F.V., Lokshtanov, D., Raman, V., Saurabh, S.: Subexponential algorithms for partial cover problems. In: IARCS Annual Conference on Foundations of Software Technology and Theoretical Computer Science. Leibniz International Proc. Informatics, vol. 4, pp. 193–201. Schloss Dagstuhl–Leibniz-Zentrum für Informatik (2009)
14. Fomin, F.V., Lokshtanov, D., Saurabh, S., Thilikos, D.M.: Bidimensionality and kernels. In: Proc. 21st Annual ACM-SIAM Symposium on Discrete Algorithms, pp. 503–510. Society for Industrial and Applied Mathematics (2010)
15. Garey, M.R., Johnson, D.S., Tarjan, R.E.: The planar hamiltonian circuit problem is np-complete. SIAM J. Comput. 5(4), 704–714 (1976)
16. Karp, R.M.: Reducibility among combinatorial problems. Complexity of Computer Computations, 85–103 (1972)

17. Kloks, T.: Treewidth. Computations and Approximations. LNCS, vol. 842. Springer, Heidelberg (1994)
18. Lipton, R.J., Tarjan, R.E.: A separator theorem for planar graphs. SIAM J. Appl. Math. 36(2), 177–189 (1979)
19. Lipton, R.J., Tarjan, R.E.: Applications of a planar separator theorem. SIAM J. Comput. 9(3), 615–627 (1980)
20. Lokshtanov, D., Nederlof, J.: Saving space by algebraization. In: STOC, pp. 321–330 (2010)
21. Nederlof, J.: Fast Polynomial-Space Algorithms Using Möbius Inversion: Improving on Steiner Tree and Related Problems. In: Albers, S., Marchetti-Spaccamela, A., Matias, Y., Nikoletseas, S., Thomas, W. (eds.) ICALP 2009. LNCS, vol. 5555, pp. 713–725. Springer, Heidelberg (2009)
22. Niedermeier, R.: Invitation to Fixed-parameter Algorithms, Oxford Lecture Series in Mathematics and its Applications, vol. 31. Oxford University Press (2006)
23. Reed, B.A.: Tree width and tangles: a new connectivity measure and some applications. In: Surveys in Combinatorics, 1997, London. London Math. Soc. Lecture Note Ser, vol. 241, pp. 87–162. Cambridge Univ. Press (1997)
24. Woeginger, G.J.: Space and Time Complexity of Exact Algorithms: Some Open Problems. In: Downey, R.G., Fellows, M.R., Dehne, F. (eds.) IWPEC 2004. LNCS, vol. 3162, pp. 281–290. Springer, Heidelberg (2004)

Approximability of the Path-Distance-Width for AT-free Graphs

Yota Otachi[1], Toshiki Saitoh[2], Katsuhisa Yamanaka[3], Shuji Kijima[4],
Yoshio Okamoto[5], Hirotaka Ono[6], Yushi Uno[7], and Koichi Yamazaki[8]

[1] Graduate School of Information Sciences, Tohoku University,
Sendai 980-8579, Japan, JSPS Research Fellow
otachi@dais.is.tohoku.ac.jp
[2] ERATO MINATO Discrete Structure Manipulation System Project,
Japan Science and Technology Agency, North 14, West 9, Sapporo, Hokkaido, 060-0814, Japan
t-saitoh@erato.ist.hokudai.ac.jp
[3] Department of Electrical Engineering and Computer Science,
Iwate University, Ueda 4-3-5, Morioka, Iwate 020-8551, Japan
yamanaka@cis.iwate-u.ac.jp
[4] Graduate School of Information Science and Electrical Engineering,
Kyushu University, Fukuoka, 819-0395, Japan
kijima@inf.kyushu-u.ac.jp
[5] Center for Graduate Education Initiative, JAIST, Asahidai 1-1,
Nomi, Ishikawa 923-1292, Japan
okamotoy@jaist.ac.jp
[6] Department of Economic Engineering, Kyushu University,
6-19-1 Hakozaki Higashi-ku, Fukuoka 812-8581, Japan
hirotaka@en.kyushu-u.ac.jp
[7] Department of Mathematics and Information Sciences, Graduate School of Science,
Osaka Prefecture University, 1-1 Gakuen-cho, Naka-ku, Sakai 599-8531, Japan
uno@mi.s.osakafu-u.ac.jp
[8] Department of Computer Science, Gunma University,
1-5-1 Tenjin-cho, Kiryu, Gunma, 376-8515 Japan
koichi@cs.gunma-u.ac.jp

Abstract. The path-distance-width of a graph measures how close the graph is to a path. We consider the problem of determining the path-distance-width for graphs with *chain-like structures* such as k-cocomparability graphs, AT-free graphs, and interval graphs. We first show that the problem is NP-hard even for a very restricted subclass of AT-free graphs. Next we present simple approximation algorithms with constant approximation ratios for graphs with chain-like structures. For instance, our algorithm for AT-free graphs has approximation factor 3 and runs in linear time. We also show that the problem is solvable in polynomial time for the class of cochain graphs, which is a subclass of the class of proper interval graphs.

1 Introduction

The path-distance-width is a graph parameter to measure how close a graph is to a path [19,18]. Roughly speaking, graphs of bounded path-distance-width, bounded bandwidth, and bounded pathwidth have *chain-like structures*. It is known that for any

P. Kolman and J. Kratochvíl (Eds.): WG 2011, LNCS 6986, pp. 271–282, 2011.

connected graph, its pathwidth is bounded by its bandwidth, and its bandwidth is less than 2 times its path-distance-width [11,19]. By these relations, many useful properties for bounded pathwidth graphs and bounded bandwidth graphs also hold for bounded path-distance-width graphs. On the other hand, even if a problem is hard for bounded pathwidth (or bandwidth) graphs, it may be tractable for bounded path-distance-width graphs. Other than these, many chain-like graphs are studied, such as interval graphs, AT-free graphs, and k-cocomparability graphs for small k. It is known that there are relationships among those graph parameters and graph classes [11,19,2,3].

This study is motivated by the research on bandwidth of AT-free graphs [13,8]. To see the motivation, let us briefly review the history of the research on bandwidth for interval graphs and AT-free graphs. One may expect that if we restrict our input graphs to interval graphs or AT-free graphs, then we would be able to easily find its chain-like structure (such as an interval representation or a dominating pair), and then from the chain-like structure we might be able to compute the bandwidth. The polynomial-time computability of the bandwidth for interval graphs was an open problem [10], but later a polynomial-time algorithm was discovered (see [12,15,17]). Since interval graphs are AT-free graphs, it would be natural to ask whether or not the bandwidth decision problem for AT-free graphs can be solved in polynomial-time. Unfortunately, the bandwidth decision problem for AT-free graphs is shown to be NP-complete [16,13]. However, it is known that for AT-free graphs, the bandwidth can be approximated within a factor 2 in $O(mn)$ time [13], where m and n denote the number of edges and the number of vertices, respectively.

From a computational view, bandwidth and path-distance-width have some similarities. For example, both problems do not admit any PTAS even for trees [1,18]. Hence it would be reasonable to ask the computational complexity of computing the path-distance-width for AT-free graphs. Unfortunately, as we will prove in this paper, the path-distance-width decision problem for AT-free graphs is also NP-complete. More precisely, we will show that the problem is NP-complete for cobipartite graphs, which is a subclass of AT-free graphs.

Although some techniques developed in the research on bandwidth can be carried over into the research on path-distance-width, path-distance-width has a serious drawback which bandwidth does not have: path-distance-width is *not* closed under edge deletion. In many cases, this drawback makes the design and analysis of algorithms very difficult. In this study, however, we find that the restriction to AT-free graphs is enough to overcome the drawback for achieving a constant-factor approximation. We first present an approximation algorithm with approximation ratio $2k + 1$ for the path-distance-width of k-cocomparability graphs. Although this algorithm is a constant-factor approximation for AT-free graphs, we present another approximation algorithm for AT-free graphs, which has a better running time and a better approximation ratio. We also consider polynomial-time solvable cases. We show that the problem is solvable in linear time for cochain graphs.

2   Preliminaries

In this paper, graphs are finite, simple, and connected. We denote by $V(G)$ and $E(G)$ the vertex set and the edge set of graph G, respectively. The *complement of graph G* is the graph $\overline{G}$ such that $V\left(\overline{G}\right) = V(G)$ and two distinct vertices are adjacent in $\overline{G}$ if and only if they are not adjacent in G.

Let G be a connected graph. The *distance between two vertices* $u, v \in V(G)$ in G, denoted $d_G(u, v)$, is the length of a shortest u–v path in G. We define the *distance between a vertex subset* $S \subseteq V(G)$ *and a vertex* $v \in V(G)$ in G as $d_G(S, v) = \min_{u \in S} d_G(u, v)$. For $S \subseteq V(G)$, we define the *diameter of* S *in* G as $\text{diam}_G(S) = \max_{u,v \in S} d_G(u, v)$. The *diameter of* G is $\text{diam}(G) = \text{diam}_G(V(G))$. The *(open) neighborhood* of a vertex v in G, denoted $N_G(v)$, is the set of vertices adjacent to v; that is $N_G(v) = \{u \mid \{u, v\} \in E(G)\}$. The *closed neighborhood* of v in G, denoted $N_G[v]$, is the set $\{v\} \cup N_G(v)$. The *(open) neighborhood* of a vertex set $S \subseteq V(G)$ in G, denoted $N_G(S)$, is the set of vertices not in S and adjacent to some vertex $u \in S$; that is $N_G(S) = \bigcup_{v \in S} N_G(v) \setminus S$.

A sequence $(L_1, \ldots, L_t)$ of subsets of vertices is a *distance structure* of a graph G if $\bigcup_{1 \le i \le t} L_i = V(G)$ and $L_i = \{v \in V(G) \mid d_G(L_1, v) = i - 1\}$ for $1 \le i \le t$. Each L_i is a *level* and especially L_1 is the *initial set*. The *width* of $(L_1, \ldots, L_t)$, denoted $\text{pdw}_{L_1}(G)$, is $\max_{1 \le i \le t} |L_i|$. The *path-distance-width* of G, denoted $\text{pdw}(G)$, is $\min_{S \subseteq V(G)} \text{pdw}_S(G)$.

If the initial set of a distance structure of G is a set which consists of only one vertex, then we say that it is a *rooted distance structure* of G. The *rooted path-distance-width* of G, denoted $\text{rpdw}(G)$, is the minimum width over all its rooted distance structures; that is, $\text{rpdw}(G) = \min_{v \in V(G)} \text{pdw}_{\{v\}}(G)$. The rooted path-distance-width can be computed in $O(mn)$ time, by running breadth-first search from every vertex.

Lemma 2.1. *The rooted path-distance-width of a connected graph G with n vertices and m edges can be computed in $O(mn)$ time.*

An *interval graph* is a graph whose vertices can be mapped to distinct intervals on the real line in such a way that two vertices are adjacent in the graph if and only if the corresponding intervals overlap. We call the set of intervals representing a graph an *interval representation* of the graph. An interval representation is *proper* if no interval properly contains other intervals in it. A graph is a *proper interval graph* if it has a proper interval representation.

An independent set of three vertices is an *asteroidal triple* if every two of them are connected by a path avoiding the neighborhood of the third. A graph is *asteroidal triple-free* (*AT-free* for short), if it contains no asteroidal triple.

A graph G is a *comparability graph* if there exists a linear ordering $<$ on $V(G)$ such that for any three vertices $u < v < w$, $\{u, v\} \in E(G)$ and $\{v, w\} \in E(G)$ imply $\{u, w\} \in E(G)$. A graph G is a *cocomparability graph* if G is the complement of a comparability graph. It is known that G is a cocomparability graph if and only if it has a *cocomparability ordering*; that is, there exists a linear order $<$ on $V(G)$ such that for any three vertices $u < v < w$, $\{u, w\} \in E(G)$ implies $\{u, v\} \in E(G)$ or $\{v, w\} \in E(G)$.

Chang, Ho, and Ko [3] generalized cocomparability graphs to k-cocomparability graphs. Let G be a graph, and let k be a positive integer. A *k-cocomparability ordering* (*k-CCPO*) of G is a linear order $<$ on $V(G)$ such that for any three vertices $u < v < w$, $d_G(u, w) \le k$ implies $d_G(u, v) \le k$ or $d_G(v, w) \le k$. A graph is a *k-cocomparability graph* if it admits a k-CCPO. Note that a 1-cocomparability ordering is just a cocomparability ordering.

A graph $G = (U, V; E)$ is a *cobipartite graph* if (U, V) is a nonempty partition of $V(G)$ and both U and V induce cliques. Thus a cobipartite graph is the complement of a bipartite graph. This implies that cobipartite graphs are cocomparability graphs, since bipartite graphs are comparability graphs. A cobipartite graph $H = (X, Y; E)$ is a *cochain graph* if the elements of X and Y can be ordered as $x_1, \ldots, x_{|X|}$ and $y_1, \ldots, y_{|Y|}$, respectively, so that $N_G[x_1] \subseteq \cdots \subseteq N_G[x_{|X|}]$ and $N_G[y_1] \subseteq \cdots \subseteq N_G[y_{|Y|}]$.

It is known that cochain graphs $\subset$ proper interval graphs $\subset$ interval graphs $\subset$ cocomparability graphs $\subset$ AT-free graphs $\subset$ 2-cocomparability graphs, and k-cocomparability graphs $\subset (k + 1)$-cocomparability graphs for any $k \geq 1$ (see [2,3]). It is easy to see that any graph G is a k_G-cocomparability graph for some $k_G \leq \mathrm{diam}(G)$.

In this paper, we present some algorithms approximating the path-distance-width for k-cocomparability graphs and their subclasses such as AT-free graphs and proper interval graphs. Each algorithm has a constant approximation ratio (if k is a fixed constant), and runs in $O(mn)$ or $O(m + n)$ time. See Fig. 1.

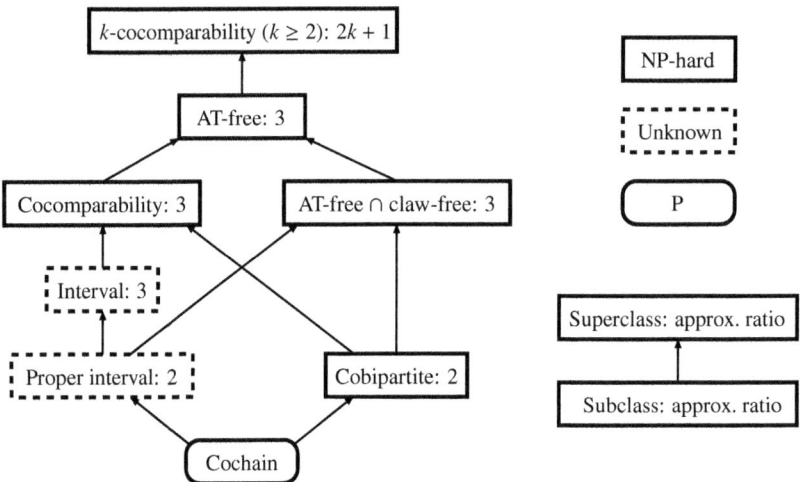

Fig. 1. Summary of results

3 NP-Hardness for Cobipartite Graphs

Before we present approximation algorithms, we show that the problem for determining the path-distance-width is NP-hard even for a very restricted graph class, the class of cobipartite graphs. To this end, we first prove the NP-completeness of an intermediate problem, by constructing a polynomial-time reduction from the following well-known NP-complete problem.

Problem: SET COVER [7, SP5]
Instance: A set $C = \{c_1, \ldots, c_n\}$, a family $\mathcal{F} = \{F_1, \ldots, F_m\} \subseteq 2^C$, and a positive integer $h \leq n$.
Question: Is there $X \subseteq \mathcal{F}$ such that $\bigcup_{F_i \in X} F_i = C$ and $|X| = h$?

In any instance of SET COVER, we can assume without loss of generality that for every $c_i \in C$, there is a subset $F_j \in \mathcal{F}$ such that $c_i \in F_j$, since otherwise the instance has no cover. We also assume $n > 1$ and $h < m$, since otherwise the problem is trivial.

Our intermediate problem is as follows.

Problem: Partial Cover in Bigraphs (PCB)
Instance: A bipartite graph $G = (U, V; E)$, and a positive integer $k \le |V|$.
Question: Is there $Y \subseteq U$ such that $|N_G(Y)| = k$?

Kobayashi [14] pointed out that PCB is NP-complete. Here, we provide a full proof.

Lemma 3.1. *PCB is NP-complete even if $|V| > k + 2$ and G has no isolated vertex.*

Proof. From an instance $(C, \mathcal{F}, h)$ of Set Cover, we first construct a bipartite graph $G = (U, V; E)$ as follows: $U = \{u_1, \ldots, u_m\}$, $V = \{v_1, \ldots, v_n\}$, and $E = \{\{u_i, v_j\} \mid c_j \in F_i\}$. The vertex sets U and V correspond to the family $\mathcal{F}$ and the ground set C, respectively. The edge set E represents the containment relation between the elements of C and the subsets in $\mathcal{F}$. Next, by adding $n + 1$ pendant vertices to each $u_i \in U$, we construct a bipartite graph $H = (U, V'; E')$. Clearly, this construction can be done in polynomial-time. Note that $|V'| = n + (n + 1)m > n + (n + 1)h + 2$ since $n > 1$ and $m > h$. Also note that H has no isolated vertex.

Let $k = n + (n + 1)h$. We shall prove that C has a cover $X \subseteq \mathcal{F}$ of size $|X| = h$ if and only if there is a set $Y \subseteq U$ such that $|N_H(Y)| = k$.

($\Longrightarrow$) Assume that there is $X \subseteq \mathcal{F}$ such that $\bigcup_{F_i \in X} F_i = C$ and $|X| = h$. We set $Y = \{u_i \mid F_i \in X\}$. Since X is a cover of C, $|N_H(Y) \cap V| = |V| = n$. Since $|N_H(Y) \setminus V| = (n + 1)h$,

$$|N_H(Y)| = |N_H(Y) \cap V| + |N_H(Y) \setminus V| = n + (n + 1)h = k.$$

($\Longleftarrow$) Assume that there exists $Y \subseteq U$ such that $|N_H(Y)| = k$. We first prove $|Y| = h$. If $|Y| \ge h + 1$, then $|N_H(Y)| \ge |N_H(Y) \setminus V| \ge (n + 1)(h + 1) > k$. If $|Y| \le h - 1$, then $|N_H(Y)| \le |V| + |N_H(Y) \setminus V| \le n + (n + 1)(h - 1) < k$. Thus $|Y| = h$. Now we have

$$|N_H(Y) \cap V| = |N_H(Y)| - |N_H(Y) \setminus V| = k - (n + 1)h = n.$$

Therefore, if we set $X = \{F_i \mid u_i \in Y\}$, then $|X| = h$ and X covers the ground set C.

From the above observation the problem is NP-hard. Since the problem clearly belongs to NP, the lemma holds. $\square$

Now we prove the NP-hardness of the path-distance-width problem for cobipartite graphs, by constructing a polynomial-time reduction from PCB. We actually prove that deciding whether $\mathrm{pdw}(G) = |V(G)|/3$ is NP-complete for cobipartite graphs with diameter 2.

Theorem 3.2. *Given a cobipartite graph H with $\mathrm{diam}(H) = 2$, it is NP-complete to decide whether $\mathrm{pdw}(H) = |V(H)|/3$.*

Proof. Clearly, the problem is in NP. Thus we prove the NP-hardness. From an instance $(G = (U, V; E), k)$ of PCB satisfying the conditions in Lemma 3.1, we construct a co-bipartite graph $H = (U', V'; E')$ as follows (see Fig. 2). Let S and T be two sets of sizes $|S| = |U| + k$ and $|T| = |U| + 2|V| - k - 2$, where S, T, U, and V are pairwise non-intersecting. We set the vertex sets as $U' = U \cup T \cup \{a\}$ and $V' = V \cup S \cup \{b\}$, where a and b are new vertices. In H, both U' and V' induce cliques. Every edge in G is also

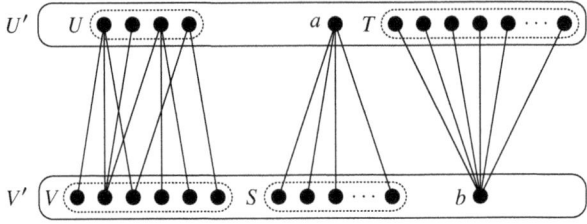

Fig. 2. Cobipartite graph $H = (U', V'; E')$

in H. Additionally, a is adjacent to all vertices in S, and b is adjacent to all vertices in T. This construction can be done in polynomial time.

Since G has no isolated vertex, diam(H) = 2. It is easy to see that $|U'| = 2|U|+2|V| - k - 1$ and $|V'| = |V| + |U| + k + 1$. Hence $|V(H)| = |U'| + |V'| = 3(|U| + |V|)$. We shall show that (G, k) is a yes instance of PCB if and only if pdw(H) = $|U| + |V|$. Note that pdw(H) $\geq |V(H)|/(\text{diam}(H) + 1) = |U| + |V|$.

($\Longrightarrow$) Assume that there exists $Y \subseteq U$ such that $|N_G(Y)| = k$. Let $X = Y \cup T'$, where T' is any subset of T such that $|T'| = |U| + |V| - |Y|$. Let $(L_1 = X, L_2, L_3)$ be the level structure with the initial set X. Clearly, $|L_1| = |X| = |U| + |V|$. The size of the second level is

$$|L_2| = |U' \setminus X| + |N_H(Y) \cap V'| + |N_H(T') \cap V'| = |U| + |V|. \tag{1}$$

This also implies $|L_3| = |V(H)| - |L_1| - |L_2| = |U| + |V|$. Therefore, pdw$_X$($H$) = $|U| + |V|$.

($\Longleftarrow$) Assume that pdw$_X$(H) = $|U| + |V|$ for some $X \subseteq V(H)$. If X intersects both U' and V', then the distance structure has at most two levels, and thus pdw$_X$(H) $\geq |V(H)|/2 > |U| + |V|$. Hence X is included in either U' or V'. Suppose $X \subseteq V'$. Since $N_H(T) \cap V' = \{b\}$, all vertices in T belong to the same level. Since $|V| > k + 2$, this implies pdw$_X$(H) $\geq |T| = |U| + 2|V| - k - 2 > |U| + |V|$, which is a contradiction. Thus we can conclude that $X \subseteq U'$.

Let $(L_1 = X, L_2, L_3)$ be the level structure with the initial set X. Since $|V(H)| = 3(|U| + |V|)$ and pdw$_X$(H) = $|U| + |V|$, each level L_i has size $|L_i| = |U| + |V|$. If $a \in X$, then $S \subseteq L_2$. This implies $|L_3| \leq |V' \setminus S| = |V| + 1 < |U| + |V|$, a contradiction. Hence $X \subseteq U \cup T$. Let $Y = X \cap U$ and $T' = X \cap T$. Clearly, $|N_H(T') \cap V'| = |\{b\}| = 1$. Since $|X| = |U| + |V|$, we have $|U' \setminus X| = |U| + |V| - k - 1$. Since Eq. (1) also holds here, we have $|N_H(Y) \cap V'| = k$. This implies $N_G(Y) = k$, and completes the proof. $\square$

Here, we note that there is a trivial factor-2 approximation algorithm for cobipartite graphs. It is easy to see that a connected cobipartite graph G has diameter at most 3, and thus pdw(G) $\geq \lceil |V(G)|/4 \rceil$. For any $S \subseteq V(G)$ with $|S| = \lceil |V(G)|/2 \rceil$, pdw$_S$($G$) = $\lceil |V(G)|/2 \rceil$. Therefore, pdw$_S$(G) $\leq \lceil |V(G)|/2 \rceil \leq 2\lceil |V(G)|/4 \rceil \leq 2 \cdot$ pdw(G).

Proposition 3.3. *For a cobipartite graph with n vertices and m edges, the path-distance-width can be approximated within a factor 2 in $O(m + n)$ time.*

4 Approximating the Path-Distance-Width

In this section, we present our main results. Namely, approximation algorithms for path-distance-width. Our algorithms are based on the following idea: bounding the diameter of each level in distance structures. This yields the approximation guarantees. The algorithms also have a special feature: we use rooted distance structures only. Thus our algorithms are very simple, and clearly run in polynomial time.

We first establish a general lower bound, which will be the main tool to guarantee the approximation ratios.

Proposition 4.1. Let $(L_1, \ldots, L_t)$ be a distance structure of G. If $u \in L_i$ and $v \in L_j$, then $d_G(u, v) \geq |i - j|$.

Proof. Assume $i \leq j$ without loss of generality. Let $(p_0, p_1, \ldots, p_\ell)$ be a shortest u–v path, where $p_0 = u$ and $p_\ell = v$. From the definition of distance structures, if $p_k \in L_h$, then $p_{k+1} \in L_{h-1} \cup L_h \cup L_{h+1}$. Since $p_0 \in L_i$, $p_\ell \in L_j$, and $i \leq j$, we need at least $j - i$ indices k such that $p_k \in L_h$ and $p_{k+1} \in L_{h+1}$. Thus $\ell \geq j - i$. $\square$

Lemma 4.2. If $S \subseteq V(G)$, then $\mathrm{pdw}(G) \geq |S|/(\mathrm{diam}_G(S) + 1)$.

Proof. Let $(L_1, \ldots, L_t)$ be an optimal distance structure of G; that is, $\mathrm{pdw}_{L_1}(G) = \mathrm{pdw}(G)$. Denote by I the set of the indices of levels having non-empty intersection with S; that is, $I = \{i \in \{1, \ldots, t\} \mid L_i \cap S \neq \emptyset\}$. By Proposition 4.1, $\max I - \min I \leq \mathrm{diam}_G(S)$. Thus the vertices of S are included in at most $\mathrm{diam}_G(S) + 1$ levels $\{L_{\min I}, L_{\min I+1}, \ldots, L_{\max I}\}$. This implies that there exists a level L_i, $i \in I$, such that $|L_i \cap S| \geq |S|/(\mathrm{diam}_G(S) + 1)$. Hence we have

$$\mathrm{pdw}(G) = \mathrm{pdw}_{L_1}(G) \geq |L_i| \geq |L_i \cap S| \geq |S|/(\mathrm{diam}_G(S) + 1),$$

as required. $\square$

4.1 Approximating the Path-Distance-Width for k-Cocomparability Graphs

By the property of k-CCPO, we are able to bound the diameter of each level in some distance structure of a k-cocomparability graph. Thus we have an approximation guarantee as follows.

Lemma 4.3. Let G be a connected k-cocomparability graph, and x be the first vertex in a k-CCPO of G. If $(L_1, \ldots, L_t)$ is the distance structure of G with the initial set $L_1 = \{x\}$, then $\mathrm{diam}_G(L_i) \leq 2k$ for all i.

Proof. Fix i arbitrarily and let $y, z \in L_i$. Without loss of generality, we may assume that $x < y < z$ in the k-CCPO. We show that $d_G(y, z) \leq 2k$. Since y and z lie in the same level, $d_G(x, y) = d_G(x, z)$. Let P be a shortest x–z path in G. Since $d_G(x, y) = d_G(x, z)$, y is not in P. Clearly, there exists an edge $\{v, w\}$ in P such that $v < y < w$. Since $d_G(v, w) = 1 \leq k$, we have $d_G(v, y) \leq k$ or $d_G(y, w) \leq k$. If $d_G(v, y) \leq k$, then $d_G(x, y) \leq d_G(x, v) + k$ and $d_G(y, z) \leq d_G(v, z) + k$. This implies

$$d_G(x, y) + d_G(y, z) \leq d_G(x, v) + d_G(v, z) + 2k = d_G(x, z) + 2k.$$

Then $d_G(y, z) \leq 2k$, since $d_G(x, y) = d_G(x, z)$. The case of $d_G(y, w) \leq k$ is almost the same. $\square$

Combining Lemmas 2.1, 4.2, and 4.3, we have the following general approximation result.

Theorem 4.4. *For a connected k-cocomparability graph G with n vertices and m edges, the path-distance-width can be approximated within a factor $2k + 1$ in $O(mn)$ time.*

4.2 Approximating the Path-Distance-Width for AT-free Graphs

Chang, Ho, and Ko [3] showed that AT-free graphs are 2-cocomparability graphs. Hence, by Theorem 4.4, the path-distance-width of a connected AT-free graph with n vertices and m edges can be approximated within a factor 5 in $O(mn)$ time. The aim of this subsection is to provide a better approximation algorithm for AT-free graphs by using some properties of AT-free graphs. More precisely, we present an $O(m + n)$-time 3-approximation algorithm for AT-free graphs. A *dominating pair* (u, v) of a graph G is a pair of vertices $u, v \in V(G)$ such that for any u–v path P in G, $V(P)$ is a dominating set of $V(G)$; that is, each vertex $v \in V(G) \setminus V(P)$ has a neighbor in $V(P)$.

Theorem 4.5 ([5,6]). *Any connected AT-free graph has a dominating pair. A dominating pair of a connected AT-free graph can be found in linear-time.*

Lemma 4.6. *Let (u, v) be a dominating pair of an AT-free graph G. If $(L_1 = \{u\}, \ldots, L_t)$ is the distance structure rooted at the vertex u, then for any i, $\mathrm{diam}_G(L_i) \leq 2$.*

Proof. Let $(p_1, \ldots, p_\ell)$ be a shortest u–v path in G, where $p_1 = u$ and $p_\ell = v$. Clearly, $p_j \in L_j$ for all j. From the definition of distance structures and dominating pairs, a vertex in a level L_i must be adjacent to at least one of p_{i-1}, p_i, and p_{i+1}, and cannot be adjacent to any other p_j, $j \notin \{i-1, i, i+1\}$. Fix i arbitrarily, and let $x, y \in L_i$. We assume $p_i \notin \{x, y\}$ since otherwise $d_G(x, y) \leq 2$. Let $(q_1, \ldots, q_i)$ be a shortest u–x path, where $q_1 = u$ and $q_i = x$. Obviously, $q_j \in L_j$ for all j. We now have three cases (see Fig. 3).

[Case 1] $\{\{x, p_{i+1}\}, \{y, p_{i+1}\}\} \cap E(G) \neq \emptyset$: By symmetry, we may assume $\{x, p_{i+1}\} = \{q_i, p_{i+1}\} \in E(G)$. Then, $(q_1, \ldots, q_i, p_{i+1}, \ldots, p_\ell)$ is a u–v path. Hence y has a neighbor in $\{q_{i-1}, q_i, p_{i+1}\}$. Since $q_i = x$ and $\{q_{i-1}, q_i\}, \{q_i, p_{i+1}\} \in E(G)$, we have $d_G(x, y) \leq 2$.

[Case 2] $\{\{x, p_i\}, \{y, p_i\}\} \cap E(G) \neq \emptyset$: By symmetry, we may assume $\{x, p_i\} = \{q_i, p_i\} \in E(G)$. Then, $(q_1, \ldots, q_i, p_i, p_{i+1}, \ldots, p_\ell)$ is a u–v path. Hence y has a neighbor in $\{q_{i-1}, q_i, p_i, p_{i+1}\}$. By Case 1, if $\{y, p_{i+1}\} \in E(G)$, then $d_G(x, y) \leq 2$. Otherwise, y has a neighbor in $\{q_{i-1}, q_i, p_i\}$. Since $q_i = x$ and $\{q_{i-1}, q_i\}, \{q_i, p_i\} \in E(G)$, we have $d_G(x, y) \leq 2$.

[Case 3] $\{\{x, p_{i-1}\}, \{y, p_{i-1}\}\} \cap E(G) \neq \emptyset$: By Cases 1 and 2, it suffices to consider the case of $\{x, p_i\}, \{x, p_{i+1}\}, \{y, p_i\}, \{y, p_{i+1}\} \notin E(G)$. Clearly, this assumption implies $\{x, p_{i-1}\}, \{y, p_{i-1}\} \in E(G)$, and hence $d_G(x, y) \leq 2$.   □

Theorem 4.5 and Lemmas 4.2 and 4.6 imply the following better approximation result for AT-free graphs.

Theorem 4.7. *For a connected AT-free graph with n vertices and m edges, the path-distance-width can be approximated within a factor 3 in $O(m + n)$ time.*

We now show that the factor 3 is the best possible even for interval graphs (thus for AT-free graphs) if we use rooted distance structures.

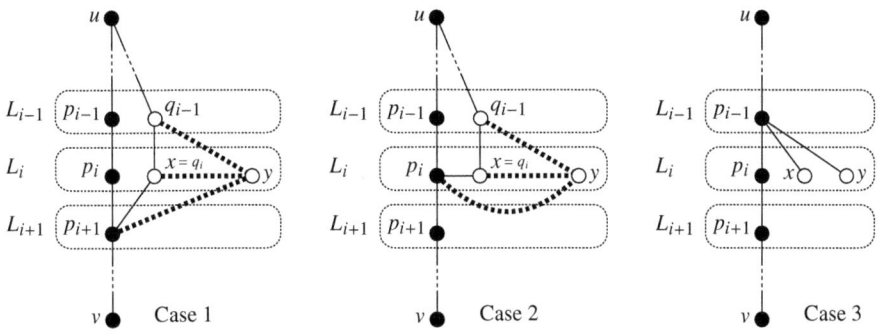

Fig. 3. The cases in the proof of Lemma 4.6

Proposition 4.8. *The approximation ratio 3 of the path-distance-width for interval graphs cannot be improved if we select only one vertex as the initial set.*

Proof. The *friendship graph* F_d is the graph with $V(F_d) = \{c\} \cup \{u_i, v_i \mid 1 \le i \le d\}$ and $E(F_d) = \{\{u_i, v_i\} \mid 1 \le i \le d\} \cup \{\{c, w\} \mid w \in V(F_d) \setminus \{c\}\}$. For any d, F_d is an interval graph (see Fig. 4).

Let c be the center of F_{3d}, and let $w \in V(F_{3d}) \setminus \{c\}$. Clearly, $\mathrm{pdw}_{\{c\}}(F_{3d}) = 6d$ and $\mathrm{pdw}_{\{w\}}(F_{3d}) = 6d - 2$. On the other hand, if $S = \{u_i \mid 1 \le i \le 2d\}$, then

$$\mathrm{pdw}_S(F_{3d}) = \max\{|\{u_i \mid 1 \le i \le 2d\}|, |\{c\} \cup \{v_i \mid 1 \le i \le 2d\}|, |\{u_i, v_i \mid 2d + 1 \le i \le 3d\}|\}$$
$$= \max\{2d, 2d + 1, 2d\} = 2d + 1.$$

Thus if we use only one vertex of F_{3d} as an initial set, then the approximation ratio is at least $(6d - 2)/(2d + 1) = 3 - 5/(2d + 1)$. Since $5/(2d + 1)$ can be arbitrarily small by increasing d, the proposition holds. □

4.3 Approximating the Path-Distance-Width for Proper Interval Graphs

Since proper interval graphs are AT-free, the result in the previous section provides an approximation algorithm for proper interval graphs as well. Fortunately, if we use proper interval representations, then we get a better approximation ratio.

Corneil, Kim, Natarajan, Olariu, and Sprague [4, Proposition 2.1(2)] showed that in the rooted distance structure of a proper interval graphs rooted at the leftmost interval, every level is a clique.

Proposition 4.9 ([4]). *Let G be a connected proper interval graph, and let $u \in V(G)$ be the vertex with the leftmost starting point in some proper interval representation of G. Let L_i be the set of vertices of distance i from u; that is, $L_i = \{v \in V(G) \mid d_G(u, v) = i\}$. Then, for any i, $\mathrm{diam}_G(L_i) = 1$ if $L_i \ne \emptyset$.*

It is known that a proper interval representation of a proper interval graph can be computed in linear time (see e.g. [4]). Thus the leftmost vertex u in the above proposition and the rooted distance structure rooted at u can be found in linear time. Therefore, by Lemma 4.2, the next theorem holds.

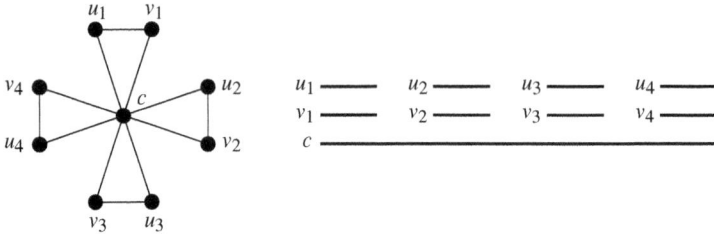

Fig. 4. Friendship graph F_4 and an interval representation of F_4

Theorem 4.10. *For a connected proper interval graph G with n vertices and m edges, the path-distance-width can be approximated within a factor 2 in $O(m + n)$-time.*

Since the complete graph K_{2n} is a proper interval graph, $pdw(K_{2n}) = n$, and $rpdw(K_{2n}) = 2n - 1$, we can conclude that the factor 2 in the above theorem cannot be improved by any algorithm using rooted distance structures only.

Proposition 4.11. *The approximation ratio 2 of the path-distance-width for proper interval graphs cannot be improved if we select only one vertex as the initial set.*

5   A Polynomial-Time Solvable Case

In this section, we show that the path-distance-width of cochain graphs can be determined in linear time. Recall that every cochain graph is a proper interval graph.

Theorem 5.1 ([9]). *Given a cochain graph G with n vertices and m edges, its bipartition (X, Y) and orderings on X and Y (which satisfies the condition in the definition) can be computed in $O(m + n)$ time.*

Theorem 5.2. *The path-distance-width of a connected cochain graph G with n vertices and m edges can be computed in $O(m + n)$ time.*

Proof. Let G be a cochain graph with a bipartition (X, Y). By Theorem 5.1, such a bipartition can be found in $O(m + n)$-time. For convenience, let $pdw(G, X) = \min\{pdw_S(G) \mid S \subseteq X\}$ and $pdw(G, Y) = \min\{pdw_S(G) \mid S \subseteq Y\}$. If $S \subseteq V(G)$ intersects both X and Y, then $pdw_S(G) \geq \lceil |V(G)|/2 \rceil$. It is easy to see that $\min\{pdw(G, X), pdw(G, Y)\} \leq \lceil |V(G)|/2 \rceil$. Therefore,

$$pdw(G) = \min\{pdw(G, X), pdw(G, Y)\}.$$

By symmetry, it is sufficient to show that $pdw(G, X)$ can be computed in $O(m + n)$ time.
   Let $X = \{x_1, \ldots, x_p\}$ and $N_G[x_1] \subseteq N_G[x_2] \subseteq \cdots \subseteq N_G[x_p]$. By Theorem 5.1, such an ordering can be computed in linear time. We also compute in linear time $|X|, |Y|$, and $\deg_G(v)$ for all $v \in V(G)$. Let $Y_0 = \{y \in Y \mid N_G(y) \cap X = \emptyset\}$. Clearly, $Y_0 = \{y \in Y \mid \deg_G(y) = |Y| - 1\}$, and thus $|Y_0|$ can be obtained in linear time.

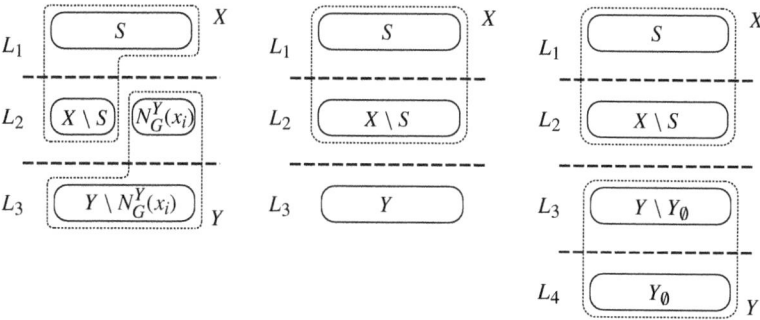

Fig. 5. Three cases in the proof of Theorem 5.2

To compute $\text{pdw}(G, X)$, we define $\text{pdw}(G, X, i)$ as follows:

$$\text{pdw}(G, X, i) = \min\{\text{pdw}_S(G) \mid S \subseteq X, \ i = \max\{j \mid x_j \in S\}\}.$$

For $x_i \in X$, we denote $N_G(x_i) \cap Y$ by $N_G^Y(x_i)$. It is easy to see that $|N_G^Y(x_i)| = \deg_G(x_i) - (|X| - 1)$. If $i = \max\{j \mid x_j \in S\}$ for some $S \subseteq X$, then $N_G(x_i) \cap Y = N_G(S) \cap Y$ since $N_G[x_j] \subseteq N_G[x_i]$ for all $j < i$. Note that $N_G^Y(x_i)$ may be empty. We will prove that $\text{pdw}(G, X, i)$ can be computed in constant time by using $|X|$, $|Y|$, $|Y_\emptyset|$, and $|N_G^Y(x_i)|$. This will imply $\text{pdw}(G, X)$ can be computed in linear time, since $\text{pdw}(G, X) = \min_{1 \le i \le p} \text{pdw}(G, X, i)$.

Let $S \subseteq \{x_1, \ldots, x_i\}$ and $x_i \in S$, and let D be the distance structure with the initial set S. We have the following three cases (see Fig. 5):

$$D = \begin{cases} (S, \ (X \setminus S) \cup N_G^Y(x_i), \ Y \setminus N_G^Y(x_i)) & \text{if } N_G^Y(x_i) \neq \emptyset, \\ (S, \ X \setminus S, \ Y) & \text{if } N_G^Y(x_i) = \emptyset \text{ and } Y_\emptyset = \emptyset, \\ (S, \ X \setminus S, \ Y \setminus Y_\emptyset, \ Y_\emptyset) & \text{if } N_G^Y(x_i) = \emptyset \text{ and } Y_\emptyset \neq \emptyset. \end{cases}$$

In any case, the average size of the first and second levels is $(|X| + |N_G^Y(x_i)|)/2$. Therefore, by setting $|S| = \min\{i, \lceil (|X| + |N_G^Y(x_i)|)/2 \rceil\}$, we can minimize the difference. One possible solution is $S = \{x_i\} \cup \{x_1, \ldots, x_{|S|-1}\}$. Since $\text{pdw}_S(G)$ can be computed in constant time with $|S|$, $|X|$, $|Y|$, $|Y_\emptyset|$, and $|N_G^Y(x_i)|$, the theorem holds. □

Note that cochain graphs, chain graphs, and threshold graphs have similar structures [9]. Thus one can design polynomial-time algorithms also for chain graphs and threshold graphs. However, achieving linear-time algorithms for them is not obvious.

6 Concluding Remarks

We have considered the problem of determining the path-distance-width of graphs in important graph classes. It turned out that the problem is NP-hard even for cobipartite graphs, and thus for cocomparability graphs and AT-free graphs. However, using their chain-like structures, we are able to present constant-factor approximation algorithms. The algorithms are very simple and fast. We also present a linear-time (exact) algorithm for cochain graphs. The computational complexity of the path-distance-width problem for interval graphs and proper interval graphs remains unsettled.

References

1. Blache, G., Karpinski, M., Wirtgen, J.: On approximation intractability of the bandwidth problem, ECCC TR98-014 (1998)
2. Brandstädt, A., Le, V.B., Spinrad, J.P.: Graph Classes: A Survey. SIAM (1999)
3. Chang, J.M., Ho, C.W., Ko, M.T.: Powers of asteroidal triple-free graphs with applications. Ars Combin. 67, 161–173 (2003)
4. Corneil, D.G., Kim, H., Natarajan, S., Olariu, S., Sprague, A.P.: Simple linear time recognition of unit interval graphs. Inform. Process. Lett. 55, 99–104 (1995)
5. Corneil, D.G., Olariu, S., Stewart, L.: Asteroidal triple-free graphs. SIAM J. Discrete Math. 10, 399–430 (1997)
6. Corneil, D.G., Olariu, S., Stewart, L.: Linear time algorithms for dominating pairs in asteroidal triple-free graphs. SIAM J. Comput. 28, 1284–1297 (1999)
7. Garey, M.R., Johnson, D.S.: Computers and Intractability: A Guide to the Theory of NP-Completeness. Freeman (1979)
8. Golovach, P., Heggernes, P., Kratsch, D., Lokshtanov, D., Meister, D., Saurabh, S.: Bandwidth on AT-Free Graphs. In: Dong, Y., Du, D.-Z., Ibarra, O. (eds.) ISAAC 2009. LNCS, vol. 5878, pp. 573–582. Springer, Heidelberg (2009)
9. Heggernes, P., Kratsch, D.: Linear-time certifying recognition algorithms and forbidden induced subgraphs. Nordic J. Comput. 14, 87–108 (2007)
10. Johnson, D.S.: The NP-completeness column: An ongoing guide. J. Algorithms 6, 434–451 (1985)
11. Kaplan, H., Shamir, R.: Pathwidth, bandwidth, and completion problems to proper interval graphs with small cliques. SIAM J. Comput. 25, 540–561 (1996)
12. Kleitman, D.J., Vohra, R.V.: Computing the bandwidth of interval graphs. SIAM J. Discrete Math. 3, 373–375 (1990)
13. Kloks, T., Kratsch, D., Müller, H.: Approximating the bandwidth for asteroidal triple-free graphs. J. Algorithms 32, 41–57 (1999)
14. Kobayashi, Y.: Private communication (September 2010)
15. Mahesh, R., Rangan, C.P., Srinivasan, A.: On finding the minimum bandwidth of interval graphs. Inform. and Comput. 95, 218–224 (1991)
16. Parra, A., Scheffler, P.: Characterizations and algorithmic applications of chordal graph embeddings. Discrete Appl. Math. 79, 171–188 (1997)
17. Sprague, A.P.: An $O(n \log n)$ algorithm for bandwidth of interval graphs. SIAM J. Discrete Math. 7, 213–220 (1994)
18. Yamazaki, K.: On approximation intractability of the path-distance-width problem. Discrete Appl. Math. 110, 317–325 (2001)
19. Yamazaki, K., Bodlaender, H.L., de Fuiter, B., Thilikos, D.M.: Isomorphism for graphs of bounded distance width. Algorithmica 24, 105–127 (1999)

Hanani-Tutte and Monotone Drawings

Radoslav Fulek[1,*], Michael J. Pelsmajer[2,**],
Marcus Schaefer[3], and Daniel Štefankovič[4]

[1] Ecole Polytechnique Fédérale de Lausanne, Lausanne, Switzerland
`radoslav.fulek@epfl.ch`
[2] Illinois Institute of Technology, Chicago, IL 60616, USA
`pelsmajer@iit.edu`
[3] DePaul University, Chicago, IL 60604, USA
`mschaefer@cs.depaul.edu`
[4] University of Rochester, Rochester, NY 14627, USA
`stefanko@cs.rochester.edu`

Abstract. A drawing of a graph is x-*monotone* if every edge intersects every vertical line at most once and every vertical line contains at most one vertex. Pach and Tóth showed that if a graph has an x-monotone drawing in which every pair of edges crosses an even number of times, then the graph has an x-monotone embedding in which the x-coordinates of all vertices are unchanged. We give a new proof of this result and strengthen it by showing that the conclusion remains true even if adjacent edges are allowed to cross oddly. This answers a question posed by Pach and Tóth. Moreover, we show that an extension of this result for graphs with non-adjacent pairs of edges crossing oddly fails even if there exists only one such pair in a graph.

1 Introduction

The classic Hanani-Tutte theorem states that if a graph can be drawn in the plane so that no pair of independent edges crosses an odd number of times, then it is planar [6,19]. (Two edges are *independent* if they do not have a shared endpoint.) There are many ways to look at this result; for example, in algebraic topology it is seen as a special case of the van Kampen-Flores theorem [9, Chapter 5] which classifies obstructions to embeddability in topological spaces. This point of view leads to challenging open questions (see, for example, [10]), but even in 2-dimensional surfaces the problem is not understood well (see [18] for a survey of what we do know).

Here, we study a variant of the problem for x-monotone drawings which was introduced by Pach and Tóth [12]. A drawing of a graph is x-*monotone* if every edge intersects every vertical line at most once and every vertical line contains

---

* The first author gratefully acknowledges support from the Swiss National Science Foundation Grant No. 200021-125287/1.
** The second author gratefully acknowledges the support from NSA Grant H98230-08-1-0043 and the Swiss National Science Foundation Grant No. 200021-125287/1.

P. Kolman and J. Kratochvíl (Eds.): WG 2011, LNCS 6986, pp. 283–294, 2011.
© Springer-Verlag Berlin Heidelberg 2011

284   R. Fulek et al.

at most one vertex. The natural analogue of the Hanani-Tutte theorem in this context would state that every x-monotone drawing in which no pair of independent edges crosses an odd number of times has an x-monotone *embedding*, that is, a crossing-free drawing—without moving the vertices. The truth of this result was left as an open problem by Pach and Tóth. We prove it as Theorem 2 in Section 3. The extension of this result in the spirit of [11,15] is not possible, which is proved in Section 4.

The weak version of the classic Hanani-Tutte theorem states that if a graph can be drawn so that *no* pair of edges crosses oddly, then it is planar. The analogue for x-monotone drawings states that there is an x-monotone embedding if there is an x-monotone drawing in which no pair of edges crosses an odd number of times. This variant of the weak Hanani-Tutte theorem was first proved by Pach and Tóth.[1] We give a new proof of this result as Theorem 1 in Section 2, which continues an elementary topological approach similar to earlier papers on the Hanani-Tutte theorem, e.g. [15].

A traditional approach to Hanani-Tutte style results is via obstructions; this sometimes leads to very slick proofs, like Kleitman's proof of the Hanani-Tutte theorem for the plane [7], but there are two drawbacks: complete obstruction sets are not always known, e.g. for the torus or, in spite of several attempts, for x-monotone embeddings (as discussed in [5]); and this approach is of little help algorithmically. Pach and Tóth took another approach, building on a proof of the weak Hanani-Tutte theorem for surfaces by Cairns and Nikolayevsky [2].

Before we begin, we introduce some basic terminology and notation. For any graph $G = (V, E)$ and $S \subseteq V(G)$, let $G[S]$ denote the *subgraph induced by S*; that is, the graph on vertex set S with edge set $\{uv \in E(G) : u \in S, v \in S\}$. By a *multigraph* we understand a graph for which the set of edges is a multiset. A *topological graph* is a graph drawn in the plane where the vertices are represented by distinct points, and edges as Jordan arcs connecting the incident vertices, but not passing through any other vertex and any pair of edges crosses a finite number of times. Throughout the paper by a *drawing* of a graph we understand its representation as a topological graph. By an *embedding* of a graph we understand its (edge) crossing-free drawing.

The *rotation* at a vertex in a drawing of a graph is the clockwise ordering of edges at that vertex. The *rotation system* of a graph is the collection of rotations at its vertices. In an x-monotone drawing, the *right (left) rotation* is the clockwise order of the edges leaving the vertex towards the right (left). So (perhaps unfortunately), the right rotation is ordered from top to bottom, and the left rotation is ordered from bottom to top. We will not carefully distinguish between an abstract graph and a topological (drawn or embedded) graph, and "vertex" and "edge" are used in both contexts. We use $x(v)$ to denote the x-coordinate of a vertex v located in the plane.

Remark 1. We are also preparing an extended version of this paper in which our results are used to solve algorithmic questions regarding level-planarity.

---

[1] There is a gap in the original argument; an updated version is now available [12,13]

2   Weak Hanani-Tutte for Monotone Drawings

An edge is *even* if it crosses every other edges an even number of times (including 0 times). A drawing is *even* if all its edges are even.

Theorem 1 (Pach, Tóth [12,13]). *If G has an x-monotone and even drawing, then G has an x-monotone embedding in which each vertex keeps its x-coordinate and the rotation system remains unchanged.*

Remark 2. The weak Hanani-Tutte theorem states that every graph with an even drawing is planar (without changing the rotation system). For background and variants of the weak Hanani-Tutte theorem, see [18].

Theorem 1 remains true if we require the resulting embedding to be straight-line. This has nothing to do with the Hanani-Tutte part of the result; it is entirely due to the fact that any x-monotone embedding can be turned into a straight-line embedding in which every vertex keeps its x-coordinate [4,12]. This redrawing can lead to an exponential blow-up in the area required for the drawing [8] (the examples in the paper allow multiple vertices in each layer, but these can be replaced by the requirement that vertices are not too close to edges they are not incident to).

Theorem 1 may prompt the reader familiar with Hanani-Tutte style results (in particular [11, Theorem 1] and [15, Theorem 2.1]) to ask whether something stronger is true: a "removing even crossings" lemma which would say that all even edges can be made crossing-free in the drawing of a graph which contains odd edges (while maintaining monotonicity). We will see in Section 4 that there cannot be any such lemma for monotone drawings.

Nearly the same result is claimed by Pach and Tóth in [12, Theorem 1.1], but instead of maintaining the rotation, Pach and Tóth state that we can find an *equivalent* x-monotone embedding for a given x-monotone and even drawing of G, where two drawings are equivalent if no edge changes whether it passes above or below a vertex. However, there are simple examples that show that one cannot hope to maintain equivalence in this sense, see Figure 1. The example can easily be turned into a 2-connected graph by replacing edges and vertices with cycles, so equivalence cannot be obtained by assuming 2-connectedness. On the other hand, see Corollary 1 for a positive result.

The original proof by Pach and Tóth contains a gap: it is not immediately clear how multiple faces that share a boundary can be embedded simultaneously.[2] Filling in the details of this gap requires dropping equivalence. Pach and Tóth have prepared an updated version of the paper that includes a more detailed argument [13].[3]

---

[2] In the text after Lemma 2.1 on page 42 of [12], D_κ cannot necessarily be glued together without changing equivalence.

[3] In this newer version, equivalence is redefined to mean having the same rotation system.

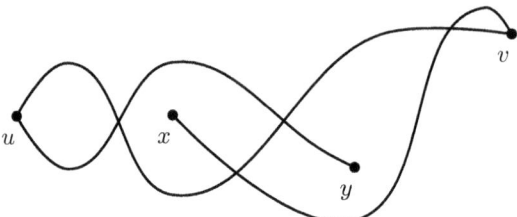

Fig. 1. y lies above xv and x lies above uv. So in any equivalent x-monotone embedding with the same relative x-order of the vertices, xv lies above uv, forcing y above uv; but this contradicts y being below uv.

We approach Theorem 1 in the spirit of earlier papers on the Hanani-Tutte theorem, e.g. [15]. The proof, which is omitted, repeatedly makes use of a simple topological observation: suppose we are given two curves (not necessarily monotone) starting at $x = x_1$ and ending at $x = x_2$ which lie entirely between x_1 and x_2. The two curves cross an even number of times if and only if they have the same vertical order at $x = x_1$ and $x = x_2$ (if they start or end in the same point the *vertical order at x* is determined by the order in which they enter x).

We will also find the following redrawing tool useful.

Lemma 1. *Suppose a multigraph G has an x-monotone embedding and let f be an inner face of the embedding, with m_f and M_f being the leftmost and the rightmost vertex of f. If we add an edge $m_f M_f$ so that $m_f M_f$ lies in f, then the resulting graph $G \cup \{m_f M_f\}$ has an x-monotone embedding in which the relative x-order of the vertices and the rotation system remain the same. Note that we do not require $m_f M_f$ to be x-monotone initially and that there may be multiple ways of inserting $m_f M_f$ into the rotations at its endpoints.*

Note that while the redrawing in Lemma 1 keeps the rotation system the same, it will destroy equivalence in the sense of Pach and Tóth. Indeed, this is necessarily the case as witnessed by a z-shaped corridor as the face in question:

Fig. 2. Adding a monotone edge $m_f M_f$ into the corridor requires destroying equivalence

Proof (of Lemma 1). If G consists of multiple components, it is sufficient to prove the result for the component containing f and shift its embedding vertically so that it does not intersect any other component. This allows us to assume that G is connected. Then every face is bounded by a closed walk.[4] The boundary of f can be broken into two m_f, M_f-walks, B_1, B_2 with B_1 starting above $m_f M_f$ in the rotation at m_f, and B_2 starting below.

Let D_f be the drawing of G intersected with $U_f := \{(x, y) \in \mathbb{R}^2 : x(m_f) < x < x(M_f)\}$. ($D_f$ is a subset of the plane, not a graph.) We will locally redraw G in U_f so that $m_f M_f$ can be inserted as a straight-line segment. For each (topologically) connected component Z of D_f, either (*i*) for every x between $x(m_f)$ and $x(M_f)$, there is a y-value of B_1 at x that is below all y-values of Z at x, or (*ii*) for any x between $x(m_f)$ and $x(M_f)$, there is a y-value of B_2 at x that is above all y-values of Z at x.

Let Z_1 be the union of all components of the first type, and Z_2 be the union of all components of the second type. Let L be the line through m_f and M_f. We will show how to move Z_1 to the half-plane above L, without changing the x-value of any point in Z_1 while fixing the points on the boundary of U_f. Let P be an x-monotone curve with endpoints m_f and M_f that lies strictly below Z_1 in U_f (note that m_f and M_f do not belong to U_f). Now move every point v of Z_1 up by the vertical distance between P and L at $x = x(v)$. We proceed similarly to move Z_2 strictly below L, at which point L is the desired embedding of $m_f M_f$. $\square$

In the proof of Theorem 1, all redrawing steps maintain equivalence except for applications of Lemma 1. This part of the proof, however, only arises in the case that $G - \{v_1, \ldots, v_i\}$, where v_j denote j-th leftmost vertex of G, is not connected. Hence, if we can make an assumption on G so that this case never occurs, we can conclude that the resulting embedding is equivalent to the original drawing in the sense of Pach and Tóth. We already saw that 2-connectedness is not sufficient, however, another notion is: a graph in which the vertices are ordered is a *hierarchy* if every vertex except the rightmost one has an edge leaving it towards the right [3].

Corollary 1. *If G has an x-monotone and even drawing and G is a hierarchy, then G has an equivalent x-monotone embedding in which each vertex keeps its x-coordinate and the rotation system remains unchanged.*

The assumption in Theorem 1 can be weakened, somewhat surprisingly, replacing x-monotonicity of edges by a weaker notion. Let us say that an edge uv in a drawing is *bounded* if every interior point p of uv satisfies $x(u) < x(p) < x(v)$. That is, an edge is bounded if it lies strictly between its endpoints; it need not be x-monotone within those bounds.

Lemma 2. *Suppose we are given a drawing of a graph G with a bounded edge e. Then e can be redrawn, without changing the remainder of the drawing or the*

---

[4] *Walks are like paths except that vertices and edges can be repeated. In a *closed walk* the last vertex is the same as the first vertex.*

position of e in the rotations of its endpoints, so that e is x-monotone and the parity of crossing between e and any other edge of G has not changed.

Proof. Suppose that $e = ab$ and let $v \in V(G)$ be an arbitrary vertex between a and b: $x(a) < x(v) < x(b)$. Now e has to cross the line $x = x(v)$ an odd number of times since it connects a to b. In particular, v splits $x = x(v)$ into two: a part which is crossed an odd number of times by e, and the other part which is crossed evenly. In a small neighborhood of $x = x(v)$ redraw G by pushing all crossings of e with $x = x(v)$ from the even side across v to the odd side. Note that the odd side of $x = x(v)$ remains odd and there are no crossing with e left on the even side. Moreover, the parity of crossing between e and any other edge does not change since e is moved an even number of times across v. Repeat this for all v between a and b; now e only passes above or below each such v, never both. We can now deform e into an x-monotone edge connecting a and b, without having the edge pass over any vertices. Therefore, this deformation does not affect the parity of crossing between e and any other edge, so we have found the redrawing required by the lemma.

In hindsight we see that the redrawing of e is quite effective: for each vertex v between a and b we only need to know whether e passes oddly above or below it, and we can build a polygonal arc from a to b that passes each vertex on the odd side.   □

3   Strong Hanani-Tutte for Monotone Drawings

Pach and Tóth [12] wrote "It is an interesting open problem to decide whether [Theorem 1] remains true under the weaker assumption that any two *non-adjacent* edges cross an an even number of times." The goal of this section is to establish this result, which was also conjectured in [18].

Theorem 2. *If G has an x-monotone drawing in which every pair of independent edges crosses evenly, then G has an x-monotone embedding in which each vertex keeps its x-coordinate.*

Remark 3. As in the case of Theorem 1, the statement of Theorem 2 remains true if we only require edges to be bounded rather than x-monotone: simply redraw edges one at a time using Lemma 2, before applying Theorem 2.

Let $G = (V, E)$ and $G' = (V', E')$ denote two graphs. We say that $G < G'$, if $|V| < |V'|$ or if $|V| = |V'|$ and $|E| < |E'|$. In the sequel we consider minimal counterexamples with respect to this relation. In a proof of the standard Hanani-Tutte theorem, it is obvious that a minimal counterexample has to be 2-connected, since embedded subgraphs can be merged at a cut-vertex. Unfortunately, the merge requires a redrawing that does not maintain monotonicity, so here we must use structural properties that are more tailored to x-monotone redrawings.

Lemma 3. *Suppose that G is a minimal counterexample to Theorem 2. Then:*

Fig. 3. Lemma 3(ii), forbidden subgraph H

Fig. 4. Lemma 3(iii), forbidden edge ac (left) and forbidden subgraph H' (right)

(i) G *is connected.*

(ii) G *has no subgraph* H *and vertices* $a, b \in V(G) \setminus V(H)$ *such that* $x(a) < x(v) < x(b)$ *for all* $v \in V(H)$, $N(H) = \{a, b\}$, *and* $V(G) \setminus (V(H) \cup \{a, b\}) \neq \emptyset$.

(iii) *If* G *has a cut-vertex* a *and* $G - \{a\}$ *has a component* H *such that* $x(a) < x(v)$ *for all* $v \in V(H)$, *then* H *has only one vertex* b, *and* G *has no edge* ac *with* $x(b) < x(c)$. *Also, in this case* G *has no induced subgraph* $H' \neq \emptyset$ *so that* $x(a) < x(v) < x(b)$ *for all* $v \in V(H')$, $a \in N(H') \neq \{a\}$, *and* $x(v) > x(b)$ *for all* $v \in N(H') \setminus \{a\}$.

Proof. If a minimal counterexample G is not connected, none of its components are counterexamples to Theorem 2. But then we could embed each component separately and stack the drawings vertically so they do not intersect each other, yielding an embedding of G. This contradiction establishes (i).

Consider case (ii) (see Figure 3). Since G is a minimal counterexample, both $G - V(H)$ and $G[V(H) \cup \{a, b\}]$ have embeddings (both graphs are smaller than G by assumption). We can deform the crossing-free drawing of $G[V(H) \cup \{a, b\}]$ so that it becomes very flat. If $ab \in E(G)$ we can then insert this drawing into the drawing of $G - V(H)$ near the edge ab, without adding crossings. This gives us a crossing-free drawing of G, which is a contradiction. If $ab \notin E(G)$ then we add ab to the drawing of $G - V(H)$ so that it has no independent odd crossings (we will presently see how this can be done); the resulting $G - V(H) \cup \{ab\}$ has fewer vertices than G so it also has an embedding, and we can proceed as in the case that $ab \in E(G)$, removing the edge ab in the end.

When $ab \notin E(G)$, here is how we draw the edge ab with no independent odd crossings: Let P be any a, b-path with interior vertices in H. By suppressing the interior vertices of P, we can consider it a bounded edge (in the sense defined earlier) between a and b, so Lemma 2 tells us that we can draw an x-monotone edge that has the same parity of crossing with all edges of G as does P.

Finally, we consider (iii) (see Figure 4), where H is a component of $G - \{a\}$ so that $x(a) < x(v)$ for all $v \in V(H)$. If $|V(H)| > 1$, let b be the vertex with the largest x-value in H and apply the part (ii) with $H := H - b$. Since the previous proof requires that $H \neq \emptyset$, we are in the case that $V(H) = \{b\}$. If G has an edge

ac with $x(b) < x(c)$, we can embed $G - \{b\}$ (since it is smaller than G), and then add ab and b to the embedding alongside of ac without crossings.

It remains to consider an induced subgraph $H' \neq \emptyset$ so that $x(a) < x(v) < x(b)$ for all $v \in V(H')$, $a \in N(H') \neq \{a\}$, and $x(v) > x(b)$ for all $v \in N(H') \setminus \{a\}$. By minimality, $G - \{b\}$ has an embedding. Consider the face which lies immediately below H'; let B be its boundary, and let c be the vertex on B with maximum x-value. B is a closed walk that intersects H', and H' has edges to the left and right, so B must contain neighbors of H' on its left and on its right. Therefore, B contains a (to the left of H'), and by the choice of H', $x(c) > x(b)$. Using Lemma 1, we can add the edge ac to the embedding of $G - \{b\}$ without introducing crossings. Since $x(a) < x(b) < x(c)$, we can instead add ab to the drawing without crossings, so G has an embedding which is a contradiction. $\square$

The proof of Theorem 2 now proceeds by induction on the number of *odd pairs* (pairs of edges that cross an odd number of times). Roughly speaking: If we encounter an odd pair (by necessity its edges are adjacent), we can either make it cross evenly or we are in a situation which has been excluded by Lemma 3. To realize this goal, we need more intermediate results. These results are not about minimal counterexamples, but are true in general.

For the lemmas we introduce some new terminology generalizing our usual notion of lying above or below a curve to curves with self-intersections: Let C be a curve in the plane with endpoints p and r so that for every point $c \in C \setminus \{p, r\}$, $x(p) < x(c) < x(r)$. (This is similar to the definition of a bounded edge except that we allow self-intersections.) Suppose that q is a point for which $x(p) \leq x(q) \leq x(r)$. Extend C via a horizontal ray from p to $x = -\infty$ and a horizontal ray from r to $x = \infty$, and consider the plane $\mathbb{R}^2$ minus that extended curve. We can 2-color its faces so that adjacent faces (faces whose boundaries intersect in a nontrivial curve) have opposite colors. We say that q is *above* (*below*) C if q lies in a face with the same color as the upper (lower) unbounded region.

In the following two lemmas, let G satisfy the assumption of Theorem 2, that is, we assume that every pair of independent edges in G crosses evenly. Both lemmas deal with the following scenario: G contains three edges $e_i = v_0 v_i$, $i \in \{1, 2, 3\}$ so that e_3 lies between e_1 and e_2 in the right rotation of v_0, with e_1 above e_2 at v_0, e_1 and e_2 cross oddly, and e_3 crosses each of the other two edges evenly.

Lemma 4. *With an arbitrary vertex $v_R > x(v_0)$ define G' as the graph induced by G on vertices v with $x(v_0) < x(v) \leq x(v_R)$. Let G'_i be the component of G' that contains v_i. (If $x(v_i) > x(v_R)$, then $G'_i = \emptyset$.)*

If G'_1, G'_2, G'_3 are pairwise disjoint and if for every i there is a path P_i from v_0 through e_i to some vertex v'_i satisfying $x(v'_i) \geq x(v_R)$ so that all vertices v of P_i satisfy $x(v) \geq x(v_0)$, then each G'_i has no neighbors (in G) to the left of $x(v_0)$, for $i \in \{1, 2, 3\}$.

Lemma 5. *Suppose that for some distinct $j, k \in \{1, 2, 3\}$, there is a cycle C that contains e_j and c_k such that every vertex v of C satisfies $x(v) \geq x(v_0)$. Let*

v_R be the vertex on C with largest x-value. Let i be the unique index such that $\{i,j,k\} = \{1,2,3\}$. Suppose that v_i is not in C.

Let G'_i be the component of $G - V(C)$ that contains v_i. Then every vertex v of G'_i satisfies $x(v_0) < x(v) < x(v_R)$.

We are finally in a position to prove Theorem 2. We need one more piece of terminology: the *distance* between two edges e, f is the number of edge ends between the ends of e, f in the right (or left) rotation. (We do not measure distance within the entire rotation; only within the right or left rotation.)

Proof (of Theorem 2). Let G be a minimal counterexample to the theorem. Fix a drawing of G which minimizes the number of odd pairs, that is, the number of pairs of edges crossing oddly. If there are no odd pairs, then Theorem 1 completes the proof.

Suppose that there are edges e_1 and e_2 that cross oddly. Then e_1 and e_2 have a shared endpoint v_0, and we may assume that v_0 is the left endpoint of e_1 and e_2. Choose e_1 and e_2 so that their ends at v_0 have minimal distance in the right rotation at v_0, with e_1 above (that is, preceding) e_2. Then e_1 and e_2 are not consecutive in the rotation at v_0; if they were, they could be redrawn so that they cross once more near v_0, by switching their order in the rotation at v_0; this contradicts the choice of drawing of G. So there is at least one edge incident to v_0 that lies between e_1 and e_2 in the rotation at v_0, and by minimality, all such edges cross each other evenly and cross both e_1 and e_2 evenly. Pick one such edge, e_3. Let v_1, v_2, v_3 be the right endpoints of e_1, e_2, e_3, respectively, and let G_0 be the subgraph of G induced by all vertices v fulfilling $x(v) \geq x(v_0)$.

Case 1. The right endpoints of e_1, e_2, e_3 are in different components of $G_0 - v_0$.

In Case 1, for each $i \in \{1,2,3\}$, consider the component of $G_0 - v_0$ that contains v_i and let v'_i be its vertex with largest x-value. Assign i,j,k so that $\{i,j,k\} = \{1,2,3\}$, and $x(v'_i)$ is smaller than $x(v'_j)$ and $x(v'_k)$. Apply Lemma 4 with $x_R = x(v'_i)$, which defines G'_i, G'_j, G'_k. By Lemma 3(iii), G'_i has only the one vertex $v_i = v'_i$, and G'_j and G'_k are non-empty because $x(v_i)$ is greater than $x(v_j)$ and $x(v_k)$ (using $a = v_0$, $c \in \{v_j, v_k\}$ and $b = v_i$). Then we can apply the second part of Lemma 3(iii) with H' equal to G'_j (or G'_k) restricted to the vertices with x-coordinate smaller than $x(v'_i)$, and we are done.

If we are not in Case 1, then let v_L be the vertex with $x(v_L)$ chosen to be smallest such that the subgraph induced by vertices v such that $x(v_0) < x(v) \leq x(v_L)$ has a component that contains at least two right endpoints of e_1, e_2, e_3. Then there is a cycle C that contains e_j and e_k for some distinct $k, j \in \{1,2,3\}$, and so that $x(v_0) \leq x(v) \leq x(v_L)$ for all $v \in V(C)$. If $vv_L \in \{e_1, e_2, e_3\}$, then we may assume that C contains vv_L.

Let i be the unique index for which $\{i,j,k\} = \{1,2,3\}$. By the previous assumption, $v_i \neq v_L$. By Lemma 5, $x(v_i) < x(v_L)$ or $v_i \in V(C) - v_L$.

Suppose that there is a path Q from v_i to C so that $x(v_0) < x(v) < x(v_L)$ for all $v \in V(Q)$. Then $Q \cup e_i \cup C - v_L$ contains a cycle C' with e_i and either e_j or e_k. But every vertex v of C' satisfies $x(v_0) \leq x(v) < x(v_L)$ for all v in C', contradicting the choice of v_L.

We can conclude that v_i is not in $V(C) - v_L$, and if we let G_i' be the component of $G - V(C)$ that contains v_i, then G_i' has no neighbors in $V(C) \setminus \{v_0, v_L\}$. By Lemma 5, G_i' lies between $x = x(v_0)$ and $x = x(v_L)$ (since $v_i \neq v_L$). Let v_i' be the vertex of G_i' with largest x-value. Apply Lemma 4 with $x_R = x(v_i')$. This defines G_i', G_j', G_k'.

Case 2. G_i' is not adjacent to v_L.

(Same as Case 1:) By Lemma 3(*iii*), G_i' has only the one vertex $v_i = v_i'$, and G_j' and G_k' are non-empty because $x(v_i)$ is greater than $x(v_j)$ and $x(v_k)$. Then we can apply Lemma 3(*iii*) with H' equal to G_j' (or G_k') restricted to the vertices with the x-coordinate smaller than $x(v_i')$, and we are done.

Case 3. There is an edge from G_i' to v_L.

Apply Lemma 3(*ii*) with $H = G_i'$. This completes the proof of the theorem.   □

4   Monotone Crossing Numbers

Our Hanani-Tutte results can be recast as results about monotone crossing numbers. For a leveled graph (G, ℓ) let mon-cr(G, ℓ) be the smallest number of crossings in any leveled drawing of (G, ℓ). Similarly, we can define mon-ocr(G, ℓ) as the smallest number of pairs of edges that cross oddly in any leveled drawing of (G, ℓ). Finally, mon-iocr(G, ℓ) is the smallest number of pairs of non-adjacent edges that cross oddly in any leveled drawing of (G, ℓ). We suppress ℓ and simply write mon-cr(G), mon-ocr(G), and mon-iocr(G). With this notation we can restate the original result by Pach and Tóth, our Theorem 1 as saying that mon-ocr$(G) = 0$ implies mon-cr$(G) = 0$. Similarly, our Theorem 2 can be restated as mon-iocr$(G) = 0$ implies mon-cr$(G) = 0$.

From this point of view we can now ask questions that parallel analogous problems for the regular (non-monotone) crossing number variants: cr, ocr, and iocr. For example, we know that ocr$(G) = $ cr(G) for ocr$(G) \le 3$ [16] and iocr$(G) = $ cr(G) for iocr$(G) \le 2$ [17]. Pach and Tóth showed that cr$(G) \le \binom{\text{ocr}(G)}{2}$ [11,15]. The core step in this result is a "removing even crossings" lemma, in this particular case: if G is drawn in the plane and E_0 is the set of its even edges, then G can be redrawn so that all edges in E_0 are free of crossings. It immediately implies cr$(G) \le \binom{\text{ocr}(G)}{2}$, since only non-even edges can be involved in crossings (and every pair of non-even edges needs to cross at most once). A similar result for monotone drawings fails dramatically. In other words: even if there are only two edges crossing oddly and all other edges are even, then any x-monotone drawing of G with the given leveling may require an arbitrary number of crossings. Thus we cannot hope to establish a "removing even crossings" lemma in the context of x-monotone drawings since it would imply a bound on mon-cr(G) in terms of mon-ocr(G).

Theorem 3. *For every n there is a graph G so that* mon-cr$(G) \ge n$ *and* mon-ocr$(G) = 1$.

5 Open Questions

We want to suggest some future avenues of research.

Monotone Crossing Numbers. The *monotone crossing number* of a leveled graph G is the smallest number of crossings in any x-monotone drawing of the leveled graph. This problem is known to be **NP**-hard (even for two levels) and the monotone crossing number can be arbitrarily large, even for a planar graph (consider nested >s). We get a more interesting question if we define the *monotone crossing number for unleveled graphs* as the smallest crossing number of any x-monotone drawing for any leveling of the graph. Is this monotone crossing number bounded in the crossing number? For comparison, $\mathrm{rcr}_2(G)$ is at most $\binom{\mathrm{cr}(G)}{2}$, where $\mathrm{rcr}_2(G)$ allows straight-line edges with one bend [1]. Pach and Tóth in [14] recently proved that monotone crossing number for unleveled graphs is at most $\binom{\mathrm{cr}(G)}{2}$. Then one can ask how far is this bound from the truth ?

Bi-monotonicity. Let us define y-monotonicity like x-monotonicity after a 90-degree rotation; not very exciting by itself, but what happens if we want embeddings that are *bi-monotone*, that is, both x- and y-monotone?

- If a graph has both an x-monotone embedding and a y-monotone embedding, does it always have a bi-monotone embedding?
- If there is a drawing of a graph which is bi-monotone, is there a straight-line drawing with the same x and y ordering?
- What about bi-level-planarity?

As far as we know, bi-monotonicity and bi-level-planarity are new concepts, however, they are quite natural: If we specify the relative locations of objects on a map, we specify them in terms of "west/east of" and "north/south of" which is exactly what bi-monotonicity models. Imagine specifying the stations for a subway map: actual distance do not matter, what matters is relative location in terms of x and y.

References

1. Bienstock, D., Dean, N.: Bounds for rectilinear crossing numbers. J. Graph Theory 17(3), 333–348 (1993)
2. Cairns, G., Nikolayevsky, Y.: Bounds for generalized thrackles. Discrete Comput. Geom. 23(2), 191–206 (2000)
3. Di Battista, G., Nardelli, E.: Hierarchies and planarity theory. IEEE Trans. Systems Man Cybernet. 18(6), 1035–1046 (1988, 1989)
4. Eades, P., Feng, Q., Lin, X., Nagamochi, H.: Straight-line drawing algorithms for hierarchical graphs and clustered graphs. Algorithmica 44(1), 1–32 (2006)
5. Estrella-Balderrama, A., Fowler, J.J., Kobourov, S.G.: On the Characterization of Level Planar Trees by Minimal Patterns. In: Eppstein, D., Gansner, E.R. (eds.) GD 2009. LNCS, vol. 5849, pp. 69–80. Springer, Heidelberg (2010)
6. Chojnacki, C., (Hanani, H.).: Über wesentlich unplättbare Kurven im drei-dimensionalen Raume. Fundamenta Mathematicae 23, 135–142 (1934)

7. Kleitman, D.J.: A note on the parity of the number of crossings of a graph. J. Combinatorial Theory Ser. B 21(1), 88–89 (1976)
8. Lin, X., Eades, P.: Towards area requirements for drawing hierarchically planar graphs. Theor. Comput. Sci. 292(3), 679–695 (2003)
9. Matoušek, J.: Using the Borsuk-Ulam theorem. Universitext. Springer, Berlin (2003); Lectures on topological methods in combinatorics and geometry, Written in cooperation with Anders Björner and Günter M. Ziegler
10. Matousek, J., Tancer, M., Wagner, U.: Hardness of embedding simplicial complexes in $\mathbb{R}^d$. In: Mathieu, C. (ed.) Proceedings of the Twentieth Annual ACM-SIAM Symposium on Discrete Algorithms, SODA 2009, New York, NY, USA, January 4-6, pp. 855–864. SIAM (2009)
11. Pach, J., Tóth, G.: Which crossing number is it anyway? J. Combin. Theory Ser. B 80(2), 225–246 (2000)
12. Pach, J., Tóth, G.: Monotone drawings of planar graphs. J. Graph Theory 46(1), 39–47 (2004)
13. Pach, J., Tóth Monotone, G.: Drawings of planar graphs. ArXiv e-prints (January 2011)
14. Pach, J., Tóth, G.: Monotone crossing number. In: Graph Drawing (to appear, 2011)
15. Pelsmajer, M.J., Schaefer, M., Štefankovič, D.: Removing even crossings. J. Combin. Theory Ser. B 97(4), 489–500 (2007)
16. Pelsmajer, M.J., Schaefer, M., Štefankovič, D.: Odd crossing number and crossing number are not the same. Discrete Comput. Geom. 39(1), 442–454 (2008)
17. Pelsmajer, M.J., Schaefer, M., Štefankovič, D.: Removing independently even crossings. SIAM Journal on Discrete Mathematics 24(2), 379–393 (2010)
18. Schaefer, M.: Hanani-Tutte and related results. To appear in Bolyai Memorial Volume
19. Tutte, W.T.: Toward a theory of crossing numbers. J. Combinatorial Theory 8, 45–53 (1970)

On Collinear Sets in Straight-Line Drawings

Alexander Ravsky and Oleg Verbitsky*

Institute for Applied Problems of Mechanics and Mathematics
Naukova St. 3-Б, Lviv 79060, Ukraine

Abstract. We consider straight-line drawings of a planar graph G with possible edge crossings. The *untangling problem* is to eliminate all edge crossings by moving as few vertices as possible to new positions. Let $fix(G)$ denote the maximum number of vertices that can be left fixed in the worst case among all drawings of G. In the *allocation problem*, we are given a planar graph G on n vertices together with an n-point set X in the plane and have to draw G without edge crossings so that as many vertices as possible are located in X. Let $fit(G)$ denote the maximum number of points fitting this purpose in the worst case among all n-point sets X. As $fix(G) \leq fit(G)$, we are interested in upper bounds for the latter and lower bounds for the former parameter.

For any $\epsilon > 0$, we construct an infinite sequence of graphs with $fit(G) = O(n^{\sigma+\epsilon})$, where $\sigma < 0.99$ is a known graph-theoretic constant, namely the shortness exponent for the class of cubic polyhedral graphs. On the other hand, we prove that $fix(G) \geq \sqrt{n/30}$ for any graph G of tree-width at most 2. This extends the lower bound obtained by Goaoc et al. [*Discrete and Computational Geometry* 42:542–569 (2009)] for outerplanar graphs. Our results are based on estimating the maximum number of vertices that can be put on a line in a straight-line crossing-free drawing of a given planar graph.

1 Introduction

1.1 Basic Definitions

Let G be a planar graph. We will denote the vertex set of G by V_G. The letter n will be reserved to always denote the number of vertices in V_G. By a *drawing* of G we mean an arbitrary injective map $\pi : V_G \to \mathbb{R}^2$. The points in $\pi(V_G)$ will be referred to as *vertices* of the drawing. For an edge uv of G, the segment with endpoints $\pi(u)$ and $\pi(v)$ will be referred to as an *edge* of the drawing. Thus, we always consider *straight-line drawings*. It is quite possible that in π we encounter edge crossings and even overlaps. A drawing is *plane* (or *crossing-free*) if this does not happen.

Given a drawing π of G, define

$$fix(G, \pi) = \max_{\pi' \text{ plane}} |\{v \in V_G : \pi'(v) = \pi(v)\}|.$$

---

* Current address: Humboldt-Universität zu Berlin, Institut für Informatik, Unter den Linden 6, D-10099 Berlin. E-mail: `verbitsk@informatik.hu-berlin.de`. Supported by the Alexander von Humboldt Foundation.

P. Kolman and J. Kratochvíl (Eds.): WG 2011, LNCS 6986, pp. 295–306, 2011.

Given an n-point set X in the plane, let

$$fix^X(G) = \min_{\pi:\,\pi(V_G)=X} fix(G,\pi).$$

Furthermore, we define

$$fix(G) = \min_X fix^X(G) = \min_\pi fix(G,\pi). \tag{1}$$

In other words, $fix(G)$ is the maximum number of vertices which can be fixed in any drawing of G while *untangling* it.

Given an n-point set X, consider now a related parameter

$$fit^X(G) = \max_{\pi \text{ plane}} |\pi(V_G) \cap X|.$$

In words, if we want to draw G allocating its vertices at points of X, then $fit^X(G)$ tells us how many points of X can fit for this purpose. To analyze the *allocation* problem in the worst case, we define

$$fit(G) = \min_X fit^X(G).$$

Note that $fit^X(G) = \max_{\pi:\,\pi(V_G)=X} fix(G,\pi)$. It follows that $fix^X(G) \le fit^X(G)$ and, therefore, $fix(G) \le fit(G)$.

1.2   Known Results on the Untangling Problem

No efficient way for evaluating the parameter $fix(G)$ is known. Note that computing $fix(G,\pi)$ is NP-hard [7,15]. Considerable effort is needed to estimate $fix(G)$ even for cycles, for which we know the bounds

$$2^{-5/3}n^{2/3} - O(n^{1/3}) \le fix(C_n) \le O((n\log n)^{2/3})$$

due to Cibulka [2] and Pach and Tardos [13], respectively. In the general case Bose et al. [1] establish a lower bound

$$fix(G) \ge (n/3)^{1/4}. \tag{2}$$

A better bound

$$fix(G) \ge \sqrt{n/2} \tag{3}$$

is proved for all trees (Bose et al. [1], Goaoc et al. [7]) and, more generally, outerplanar graphs (Goaoc et al. [7], cf. Corollary 2 below).

On the other hand, [1,7,11] provide examples of planar graphs (even acyclic ones) with

$$fix(G) = O(\sqrt{n}). \tag{4}$$

In particular, for the fan graphs F_n we have

$$fix^X(F_n) \le (2\sqrt{2} + o(1))\sqrt{n} \text{ for every } X, \tag{5}$$

see [11]. Cibulka [2] establishes some general upper bounds, namely $fix(G) = O(\sqrt{n}(\log n)^{3/2})$ for graphs whose maximum degree and diameter are bounded by a logarithmic function of n and $fir(G) = O((n\log n)^{2/3})$ for 3-connected graphs.

1.3   Known Results on the Allocation Problem

The question whether or not $fit^X(G) = n$ has been studied in the literature, especially for X in general position. If X is in convex position, any triangulation on X is outerplanar. By this reason, $fit^X(G) < n$ for all non-outerplanar graphs G and all sets X in convex position. On the other hand, Gritzmann et al. [8] proved that $fit^X(G) = n$ for all outerplanar G and all X in general position. Other results and references on this subject can be found, e.g., in [5].

It is known that there are $\Omega(27.22^n)$ unlabeled planar graphs with n vertices [6], while a set of n points in convex position admits no more than $O(11.66^n)$ plane drawings [4]. Combining the two results, we see that, if X is in convex position, then $fit^X(G) < n$ for almost all planar G.

1.4   Our Present Contribution

We aim at proving upper bounds for $fit(G)$ and lower bounds for $fix(G)$. Our approach to both problems is based on an analysis of collinear sets of vertices in straight-line graph drawings. We show the relevance of the following questions. How many collinear vertices can occur in a plane drawing of a graph G? If there is a large collinear set, which useful features can it have?

Suppose that π is a crossing-free drawing of a graph G. A set $S \subseteq \pi(V_G)$ of vertices in π is *collinear* if all of them lie on a common line ℓ. By a *displacement* of S we mean a relocation $\delta : S \to \ell$ preserving the relative order in which the vertices in S lie on ℓ. We call S *free* if every displacement $\delta : S \to \ell$ is extendable to a mapping $\delta : \pi(V_G) \to \mathbb{R}^2$ so that $\delta \circ \pi$ is a crossing-free drawing of G (i.e., whenever we shift vertices in S along ℓ without breaking their relative order, then all edge crossings that may arise can be eliminated by subsequently moving the vertices in $\pi(V_G) \setminus S$). Let $\bar{v}(G, \pi)$ denote the maximum size of a collinear set in π and $\tilde{v}(G, \pi)$ the maximum size of a free collinear set in π. Define

$$\bar{v}(G) = \max_{\pi \text{ plane}} \bar{v}(G, \pi) \text{ and } \tilde{v}(G) = \max_{\pi \text{ plane}} \tilde{v}(G, \pi).$$

Obviously, $\tilde{v}(G) \leq \bar{v}(G)$. These parameters have a direct relation to $fix(G)$ and $fit(G)$, namely

$$\sqrt{\tilde{v}(G)} \leq fix(G) \leq fit(G) \leq \bar{v}(G). \tag{6}$$

The last inequality follows directly from the definitions. The first inequality is proved as Theorem 2 below.

In Section 3 we construct, for any $\epsilon > 0$, an infinite sequence of graphs with $\bar{v}(G) = O(n^{\sigma + \epsilon})$ where $\sigma \leq \frac{\log 26}{\log 27}$ is a known graph-theoretic constant, namely the shortness exponent for the class of cubic polyhedral graphs (see Section 2 for the definition). To the best of our knowledge, this gives us the first example of graphs with $fit(G) = o(n)$. While the known upper bounds (4) for $fix(G)$ are still better, note that the problems of bounding $fix(G)$ and $fit(G)$ from above are inequivalent. The two parameters can be far away from one another: for example, in contrast to (5) we have $fit^X(F_n) \geq n - 1$ for any X.

By the lower bound in (6), we have $\mathit{fix}(G) = \Omega(\sqrt{n})$ whenever $\tilde{v}(G) = \Omega(n)$. Therefore, identification of classes of planar graphs with linear $\tilde{v}(G)$ is of big interest. In Section 4 we show that $\tilde{v}(G) \geq n/2$ for every outerplanar graph G. This gives us another proof of the bound $\mathit{fix}(G) \geq \sqrt{n/2}$ proved by Goaoc et al. [7][1] for outerplanar graphs.

Furthermore, we consider the broader class of graphs with tree-width at most 2. It coincides with the class of partial 2-trees and contains also all series-parallel graphs. For any graph G in this class, we prove that $\tilde{v}(G) \geq n/30$ and, therefore, $\mathit{fix}(G) \geq \sqrt{n/30}$.

2   Preliminaries

Given a planar graph G, we denote the number of vertices, edges, and faces in it by $v(G)$, $e(G)$, and $f(G)$, respectively. The number of faces does not depend on a particular plane embedding of G and hence is well defined.

A graph is k-*connected* if it has more than k vertices and stays connected after removal of any set of at most $k-1$ vertices. 3-connected planar graphs are called *polyhedral* as, according to Steinitz's theorem, these graphs are exactly the 1-skeletons of convex polyhedra. By Whitney's theorem, all plane embeddings of a polyhedral graph G are equivalent, that is, obtainable from one another by a plane homeomorphism up to the choice of outer face. In particular, the set of facial cycles (i.e., boundaries of faces) of G does not depend on a particular plane embedding.

A planar graph G is *maximal* if adding an edge between any two non-adjacent vertices of G violates planarity. Since all facial cycles in maximal planar graphs on at least 3 vertices have length 3, such graphs are also called *triangulations*. Every triangulation on more than 3 vertices is 3-connected, see, e.g., [12, Section 2.3].

The *dual* of a polyhedral graph G is a graph G^* whose vertices are the faces of G (represented by their facial cycles). Two faces are adjacent in G^* if they share a common edge. G^* is also a polyhedral graph. If we consider $(G^*)^*$, we obtain a graph isomorphic to G. In a *cubic* graph every vertex is incident to exactly 3 edges. As easily seen, the dual of a triangulation is a cubic graph. Conversely, the dual of any cubic polyhedral graph is a triangulation.

The *circumference* of a graph G, denoted by $c(G)$, is the length of a longest cycle in G. The *shortness exponent* of a class $\mathcal{G}$ of graphs is the limit inferior of quotients $\log c(G) / \log v(G)$ over all $G \in \mathcal{G}$. Let σ denote the shortness exponent for the class of cubic polyhedral graphs. It is known that

$$0.694\ldots = \log_2(1 + \sqrt{5}) - 1 \leq \sigma \leq \frac{\log 26}{\log 27} = 0.988\ldots$$

(see [10] for the lower bound and [9] for the upper bound).

---

[1] A preliminary version of [7] gave a somewhat worse bound of $\mathit{fix}(G) > \sqrt{n/3}$ An improvement to $\sqrt{n/2}$ was made in the early version of the present paper independently of [7].

3 Graphs with Small Collinear Sets

We now construct a sequence of triangulations G with $\bar{v}(G) = o(v(G))$. For our analysis, we need another parameter of a straight-line drawing. Given a crossing-free drawing π of a graph G, let $\bar{f}(G, \pi)$ denote the maximum number of faces of π whose interiors can be cut by a line. Furthermore, let $\bar{f}(G) = \max_\pi \bar{f}(G, \pi)$.

For the triangulations constructed below, we will show that $\bar{v}(G)$ is small with respect to $v(G)$ because $\bar{f}(G)$ is small with respect to $f(G)$ (though we do not know any relation between $\bar{v}(G)$ and $\bar{f}(G)$ in general). Our construction can be thought of as a recursive procedure for essentially decreasing the ratio $\bar{f}(G)/f(G)$ at each recursion step provided that we initially have $\bar{f}(G) < f(G)$.

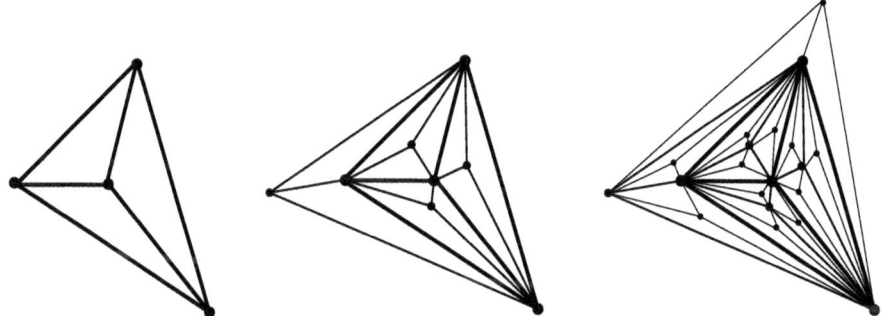

Fig. 1. An example of the construction: $G_1 = K_4$, G_2, G_3

Starting from an arbitrary triangulation G_1 with at least four vertices, we recursively define a sequence of triangulations $G_1, G_2, \ldots$. To define G_k, we describe a spherical drawing δ_k of this graph. Let δ_1 be an arbitrary drawing of G_1 on a sphere. Furthermore, δ_{i+1} is obtained from δ_i by triangulating each face of δ_i so that this triangulation is isomorphic to G_1. An example is shown in Fig. 1. In general, upgrading δ_i to δ_{i+1} can be done in different ways, that may lead to non-isomorphic versions of G_{i+1}. We make an arbitrary choice and fix the result. The following lemma points out a property of the initial triangulation G_1 that ensures the desired relation $\bar{v}(G_k) = o(v(G_k))$.

Lemma 1. *Denote* $f = f(G_1)$, $\bar{f} = \bar{f}(G_1)$, *and* $\alpha = \dfrac{\log(\bar{f} - 1)}{\log(f - 1)}$.

1. $f(G_k) = (f - 1)^{k-1} f$.
2. $\bar{f}(G_k) \le (\bar{f} - 1)^{k-1} \bar{f}$.
3. $\bar{v}(G_k) < c\, v(G_k)^\alpha$, *where* c *is a constant depending only on* G_1.

Proof. The first part follows from the obvious recurrence $f(G_{i+1}) = f(G_i)(f-1)$. We have to prove the other two parts.

Consider an arbitrary crossing-free straight-line drawing π_k of G_k. Recall that, by construction, $G_1, \ldots, G_{k-1}$ is a chain of subgraphs of G_k with

$$V_{G_1} \subset V_{G_2} \subset \ldots \subset V_{G_{k-1}} \subset V_{G_k}.$$

Let π_i be the part of π_k that is a drawing of the subgraph G_i. By the Whitney theorem, π_k can be obtained from δ_k (the spherical drawing defining G_k) by an appropriate stereographic projection of the sphere to the plane combined with a homeomorphism of the plane onto itself. It follows that, like δ_{i+1} and δ_i, drawings π_{i+1} and π_i have the following property: the restriction of π_{i+1} to any face of π_i is a drawing of G_1. Given a face F of π_i, the restriction of π_{i+1} to F (i.e., a plane graph isomorphic to G_1) will be denoted by $\pi_{i+1}[F]$.

Consider now an arbitrary line ℓ. Let $\bar{f}_i$ denote the number of faces in π_i cut by ℓ. By definition, we have

$$\bar{f}_1 \leq \bar{f}. \tag{7}$$

For each $1 \leq i < k$, we claim that

$$\bar{f}_{i+1} \leq \begin{cases} \bar{f} & \text{if } \bar{f}_i = 1, \\ (\bar{f} - 1)\bar{f}_i & \text{if } \bar{f}_i > 1. \end{cases} \tag{8}$$

Indeed, let K denote the outer face of π_i. Equality $\bar{f}_i = 1$ means that, of all faces of π_i, ℓ cuts only K. Within K, ℓ can cut only faces of $\pi_{i+1}[K]$ and, therefore, $\bar{f}_{i+1} \leq \bar{f}$.

Assume that $\bar{f}_i > 1$. Within K, ℓ can now cut at most $\bar{f} - 1$ faces of π_{i+1} (because ℓ cuts $\mathbb{R}^2 \setminus K$, a face of $\pi_{i+1}[K]$ outside K). Within any inner face F of π_i, ℓ can cut at most $\bar{f} - 1$ faces of π_{i+1} (the subtrahend 1 corresponds to the outer face of $\pi_{i+1}[F]$, which surely contributes to $\bar{f}$ but is outside F). The number of inner faces F cut by ℓ is equal to $\bar{f}_i - 1$ (again, the subtrahend 1 corresponds to the outer face of π_i). We therefore have $\bar{f}_{i+1} \leq (\bar{f}-1) + (\bar{f}_i - 1)(\bar{f} - 1) = (\bar{f} - 1)\bar{f}_i$, completing the proof of (8).

Using (7) and (8), a simple inductive argument gives us

$$\bar{f}_i \leq (\bar{f} - 1)^{i-1}\bar{f} \tag{9}$$

for every $i \leq k$. As π_k and ℓ are arbitrary, part 2 of the lemma is proved by setting $i = k$ in (9).

To prove part 3, we have to estimate from above $\bar{v}_k = |\ell \cap V(\pi_k)|$, the number of vertices of π_k on the line ℓ. Put $\bar{w}_1 = |\ell \cap V(\pi_1)|$ and $\bar{w}_i = |\ell \cap (V(\pi_i) \setminus V(\pi_{i-1}))|$ for $1 < i \leq k$. Clearly, $\bar{v}_k = \sum_{i=1}^{k} \bar{w}_i$. Abbreviate $v = v(G_1)$. It is easy to see that $\bar{w}_1 \leq v - 2$ and $\bar{w}_i \leq \bar{f}_{i-1}(v - 3)$ for all $1 < i \leq k$. It follows that

$$\bar{v}_k \leq (v - 2) + (v - 3) \sum_{i=1}^{k-1} \bar{f}_i \leq \frac{(v-3)\bar{f}}{\bar{f} - 2}(\bar{f} - 1)^{k-1}, \tag{10}$$

where we use (9) for the latter estimate. It remains to express the obtained bound in terms of $v(G_k)$. By part 1 of the lemma, we have $(f-1)^{k-1}f = f(G_k) = 2v(G_k) - 4$ and, therefore,

$$(\bar{f}-1)^{k-1} = (f-1)^{\alpha(k-1)} < (2/f)^{\alpha} v(G_k)^{\alpha}.$$

Plugging this in to (10), we arrive at the desired bound for $\bar{v}_k$ and hence for $\bar{v}(G_k)$.   □

We now need an initial triangulation G_1 with $\bar{f}(G_1) < f(G_1)$. The following lemma shows a direction where one can seek for such triangulations.

Lemma 2. *For every triangulation G with more than 3 vertices, we have*

$$\bar{f}(G) \le c(G^*).$$

Proof. Given a crossing-free drawing π of G and a line ℓ, we have to show that ℓ crosses no more than $c(G^*)$ faces of π. Shift ℓ a little bit to a new position ℓ' so that ℓ' does not go through any vertex of π and still cuts all the faces that are cut by ℓ. Thus, ℓ' crosses boundaries of faces only via inner points of edges. Each such crossing corresponds to transition from one vertex to another along an edge in the dual graph G^*. Note that this walk is both started and finished at the outer face of π. Since all faces are triangles, each of them is visited at most once. Therefore, ℓ' determines a cycle in G^*, whose length is at least the number of faces of π cut by ℓ.   □

Lemma 2 suggests the following choice of G_1: Take a cubic polyhedral graph H getting close to the infimum of the set of quotients $\log(c(G)-1)/\log(v(G)-1)$ over all cubic polyhedral graphs G and set $G_1 = H^*$. In particular, we can get arbitrarily close to the shortness exponent σ defined in Section 2. By Lemma 1.3, we arrive at the main result of this section.

Theorem 1. *Let σ denote the shortness exponent of the class of cubic polyhedral graphs. Then for any $\alpha > \sigma$, there is a sequence of triangulations G with*

$$\bar{v}(G) = O(v(G)^{\alpha}).$$

Corollary 1. *For infinitely many n there is a planar graph G on n vertices with*

$$fix(G) = O(n^{0.99}).$$

4   Graphs with Large Free Collinear Sets

Let π be a crossing-free drawing of a graph G, and let ℓ be a line. Recall that a set $S \subset \pi(V_G) \cap \ell$ is called *free* if, whenever we displace the vertices in S along ℓ without violating their mutual order (thereby introducing edge crossings), we are able to untangle the modified drawing by only moving the vertices in $\pi(V_G) \setminus S$. By $\tilde{v}(G)$ we denote the largest size of a free collinear set maximized over all drawings of a graph G.

Theorem 2. $fix(G) \ge \sqrt{\tilde{v}(G)}.$

Proof. Let $fix^-(G)$ be defined similarly to (1) but with minimization over all collinear X (or over π such that $\pi(V_G)$ is collinear). Obviously, $fix(G) \leq fix^-(G)$. As proved by Kang et al. [11] (based on [1, Lemma 1]), we actually have

$$fix(G) = fix^-(G). \qquad (11)$$

We use this equality here.

Choose an integer k such that $(k-1)^2 < \tilde{v}(G) \leq k^2$. By (11), it suffices to show that any drawing $\pi : V_G \to \ell$ of G on a line ℓ can be made crossing-free while keeping k vertices fixed. Let ρ be a crossing-free drawing of G such that, for some $S \subset V_G$ with $|S| > (k-1)^2$, $\rho(S)$ is a free collinear set on ℓ. By the Erdős-Szekeres theorem, any two orderings of a set with cardinality more than $(k-1)^2$ agree, up to reversal, on at least k elements. We, therefore, can find a set $F \subset S$ of k vertices such that $\pi(F)$ and $\rho(F)$ lie on ℓ in the same order (up to reversal). By the definition of a free set, there is a crossing-free drawing ρ' of G with $\rho'(F) = \pi(F)$. Thus, we can come from π to ρ' with F staying fixed.   □

Theorem 2 sometimes gives a short way of proving bounds of the kind $fix(G) = \Omega(\sqrt{n})$. For example, for the wheel graph W_n we immediately obtain $fix(W_n) > \sqrt{n} - 1$ from the easy observation that $\tilde{v}(W_n) = n - 2$ (in fact, this repeats the argument of Pach and Tardos for cycles [13]). The classes of graphs with linear $\tilde{v}(G)$ are therefore of big interest in the context of disentanglement of drawings. One of these classes is addressed below.

Given a drawing π, we call it a *track drawing* if there are parallel lines, called *tracks*, such that every vertex of π lies on one of the layers and every edge either lies on one of the layers or connects endvertices lying on two consecutive layers. We call a graph *track drawable* if it has a crossing-free track drawing. Note that any displacement of the vertices on a track does not introduce edge crossings.

An obvious example of a track drawable graph is a grid graph $P_s \times P_s$. It is also easy to see that any tree is track drawable: two vertices are to be aligned on the same layer if they are at the same distance from an arbitrarily assigned root. The latter example can be considerably extended.

Call a drawing *outerplanar* if all the vertices lie on the outer face. An *outerplanar graph* is a graph admitting an outerplanar drawing (this definition does not depend on whether straight-line or curved drawings are considered). The following fact is illustrated by Fig. 2.

Lemma 3 (Felsner, Liotta, and Wismath [3]). *Outerplanar graphs are track drawable.*

Lemma 4. *For any track drawable graph G on n vertices, it holds that*

$$\tilde{v}(G) \geq n/2.$$

Proof. Let π be a track drawing of G with tracks $t_1, \ldots, t_s$, lying in the plane in this order. It is practical to assume that $t_1, \ldots, t_s$ are parallel straight-line segments (rather than unbounded lines) containing all the vertices of π. Let ℓ be a horizontal line. Consider two redrawings of π.

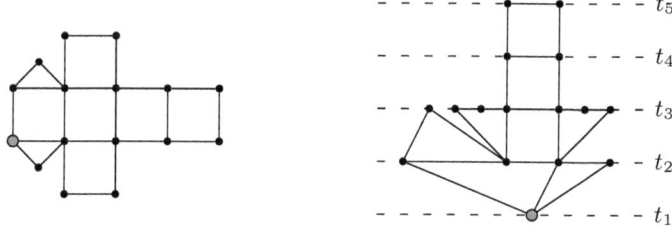

Fig. 2. An outerplanar graph and its track drawing

To make a redrawing π', we put $t_1, t_3, t_5, \ldots$ next to each other on ℓ. For each even index $2i$, we drop a perpendicular p_{2i} to ℓ between the segments t_{2i-1} and t_{2i+1}. We then put each t_{2i} on p_{2i} so that t_{2i} is in the upper half-plane if i is odd and in the lower half-plane if i is even. It is clear that such a relocation can be done so that π' is crossing-free (the whole procedure can be thought of as sequentially unfolding each strip between consecutive layers to a quadrant of the plane, see the left side of Fig. 3).

It is clear that the vertices on ℓ form a free collinear set: if the neighboring vertices of t_{2i-1} and t_{2i+1} are displaced, then p_{2i} is to be shifted appropriately.

In the redrawing π'' the roles of odd and even indices are interchanged, that is, $t_2, t_4, t_6, \ldots$ are put on ℓ and $t_1, t_3, t_5, \ldots$ on perpendiculars (see the right side of Fig. 3). It remains to observe that at least one of the inequalities $\tilde{v}(G, \pi') \geq n/2$ and $\tilde{v}(G, \pi'') \geq n/2$ must be true.   □

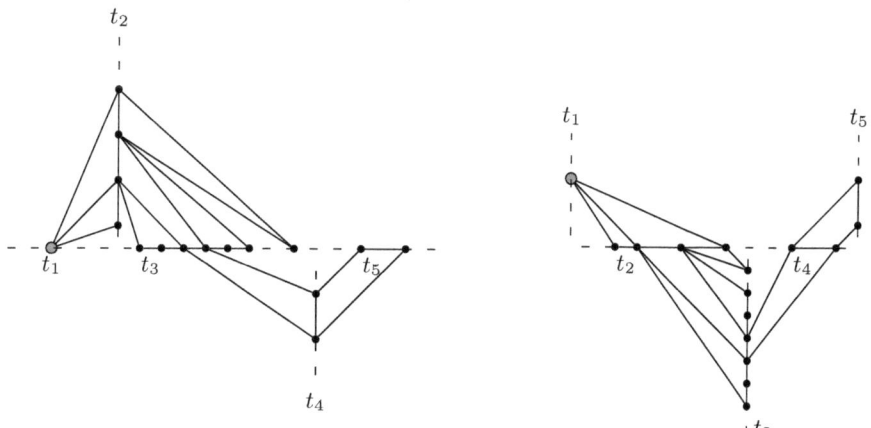

Fig. 3. Proof of Lemma 4: two redrawings of the graph from Fig. 2

Combining Lemmas 3 and 4 with Theorem 2, we obtain another proof for the following result.

Corollary 2 (Goaoc et al. [7]). *For any outerplanar graph G with n vertices, it holds that* $fix(G) \geq \sqrt{n/2}$.

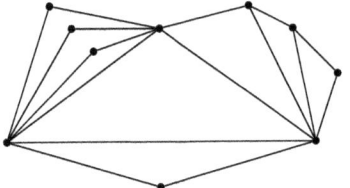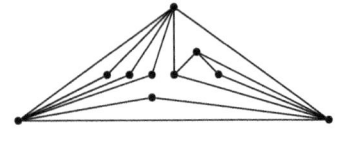

Fig. 4. A 2-tree and its folded drawing

In fact, Theorem 2 has a much broader range of application. The class of *2-trees* is defined recursively as follows:

- the graph consisting of two adjacent vertices is a 2-tree;
- if G is a 2-tree and H is obtained from G by adding a new vertex and connecting it to two adjacent vertices of G, then H is a 2-tree.

A graph is a *partial 2-tree* if it is a subgraph of a 2-tree. It is well known that the class of partial 2-trees coincides with the class of graphs with treewidth at most 2. Any outerplanar graph is a partial 2-tree, and the same holds for series-parallel graphs.

Theorem 3. *If G is a partial 2-tree with n vertices, then $\tilde{v}(G) > n/30$.*

Corollary 3. *For any partial 2-tree G with n vertices, it holds that*
$$fix(G) \geq \sqrt{n/30}.$$

In the rest of this section we briefly discuss the proof of Theorem 3. It is not hard to show that any partial 2-tree G with more than one vertex is a spanning subgraph of some 2-tree H, that is, $V_G = V_H$. The following lemma, therefore, shows that it is enough to prove Theorem 3 for 2-trees.

Lemma 5. *If G is a spanning subgraph of a planar graph H, then $\tilde{v}(G) \geq \tilde{v}(H)$.*

Proof. Any drawing of H specifies a drawing of G if we just ignore the edges of H absent in G. It suffices to note that, if X is a free set of collinear vertices in a drawing of H, then X stays free in the corresponding drawing of G.   □

Thus, from now on we suppose that G is a 2-tree. Call two triangles of G *neighbors* if they share an edge. We consider plane drawings of G having a special shape. Specifically, we call a drawing *folded* if for any two neighboring triangles, one contains the other. Thus, a sequence of triangles sharing an edge forms a containment chain, see Fig. 4. Folded drawings can be obtained by the following recursive procedure. Suppose that G is obtained from a 2-tree G' by attaching a new vertex v to an edge e of G'. Then, once G' is drawn, we put v inside that triangular face of the current drawing of G' whose boundary contains e. The procedure can be implemented in many ways giving different outputs; all of them are called folded drawings.

Using an inductive argument on the "depth" of a folded drawing (i.e., on the maximum number of triangles in a containment chain), we can prove the following fact.

Lemma 6. *Every collinear set of vertices in a folded drawing is free.*

Lemma 6 reduces our task to constructing folded drawings with linear number of collinear vertices. We will need some notions.

Call $v \in V_G$ a *leaf vertex* if it has degree 2. The number of leaf vertices in G will be denoted by $t_1(G)$. Call a triangle T in G *linking* if it has exactly two neighbors and they are edge-disjoint. In other words, T shares one of its edges with no other triangle and shares each of the other two edges with exactly one triangle. The number of linking triangles in G will be denoted by $t_2(G)$. For the graph G in Fig. 4, $t_1(G) = 5$ and $t_2(G) = 2$. Call a vertex *interposed* if it belongs only to linking triangles.

Lemma 7. *If G is a 2-tree with $n \geq 3$ vertices, then $4t_1(G) + t_2(G) > n$.*

Lemma 7 implies that we always have $t_1(G) = \Omega(n)$ or $t_2(G) = (1 - o(1))n$. To prove that $\tilde{v}(G) = \Omega(n)$, it now suffices to show that we can always make a constant fraction of all leaf vertices collinear and we also can make at least $t_2(G) - \epsilon n$ interposed vertices collinear. This is ensured by the next two lemmas.

Lemma 8. *Every 2-tree G has a folded drawing with at least $\frac{2}{3}t_1(G)$ collinear leaf vertices.*

Lemma 9. *Every 2-tree G with n vertices has a folded drawing with more than $t_2(G) - \frac{4}{5}n$ collinear interposed vertices.*

As a direct consequence of Lemmas 6–9, we obtain a bound $\tilde{v}(G) \geq n/35$, which is weaker than the bound $\tilde{v}(G) \geq n/30$ stated in Theorem 3. The latter bound is based on a hybrid of the drawings provided by Lemmas 8 and 9. The details will appear in a full version of the paper and can currently be found in [14].

5   Further Questions

1. How far or close are parameters $\tilde{v}(G)$ and $\bar{v}(G)$? It seems that a priori we cannot even exclude equality.

2. Are there graphs with $\bar{v}(G) = O(\sqrt{n})$? If so, this could be considered a strengthening of the examples of graphs with $fix(G) = O(\sqrt{n})$ given in [1,7,11]. Are there graphs with, at least, $\tilde{v}(G) = O(\sqrt{n})$? If not, by Theorem 2 this would lead to an improvement of Bose et al.'s bound (2).

3. By Theorem 3, we have $\tilde{v}(G) \geq n/30$ for any graph G of tree-width no more than 2. For Halin graphs, whose tree-width can attain 3, we can show that $\tilde{v}(G) \geq n/2$. For which other classes of graphs do we have $\tilde{v}(G) = \Omega(n)$ or, at least, $\bar{v}(G) = \Omega(n)$? Classes of planar graphs with bounded tree-width or with bounded vertex degrees are of especial interest.

Acknowledgement. We thank anonymous referees for their useful comments.

References

1. Bose, P., Dujmovic, V., Hurtado, F., Langerman, S., Morin, P., Wood, D.R.: A polynomial bound for untangling geometric planar graphs. Discrete and Computational Geometry 42, 570–585 (2009)
2. Cibulka, J.: Untangling polygons and graphs. Discrete and Computational Geometry 43, 402–411 (2010)
3. Felsner, S., Liotta, G., Wismath, S.K.: Straight-line drawings on restricted integer grids in two and three dimensions. J. Graph Algorithms Appl. 7, 363–398 (2003)
4. Flajolet, P., Noy, M.: Analytic combinatorics of non-crossing configurations. Discrete Mathematics 204, 203–229 (1999)
5. García, A., Hurtado, F., Huemer, C., Tejel, J., Valtr, P.: On triconnected and cubic plane graphs on given point sets. Comput. Geom. 42, 913–922 (2009)
6. Giménez, O., Noy, M.: Counting planar graphs and related families of graphs. In: Surveys in Combinatorics 2009, pp. 169–210. Cambridge University Press, Cambridge (2009)
7. Goaoc, X., Kratochvíl, J., Okamoto, Y., Shin, C.S., Spillner, A., Wolff, A.: Untangling a planar graph. Discrete and Computational Geometry 42, 542–569 (2009)
8. Gritzmann, P., Mohar, B., Pach, J., Pollack, R.: Embedding a planar triangulation with vertices at specified points. Amer. Math. Monthly 98, 165–166 (1991)
9. Grünbaum, B., Walther, H.: Shortness exponents of families of graphs. J. Combin. Theory A 14, 364–385 (1973)
10. Jackson, B.: Longest cycles in 3-connected cubic graphs. J. Combin. Theory B 41, 17–26 (1986)
11. Kang, M., Pikhurko, O., Ravsky, A., Schacht, M., Verbitsky, O.: Untangling planar graphs from a specified vertex position — Hard cases. Discrete Applied Mathematics 159, 789–799 (2011)
12. Mohar, B., Thomassen, C.: Graphs on surfaces. The John Hopkins University Press, Baltimore (2001)
13. Pach, J., Tardos, G.: Untangling a polygon. Discrete and Computational Geometry 28, 585–592 (2002)
14. Ravsky, A., Verbitsky, O.: On collinear sets in straight-line drawings. E-print (2011), http://arxiv.org/abs/0806.0253v3
15. Verbitsky, O.: On the obfuscation complexity of planar graphs. Theoretical Computer Science 396, 294–300 (2008)

From Few Components to an Eulerian Graph by Adding Arcs

Manuel Sorge[*], René van Bevern[**], Rolf Niedermeier, and Mathias Weller[***]

Institut für Softwaretechnik und Theoretische Informatik,
TU Berlin, Berlin, Germany
{manuel.sorge,rene.vanbevern,
rolf.niedermeier,mathias.weller}@tu-berlin.de

Abstract. EULERIAN EXTENSION (EE) is the problem to make an arc-weighted directed multigraph Eulerian by adding arcs of minimum total cost. EE is NP-hard and has been shown fixed-parameter tractable with respect to the number of arc additions. Complementing this result, on the way to answering an open question, we show that EE is fixed-parameter tractable with respect to the combined parameter "number of connected components in the underlying undirected multigraph" and "sum of $\mathrm{indeg}(v) - \mathrm{outdeg}(v)$ over all vertices v in the input multigraph where this value is positive." Moreover, we show that EE is unlikely to admit a polynomial-size problem kernel for this parameter combination and for the parameter "number of arc additions".

1 Introduction

A directed (multi-)graph G is called *Eulerian* if it contains a tour that traverses every arc in G exactly once. We study the following NP-complete decision problem:

EULERIAN EXTENSION (EE)
Input: A directed multigraph $G = (V, A)$, a positive integer $\omega_{\max}$, and a weight function $\omega : V \times V \to [0, \omega_{\max}] \cup \{\infty\}$.
Question: Is there an Eulerian extension for G of weight at most $\omega_{\max}$?

Herein, an *Eulerian extension* is a multiset E over $V \times V$ such that $G' = (V, A \cup E)$ is Eulerian. In the weight of an Eulerian extension, multiple arcs are counted multiple times, according to their multiplicity. Recently, there has been renewed interest in EULERIAN EXTENSION for at least two reasons. First, there are interesting applications (of special cases) of EULERIAN EXTENSION for sequencing problems [10]. Second, it has been pointed out that EULERIAN EXTENSION is "parameterized equivalent" to RURAL POSTMAN [4], a famous arc routing problem in combinatorial optimization [6, 12]. The main focus of this paper is on

---

[*] Supported by the DFG, project AREG, NI 369/9 and project PABI, NI 369/7.
[**] Supported by the DFG, project AREG, NI 369/9.
[***] Supported by the DFG, project DARE, NI 369/11.

P. Kolman and J. Kratochvíl (Eds.): WG 2011, LNCS 6986, pp. 307–318, 2011.

assessing the parameterized computational complexity of EE with respect to the parameter c denoting the number of connected components of the underlying undirected multigraph.

We are aware of only two papers explicitly dedicated to studying the parameterized complexity of EULERIAN EXTENSION [4, 17]. Dorn et al. [4] studied the "standard parameterization" by the number k of extension arcs and their main result was to show that EULERIAN EXTENSION is fixed-parameter tractable with respect to k. Motivated by early work of Orloff [14, 15] and Frederickson [8, 9], we complement the previous considerations and start a deeper study of EULERIAN EXTENSION parameterized by the component parameter c. Frederickson [8, 9] showed that EULERIAN EXTENSION is polynomial-time solvable for constant c. However, in his algorithm c influences the degree of the polynomial and so this only shows containment in the parameterized complexity class XP; it does not yield fixed-parameter tractability with respect to parameter c. Already in the 1970's Lenstra and Kan [12] and Orloff [15] pointed out the importance of the "complexity parameter" c. Indeed, to date, it is open whether EULERIAN EXTENSION can be solved in (fixed-parameter tractable) time $f(c) \cdot n^{O(1)}$ for an n-vertex multigraph, f being an arbitrary computable function exclusively depending on c. In companion work [17], we related EULERIAN EXTENSION to a matching variant and derived fixed-parameter tractability with respect to c on some special graph classes.

Our Results. We make some partial progress on resolving the complexity question for EULERIAN EXTENSION with respect to the parameter c. More specifically, we show that EULERIAN EXTENSION is fixed-parameter tractable with respect to the combined parameter (b, c), where b denotes the sum of all positive values indeg(v) − outdeg(v) over all vertices in the multigraph. See Orloff [15] for an early indication towards the relevance of this combined parameter. Notably, both b and c are upper-bounded by k and should typically be significantly smaller than k for most input multigraphs. In addition, we show that there is no polynomial-size problem kernel for the single parameters b, c, or k, or the combined parameter (b, c), unless coNP $\subseteq$ NP/poly. To this end, we introduce the NP-hard SWITCH SET COVER, a combinatorial problem of potentially independent interest. Due to the lack of space, many details are deferred to a full version of the paper.[1]

Preliminaries. We consider directed *multigraphs* $G = (V, A)$, where $V(G) := V$ is the set of vertices and $A(G) := A$ is the multiset of arcs. We use $n := |V|$ and $m := |A|$. A *trail* W in G is a sequence of arcs in G such that each arc ends in the same vertex as the next arc starts in and such that no arc is used more often than it is present in G. We use $V(W)$ and $A(W)$ to refer to the set of vertices in which arcs of W start or end, and the multiset of arcs of W, respectively. A *path* in the multigraph G is a trail that traverses every vertex of G at most once. A closed trail that traverses its initial and terminal vertex exactly twice and every other vertex of G at most once is called a *cycle*. A directed multigraph G is said

---

[1] Further details can also be found in the first author's diploma thesis [16].

to be *weakly connected* if every pair of vertices $u, v \in V(G)$ is *weakly connected*, that is, there is a path with the endpoints u, v in the underlying undirected multigraph of G. For brevity, we also write *connected* instead of weakly connected. A *connected component* is a maximal vertex set C such that $G[C]$ is connected. We use balance$(v) := \text{indeg}(v) - \text{outdeg}(v)$ to denote the *balance* of a vertex v in G and I_G^+ and I_G^- to denote the set of all vertices v in G with balance$(v) > 0$ and balance$(v) < 0$, respectively. A vertex v is *balanced* if balance$(v) = 0$.

We use the following characterization of Eulerian multigraphs, due to Euler: A multigraph is Eulerian if and only if all arcs are contained in the same connected component and all vertices are balanced.

Our results are in the context of parameterized complexity, which is a two-dimensional framework for studying computational complexity [5, 7, 13]. A *parameterized problem* $L \subseteq \Sigma^* \times \mathbb{N}$ is called *fixed-parameter tractable (FPT)* with respect to a parameter k if $(x, k) \in L$ is decidable in $f(k) \cdot |x|^{O(1)}$ time, where f is a computable function depending only on k. For a language $L \subseteq \Sigma^* \times \mathbb{N}$, a *reduction to a problem kernel* is a function that takes as input an instance (x, k) and, in time polynomial in $|x| + k$, outputs an instance (x', k') such that $(x', k') \in L \Leftrightarrow (x, k) \in L$, $|x'| \le f(k)$, and $k \le g(k)$. Here, f and g are computable functions solely depending on k; f is called the *size* of the problem kernel. A *polynomial-parameter polynomial-time many-one reduction* ($\le_m^{PPP}$-reduction) from a parameterized problem L to a parameterized problem L' is a polynomial-time computable function f such that $(x, k) \in L \Leftrightarrow (x', k') \in L'$, where $(x', k') := f(x, k)$, $k' \le p(k)$, and p is a polynomial depending only on k. If such a reduction exists, we write $L \le_m^{PPP} L'$.

2   Limiting Imbalance Helps

The fixed-parameter tractability of EULERIAN EXTENSION with respect to the number c of connected components in the input multigraph is an open question that arose implicitly in work of Frederickson [8, 9]. While we cannot answer this question, this section presents a fixed-parameter algorithm for EE that, *additionally* to the parameter c, uses the sum b of all positive balances of vertices as parameter. An early indication that both parameters influence the complexity of EE was given by Orloff [15].

Theorem 1. EULERIAN EXTENSION *is solvable in* $O(4^{c \log(bc^2)} n^2 (b^2 + n \log n) + n^2 m)$ *time. Here, c is the number of connected components in the input multigraph, and b is the sum of all positive balances of vertices in the input multigraph.*

To prove Theorem 1, we consider a restricted version of EE that takes as input, additionally to a regular EE-instance, an "advice" that determines how components in the input multigraph are to be connected. Then, we will see that the number of ways to connect two given components is upper-bounded in terms of b.

More details follow. Let $G = (V, A)$ be a directed multigraph. By $\mathbb{C}_G$ we denote the *component graph*, which is a clique whose vertices one-to-one correspond to the weakly connected components of G. A *hint* for G is an undirected path or cycle t of length at least one in the component graph $\mathbb{C}_G$ together with the information whether t shall form a cycle or a path in an Eulerian extension of G.[2] We call the corresponding hints *cycle hints* and *path hints*, respectively. We say a set of hints P is an *advice* to the multigraph G if the hints are edge-disjoint.[3] For a trail t in the graph $(V, V \times V)$, we define $\mathbb{C}_G(t)$ as the trail in $\mathbb{C}_G$ that is obtained by making t undirected and, for every connected component C of G, substituting every maximum length subtrail t' of t with $V(t') \subseteq C$ by the vertex in $\mathbb{C}_G$ corresponding to C. We say that a path p in the graph $(V, V \times V)$ *realizes* a path hint h if $\mathbb{C}_G(p) = h$ and the initial vertex of p has positive balance and the terminal vertex has negative balance in G. We say that a cycle c in the graph $(V, V \times V)$ realizes a cycle hint h if $\mathbb{C}_G(c) = h$. We say that an Eulerian extension E *heeds the advice* P if it can be decomposed into a number of paths and cycles that realize all hints in P. An advice P is *connecting* if the hints in P connect every pair of vertices in $\mathbb{C}_G$. Now consider the following restricted version of EE:

> EULERIAN EXTENSION WITH MINIMAL CONNECTING ADVICE (EEA)
> *Input:* A directed multigraph $G = (V, A)$, an integer $\omega_{\max}$, a weight function $\omega : V \times V \to [0, \omega_{\max}] \cup \{\infty\}$, and a minimal connecting advice P.
> *Question:* Is there an Eulerian extension E of G that is of weight at most $\omega_{\max}$ and heeds the advice P?

Section 2.1 first shows how to solve EE with help of an algorithm for EEA:

Lemma 1. EE *can be solved by solving* $O(c^{4c-2})$ *instances of* EEA, *where each instance is computable in* $O(c^3 + n + m)$ *time.*

Then, Section 2.2 shows an algorithm that solves EEA in the following time:

Proposition 1. EEA *is solvable in* $O(4^{c \log b} n^2 (b^2 + n \log n) + n^2 m)$ *time.*

Then, Theorem 1 follows by combining Lemma 1 and Proposition 1. In order to prove these, we first present two transformations that, applied to the input multigraph, allow us to assume that each vertex has balance between -1 and 1 and to assume that weight functions abide the triangle inequality. For each given transformation, we show that it is *correct*, that is, it transforms yes-instances and only yes-instances to yes-instances.

Transformation 1 (Splitting Vertices). Let the multigraph $G = (V, A)$, the weight function ω, and the maximum weight $\omega_{\max}$ constitute an instance of EE. Compute a new instance as follows: Search for a vertex v with $|\,\text{balance}(v)| > 1$, and introduce a new vertex u. If balance$(v) > 0$, choose an arbitrary arc (w, v), delete it, and add the arc (w, u). Proceed analogously if balance$(v) < 0$. Add

---

[2] The extra information is necessary because a hint to a path may be a cycle in $\mathbb{C}_G$.
[3] Note the difference between advice in our sense and advice in computational complexity theory. There, a piece of advice applies to every instance of a specific length.

the arcs $(u, v), (v, u)$. Finally, define a new weight function ω' for each pair of vertices $x, y \in V$ as follows.

$$\omega'(x, y) = \begin{cases} \infty & \text{if } (x = u \wedge y = v) \vee (x = v \wedge y = u), \\ \omega(v, y) & \text{if } x = u, \\ \omega(x, v) & \text{if } y = u, \\ \omega(x, y) & \text{otherwise} \end{cases}$$

Lemma 2. *Transformation 1 is correct. It can exhaustively be applied in* $O(n^2 m)$ *time. The resulting instance only contains vertices v with* $|\text{balance}(v)| \leq 1$.

A further preprocessing routine allows us to assume that weight functions abide the triangle inequality.

Transformation 2 (Shortest-Path Preprocessing). For an input instance of EE consisting of the directed multigraph $G = (V, A)$, the weight function ω and the maximum weight ω_{max}, derive a new instance by computing a new weight function ω', where for each $u, v \in V$, $\omega'(u, v)$ is the weight of a shortest path from u to v in the graph $(V, V \times V)$.

Lemma 3. *Transformation 2 is correct and can be applied in* $O(n^3)$ *time.*

Observe that the number of components and the sum of all positive balances of vertices in an instance of EE are invariant under Transformation 1 and Transformation 2. In the following we assume all instances of EE and EEA to be exhaustively preprocessed using Transformation 1 and Transformation 2.

2.1   Generating Advice for Eulerian Extension

This section shows how to generate advice for EE in order to solve EE with the help of an algorithm for EEA, thus proving Lemma 1. To prove Theorem 1, it then remains to prove Proposition 1, that is, to present an algorithm for EEA; this is done in Section 2.2. To solve EE using an algorithm for EEA, we simply try every minimal connecting advice and solve the resulting instances of EEA. Exploiting the structure obtained by Transformations 1 and 2, one can show that we only have to try very restricted forms of advice:

Lemma 4. *Let G be a directed multigraph and let E be a minimum-weight Eulerian extension with respect to a weight function $\omega : V \times V \to [0, \omega_{max}] \cup \{\infty\}$ for G. There is a minimal connecting advice $P = \{h_1, \ldots, h_i\}$ such that*
  (i) *E heeds P and*
  (ii) *the graph defined by the union of all trails $h_1, \ldots, h_i$ without their initial vertices does not contain a cycle.*

Using this restriction, we can enumerate all forests in $\mathbb{C}_G$ and try all possibilities to extend them to an advice. Thus, we can prove Lemma 1.

Proof (Lemma 1). We give an algorithm that decides EE by solving $O(4^{c \log(c^2)})$ instances of EEA, each of which can be generated in $O(c^3 + n + m)$ time. Let the directed multigraph $G = (V, A)$ with c connected components and the weight function $\omega : V \times V \to [0, \omega_{max}] \cup \{\infty\}$ constitute an instance of EE.

We simply generate all possible pieces of advice and solve the so-obtained EEA-instances. If one of these instances is a yes-instance, then, clearly, the original instance is a yes-instance. Also, for every yes-instance of EE, there is an advice derivable from a solution to the instance because of Lemma 4.

Concerning the generation of the advices, by Lemma 4 we may assume that the hints without their initial vertices form a forest in $\mathbb{C}_G$. Thus, we may simply enumerate all forests contained in $\mathbb{C}_G$, partition their edges into at most c hints and try all possibilities to reinsert the initial vertices back onto the hints. To enumerate all forests, we first partition the vertices into at most c cells (there are at most c^c such partitions), then enumerate all spanning trees in each cell (in each cell there are at most c^{c-2} spanning trees [2] that can be numerated in $O(c^{c-2} + c^2)$ time [11]).

Hence, in total, $O(c^c c^{c-2})$ forests are computed. We partition the edges of each forest into at most c hints (there are at most c^c partitions for each forest), extend every hint by adding an initial vertex (for each of the c hints, there are c possibilities, yielding c^c possibilities in total) and check whether this yields a valid advice—that is, we check whether the hints are paths or cycles and whether the advice is connecting. In total, we generated $O(c^c c^{c-2} c^c c^c) \subseteq O(c^{4c}) = O(4^{c \log(c^2)})$ advices. The validity check for each advice can be carried out in $O(c^3)$ time. Hence, each instance can be generated in $O(c^3 + n + m)$ time.
□

2.2   Solving Eulerian Extension with Advice

Having shown how EE can be solved using EEA, it remains to present an algorithm for EEA to solve EE. To this end, this section proves Proposition 1. This will conclude the proof of Theorem 1. To obtain an algorithm for EEA, we use the fact that EE on connected multigraphs is solvable in $O(n^3 \log n)$ time [4]. Hence, given an instance of EEA, we can solve it by realizing all hints given in the given minimal connecting advice and then solving EE on the resulting connected multigraph. In this approach, the parameter b helps to bound the number of possible ways we have to try to realize each hint. An algorithm that achieves the running time given in Proposition 1 can simply try each combination of optimal realizations of each hint in the given advice and then solve the resulting instance comprising a connected multigraph via the polynomial-time algorithm of Dorn et al. [4]. We denote a call to this algorithm by solve_connected(G, ω), where G is a connected multigraph and $\omega : V \times V \to [0, \omega_{max}] \cup \{\infty\}$ is a weight function.

Realizing Hints. First, we show how to realize a given path hint using a path between two given vertices. Then, we can try all possible initial and terminal vertices of such a path in order to optimally realize a path hint. For a directed multigraph $G = (V, A)$ and a weight function $\omega : V \times V \to [0, \omega_{max}] \cup \{\infty\}$, let p

---

Algorithm SolveEEA. Solving EEA.

Input: A directed multigraph $G = (V, A)$, a weight function $\omega : V \times V \to [0, \omega_{\max}] \cup \{\infty\}$, a cycle-less advice P, and an arc-set E.
Output: A minimum-weight Eulerian extension for G that heeds the advice P.

1   **if** $P = \varnothing$ **then return** $E \cup \text{solve\_connected}(G, \omega)$;
2   **else**
3      $h \leftarrow$ a hint in P;
4      $C_A \leftarrow$ connected component of G corresponding to the initial vertex of h;
5      $C_\Omega \leftarrow$ connected component of G corresponding to the terminal vertex of h;
6   MinEE $\leftarrow \varnothing$;
7   found_solution $\leftarrow$ false;
8   **for** $(u, v) \in I_G^+ \times I_G^-$ such that $u \in C_A \wedge v \in C_\Omega$ or vice versa **do**
9          $p \leftarrow \text{minpath}(G, \omega, h, u, v)$;
10   ActEE $\leftarrow$ SolveEEA $(G + p, \omega, P \setminus \{h\}, E \cup A(p))$;
11   **if** $(\omega(\text{ActEE}) < \omega(\text{MinEE})) \vee (\text{found\_solution} = \text{false})$ **then**
12   found_solution $\leftarrow$ true;
13   MinEE $\leftarrow$ ActEE;

14   **return** MinEE;

---

be a path in $\mathbb{C}_G$, let u be a vertex in the component of G that corresponds to the initial vertex of p, and let v be a vertex in the component that corresponds to the terminal vertex of p. Define $\text{minpath}(G, \omega, p, u, v)$ as a shortest path s from u to v in the complete graph $(V, V \times V)$ such that $\mathbb{C}_G(s) = p$.

In the following, we show that the minpath function indeed yields realizations of hints that can be assumed to be part of an optimal Eulerian extension and how it can be computed in $O(n^2)$ time.

Observation 1. *Let E be an Eulerian extension for the multigraph G that heeds the advice P, let P contain a path-hint h, and let ω be a weight function $V \times V \to [0, \omega_{\max}] \cup \{\infty\}$. Then, there is an Eulerian extension E' such that E' heeds the advice P, $\omega(E') \leq \omega(E)$, and $A(\text{minpath}(G, \omega, h, u, v)) \subseteq E'$. Here, u, v are vertices contained in the connected components of G that correspond to the initial and terminal vertices of h, respectively.*

Exploiting the structure obtained by Transformations 1 and 2, we can show:

Lemma 5. *$\text{minpath}(G, \omega, p, u, v)$ is computable in $O(n^2)$ time.*

Having shown how to realize path hints, we can show how to realize all cycle hints in an advice P in $O(|P|n^3)$ time.

Lemma 6. *For a given directed multigraph G, a weight function $\omega : V \times V \to [0, \omega_{\max}] \cup \{\infty\}$, and an advice P, realizations of all cycle hints in P are computable in $O(|P|n^3)$ time.*

Solution Algorithm. We now give an algorithm for EULERIAN EXTENSION WITH MINIMAL CONNECTING ADVICE, thus proving Proposition 1 and, therefore, proving Theorem 1.

Proof (Proposition 1). Given an instance of EEA, we first compute realizations of all cycle hints in the advice P in $O(|P|n^3)$ time (see Lemma 6), add them to G and remove all cycle hints from P. Hence, in the following, we assume that P only contains path hints. Then, we apply Algorithm SolveEEA that solves instances of EEA whose advice does not contain cycle hints. We first look at the correctness of Algorithm SolveEEA and then analyze the overall running time.

Consider the return value E' of Algorithm SolveEEA when called with an initially empty arc set E and an instance of EEA consisting of the multigraph G, the weight function ω, and minimal connecting advice P without cycle hints. For every hint in P there is realization in E', that is, E' connects all connected components of G. Because of the call to solve_connected(), the set E' also makes every vertex in G balanced. Hence, E' is an Eulerian extension for G that heeds P. Also, E' is of minimum weight among all Eulerian extensions for G that heed the advice P. This is because, first, the solution of solve_connected() is weight-minimal and, second, because, by Observation 1, we may assume that in a minimum-weight Eulerian extension all path hints h are realized by minpath(G, ω, h, u, v) for appropriate vertices u, v.

Concerning the running time of the overall procedure, we have to preprocess the input instance using Transformation 1 and Transformation 2 (recall that we assume that all instances are preprocessed accordingly). By Lemmas 2 and 3 this takes $O(n^3 + n^2 m)$ time. Next, all cycle hints are realized. By Lemma 6, this is possible in $O(|P|n^3)$ time. Finally, we apply Algorithm SolveEEA: Obviously its recursion depth is at most $|P|$. Because of $b \geq |I_G^+| = |I_G^-|$, every call of Algorithm SolveEEA yields at most b^2 recursive calls. This means that the sum of all calls is $b^{2|P|}$. The running time of one call is dominated by either the computation of b^2 minpath instances which takes $O(b^2 n^2)$ time (Lemma 5) or the computation of solve_connected() which takes $O(n^3 \log n)$ time [4]. Thus, Algorithm SolveEEA can be executed in $O(b^{2|P|}(b^2 n^2 + n^3 \log n)) = O(2^{2|P| \log(b)} n^2 (b^2 + n \log n))$ time. Since P is a *minimal* connecting advice, we have $|P| \leq c$, and thus the overall procedure runs in $O(2^{2c \log(b)} n^2 (b^2 + n \log n) + cn^3 + n^2 m) = O(4^{c \log(b)} n^2 (b^2 + n \log n) + n^2 m)$ time.   □

3   Non-existence of Polynomial-Size Problem Kernels

In this section, we prove the following theorem.

Theorem 2. EULERIAN EXTENSION *does not admit a polynomial-size problem kernel with respect to the parameters*
  (i) *minimum number k of arcs in an Eulerian extension E with $\omega(E) \leq \omega_{\max}$,*
  (ii) *sum b of positive balances,*
  (iii) *number c of connected components in the input multigraph, or*
  (iv) *the combined parameter (b, c),*
unless coNP $\subseteq$ NP/poly.

Since parameters b and c are upper-bounded by k, a polynomial-size problem kernel with respect to the parameters b, c, or the combined parameter (b, c) would imply a polynomial-size problem kernel with respect to the parameter k. Thus, we only have to show the theorem for the parameter k. Instead of considering the general problem EE, we show that Theorem 2 holds even for the following, more restricted problem variant. Since this is a special case of EE, the hardness result also holds for EE.

2-DIMENSIONAL EULERIAN EXTENSION (2DEE)
Input: A directed graph $G = (V, A)$ with $V \subseteq \mathbb{N} \times \mathbb{N}$.
Question: Is there an Eulerian extension E for G with $\forall (u, v) \in E : u \preceq v$?

Herein, $(u_1, u_2) \preceq (v_1, v_2)$ means $u_1 \leq v_1$ and $u_2 \leq v_2$ and we call $(u, v) \in V \times V$ an *allowed* arc if $u \preceq v$. The 2DEE problem stems from an application in sequencing [10]. In order to prove that 2DEE does not admit a polynomial-size problem kernel, we use an intermediate problem called SWITCH SET COVER (SSC). To define SSC, we use the following terminology. Let U be a non-empty set. A U-*position* is a multiset with elements drawn from the *universe* U. A U-*switch* is a multiset whose elements are U-positions. When the set U is clear from the context, we simply speak of positions and switches.

SWITCH SET COVER (SSC)
Input: A set U and s switches each containing a number of positions.
Question: Is it possible to choose exactly one position in each switch such that each element of U is contained in at least one of the chosen positions?

We show that we can derive a polynomial-size problem kernel for SSC with respect to the combined parameter $(s, |U|)$ from a polynomial-size problem kernel for 2DEE with respect to the parameter k. This is done using a $\leq_{\mathrm{m}}^{\mathrm{PPP}}$-reduction.

Proposition 2. SSC *parameterized by* $(s, |U|)$ *is* $\leq_{\mathrm{m}}^{\mathrm{PPP}}$-*reducible to* 2DEE *parameterized by* k.

We also use the fact that both 2DEE and SSC are NP-complete. For 2DEE this is proven by Höhn et al. [10]. The same can be shown for SSC using a simple reduction from the well-known NP-hard SET COVER problem. Given a polynomial-size problem kernel for 2DEE, and an instance of SSC, we derive a problem kernel for SSC as follows. First, we construct an equivalent instance of 2DEE using the reduction from Proposition 2. Then, we reduce this instance of 2DEE to a polynomial-size problem kernel and transform the kernel-instance back to an equivalent instance of SSC using one of its NP-hardness reductions. This procedure yields a polynomial-size problem kernel for SSC because the first reduction increases the parameter at most polynomially, and the second reduction increases the instance size at most polynomially. However, we also prove that a polynomial-size problem kernel is unlikely to exist for SSC.

Proposition 3. SWITCH SET COVER *does not admit a polynomial-size problem kernel with respect to the combined parameter* $(s, |U|)$, *unless* $coNP \subseteq NP/poly$.

Consequently, in order to prove Theorem 2, we have to prove Proposition 2 and Proposition 3. To prove Proposition 3, we use a framework introduced by

Bodlaender et al. [1]: An *or-composition algorithm* for a parameterized problem $L \subseteq \Sigma^* \times \mathbb{N}$ is an algorithm that

(1) receives a number of instances $(I_1, k), \ldots, (I_m, k)$,
(2) runs in time that is polynomial in $\sum_{i=1}^{m} |I_i| + k$, and
(3) outputs an instance (I^*, k'), such that k' is bounded by a polynomial in k

and $(I^*, k') \in L$ if and only if $(I_j, k) \in L$ for some $1 \leq j \leq m$.
A parameterized problem is called *or-compositional* if there is an or-composition algorithm for it. Bodlaender et al. [1] showed that if an or-compositional parameterized problem admits a polynomial-size problem kernel, then coNP $\subseteq$ NP/poly.

To prove that SSC is or-compositional with respect to the parameter $(s, |U|)$, we employ the following strategy, introduced by Dom et al. [3]: First, we prove that SWITCH SET COVER is fixed-parameter tractable. More specifically, we show that it is solvable in $O^*(2^{s|U|})$ time.[4] Then, in the composition algorithm, if there are $m \geq 2^{s|U|}$ input instances, then we can directly solve all the instances in $O^*(m2^{s|U|}) \subseteq O^*(m^2)$ time and return a trivial yes- or no-instance depending on whether any of the m instances is a yes-instance. Thus, we may then assume that $m \leq 2^{s|U|}$ and, hence, $\log m \leq s|U|$. We exploit this to create an instance-chooser gadget by introducing $\log m$ new switches and $s \log m$ new elements into the output instance, increasing the parameter at most polynomially since $\log m \leq s|U|$. Every possible way to choose positions in these new switches will correspond to exactly one original instance that then is a yes-instance if the composite instance is a yes-instance. If, however, there is a yes-instance among the original instances, then the composite instance can be solved by simply configuring the chooser for this instance. We obtain the following result:

Observation 2. SWITCH SET COVER *is or-compositional with respect to the combined parameter $(s, |U|)$.*

In order to prove that EE has no polynomial-size problem kernel, according to our strategy, it remains to show the following:

Observation 3. SWITCH SET COVER *can be solved in $O^*(2^{s|U|})$ time.*

Proof. An algorithm to solve SSC may simply try each combination of positions for all the switches: We may assume that in every switch there are at most $2^{|U|}$ positions because positions containing the same elements as other positions may be deleted and multiple copies of one element in one position may also be deleted. Thus, there are at most $(2^{|U|})^s$ combinations of positions.   □

Next, we prove Proposition 2 by briefly sketching a $\leq_m^{PPP}$-reduction from SSC parameterized by the number of elements $|U|$ and the number of switches s to 2DEE parameterized by the number of extension arcs k. Since switches and positions are multisets, we can assume without loss of generality that all positions contain exactly $|U|$ elements. If this is not already the case, we can simply repeat elements or delete repeated elements.

---

[4] Here, $\tilde{O}^*$ suppresses polynomial factors.

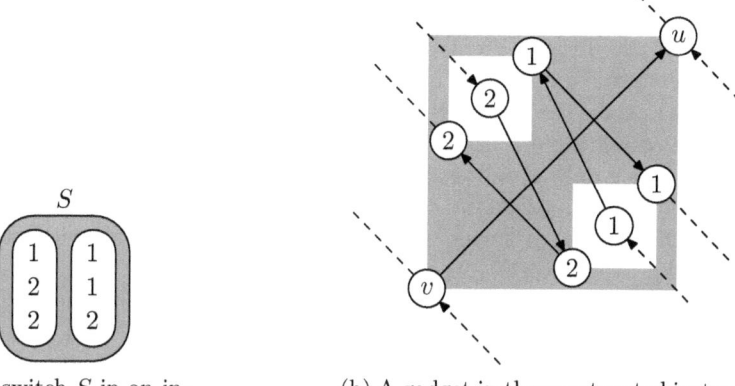

(a) A switch S in an instance of SSC

(b) A gadget in the constructed instance
of 2DEE corresponding to S

Fig. 1. The imbalanced vertices u and v are part of the frame. 2DEE only allows inserting arcs pointing to the lower left. Therefore, vertices in distinct white squares are independent in the sense that no arcs may be inserted between them. Hence, only one of the groups of three vertices in the white squares can be connected to the frame and an algorithm for 2DEE has to choose which group to connect, thus choosing a position for the switch S. Note that, since u and v are imbalanced, this gadget requires adding at least one arc. Similarly to the pairwise independence of the white squares, we ensure that all switch gadgets are pairwise independent.

The reduction uses the fact that, in 2DEE, all components of the input have to be connected. Analogously, all elements of an SSC instance have to be covered. We exploit this analogy by modeling elements as connected components. We represent each of the elements in U by a different component and introduce a special component, the "frame", to which the element-components can be connected. We use the geometrical restrictions of 2DEE to only allow connecting elements of exactly one position for each switch. To this end, consider a switch S. We introduce a gadget for S that allows connecting all elements of exactly one position of S to the frame (see Figure 1). Each switch is represented by one of these gadgets and we use the restrictions of 2DEE to ensure that each of these gadgets is extended independently, thus representing the choice of a position for each of the switches in the original instance. We note that a similar notion of independence is also used in the NP-hardness proof for 2DEE by Höhn et al. [10].

The described reduction runs in polynomial time. Furthermore, since each position contains at most $|U|$ elements, each gadget representing a switch allows for at most $|U| + 1$ arcs added. Hence, at most $s|U| + s$ arcs have to be added. Hence, the presented construction constitutes a $\leq_{\mathrm{m}}^{\mathrm{PPP}}$-reduction from SSC to 2DEE. Together with Proposition 3 and the NP-hardness of SSC, Theorem 2 follows.

4   Conclusion

The most important remaining open question is to determine whether EULERIAN EXTENSION is fixed-parameter tractable solely with respect to the number of weakly connected components. Furthermore, it seems worthwhile to search for more efficient algorithms for the special case 2-DIMENSIONAL EULERIAN EXTENSION (see Section 3). It would also be interesting to see whether the newly introduced SWITCH SET COVER turns out be useful in other contexts. Having (almost) excluded the possibility of polynomial-size many-one problem kernels, it seems tempting to analyze the potential existence of Turing problem kernels. Finally, the algorithms presented and those of previous work [4] appear to be not only of theoretical interest, making it promising to work on implementations and experiments with these algorithms.

References

[1] Bodlaender, H.L., Downey, R.G., Fellows, M.R., Hermelin, D.: On problems without polynomial kernels. J. Comput. System Sci. 75(8), 423–434 (2009)
[2] Cayley, A.: A theorem on trees. Quart. J. Math. 23, 376–378 (1889)
[3] Dom, M., Lokshtanov, D., Saurabh, S.: Incompressibility Through Colors and IDs. In: Albers, S., Marchetti-Spaccamela, A., Matias, Y., Nikoletseas, S., Thomas, W. (eds.) ICALP 2009. LNCS, vol. 5555, pp. 378–389. Springer, Heidelberg (2009)
[4] Dorn, F., Moser, H., Niedermeier, R., Weller, M.: Efficient Algorithms for Eulerian Extension. In: Thilikos, D.M. (ed.) WG 2010. LNCS, vol. 6410, pp. 100–111. Springer, Heidelberg (2010)
[5] Downey, R.G., Fellows, M.R.: Parameterized Complexity. Springer, Heidelberg (1999)
[6] Eiselt, H.A., Gendreau, M., Laporte, G.: Arc routing problems, part II: The rural postman problem. Oper. Res. 43(3), 399–414 (1995)
[7] Flum, J., Grohe, M.: Parameterized Complexity Theory. Springer, Heidelberg (2006)
[8] Frederickson, G.N.: Approximation Algorithms for NP-hard Routing Problems. PhD thesis, Faculty of the Graduate School of the University of Maryland (1977)
[9] Frederickson, G.N.: Approximation algorithms for some postman problems. J. ACM 26(3), 538–554 (1979)
[10] Höhn, W., Jacobs, T., Megow, N.: On Eulerian extensions and their application to no-wait flowshop scheduling. J. Sched. (to appear, 2011)
[11] Kapoor, S., Ramesh, H.: Algorithms for enumerating all spanning trees of undirected and weighted graphs. SIAM J. Comput. 24, 247–265 (1995)
[12] Lenstra, J.K., Kan, A.H.G.R.: On general routing problems. Networks 6(3), 273–280 (1976)
[13] Niedermeier, R.: Invitation to Fixed-Parameter Algorithms. Oxford University Press (2006)
[14] Orloff, C.S.: A fundamental problem in vehicle routing. Networks 4(1), 35–64 (1974)
[15] Orloff, C.S.: On general routing problems: Comments. Networks 6(3), 281–284 (1976)
[16] Sorge, M.: On Making Directed Graphs Eulerian. Diplomarbeit, Institut für Informatik, Friedrich-Schiller-Universität Jena, Available electronically. arXiv:1101.4283 [cs.DM] (2011)
[17] Sorge, M., van Bevern, R., Niedermeier, R., Weller, M.: A New View on Rural Postman Based on Eulerian Extension and Matching. In: Iliopoulos, C.S. (ed.) IWOCA 2011. LNCS, vol. 7056, pp. 310–322. Springer, Heidelberg (2011)

Recognizing Some Subclasses of Vertex Intersection Graphs of 0-Bend Paths in a Grid[*]

Steven Chaplick[1], Elad Cohen[2], and Juraj Stacho[2]

[1] Department of Computer Science, University of Toronto, 10 Kings College Road,
Toronto, Ontario M5S 3G4, Canada
chaplick@cs.toronto.edu

[2] Caesarea Rothschild Institute, University of Haifa, Mt. Carmel, Haifa, Israel 31905
eladdc@gmail.com, stacho@cs.toronto.edu

Abstract. We investigate graphs that can be represented as vertex intersections of horizontal and vertical paths in a grid, known as B_0-VPG graphs. Recognizing these graphs is an NP-hard problem. In light of this, we focus on their subclasses. In the paper, we describe polynomial time algorithms for recognizing chordal B_0-VPG graphs, and for recognizing B_0-VPG graphs that have a representation on a grid with 2 rows.

1 Introduction

A *VPG representation*[1] of a graph G is a collection of paths of the two-dimensional grid where the paths represent the vertices of G in such a way that two vertices of G are adjacent if and only if the corresponding paths share at least one vertex. We focus on a special subclass of VPG representations.

A *B_0-VPG representation* of G is a VPG representation in which all paths in the collection have no bends. In other words, it is a representation of G by the intersections of orthogonal segments of the plane. Here, we emphasize the grid-based definition in order to focus on some properties of the underlying grid (e.g. size). A graph is a *B_0-VPG graph* if it has a B_0-VPG representation.

Intersection representations of paths on grids arise naturally in the context of circuit layout problems and layout optimization [18] where a layout is modeled as paths (wires) on a grid. Often one seeks to minimize the number of times a wire is bent [3,17] in order to minimize the cost or difficulty of production. Other times layouts may consist of several layers where the wires on each layer are not allowed to intersect. This is naturally modeled as the colouring problem on the corresponding intersection graph.

VPG graphs were defined in [1,2] where, in particular, the subclasses with bounded number of bends are studied. These classes are shown to have many connections to other, more traditional graph classes such as interval graphs, planar graphs, string graphs, segments graphs, circle graphs and circular-arc graphs. Unfortunately, due to these connections, many natural problems for VPG graphs are hard. For instance, colouring is NP-hard even for B_0-VPG graphs,

---

[*] The authors wish to thank Krishnam Raju Jampani, Therese Biedl, and Martin Charles Golumbic for fruitful discussions in the early stages of this work.

[1] Representation by <u>V</u>ertex intersection of <u>P</u>aths in a <u>G</u>rid.

P. Kolman and J. Kratochvíl (Eds.): WG 2011, LNCS 6986, pp. 319–330, 2011.
© Springer-Verlag Berlin Heidelberg 2011

and recognition is NP-hard for both VPG and B_0-VPG graphs (for more details about these and related results, see [2,13,14]).

Thus, in the quest for polynomial algorithms, we need to restrict our attention further to specific cases with (potentially) useful structure. In this respect, in [8], certain subclasses of B_0-VPG graphs have been characterized and shown to admit polynomial time recognition; namely split, chordal claw-free, and chordal bull-free B_0-VPG graphs are discussed in [8].

In this paper, we continue this line of research by further investigating two other subclasses of B_0-VPG graphs. In particular, we describe a polynomial time algorithm for recognizing chordal B_0-VPG graphs, and a polynomial time algorithm for recognizing 2-row B_0-VPG graphs, i.e., B_0-VPG graphs that can be represented on a grid with just 2 rows (and arbitrary number of columns). Note that the former generalizes [8] and one can easily verify that the underlying grid graph induced by the paths of a chordal B_0-VPG graph is, in fact, a tree.

Studying B_0-VPG representation of chordal graphs is a natural choice as they are precisely the intersection graphs of subtrees of a tree, and can be also seen as the intersection graphs of leaf generated subtrees of a complete binary host tree [11,16] which by [10] has a near optimal embedding on a grid. Moreover, the colouring problem can be solved in linear time on chordal graphs (see [7]). Similarly, the choice of 2-row representation of B_0-VPG graphs is a natural one since when considering embeddings of graphs in grids, one objective is to utilize as little space as possible; in this context, 2-row representations constitute the smallest non-trivial case one can study and one that has not been considered before this work. In the conclusion of the paper, we discuss the complexity of the colouring problem on such representations (with bounded number of rows).

Both our recognition algorithms are based on essentially the same idea which follows from the realization that the rows and columns of the grid induce interval representations. That is, a graph G is a B_0-VPG graph if there is a partition of its vertex set such that each class of the partition induces in G an interval graph and the connections between the classes follow "certain" structure. There are two specific problems related to this approach: we need to find the partition (it is easy to test if each class is an interval graph), and we need an efficient way to construct a representation of the connection graph of the classes.

We deal with these questions differently in each case. In the case of chordal B_0-VPG graphs, the classes are based on an equivalence relation on the vertices. The connection graph turns out to be a tree, and we describe an algorithm to find a layout of this tree which yields a representation of G. In the case of 2-row B_0-VPG graphs, the classes are found by splitting *bisimplicial* vertices (vertices whose neighbourhood induces two cliques with no edges between them). The connection graph of the classes is a planar graph that can be drawn on two layers without crossings, and this drawing satisfies additional conditions.

The two cases are discussed separately in the following sections. We mostly follow the definitions and notation from [7] and [19]. In the interest of brevity, we shall assume some familiarity with *chordal graphs*, *interval graphs*, *clique paths* (linear orderings of maximal cliques [6]), *planar drawings*, and *PQ-trees* [4].

A vertex of G is *horizontal*, resp. *vertical* in a B_0-VPG representation of G if it is represented by a horizontal, resp. vertical path.

2   Chordal B_0-VPG Graphs

In this section, we describe a polynomial time algorithm for recognizing chordal B_0-VPG graph. First, we recall the following lemma from [8].

(2.1) Diamond rule. *Let G be the graph with $V(G) = \{u, v, x, y\}$ and $E(G) = \{uv, ux, uy, vx, vy\}$. Then in every B_0-VPG representation of G, the two paths representing u and v use a common horizontal or a common vertical line.*

This inspires the following definition.

Let G be a graph. The binary relation $\sim_0$ on $V(G)$ is defined as follows:

$$u \sim_0 v \iff uv \in E(G) \text{ and } \exists x, y \in N(u) \cap N(v) \text{ with } xy \notin E(G)$$

In other words, u and v are related by $\sim_0$ if they form the diagonal of some diamond. Let $\sim$ denote the transitive closure of $\sim_0$.

(2.2) *Let G be a chordal graph, S be an equivalence class of $\sim$, and K be a connected component of $G - S$. Then $(N(S) \cap K) \cup (N(K) \cap S)$ is a clique of G.*

Proof (Sketch). If there are no edges between K and S we are done. Otherwise, suppose that there is $x \in N(K) \cap S$ and $y \in N(S) \cap K$ such that $xy \notin E(G)$. Since $x \in N(K)$, there exists $y' \in K$ with $xy' \in E(G)$, and since $y \in N(S)$, there is $x' \in S$ where $x'y \in E(G)$. Choose x, y, x', y' so that $d_S(x, x') + d_K(y, y')$ is minimized where $d_S(x, x')$ is the distance between x and x' in $G[S]$, and $d_K(y, y')$ is the distance between y and y' in $G[K]$. By the minimality of the choice and chordality of G, we conclude that $xx', yy' \in E(G)$ and $x'y' \in E(G)$. This shows that $x' \sim y'$ which is impossible, since $x' \in S$, $y' \notin S$ and S is an equivalence class of $\sim$. So, there is no such x, y and the rest of the claim follows easily.   □

The *clique-class intersection graph* (see Fig. 1) of G is the bipartite graph whose vertices are the maximal cliques of G and the equivalence classes of $\sim$, where a clique Q is adjacent to a class S just if they share at least one vertex.

(2.3) *If G is a connected chordal B_0-VPG graph, then the clique-class intersection graph of G is a tree.*

(Note that the clique-class intersection graph of G is precisely the block-cutpoint tree of the graph we obtain from G by contracting all equivalence classes of $\sim$.)

The proof of the following claim follows directly from (2.2).

(2.4) *Let G be a chordal B_0-VPG graph, and S be an equivalence class of $\sim$. Then the closed neighbourhood $N[S]$ of S induces in G an interval graph.*

Consider a B_0-VPG representation of G, and let Q be a maximal clique of G. Let I denote the intersection of the paths representing the vertices in Q. By the Helly property [2], the set I is non-empty. Let S be an equivalence class of $\sim$ such that $Q \cap S \neq \emptyset$. We say that Q is an *end* of S in this representation, if the vertices of $S \setminus Q$ are represented by paths that are either all to the right, or all to the left, or all above, or all below all points in I.

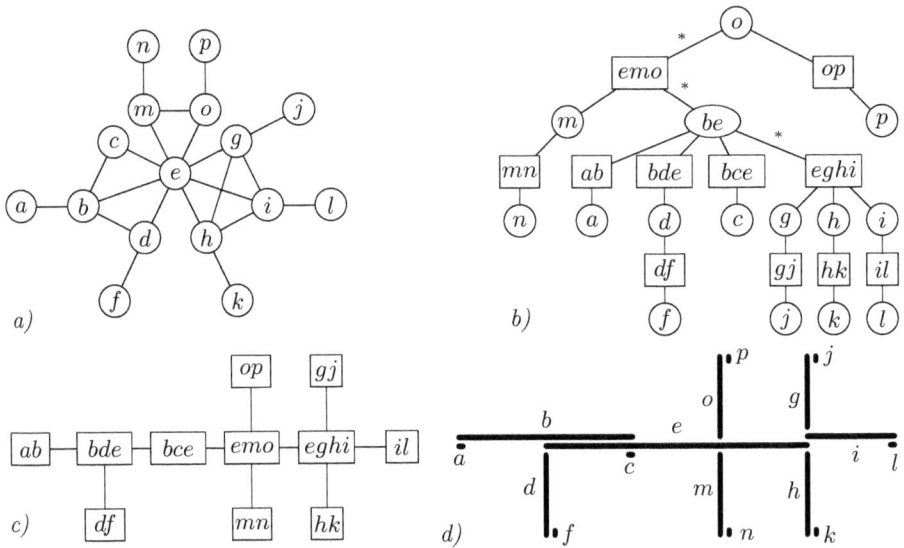

Fig. 1. *a)* Example chordal graph G, *b)* the clique-class intersection graph H of G (edges with * are those marked by the algorithm), *c)* the graph T=the union of the clique paths P_S after identifying cliques, *d)* the corresponding B_0-VPG representation

We say that Q is a *forced end* of S if Q is an end of S in every B_0-VPG representation of G. We say that Q is a *forced midpoint* of S if there is no B_0-VPG representation of G in which Q is an end of S.

The following two claims explain the role of forced ends and midpoints in chordal B_0-VPG graphs. They are simple consequences of (2.1), (2.2), and (2.4).

(2.5) *Let G be a chordal B_0-VPG graph. Let Q be a maximal clique of G, and let $S_1, \ldots, S_t$ be all the equivalence classes of $\sim$ that have non-empty intersection with Q. Let r be the number of indices $i \in \{1 \ldots t\}$ for which Q is a forced midpoint of S_i. Let s be the number of indices $i \in \{1 \ldots t\}$ with $Q \supseteq S_i$.*

Then $r + t - s \leq 4$. Moreover, if there exists $i \in \{1 \ldots t\}$ such that Q is neither a forced end of S_i nor a forced midpoint of S_i, then $r + t - s \leq 3$.

(2.6) *Let G be a chordal B_0-VPG graph. Let S be an equivalence class of $\sim$, and let $Q_1, \ldots, Q_t$ be the maximal cliques of G with non-empty intersection with S.*

Construct a graph G' starting from $G[N[S]]$ as follows: for each $i \in \{1 \ldots t\}$ such that Q_i is a forced end of S, add a new vertex u_i and make it adjacent to each vertex in $Q_i \setminus S$. Then G' is an interval graph.

Moreover, if for $i \in \{1 \ldots t\}$ adding to G' a vertex u_i adjacent to each vertex in $Q_i \setminus S$ results in a non-interval graph, then Q_i is a forced midpoint of S.

2.1   Algorithm

Now, we are ready to describe our algorithm for recognition of chordal B_0-VPG graphs. Let G be a graph given as input. We may assume that G is connected. Otherwise, we obtain a representation of G by finding a representation for each of its connected components, and by putting the representations side-by-side.

First, we check if G is chordal (see [7]). If not, we reject G. Otherwise, we compute the maximal cliques of G and the equivalence classes of $\sim$. We test if the closed neighbourhood of each equivalence class induces in G an interval graph. If not, we reject G based on (2.4). Otherwise, we construct the clique-class intersection graph H. By (2.3), the graph H is in fact a tree. The algorithm uses dynamic programming on H to find out which cliques of G are forced ends or forced midpoints of equivalence classes. To test this, we use (2.5) and (2.6).

We start by rooting H at an arbitrary node. The nodes of H are processed bottom-up, processing a node only after all its descendants are processed. We mark some edges of H in this process; initially, no edges are marked, and once an edge becomes marked, it remains marked. The meaning of a marked edge e between a clique Q and class S is the following. If Q is the parent of S, and e becomes marked when processing S, then Q is a forced midpoint of S (otherwise, Q is not is not a forced midpoint of S). Similarly, if Q is a child of S, and the edge e becomes marked when processing Q, then Q is a forced end of S.

When processing a clique Q, we count the number of children S such that the edge between Q and S is marked. Thus, Q is a forced midpoint in each such S and is not a forced midpoint in all other children. If Q is the root of H, we apply the test from (2.5). If this fails, we reject G. If Q has a parent S^*, we test (2.5) assuming that Q is not a forced midpoint of S^*. If this fails, we reject G. If only the second part of (2.5) fails, then Q is necessarily a forced end of S^*, and we mark the edge between Q and S^*. Otherwise, we do not mark the edge.

Similarly, we process each class S using (2.6). First, we look at the marked children Q of S; each such Q is a forced end of S, and we shall assume that all other children are not. If S is the root of H, we just perform the first test from (2.6), and if it fails, we reject G. Otherwise, if S has a parent Q^*, we use (2.6) to determine whether or not Q^* is a forced midpoint of S. If so, we mark the edge between S and Q^*. If not, we conclude that Q^* is not a forced midpoint of S (by providing a representation), and thus we do not mark the edge.

It remains to explain how we obtain a representation of G if this process finishes without rejecting G. For each class S, we assign to S the interval representation of G' guaranteed by (2.6); in this representation every forced end of S is necessarily an end of S. If the parent of S is not a forced midpoint, we assign to S the representation from the second part of (2.6); in this representation, the parent of S is an end of S. From this representation, we remove the vertices that are not in G (the vertices u_i added in the process of creating G'), and consider the resulting representation as an equivalent clique path that we denote by P_S. Observe that the cliques on this path are maximal cliques in G.

Now, consider the graph obtained by taking the disjoint union of the above clique paths; each vertex corresponds to some maximal clique of G, and each connected component is the path P_S for some S. In this graph, for each clique Q, we find all vertices that correspond to Q, and identify them to a single node. This results in a graph T. We observe that T is a tree, since H is a tree. In fact, T is a clique tree, since the paths P_S are clique paths. By the choice of the paths using (2.6) and since each clique Q satisfies (2.5), the tree T has maximum degree four, and hence, can be drawn in the plane so that the edges are represented by horizontal or vertical segments. In fact (by appropriately permuting neighbors), we can draw T so that each path P_S is horizontal or vertical in the drawing. Since each vertex of G belongs to exactly one equivalence class S, we conclude that it appears only in cliques on the path P_S, and we represent it by a path connecting all such cliques on P_S. This yields a B_0-VPG representation of G.

That concludes the description of our algorithm. We now briefly analyze its running time. Testing for chordality takes linear time (see [7]), so does computing all maximal cliques (there is $O(n)$ of them). Finding all diamonds and thus computing the equivalence classes takes $O(nm)$ time. Having done that, the construction of the clique-class intersection graph H takes linear time. Finally, since H has $O(n)$ nodes, the dynamic programming takes $O(nm)$ time.

Thus, the recognition algorithm runs in $O(nm)$ time.

3   2-Row B_0-VPG Graphs

A *2-row B_0-VPG representation* of G is a B_0-VPG representation where the underlying grid has two rows (and arbitrary number of columns). We call the two rows *layers* and distinguish the *top* and the *bottom* layer. For simplicity, any path of the representation that is a single grid-point is considered to be horizontal. Thus, a vertical path always consists of exactly two points of the grid (in the same column), one on the top and one on the bottom layer.

A graph is a *2-row B_0-VPG graph* if it has a 2-row B_0-VPG representation. In this section, we describe a polynomial time algorithm for recognizing 2-row B_0-VPG graphs. It suffices to focus on connected graphs. Also, it suffices to consider graphs with no true twins, where u, v are *true twins* if $N[u] = N[v]$. Clearly, if u, v are true twins in G, then a representation of G can be obtained from a representation of $G - v$ by assigning to v the path corresponding to u.

A vertex v of G is *bisimplicial* if the neighbourhood of v in G induces a disjoint union of two non-empty cliques. In other words, the set $N(v)$ induces in G a graph consisting of two non-empty cliques with no edges between them. (See Fig. 2 for an illustration of this and subsequent notions.)

A B_0-VPG representation of G is *proper* if for every path of the representation, each of its endpoints belongs to at least one other path. It suffices to consider only proper representations. We note the following properties.

(3.1) *If v is a bisimplicial vertex of G, then in every B_0-VPG representation of G, some edge of the underlying grid belongs only to the path representing v.*

(3.2) *In each proper 2-row B_0-VPG representation of a graph G with no true twins, every vertical vertex is a bisimplicial vertex of G.*

Proof. If v is a vertical vertex, then, since the representation is proper, the vertical path P representing v intersects both a horizontal path on the top layer and a horizontal path on the bottom layer of the representation. Thus the horizontal paths intersecting P on the top layer and the horizontal paths intersecting P at the bottom layer yield the two cliques in $N(v)$ as required. □

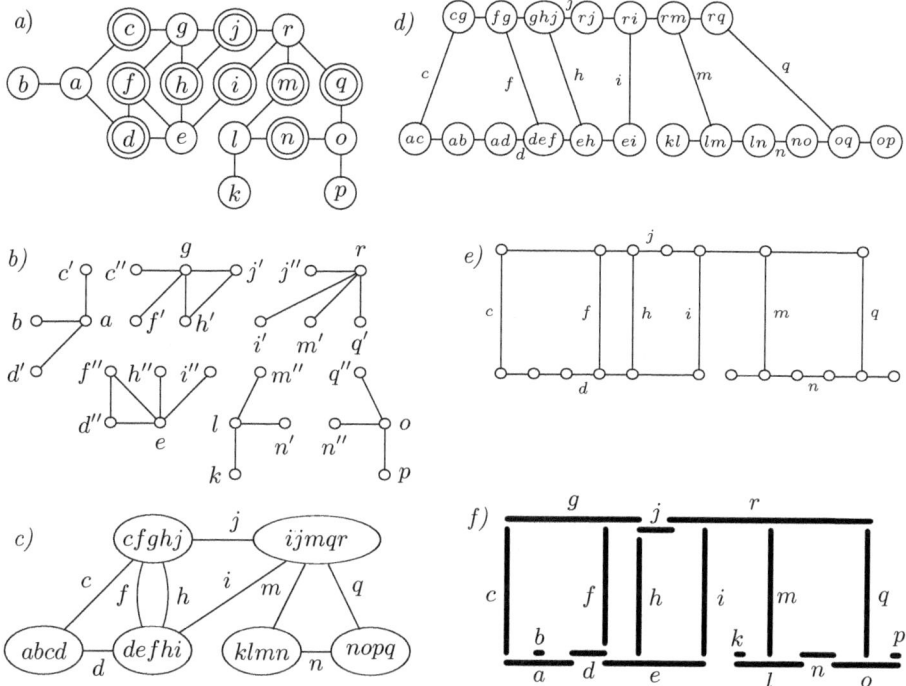

Fig. 2. *a)* Example graph G (bisimplicial vertices shown in double circles), *b)* the bisplitting of G, *c)* the bicontraction H of G, *d)* the graph $H' =$ substituting clique paths to H, *e)* orthogonal drawing of H', *f)* the corresponding 2-row B_0-VPG representation

The converse of (3.2) is false, i.e., not all bisimplicial vertices are necessarily represented by vertical paths. To find out which are, we use two auxiliary graphs.

Let G be a graph. The *bisplitting* (see Fig. 2) of G is the graph obtained from G as follows. We consider each bisimplicial vertex v of G and let Q_1, Q_2 denote the two cliques induced by $N(v)$. We remove v, add new vertices v', v'', make v' adjacent to each vertex in Q_1, and make v'' adjacent to each vertex in Q_2.

(3.3) *If G is a 2-row B_0-VPG graph with no true twins, then the bisplitting of G is an interval graph.*

Proof. Consider a proper 2-row B_0-VPG representation of G. By (3.1), each bisimplicial vertex v of G has an edge e of the underlying grid that only belongs to v's path P. We remove e from P to obtain two subpaths P_1 and P_2. Since the representation is proper, the paths representing the vertices of one of the two cliques in $N(v)$ intersect P_1, while the paths for the other clique intersect P_2. We represent v' by P_1 and v'' by P_2. Now, by (3.2), no vertical paths remain in the representation. Thus the result is an interval representation.   □

The *bicontraction* (see Fig. 2) of G is the multigraph H described as follows. The vertex set of H is the set of connected components of the bisplitting of G. There is an edge e with label v between connected components C', C'' if and only if if v is a bisimplicial vertex of G that was "split" into v' and v'' where $v' \in C'$ and $v'' \in C''$. We allow possibly $C' = C''$ in which case e is a loop.

For further use, we denote by $Q_e^{C'}$ the closed neighbourhood of v' in C' and denote by $Q_e^{C''}$ the closed neighbourhood of v'' in C''. Note that, by the definition of bisplitting, $Q_e^{C'}$ and $Q_e^{C''}$ are maximal cliques in C' and C'', respectively.

(3.4) *If G is a 2-row B_0-VPG graph, then the bicontraction of G contains no loops, and no two parallel edges have labels that are adjacent in G.*

3.1   LL-Drawings

A *planar LL-drawing* of a multigraph H is a planar embedding of H on two layers. In other words, the vertices are placed on two horizontal lines such that every edge is drawn (without crossing another edge) either between consecutive vertices on a layer or between the two layers, and there are no parallel edges between vertices on the same layer. A linear time algorithm for finding planar LL-drawings is discussed in [5]. An edge of a planar LL-drawing is *horizontal* (respectively *vertical*) if its endpoints are on one (respectively two) layers.

A vertical edge e of a planar LL-drawing $\mathcal{E}$ splits the drawing into two areas, the area L_e to left of e, and the area R_e to the right e. For any edge e', we write $e' \prec_{\mathcal{E}} e$ (respectively $e' \succ_{\mathcal{E}} e$) if the interior of e' belongs to L_e (respectively R_e). For all other pairs of edges, we derive $\prec_{\mathcal{E}}$ by transitivity. Since there are no crossings, the resulting relation $\prec_{\mathcal{E}}$ is a partial order on the edges of $\mathcal{E}$; it is a total order on the vertical edges of $\mathcal{E}$. We refer to $\prec_{\mathcal{E}}$ as the $\mathcal{E}$-*order* of the edges.

Using the proof of (3.3), we now *associate* to each 2-row B_0-VPG representation of G with no true twins a planar LL-drawing of the bicontraction of G.

We proceed as follows. Starting with a 2-row B_0-VPG representation of G, we denote by H_0 the subgraph of the underlying grid obtained by taking the union of all paths of the representation. Using the proof of (3.3), we obtain a representation of the bisplitting G' of G. During this process, a set F of edges is removed from the paths of the representation. By definition, each edge in F is also an edge of H_0. Further, by the construction, each connected component K of $H_0 - F$ corresponds to some connected component C of G' in that K is a horizontal path that includes all paths representing the vertices of C.

Now, label each edge $e \in F$ with the bisimplicial vertex of G from whose path we removed e. After that, consider the multigraph H obtained by contracting the edges of H_0 that are not in F (while keeping parallel edges). Notice that each connected component of G' shrinks to one point and F is the edge-set of H. Thus, we conclude that H is the bicontraction of G.

Finally, note that H_0 itself (as a subgraph of the underlying grid) is a planar LL-drawing, and we obtained H by contracting horizontal edges in this drawing. Thus, the result is a planar LL-drawing of H, the bicontraction of G, and we associate it to the 2-row B_0-VPG representation of G we started with.

Notice that this associated LL-drawing has some special properties. They are summarized in the following definition. Again, let H be the bicontraction of G, and let $\mathcal{E}$ be a planar LL-drawing of H. Let C be a vertex of H, and let $e_1, \ldots, e_t$ be the edges incident to C in H such that $e_1 \prec_{\mathcal{E}} \ldots \prec_{\mathcal{E}} e_t$.

We say that C is _good_ in $\mathcal{E}$ if there exists a clique path $\mathcal{P}$ of C such that

(L1) for all $i < j$, if $Q^C_{e_i} \neq Q^C_{e_j}$, then $Q^C_{e_i}$ appears before $Q^C_{e_j}$ on $\mathcal{P}$,
(L2) for all $i < j$, if $Q^C_{e_i} = Q^C_{e_j}$, then at least one of the following holds:
    (L2a) $i = 1$, $j = 2$, and e_1 is a horizontal edge,
    (L2b) $i = (t-1)$, $j = t$, and e_t is a horizontal edge,
    (L2c) $i = 1$, $j = t = 3$, and $Q^C_{e_1} = Q^C_{e_2} = Q^C_{e_3}$,
(L3) if e_1 is a horizontal edge, then $Q^C_{e_1}$ is the first clique on $\mathcal{P}$, and
    if e_t is a horizontal edge and $t \geq 2$, then $Q^C_{e_t}$ is the last clique on $\mathcal{P}$.

We say that $\mathcal{E}$ is a _good LL-drawing_ if every vertex of H is good in $\mathcal{E}$.

It is not difficult to see that the associated planar LL-drawing of a 2-row B_0-VPG representation of G is a good LL-drawing. Namely, for each vertex C, we use the clique path $\mathcal{P}$ corresponding to the interval representation of C induced by the representation of G. Then $\mathcal{P}$ (or its reverse[4]) satisfies the above conditions. It turns out that the converse is also true yielding the following characterization.

(3.5) _A graph G has a 2-row B_0-VPG representation if and only if there exists a good LL-drawing of the bicontraction of G._

Proof. The forward direction is discussed above the claim. For the backward direction, we consider a good LL-drawing of the bicontraction H of G. In the drawing, we replace each vertex C by the clique path $\mathcal{P}$ of C satisfying (L1)-(L3). We arrange the vertices of $\mathcal{P}$ from left to right in the order given[4] by $\mathcal{P}$. For each edge e incident to C, we reattach e from C to Q^C_e (which is one of the vertices on $\mathcal{P}$). From (L1)-(L3) we conclude that the result of this process is a planar LL-drawing of a graph H'. In particular, the vertices of H' correspond precisely to the maximal cliques of G. By (possibly) adding gaps between consecutive vertices on layers, we modify the drawing so that every vertical edge of the drawing is drawn as a vertical segment, and vertices are on integer coordinates. Finally, we assign to each vertex v of G the path in the drawing connecting all cliques of G that contain v. This yields a 2-row B_0-VPG representation of G. $\square$

3.2 PQ-Trees

To find a good LL-drawing of the bicontraction of G, we use the well-known concept of a PQ-tree [4]. We briefly review some key properties of such trees.

A PQ-tree is a _rooted ordered_ tree (children of each node are totally ordered) where each internal node is either a P-node or a Q-node, and whose _leaf order_ is defined as the ordering of leaves in the traversal that visits the children of each node in the given total order. Two PQ-trees are _equivalent_ if one can be obtained from the other by possibly permuting the children of some P-nodes and reversing the order of children of some Q-nodes. A permutation π of a set is _consistent_ with a PQ-tree if there exists an equivalent PQ-tree whose leaf order is π.

If T, T' are PQ-trees with the same leaf-set, then the _intersection_ of T and T' is the PQ-tree whose consistent permutations are precisely those that are consistent with both T and T'. If no such permutations exist, the intersection is the _null tree_. The intersection takes linear time to construct (for instance, see [9]).

---

[4] If C belongs to only one horizontal edge, we may need to reverse $\mathcal{P}$.

3.3 Algorithm

Now, we are ready to describe our recognition algorithm for 2-row B_0-VPG graphs. By (3.5), it suffices for the algorithm to find a good LL-drawing of the bicontraction of the given graph G. This, up to minor technical details, boils down to the following problem: given a multigraph H with a collection of PQ-trees $\{T_v\}_{v \in V(H)}$, find a planar LL-drawing of H such that the clockwise ordering of edges around each $v \in V(H)$ is consistent with T_v. Our algorithm follows this idea. We assign to each vertex C of the bicontraction H of G a PQ-tree T_C built as follows. Starting with the PQ-tree representing the clique paths of C, we replace the clique Q_e^C by e for each edge e incident to C in H (in case two such cliques coincide we introduce a P-node). We further reduce this tree to account for blocks and parallel edges in H (for brevity, we omit these technical details).

First, suppose that H is a 2-connected simple graph. Then, by [5], H has a planar LL-drawing if and only if H is an outerplanar graph and in every outerplanar embedding of H the bounded faces form a path in the dual graph. As a planar LL-drawing, we notice that every face has exactly two vertical edges. Further, every edge that belongs to two faces is vertical. This fixes the drawing of every inner face on the path of the dual. For the end-faces, we try all possible choices for the vertical edges. This results in $O(|V(H)|^2)$ choices that completely cover all possible drawings. For each such choice, we use the PQ-trees T_C to test whether or not it corresponds to a good LL-drawing of H.

Next, if H is a simple graph but not 2-connected, we use dynamic programming to process the blocks of H. For this, similarly to [5], we distinguish special blocks in H. Let C be a vertex of H, and let K be a connected component of $H - C$. Let $\{e_1, \ldots, e_t\}$ be the edges of H between C and its neighbours in K.

We say that K is a *fan* of H *attached to C* if $Q_{e_1}^C, \ldots, Q_{e_t}^C$ are distinct cliques, and the subgraph of $\overline{G}$ corresponding to the union of $C' \in K$ is an interval graph. A fan K is a *tail* of H if the subgraph of G corresponding to the union of C and all $C' \in K$ is an interval graph. A fan is a *proper fan* if it is not a tail.

To illustrate these concepts, note that in Fig. 2 for $C = ijmqr$, the connected component $K = \{klmn, nopq\}$ of $H - C$ is a fan attached to C but it is not a tail, since $G[\{i, \ldots, r\}]$ is not an interval graph while $G[\{k, \ldots, q\}]$ is.

(**3.6**) *Let K be a fan of H. If H has a good LL-drawing $\mathcal{E}$, then there is a good LL-drawing $\mathcal{E}'$ of H in which all vertices of K are on the same layer.*

Let H' be obtained from H by removing all fans of H. We say that a block of H is a *proper block* of H if it is also a block of H'. A cutpoint C of H is a *proper cutpoint* of H if it belongs to two proper blocks of H, or some component of $H - C$ is a proper fan, or if at least three components of $H - C$ are tails. For blocks B, B', we write $B \prec_{\mathcal{E}} B'$ if $e \prec_{\mathcal{E}} e'$ for all $e \in E(B)$, $e' \in E(B')$.

(**3.7**) *The proper blocks and proper cutpoints induce a path in the block-cutpoint tree of H. If $B_1, \ldots, B_k$ are the proper blocks on this path in this order, then either $B_1 \prec_{\mathcal{E}} \ldots \prec_{\mathcal{E}} B_k$ or $B_k \prec_{\mathcal{E}} \ldots \prec_{\mathcal{E}} B_1$ for every good LL-drawing $\mathcal{E}$ of H.*

We process the proper blocks and cutpoints of H in the order $B_1, \ldots, B_k$ given by the above claim. For each proper block B_i, we consider B_i together with all tails attached to it via non-proper cutpoints. We try all possible arrangements.

It can be shown that this results in a polynomial number of choices, and we test each of them using the PQ-trees T_C, and discard those that fail the test.

Next, we test consecutive blocks B_i, B_{i+1}; let C be the cutpoint they share. We consider all possible good LL-drawings $\mathcal{E}_i$ of B_i and $\mathcal{E}_{i+1}$ of B_{i+1} that are compatible (i.e., C is on the same layer in both drawings and is the rightmost, resp. leftmost in $\mathcal{E}_i$, resp. $\mathcal{E}_{i+1}$). We try all such feasible drawings together with all attached fans and tails. In this case, we do not try all possibilities directly (since there may be exponentially many of them), but instead use the PQ-tree T_C to try them indirectly using an intersection with another PQ-tree representing our choices. As discussed earlier, this can be done in linear time. In a similar fashion, we deal with proper cutpoints that are not between proper blocks.

Finally, we deal with parallel edges which is done by incorporating additional tests that do not increase the complexity by more than a constant factor.

Now, to obtain a good LL-drawing of H, we combine good LL-drawings of compatible pairs of proper blocks and cutpoints (if possible). Since there are polynomially many choices for each block, the resulting algorithm is polynomial. We remark that the number of choices for each block can be further reduced to a constant by additional tests. This produces a procedure whose complexity is linear in the size of G. (For complete details, see the full version of the paper.)

We conclude by analyzing the total complexity of our algorithm. First, finding all bisimplicial vertices of G can be done in $O(nm)$ time by examining the neighbourhood of each vertex in G. From this, the bisplitting and the bicontraction H of G can be constructed in linear time. Also, checking for true twins in G and for loops in H is a linear time procedure, and so is [4] the intervality test on H. Similarly, constructing the PQ-trees for all vertices of H takes linear time [4], and so do all other necessary operations on these PQ-trees. Finally, our dynamic programming as described above can be also implemented to run in linear time. Hence, the overall complexity of the algorithm is $O(nm)$.

4   Conclusion

We studied recognition algorithms for special cases of B_k-VPG graphs, namely chordal B_0-VPG and 2-row B_0-VPG graphs. For both cases, we described $O(nm)$ time algorithms for recognition. The interest in these types of representations comes from applications in VLSI where they can be used to model some aspects of electrical circuits. In particular, solving the colouring problem is of interest but is unfortunately NP-complete [2] on B_k-VPG graphs for every fixed $k \geq 0$. It turns out that the problem is NP-complete already on ℓ-row B_0-VPG graphs (B_0-VPG graphs that have a representation with ℓ rows) for every fixed $\ell \geq 2$. In contrast, one can decide in linear time if an ℓ-row B_0-VPG graph can be properly coloured with t colours, when t is fixed, since in this case all yes-instances have bounded *pathwidth*, while the problem is NP-complete on B_0-VPG graphs for every fixed $t \geq 3$. In a similar vein, the independent set problem is NP-complete on B_0-VPG graphs [15] but can be solved in polynomial time on ℓ-row B_k-VPG graphs for all fixed k, ℓ. Detailed proofs are in the full version of this paper.

As a continuation of this work, it may be interesting to look at other cases where representation has bounded number of rows (three or more) or other structural restriction in order to overcome hardness of optimization problems on these representations. It should be noted that the recognition of B_k-VPG graphs for every k was recently proved to be NP-complete [12]. In that respect, it seems natural to study as well special cases of these graphs, where for instance one only allows certain types of paths in the representation. Specifically the case $k = 1$ is already of interest and is currently a subject of our ongoing research.

References

1. Asinowski, A., et al.: String graphs of k-bend paths on a grid. Electronic Notes in Discrete Mathematics 37, 141–146 (2011); LAGOS 2011 - VI Latin-American Algorithms, Graphs and Optimization Symposium,
   http://dx.doi.org/10.1016/j.endm.2011.05.025
2. Asinowski, A., Cohen, E., Golumbic, M.C., Limouzy, V., Lipshteyn, M., Stern, M.: Intersection graphs of paths on a grid, technical report (2010)
3. Bandy, M., Sarrafzadeh, M.: Stretching a knock-knee layout for multilayer wiring. IEEE Trans. Computing 39, 148–151 (1990)
4. Booth, K.S., Lueker, G.S.: Testing for the consecutive ones property, interval graphs, and graph planarity using PQ-tree algorithms. Journal of Computer and System Sciences 13, 335–379 (1976)
5. Cornelsen, S., Schank, T., Wagner, D.: Drawing graphs on two and three lines. Journal of Graph Algorithms and Applications 8, 161–177 (2004)
6. Fulkerson, D.R., Gross, O.A.: Incidence matrices and interval graphs. Pacific Journal of Mathematics 15, 835–855 (1965)
7. Golumbic, M.C.: Algorithmic graph theory and perfect graphs, 2nd edn. North-Holland (2004)
8. Golumbic, M.C., Ries, B.: On the intersection graphs of orthogonal line segments in the plane: characterizations of some subclasses of chordal graphs (submitted manuscript, 2011)
9. Haeupler, B., Jampani, K.R., Lubiw, A.: Testing simultaneous planarity when the common graph is 2-connected. In: Cheong, O., Chwa, K.-Y., Park, K. (eds.) ISAAC 2010, Part II. LNCS, vol. 6507, pp. 410–421. Springer, Heidelberg (2010)
10. Heckmann, R., Klasing, R., Monien, B., Unger, W.: Optimal Embedding of Complete Binary Trees into Lines and Grids. In: Schmidt, G., Berghammer, R. (eds.) WG 1991. LNCS, vol. 570, pp. 25–35. Springer, Heidelberg (1992)
11. Jamison, R.E., Mulder, H.M.: Constant tolerance intersection graphs of subtrees of a tree. Discrete Mathematics 290, 27–46 (2005)
12. Kratochvíl, J.: Personal communication
13. Kratochvíl, J.: String graphs II. Recognizing string graphs is NP-hard. Journal of Combinatorial Theory B 52, 67–78 (1991)
14. Kratochvíl, J., Matoušek, J.: Intersection graphs of segments. Journal of Combinatorial Theory Series B 62, 289–315 (1994)
15. Kratochvíl, J., Nešetřil, J.: Independent set and clique problems in intersection-defined classes of graphs. Commentationes Mathematicae Universitatis Carolinae 31, 85–93 (1990)
16. McMorris, F.R., Scheinerman, E.R.: Connectivity threshold for random chordal graphs. Graphs and Combinatorics 7, 177–181 (1991)
17. Molitor, P.: A survey on wiring. EIK Journal of Information Processing and Cybernetics 27, 3–19 (1991)
18. Sinden, F.: Topology of thin film circuits. Bell System Technical Journal 45, 1639–1662 (1966)
19. West, D.B.: Introduction to graph theory, 2nd edn. Prentice-Hall (2000)

A Polynomial Time Algorithm for Bounded Directed Pathwidth

Hisao Tamaki

Department of Computer Science, Meiji University
Kawasaki, Japan 214-8571
tamaki@cs.meiji.ac.jp

Abstract. We give a polynomial time algorithm for bounded directed pathwidth. Given a positive integer k and a digraph G with n vertices and m edges, it runs in $O(mn^{k+1})$ time and constructs a directed path-decomposition of G of width at most k if one exists and otherwise reports the non-existence.

1 Introduction

According to Barát [3], the notion of directed pathwidth of digraphs was introduced by Reed, Thomas, and Seymour around 1995. It is a generalization of pathwidth [17], which is defined for undirected graphs, in the sense that if G is an undirected graph and G' is a digraph obtained from G by replacing each edge by a pair of directed edges in both directions, then the directed pathwidth of G' equals the pathwidth of G.

Following the tremendous success of the notion of treewidth [18] of undirected graphs, as a key tool for the graph minor theory [19] and for designing efficient algorithm [2], several authors have proposed extensions of this notion to digraphs. Johnson, Robertson, Seymour, and Thomas introduced directed treewidth [11], and showed that some NP-hard problems on digraphs including the directed Hamilton cycle problem can be solved in polynomial time if the given digraph has bounded directed treewidth. Since then, several variants of directed treewidth have been proposed: D-width [20], DAG-width [16,4], and Kelly-width [10]. It is the subject of ongoing active research to compare respective power of these variants and other related digraph measures [9].

In contrast, the extension of the notion of pathwidth to digraphs seems stable. Only one parameter, the directed pathwidth, has been proposed, which enjoys several equivalent formulations just as undirected treewidth and pathwidth do.

Although the applicability of these digraph parameters in designing efficient algorithms is provably limited in the sense that directed graph counterparts of some fixed parameter tractable problems on undirected graphs are hard to solve when parameterized by these width parameters [14], they are nonetheless fundamental digraph parameters that deserve further explorations for algorithmic applications. For example, in [21], the present author used directed pathwidth in a heuristic algorithm for exactly identifying the set of attractors of a given boolean network and experimentally showed the effectiveness of the approach.

P. Kolman and J. Kratochvíl (Eds.): WG 2011, LNCS 6986, pp. 331–342, 2011.

Since it is NP-complete to decide, given a positive integer k and an undirected graph G, whether the pathwidth of G is at most k [12], the same holds for the directed pathwidth. The situation is quite different between these problems if k is fixed. In this case, the problem of deciding if an undirected graph G has pathwidth at most k (and of constructing the associated path-decomposition) can be solved in linear time [6,5]. In contrast, no polynomial time algorithm for fixed k (even for $k = 2$) was previously known, that decides whether the directed pathwidth of a given digraph is at most k.

In the undirected case, the fact that there is a polynomial time algorithm for fixed k that decides whether a given graph has pathwidth at most k is an immediate consequence of the graph minor theorem due to Robertson and Seymour [19]: since the class of graphs with pathwidth k or smaller is closed under taking minors, that class is characterized by a fixed set of forbidden minors and therefore the membership to that class can be tested by checking if the given graph contains any of the forbidden minors. This does not hold for the directed case. Although the class of digraphs with directed pathwidth at most k, for any fixed k, is closed under taking minors, with a suitable definition of digraph minors [11], no counterpart of the graph minor theorem is known for digraphs.

The standard algorithmic approach for undirected pathwidth for fixed k that leads to the linear time algorithm mentioned above is to first obtain a tree-decomposition of width $O(k)$ of the given graph and then perform a dynamic programming on this tree-decomposition to optimally solve the problem. There are again difficulties in extending this approach to the directed case. Although there is a fast approximation algorithm [11] to obtain a directed tree-decomposition of G of width $O(k)$, given that G has directed pathwidth at most k, directed tree-decompositions do not seem to support dynamic programming solutions to the problem of exactly determining the directed pathwidth. We may try to use a tree-decomposition of the underlying undirected graph, but since the treewidth of the underlying undirected graph is not bounded by any function of the directed pathwidth of the original digraph, we do not obtain a time bound that is polynomial in the size of the digraph even if the directed pathwidth is bounded.

In this paper, we show that the directed pathwidth problem for fixed k can be solved in polynomial time. We denote the directed pathwidth of digraph G by $\mathrm{dpw}(G)$.

Theorem 1. *Given a positive integer k and a digraph G of n vertices and m edges, it can be decided in $O(mn^{k+1})$ time whether $\mathrm{dpw}(G) \leq k$. Moreover, if $\mathrm{dpw}(G) \leq k$, a directed path-decomposition of width at most k can be constructed in the same amount of time.*

Our algorithm is based on a lemma (Lemma 1), which enables us to prune the natural search tree of factorial size into one of polynomial size. This lemma, which we call the commitment lemma, asserts that if a descendant of a node satisfies certain conditions then all other descendants of the node in the same generation can be safely removed from the search tree.

Our algorithm is extremely simple and easy to implement. We remark that even for undirected pathwidth, for which a fixed parameter linear-time algorithm is known [6], our algorithm is a strong alternative for practical use, as the linear-time algorithm depends exponentially on k^3 and is considered highly impractical. To the best of the present author's knowledge, an explicit and implementable $n^{O(k)}$ time algorithm has been known for treewidth [1] but not for pathwidth for general fixed k. We also remark that, even for the ranges of n and k where the time bound in Theorem 1 is practically useless, the commitment lemma would be useful in designing heuristic algorithms.

The rest of this paper is organized as follows. After some preliminaries in Section 2, we describe some basic ideas underlying the pruning of search trees in Section 3, assuming the commitment lemma. The proof of this lemma is given in Section 4. Section 5 provides some details of the algorithm which are necessary to establish the exact running time bound stated in Theorem 1.

2 Preliminaries

Let G be a digraph. For each subset U of $V(G)$, we denote by $N_G^-(U)$ the set of in-neighbors of U, i.e., $N_G^-(U) = \{v \in V(G) \setminus U \mid \exists u \in U : (v, u) \in E(G)\}$, and $d_G^-(U) = |N_G^-(U)|$ the number of in-neighbors of U.

Rather than giving the standard definition of the directed pathwidth, we use an alternative formulation called the directed vertex separation number, defined below.

We call a sequence σ of vertices of G non-duplicating if each vertex of G occurs at most once in σ. We denote by $\Sigma(G)$ the set of all non-duplicating sequences of vertices of G. For each sequence $\sigma \in \Sigma(G)$, we denote by $V(\sigma)$ the set of vertices constituting σ and by $|\sigma| = |V(\sigma)|$ the length of σ. For brevity, we write $d_G^-(\sigma)$ and $N_G^-(\sigma)$ for $d_G^-(V(\sigma))$ and $N_G^-(V(\sigma))$, respectively.

For each pair of sequences $\sigma, \tau \in \Sigma(G)$ such that $V(\sigma) \cap V(\tau) = \emptyset$, we denote by $\sigma\tau$ the sequence in $\Sigma(G)$ that is σ followed by τ. If $\sigma' = \sigma\tau$ for some τ, then we say that σ is a *prefix* of σ' and that σ' is an *extension* of (or *extends*) σ; we say that σ is a *proper* prefix of σ' and that σ' is a *proper* extension of σ if τ is nonempty. For each non-empty sequence $\sigma \in \Sigma(G)$, we denote by $\pi(\sigma)$ the prefix of σ with length $|\sigma| - 1$.

For $\sigma, \tau \in \Sigma(G)$, we say σ is a *subsequence* of τ if $V(\sigma) \subseteq V(\tau)$ and, for each pair of distinct vertices u and v in $V(\sigma)$, u occurs before v in σ if and only if u occurs before v in τ.

Let G be a digraph and k a positive integer. We say $\sigma \in \Sigma(G)$ is *k-feasible* for G if $d_G^-(\sigma') \leq k$ for every prefix σ' of σ. We may drop the reference to G and say σ is k-feasible when G is clear from the context.

Definition 1. *The* directed vertex separation number *of digraph G, denoted by* dvsn(G), *is the minimum integer k such that there is a k-feasible sequence $\sigma \in \Sigma(G)$ with $V(\sigma) = V(G)$.*

Note that, because of the equivalence of the directed vertex separation number to the directed pathwidth stated below, this parameter is invariant under the simultaneous reversal of all the edges.

It is known that $\mathrm{dvsn}(G) = \mathrm{dpw}(G)$ for every digraph G [22] (see also [13] for the undirected case) and the conversions between the sequences achieving the directed vertex separation number and the optimal directed path decompositions are simple. In particular, the conversion from the former to the latter can be done in $O(m + kn)$ time, where $n = |V(G)|$, $m = |E(G)|$, and $k = \mathrm{dpw}(G)$. Based on these equivalence and conversion, we focus on computing the directed vertex separation number and the corresponding sequence in the following sections.

3   Search Tree Pruning

Let digraph G be fixed and let $n = |V(G)|$. A straightforward exponential time algorithm for deciding if $\mathrm{dvsn}(G) \leq k$ constructs a search tree in which each node at level i of the tree is a k-feasible sequence of length i and the parent of a non-empty sequence σ is $\pi(\sigma)$, the prefix of σ with length $|\sigma| - 1$. We show that this search tree can be pruned into one with $O(n^{k+1})$ nodes.

The key to this pruning is the notion of non-expanding extensions. We say that an extension τ of $\sigma \in \Sigma(G)$ is *non-expanding* if τ is a proper extension of σ and $d_G^-(\tau) \leq d_G^-(\sigma)$. Suppose σ is k-feasible and has an immediate non-expanding extension σv, where $v \in V(G) \setminus V(\sigma)$. Then it appears plausible to hope that committing to this child of σ in the search tree, discarding all the other children, is safe in the sense that if σ has a k-feasible extension of length n then so does σv and therefore we do not lose completeness of the search through this commitment. The following lemma states that this hope is true in a more general manner: we may safely commit not only to an immediate non-expanding extension but also to any shortest non-expanding extension. We say that an element of $\Sigma(G)$ is *strongly k-feasible* if it has a k-feasible extension of length n.

Lemma 1. *(Commitment Lemma) Let σ be a strongly k-feasible sequence in $\Sigma(G)$ and let τ be a shortest non-expanding k-feasible extension of σ, that is,*

1. *$d_G^-(\tau) \leq d_G^-(\sigma)$, and*
2. *$d_G^-(\tau') > d_G^-(\sigma)$ for every k-feasible proper extension τ' of σ with $|\tau'| < |\tau|$.*

Then, τ is strongly k-feasible.

The proof of this lemma is given in the next section.

In the following, we assume a fixed total ordering $<$ on $V(G)$ and use a standard lexicographic ordering $<$ on $\Sigma(G)$ based on this total ordering. Let σ and τ be sequences of equal length in $\Sigma(G)$. We say that σ is *preferable to τ*, if either $d_G^-(\sigma) < d_G^-(\tau)$ or $d_G^-(\sigma) = d_G^-(\tau)$ and $\sigma < \tau$. Clearly, this preferable-to relation is a total ordering on the subset of $\Sigma(G)$ consisting of sequences of length i, for each $0 \leq i \leq n$.

Let σ and τ be k-feasible sequences of equal length. We say that σ *suppresses* τ, if σ and τ has a common prefix σ' such that σ is a shortest non-expanding k-feasible extension of σ' and σ is preferable to τ.

Proposition 1. *Let σ, τ, and η be k-feasible sequences of equal length. If σ suppresses τ and τ suppresses η, then σ suppresses η.*

Proof. Under the assumptions of the lemma, σ is preferable to η, since σ is preferable to τ and τ is preferable to η. Therefore, it suffices to show that σ and η has a common prefix α such that σ is a shortest non-expanding k-feasible extension of α.

Since σ suppresses τ, there is a common prefix β of σ and τ such that σ is a shortest non-expanding k-feasible extension of β. Similarly, there is a common prefix γ of τ and η such that τ is a shortest non-expanding k-feasible extension of γ. Since both β and γ are prefixes of τ, one is a prefix of the other. If β is a prefix of γ, then we are done with $\alpha = \beta$. If γ is a prefix of β, then γ is a common prefix of σ and η. Since σ is preferable to τ, we have $d_G^-(\sigma) \leq d_G^-(\tau)$. This, together with the assumption that τ is a shortest non-expanding k-feasible extension of γ implies that σ is also a shortest non-expanding k-feasible extension of γ. We are done with $\alpha = \gamma$.   □

It should be intuitively clear that suppressed sequences are not necessary in the search tree, as a consequence of the commitment lemma. To formalize this intuition, we define the set S_i of unsuppressed k-feasible sequences of length i, for each $0 \leq i \leq n$, inductively as follows.

1. S_0 consists of the empty sequence.
2. A k-feasible sequence σ of length $i > 0$ is in S_i if and only if $\pi(\sigma) \in S_{i-1}$ and there is no k-feasible sequence τ of length i such that $\pi(\tau) \in S_{i-1}$ and τ suppresses σ.

Lemma 2. *If there is a k-feasible sequence of length n in $\Sigma(G)$, then there is at least one such sequence in S_n.*

Proof. For each k-feasible sequence σ of length n, let i_σ denote the largest i, $0 \leq i \leq n$, such that the prefix of σ of length i is in S_i. If there is some k-feasible σ of length n with $i_\sigma = n$, then we are done. So, suppose otherwise and fix k-feasible σ of length n so that i_σ is the largest over all choices of σ. Let σ' be the prefix of σ of length $i_\sigma + 1$. Then, since $\sigma' \notin S_{i_\sigma+1}$ and $\pi(\sigma') \in S_{i_\sigma}$, σ' must be suppressed by some k-feasible sequence τ of length $i_\sigma + 1$ such that $\pi(\tau) \in S_{i_\sigma}$. Choose τ so that it is the most preferable among all the candidates. Then, τ is not suppressed by any τ' with $\pi(\tau') \in S_{i_\sigma}$, since otherwise τ' suppresses σ' by Proposition 1 and is preferable to τ, contradicting the choice of τ. Therefore $\tau \in S_{i_\sigma+1}$. But since τ is a shortest non-expanding k-feasible extension of some prefix of σ', which is strongly k-feasible because of its extension σ, τ is strongly k-feasible by Lemma 1. This contradicts the choice of σ, since $i_\eta \geq i_\sigma + 1$, where η is a k-feasible extension of τ with length n.   □

Thus, in our pruned search, we need only to generate k-feasible sequences in S_i, for $1 \leq i \leq n$.

To analyze the size of each set S_i, we assign a *signature* $\text{sgn}(\sigma) \in \Sigma(G)$ to each k-feasible sequence $\sigma \in \Sigma(G)$ as follows. Call a non-expanding k-feasible extension τ of σ *locally shortest*, if no proper prefix of τ is a non-expanding extension of σ. We define $\text{sgn}(\sigma)$ inductively as follows.

1. If σ is empty then $\text{sgn}(\sigma)$ is empty.
2. If σ is non-empty and is a locally shortest non-expanding extension of some prefix of σ, then $\text{sgn}(\sigma) = \text{sgn}(\tau)$, where τ is the shortest prefix of σ such that σ is a locally shortest non-expanding k-feasible extension of τ.
3. Otherwise $\text{sgn}(\sigma) = \text{sgn}(\pi(\sigma))v$, where v is the last vertex of σ (and hence $\sigma = \pi(\sigma)v$).

Proposition 2. *For each k-feasible sequence $\sigma \in \Sigma(G)$, we have $|\text{sgn}(\sigma)| \leq d_G^-(\sigma)$.*

Proof. The proof is by induction on the length of σ. The base case where σ is empty is trivial. Suppose rule 2 of the definition of signatures applies to σ: $\text{sgn}(\sigma) = \text{sgn}(\tau)$, where τ is the shortest prefix of σ such that σ is a locally shortest non-expanding k-feasible extension of τ. If $d_G^-(\tau) = d_G^-(\sigma)$ then we are done, since we have $|\text{sgn}(\sigma)| = |\text{sgn}(\tau)| \leq d_G^-(\tau)$ by the induction hypothesis. So suppose $d_G^-(\tau) > d_G^-(\sigma)$. Let τ' be the shortest prefix of τ such that $d_G^-(\tau'') = d_G^-(\tau)$ for every prefix τ'' of τ that is an extension of τ', including τ' itself. Then, we have $\text{sgn}(\sigma) = \text{sgn}(\tau) = \text{sgn}(\tau')$ by a repeated application of rule 2. Since $d_G^-(\tau') > 0$, τ' is non-empty and we have $d_G^-(\pi(\tau')) < d_G^-(\sigma)$ since σ is not a locally shortest non-expanding extension of $\pi(\tau')$ by the choice of τ. In this case, τ' cannot be a locally shortest non-expanding extension of any of its prefixes because $d_G^-(\pi(\tau')) < d_G^-(\tau)$. Therefore, rule 3 applies to τ' and we have $|\text{sgn}(\tau')| = |\text{sgn}(\pi(\tau'))| + 1 \leq d_G^-(\pi(\tau')) + 1$ by the induction hypothesis and therefore $|\text{sgn}(\sigma)| = |\text{sgn}(\tau')| \leq d_G^-(\sigma)$. Finally suppose that rule 3 applies to σ: $\text{sgn}(\sigma) = \text{sgn}(\pi(\sigma))v$, where v is the last vertex of σ. Since σ is not a non-expanding extension of $\pi(\sigma)$, we have $d_G^-(\sigma) > d_G^-(\pi(\sigma))$ and therefore $|\text{sgn}(\sigma)| \leq d_G^-(\sigma)$ follows from the induction hypothesis on $\pi(\sigma)$. $\square$

The following observation is straightforward.

Proposition 3. *Let σ be a k-feasible sequence of length i that belongs to S_i. Then $v \in V(\sigma)$ does not appear in $\text{sgn}(\sigma)$ if and only if there are prefixes σ_1 and σ_2 of σ such that $v \notin V(\sigma_1)$, $v \in V(\sigma_2)$, and σ_2 is a locally shortest non-expanding k-feasible extension of σ_1.*

Lemma 3. *Let i, $1 \leq i \leq n$, be arbitrary. If σ and τ are distinct elements of S_i then neither $\text{sgn}(\sigma)$ nor $\text{sgn}(\tau)$ is a prefix of the other.*

Proof. Let $\sigma, \tau \in S_i$ be distinct. For each j, $0 \leq j \leq i$, let σ_j (τ_j, resp.) denote the prefix of σ (τ, resp.) of length j. Let j_0 be the smallest integer such that $\sigma_{j_0} \neq \tau_{j_0}$. Let u_0 be the last vertex of σ_{j_0} and v_0 the last vertex of τ_{j_0}. We claim that there is no pair of integers j_1 and j_2 such that $0 \leq j_1 < j_0 \leq j_2 \leq i$ and σ_{j_2} is a locally shortest non-expanding extension of σ_{j_1}. To see this, suppose such a pair of integers j_1 and j_2 exists. If there is a non-expanding k feasible extension of σ_{j_1} shorter than σ_{j_2} then this extension is not a prefix of σ_{j_2} since σ_{j_2} is a locally shortest non-expanding k-feasible extension of σ_{j_1}. But this is impossible because then a prefix of σ would be suppressed and σ would not be in S_i. Therefore, σ_{j_2} is a shortest non-expanding k-feasible extension of σ_{j_1}. Since σ_{j_1} is a common prefix of σ_{j_2} and τ_{j_2}, τ_{j_2} is suppressed by σ_{j_2} if σ_{j_2} is preferable to τ_{j_2}. On the other hand, if τ_{j_2} is preferable to σ_{j_2}, then $d_G^-(\tau_{j_2}) \leq d_G^-(\sigma_{j_2}) \leq d_G^-(\sigma_{j_1})$ and, noting that $\sigma_{j_1} = \tau_{j_1}$ because $j_1 < j_0$, we see that τ_{j_2} is also a shortest non-expanding k-feasible extension of σ_{j_1} and hence suppresses σ_{j_2}. In either case, we have a contradiction to the fact that both σ_{j_2} and τ_{j_2} are in S_{j_2}. This verifies the claim that there is no such pair j_1, j_2.

It follows from this claim and Proposition 3 that:

1. u_0 appears in $\mathrm{sgn}(\sigma)$ and
2. each vertex in $V(\sigma_{j_0-1})$ appears in $\mathrm{sgn}(\sigma)$ if and only if it appears in $\mathrm{sgn}(\sigma_{j_0-1})$.

Similarly, we have:

1. v_0 appears in $\mathrm{sgn}(\tau)$ and
2. each vertex in $V(\tau_{j_0-1})$ appears in $\mathrm{sgn}(\tau)$ if and only if it appears in $\mathrm{sgn}(\tau_{j_0-1})$.

Thus, $\mathrm{sgn}(\sigma)$ and $\mathrm{sgn}(\tau)$ have a common prefix $\mathrm{sgn}(\sigma_{j_0-1}) = \mathrm{sgn}(\tau_{j_0-1})$, which is followed by u_0 in $\mathrm{sgn}(\sigma)$ and by v_0 in $\mathrm{sgn}(\tau)$. Since $u_0 \neq v_0$, neither $\mathrm{sgn}(\sigma)$ nor $\mathrm{sgn}(\tau)$ is a prefix of the other. $\qquad\square$

Our desired bound on $|S_i|$ immediately follows from this lemma and Proposition 2.

Corollary 1. $|S_i| \leq n^k$ holds for $0 \leq i \leq n$.

From this corollary, it is clear that the directed pathwidth problem can be solved in $n^{k+O(1)}$ time. Some implementation details needed to obtain the specific time bound stated in Theorem 1 are given in Section 5.

4   Proof of the Commitment Lemma

The following observation that the function d_G^- is submodular is straightforward. For self-containedness, we include a proof.

Proposition 4. Let G be a digraph and let X and Y be two arbitrary subsets of $V(G)$. Then, we have

$$d_G^-(X) + d_G^-(Y) \geq d_G^-(X \cap Y) + d_G^-(X \cup Y). \tag{1}$$

Proof. For each vertex $v \in V(G)$, we show that the number of times v is counted in the right-hand side of (1) does not exceed the number of times it is counted in the left-hand side of (1). If v is counted both in $d_G^-(X \cap Y)$ and $d_G^-(X \cup Y)$ then $v \notin X \cup Y$ and v has an out-neighbor in $X \cap Y$ and, therefore, v is counted both in $d_G^-(X)$ and $d_G^-(Y)$. If v is counted in $d_G^-(X \cup Y)$ then $v \notin X \cup Y$ and

v has an out-neighbor in $X \cup Y$ and, therefore, v is counted either in $d_G^-(X)$ or in $d_G^-(Y)$. If v is counted in $d_G^-(X \cap Y)$ then either $v \notin X$ or $v \notin Y$ and v has an out-neighbor in $X \cap Y$ and, therefore, v is counted either in $d_G^-(X)$ or in $d_G^-(Y)$.   □

Lemma 1 is a direct consequence of the following two lemmas.

Lemma 4. *Let G be a directed graph and k a positive integer. Let σ be a strongly k-feasible sequence in $\Sigma(G)$ and τ a k-feasible proper extension of σ such that $d_G^-(X) \geq d_G^-(\tau)$ for every X with $V(\sigma) \subseteq X \subseteq V(\tau)$. Then, τ is strongly k-feasible.*

Proof. Let σ and τ be as in the statement of the lemma and σ' a k-feasible extension of σ of length n. Let α be the subsequence of σ' such that $V(\alpha) = V(G) \setminus V(\tau)$. Let $\tau' = \tau\alpha$. Note that $\tau' \in \Sigma(G)$ and $V(\tau') = V(G)$. We claim that τ' is k-feasible and therefore τ is strongly k-feasible.

Since the prefix τ of τ' is k-feasible, we only need to show that, for $1 \leq i \leq |\alpha|$, $d_G^-(V(\tau) \cup V_i(\alpha)) \leq k$, where we denote by $V_i(\alpha)$ the set of first i vertices of α.

For each i, $1 \leq i \leq |\alpha|$, let σ_i denote the minimal prefix of σ' such that $V(\sigma_i) \setminus V(\tau) = V_i(\alpha)$. Since each member of σ precedes each member of α in σ', σ is a prefix of σ_i for $1 \leq i \leq |\alpha|$. Fix i, $1 \leq i \leq |\alpha|$. By the submodularity of d_G^-, we have

$$d_G^-(\tau) + d_G^-(\sigma_i) \geq d_G^-(V(\tau) \cap V(\sigma_i)) + d_G^-(V(\tau) \cup V(\sigma_i)).$$

Since σ' is k-feasible, we have $d_G^-(\sigma_i) \leq k$. By the assumption on τ in the statement of the lemma, we also have $d_G^-(V(\tau) \cap V(\sigma_i)) \geq d_G^-(\tau)$ as $V(\sigma) \subseteq V(\tau) \cap V(\sigma_i) \subseteq V(\tau)$. Therefore we have $d_G^-(V(\tau) \cup V_i(\alpha)) = d_G^-(V(\tau) \cup V(\sigma_i)) \leq k$, which proves the claim.   □

Lemma 5. *Let G be a directed graph and k a positive integer. Let σ be a k-feasible sequence in $\Sigma(G)$ and τ a shortest non-expanding k-feasible extension of σ. Then, for every X such that $V(\sigma) \subseteq X \subseteq V(\tau)$, we have $d_G^-(X) \geq d_G^-(\tau)$.*

Proof. Suppose to the contrary that there is some X, $V(\sigma) \subseteq X \subseteq V(\tau)$, such that $d_G^-(X) < d_G^-(\tau)$. Since $d_G^-(\sigma) \geq d_G^-(\tau)$, we have $V(\sigma) \subsetneq X \subsetneq V(\tau)$. We show that there is some non-expanding k-feasible extension η of σ that is shorter than τ. This contradicts the assumption that τ is a shortest such extension, and therefore we will be done.

Let α be the subsequence of τ such that $V(\alpha) = X$. Note that α extends σ since $V(\sigma) \subseteq X$. Let h be an integer, $|\sigma| < h \leq |X|$, such that $d_G^-(V_h(\alpha))$ is the largest, where we denote by $V_h(\alpha)$ the set of first h vertices of α. If $d_G^-(V_h(\alpha)) \leq k$ then α is k-feasible and we are done with $\eta = \alpha$: $|\alpha| < |\tau|$ holds since $V(\alpha) = X$ is a proper subset of $V(\tau)$.

Suppose $d_G^-(V_h(\alpha)) > k$. Since $d_G^-(X) < k$, we have $h < |X|$. For each i, $0 \leq i \leq X$, let τ_i denote the minimal prefix of τ such that $V(\tau_i) \cap X = V_i(\alpha)$. Since $V(\sigma) \subseteq X$, we have $\tau_{|\sigma|} = \sigma$.

We set $\eta = \tau_h \alpha'$, where α' is the subsequence of α consisting of its last $|X| - h$ elements, and verify that η is a non-expanding k-feasible extension of σ and is shorter than τ. Let i be an integer, $h \leq i \leq |X|$. By the submodularity of d_G^-, we have

$$d_G^-(\tau_h) + d_G^-(V_i(\alpha)) \geq d_G^-(V_h(\alpha)) + d_G^-(V(\tau_h) \cup V_i(\alpha)), \qquad (2)$$

where we have used $V(\tau_h) \cap V_i(\alpha) = V_h(\alpha)$. We have $d_G^-(V_i(\alpha)) \leq d_G^-(V_h(\alpha))$ by the choice of h and moreover $d_G^-(\tau_h) \leq k$ since τ is k-feasible. Therefore, we have $d_G^-(V(\tau_h) \cup V_i(\alpha)) \leq k$. Since this holds for every i, $h \leq i \leq |X|$, η is k-feasible. Since $d_G^-(\tau_h) \leq k < d_G^-(V_h(\alpha))$, (2) also implies $d_G^-(V(\tau_h) \cup V_i(\alpha)) < d_G^-(V_i(\alpha))$. Letting $i = |X|$, we have $d_G^-(\eta) = d_G^-(V(\tau_h) \cup V(\alpha)) < d_G^-(\alpha) = d_G^-(X) < d_G^-(\tau) \leq d_G^-(\sigma)$. Thus, η is a non-expanding extension of σ. Finally, the inclusion $V(\eta) \subseteq V(\tau)$ and the strict inequality $d_G^-(\eta) < d_G^-(\tau)$ imply that η is shorter than τ. □

Proof. (of Lemma 1.) Let σ be a strongly k-feasible sequence in $\Sigma(G)$ and τ a shortest non-expanding k-feasible extension of σ. Then, by Lemma 5, we have $d_G^-(X) \geq d_G^-(\tau)$ for every X such that $V(\sigma) \subseteq X \subseteq V(\tau)$. Lemma 4 applies and τ is strongly k-feasible. □

5 Implementation Details

In this section, we verify that our algorithm can be implemented to run in the time bound of $O(mn^{k+1})$ stated in Theorem 1, where $n = |V(G)|$ and $m = |E(G)|$. We assume that G is strongly connected and hence $m \geq n$.

Data Structures

We represent each nonempty sequence $\sigma \in \Sigma(G)$ by a pair consisting of the last vertex of σ and a pointer to the prefix $\pi(\sigma)$ of σ of length $|\sigma| - 1$. Thus, the elements of the sets S_i, $0 \leq i \leq i$, naturally form a rooted tree in which the parent of each non-root node σ is $\pi(\sigma)$ and the set of nodes at the ith level is S_i. In addition, we represent the set S_i, for each $0 \leq i \leq n$, as a list sorted in the lexicographic ordering.

We assume the input digraph G is given in the in-neighbor list representation: each vertex v has a list $\mathrm{in}(v)$ of its in-neighbors ordered in the assumed total ordering $<$ on $V(G)$.

Constructing Immediate Extensions

In this step, we generate k-feasible extensions of each element of S_{i-1} and let the set of all those extensions be T_i. Let σ be an element of S_{i-1} being processed. We first construct the bit-vector representation of $N_G^-(\sigma)$ in $O(n)$ time. Then, we iterate through all the vertices in $V(G)$. If $v \in V(G)$ is not in σ, we compute $d_G^-(\sigma v)$ in $O(d_G^-(v))$ time, using the bit-vector for $N_G^-(\sigma)$. If $d_G^-(\sigma v) \leq k$ then we add σv to our list of feasible extensions. Doing this for all elements of S_{i-1} in the sorted order, we obtain the set T_i in the form of a sorted list. The time required for this step is $O(mn^k)$.

Identifying Shortest Non-expanding k-Feasible Extensions and Inheritors

In this step, for each pair (τ, σ) such that $\sigma \in T_i$ and σ is the most preferable shortest non-expanding k-feasible extension of τ, we register σ as the *inheritor* of τ.

We first observe that $\sigma \in T_i$ can be a shortest non-expanding k-feasible extension of some proper prefix of σ only if $d_G^-(\sigma) \leq d_G^-(\pi(\sigma))$. Moreover, for each $\eta \in S_{i-1}$, among the extensions of η in T_i satisfying $d_G^-(\sigma) \leq d_G^-(\eta)$, only the most preferable one can be the most preferable shortest non-expanding k-feasible extension of some sequence. Based on this observation, we collect, for each $\eta \in S_{i-1}$, at most one extension $\sigma \in T_i$ of η: σ satisfies $d_G^-(\sigma) \leq d_G^-(\eta)$ and is the most-preferable over all extensions of η in T_i. We let the resulting set T_i' and scan its elements in the lexicographic ordering.

Let σ be an element of T_i'. For each proper prefix τ of σ, σ is a shortest non-expanding k-feasible extension of τ if and only if σ is a locally shortest non-expanding k-feasible extension of τ. The "only if" part is obvious. For the "if" part, suppose τ has a non-expanding k-feasible extension τ' that is shorter than σ but is not a prefix of σ. We assume τ' is the shortest among such and hence is a shortest non-expanding k-feasible extension of τ. Let τ'' be a prefix of σ of length $|\tau'|$. Since the presence of $\pi(\sigma)$ in S_{i-1} implies that τ' does not suppress τ'', it must hold that $d_G^-(\tau'') \leq d_G^-(\tau') \leq d_G^-(\tau)$ and therefore σ is not a locally shortest non-expanding k-feasible extension of τ. Since $d_G^-(\tau)$ has been calculated for every $\tau \in \bigcup_{j \leq i} S_j$, the above condition can be tested in $O(n)$ total time for all prefixes τ of σ.

When we find a prefix τ of σ such that σ is a shortest k-feasible non-expanding extension of τ, we check whether the inheritor of τ is already registered. If not, then register σ as such. Otherwise, let σ' be the registered extension. If $d_G^-(\sigma) < d_G^-(\sigma')$ then we replace σ' with σ; otherwise we retain σ'. Since we are processing the elements of T_i' in the lexicographic order, the registered inheritor is correctly the most-preferable shortest k-feasible non-expanding extension after all the elements of T_i' are processed. The time required for this registering process is also $O(n)$ for each $\sigma \in T_i'$. The overall processing time for this step is $O(n^{k+1})$.

Filtering Out Suppressed Elements

In this step, we collect those elements of T_i that are not suppressed, obtaining the set S_i.

Let $\eta \in S_{i-1}$. Suppose first that η does not have an extension in T_i', that is, $d_G^-(\sigma) > d_G^-(\eta)$ for every extension σ of η in T_i. In this case, if some prefix of η has some inheritor registered then all extensions of η in T_i are suppressed; otherwise, no prefix of η has a non-expanding k-feasible extension in T_i and therefore none of the extensions of η in T_i is suppressed. Suppose next that η has an extension σ in T_i' (which is unique). Then all extensions of η in T_i but σ are suppressed by σ. This extension σ is suppressed if and only if some prefix of η has an inheritor other than σ registered.

In either case, the processing time for each $\eta \in S_{i-1}$ is $O(n)$ and therefore the total time for this step is $O(n^{k+1})$.

Overall Running Time

We repeat the above construction of S_i for $i = 1, 2, \ldots, n$ in $O(mn^{k+1})$ total time. Checking whether S_n is empty is trivial. If it is not empty, any element of S_n achieves the directed vertex separation number at most k.

6   Concluding Remarks

In the terminology of parameterized complexity theory [7,8,15], the result of this paper puts the problem of deciding the directed pathwidth in class XP. It is open whether it is in FPT, that is, if there is an algorithm that, given positive integer k and digraph G, decides if $\mathrm{dpw}(G) \leq k$ in time $f(k)n^{O(1)}$ where f is some function independent of n.

It was pointed out, at the workshop site, by Sang-il Oum and by Hiroshi Nagamochi that the commitment lemma holds in a more general setting, where the in-degree function d_G^- is replaced by an arbitrary submodular function, and thus may be useful in other contexts. Exploring such applications of the lemma and the techniques in this work is also an attractive avenue of future research.

Acknowledgment. The author would like to thank Yuichiro Miyamoto, Ryuhei Uehara, Hirotaka Ono, Takehiro Ito, Katsuhisa Yamanaka, Yasuaki Kobayashi, and Fumihito Ōhtaki for useful discussions. Thanks are also due to Hiroshi Nagamochi who read the submission version carefully and helped improve the presentation.

References

1. Arnborg, S., Corneil, D., Proskurowski, A.: Complexity of finding embeddings in a k-tree. SIAM Journal on Matrix Analysis and Applications 8(2), 277–284
2. Arnborg, S., Proskurowski, A.: Linear time algorithms for NP-hard problems restricted to partial k-trees. Discrete Applied Mathematics 23(1), 11–24
3. Barát, J.: Directed path-width and monotonicity in digraph searching. Graphs and Combinatorics 22(2), 161–172 (2006)
4. Berwanger, D., Dawar, A., Hunter, P., Kreutzer, S.: DAG-Width and Parity Games. In: Durand, B., Thomas, W. (eds.) STACS 2006. LNCS, vol. 3884, pp. 524–536. Springer, Heidelberg (2006)
5. Bodlaender, H.L., Kloks, T.: Efficient and constructive algorithms for the pathwidth and treewidth of Graphs. Journal of Algorithms 21, 358–402 (1996)
6. Bodlaender, H.L.: A linear-time algorithm for finding tree-decompositions of small treewidth. SIAM Journal on Computing 25(6), 1305–1317 (1996)
7. Downey, R.G., Fellows, M.R.: Parameterized Complexity. Springer, Heidelberg (1999)

8. Flum, J., Grohe, M.: Parameterized Complexity Theory. Springer, Heidelberg (2006)
9. Ganian, R., Hliněný, P., Kneis, J., Langer, A., Obdržálek, J., Rossmanith, P.: On Digraph Width Measures in Parameterized Algorithmics. In: Chen, J., Fomin, F.V. (eds.) IWPEC 2009. LNCS, vol. 5917, pp. 185–197. Springer, Heidelberg (2009)
10. Hunter, P., Kreutzer, S.: Digraph Measures: Kelly Decompositions, Games, and Orderings. In: Proc. the 18th Annual ACM-SIAM Symposium on Discrete Algorithms, pp. 637–644 (2007)
11. Johnson, T., Robertson, N., Seymour, P.D., Thomas, R.: Directed tree-width. Journal of Combinatorial Theory Series B 82(1), 138–154 (2001)
12. Kashiwabara, T., Fujisawa, T.: NP-completeness of the problem of finding a minimum-clique-number interval graph containing a given graph as a subgraph. In: Proc. International Symposium on Circuits and Systems, pp. 657–660 (1979)
13. Kinnersley, N.G.: The vertex separation number of a graph equals its path-width. Information Processing Letters 42, 345–350 (1992)
14. Lampis, M., Kaouri, G., Mitsou, V.: On the Algorithmic Effectiveness of Digraph Decompositions and Complexity Measures. In: Proc. of 19th International Symposium on Algorithms and Computation, pp. 220–231 (2008)
15. Niedermeier, R.: Invitation to Fixed-Parameter Algorithms. Oxford University Press (2006)
16. Obdržálek, J.: DAG-width - Connectivity Measure for Directed Graphs. In: Proc. the 17th Annual ACM-SIAM Symposium on Discrete Algorithms, pp. 814–821 (2006)
17. Robertson, N., Seymour, P.: Graph minors. I. Excluding a forest. Journal of Combinatorial Theory, Series B 35(1), 39–61 (1983)
18. Robertson, N., Seymour, P.: Graph minors III: Planar tree-width. Journal of Combinatorial Theory, Series B 36(1), 49–64 (1984)
19. Robertson, N., Seymour, P.: Graph Minors. XX. Wagner's conjecture. Journal of Combinatorial Theory, Series B 92(2), 325–335 (2004)
20. Safari, M.A.: D-Width: A More Natural Measure for Directed Tree Width. In: Jedrzejowicz, J., Szepietowski, A. (eds.) MFCS 2005. LNCS, vol. 3618, pp. 745–756. Springer, Heidelberg (2005)
21. Tamaki, H.: A directed path-decomposition approach to exactly identifying attractors of boolean networks. In: Proc. 10th International Symposium on Communications and Information Technologies, pp. 844–849 (2010)
22. Yang, B., Cao, Y.: Digraph searching, directed vertex separation and directed pathwidth. Discrete Applied Mathematics 156(10), 1822–1837 (2008)

Author Index

Alcón, Liliana   11
Arends, Felix   23
Auer, Christopher   35

Belmonte, Rémy   47
Bílka, Ondřej   83
Bodlaender, Hans L.   59
Bodlaender, Marijke H.L.   71

Cechlárová, Katarína   95
Chaplick, Steven   319
Cheng, Christine   107
Cohen, Elad   319
Couturier, Jean-François   119
Cygan, Marek   131

Das, Shantanu   143

Faria, Luerbio   11
Feldmann, Andreas Emil   143
Figueiredo, Celina M.H. de   11
Flier, Holger   155
Fulek, Radoslav   283

Galčík, František   250
Gaspers, Serge   167
Gleißner, Andreas   35
Golovach, Petr A.   119
Gutierrez, Marisa   11

Halldórsson, Magnús M.   191
Hasunuma, Toru   203
Hermelin, Danny   215
Huang, Chien-Chung   215
Hurkens, Cor A.J.   71

Jelínková, Eva   95
Jirásek, Jozef   83
Junosza-Szaniawski, Konstanty   227

Kanj, Iyad A.   238
Katrenič, Ján   250
Kijima, Shuji   271
Kitaev, Sergey   191

Klavík, Pavel   83
Kratsch, Dieter   59, 119
Kratsch, Stefan   215

Liedloff, Mathieu   167
Lokshtanov, Daniel   262
Lonc, Zbigniew   227

Marchetti-Spaccamela, Alberto   1
Marx, Dániel   5, 131
McDermid, Eric   107
McGrae, Andrew R.A.   179
Mihalák, Matúš   155
Mnich, Matthias   262

Nagamochi, Hiroshi   203
Niedermeier, Rolf   307

Okamoto, Yoshio   271
Ono, Hirotaka   271
Otachi, Yota   271
Ouaknine, Joël   23

Paulusma, Daniël   119
Pelsmajer, Michael J.   283
Pilipczuk, Marcin   131
Pilipczuk, Michał   131
Pyatkin, Artem   191

Ravsky, Alexander   295

Saitoh, Toshiki   271
Saurabh, Saket   262
Schaefer, Marcus   283
Schlotter, Ildikó   131
Semanišin, Gabriel   250
Sorge, Manuel   307
Stacho, Juraj   319
Štefankovič, Daniel   283
Stein, Maya   167
Suchan, Karol   167
Suzuki, Ichiro   107

Tamaki, Hisao   331
Tancer, Martin   83
Tuczyński, Michał   227

344   Author Index

Uno, Yushi   271

van Bevern, René   307
Vatshelle, Martin   47
Verbitsky, Oleg   295
Volec, Jan   83

Wahlström, Magnus   215
Wampler, Charles W.   23
Weller, Mathias   307

Widmayer, Peter   143, 155
Woeginger, Gerhard J.   71

Yamanaka, Katsuhisa   271
Yamazaki, Koichi   271

Zhang, Fenghui   238
Zito, Michele   179
Zych, Anna   155